Communication
Electronics

Communication Electronics

SECOND EDITION

Louis E. Frenzel

GLENCOE

Macmillan/McGraw-Hill

New York, New York   Columbus, Ohio   Mission Hills, California   Peoria, Illinois

Cover photograph: COMSTOCK Inc./Michael Stuckey. Safety section photographs: *Top left:* Charles Thatcher/Tony Stone Worldwide; *Top right:* Lou Jones/The Image Bank; *Lower right:* © Cindy Lewis.

ACKNOWLEDGMENTS

The *Basic Skills in Electricity and Electronics* series was conceived and developed through the talents and energies of many individuals and organizations.

The original, on-site classroom testing of the texts and manuals in this series was conducted at the Burr D. Coe Vocational Technical High School, East Brunswick, New Jersey; Chantilly Secondary School, Chantilly, Virginia; Nashoba Valley Technical High School, Westford, Massachusetts; Platt Regional Vocational Technical High School, Milford, Connecticut; and the Edgar Thomson, Irvin Works of the United States Steel Corporation, Dravosburg, Pennsylvania. Postpublication testing took place at the Alhambra High School, Phoenix, Arizona; St. Helena High School, St. Helena, California; and Addison Trail High School, Addison, Illinois.

Early in the publication life of this series, the appellation "Rainbow Books" was used. The name stuck and has become a point of identification ever since.

In the years since the publication of this series, extensive follow-up studies and research have been conducted. Thousands of instructors, students, school administrators, and industrial trainers have shared their experiences and suggestions with the authors and publishers. To each of these people we extend our thanks and appreciation.

Library of Congress Cataloging-in-Publication Data

Frenzel, Louis E.
    Communication electronics / Louis E. Frenzel—2nd ed.
        p. cm. — (Basic skills in electricity and electronics)
    Includes index.
    ISBN 0-02-801842-7
    1. Electronics. 2. Radio—Receivers and reception.
  3. Telecommunication. I. Title. II. Series.
TK7816.F678 1994
  621.382—dc20   93-34935
                                          CIP

Communication Electronics, Second Edition

Send all inquiries to:
GLENCOE DIVISION
Macmillan/McGraw-Hill
936 Eastwind Drive
Westerville, Ohio 43081

ISBN 0-02-801842-7

Printed in the United States of America

1 2 3 4 5 6 7 8 9   A-KP   01 00 99 98 97 96 95 94 93

Contents

■

Editor's Foreword

The Glencoe *Basic Skills in Electricity and Electronics* series has been designed to provide entry-level competencies in a wide range of occupations in the electrical and electronic fields. The series consists of coordinated instructional materials designed especially for the career-oriented student. Each major subject area covered in the series is supported by a textbook, an activities manual, and a teacher's manual. All the materials focus on the theory, practices, applications, and experiences necessary for those preparing to enter technical careers.

There are two fundamental considerations in the preparation of materials for such a series: the needs of the learner and needs of the employer. The materials in this series meet these needs in an expert fashion. The authors and editors have drawn upon their broad teaching and technical experiences to accurately interpret and meet the needs of the student. The needs of business and industry have been identified through questionnaires, surveys, personal interviews, industry publications, government occupational trend reports, and field studies.

The processes used to produce and refine the series have been ongoing. Technological change is rapid, and the content has been revised to focus on current trends. Refinements in pedagogy have been defined and implemented based on classroom testing and feedback from students and teachers using the series. Every effort has been made to offer the best possible learning materials.

The widespread acceptance of the *Basic Skills in Electricity and Electronics* series and the positive responses from users confirm the basic soundness in content and design of these materials as well as their effectiveness as learning tools. Teachers will find the texts and manuals in each of the subject areas logically structured, well-paced, and developed around a framework of modern objectives. Students will find the materials to be readable, lucidly illustrated, and interesting. They will also find a generous amount of self-study and review materials to help them determine their own progress.

The publisher and editor welcome comments and suggestions from teachers and students using the materials in this series.

Charles A. Schuler
Project Editor

Basic Skills in Electricity and Electronics

Charles A. Schuler, Project Editor

BOOKS IN THIS SERIES:

Introduction to Television Servicing by Wayne C. Brandenburg

Electricity: Principles and Applications by Richard J. Fowler

Communication Electronics by Louis E. Frenzel

Instruments and Measurements: A Text-Activities Manual by Charles M. Gilmore

Microprocessors: Principles and Applications by Charles M. Gilmore

Small Appliance Repair by Phyllis Palmore and Nevin E. André

Electronics: Principles and Applications by Charles A. Schuler

Digital Electronics by Roger L. Tokheim

Industrial Electricity and Electronics by Frank D. Petruzella

Mathematics for Electronics by Harry Forster, Jr.

Preface

This book is written to accompany a first course in electronic communications. To get the most from this text, students should have the appropriate prerequisites. These include an understanding of ac circuits including resonance and filters, basic electronic circuits, and digital techniques.

In recent years, the study of microprocessors and personal computers has overshadowed study in other areas of electronics, including communications. Yet communications plays a vital role in our lives. Developments in communications technology have increased its application in allied fields of electronics including computers and industrial control. Regardless of a student's ultimate area of specialization, knowledge of communication concepts and applications is no longer optional; it is essential to understanding today's multidisciplinary applications.

This book begins by introducing basic communication concepts and circuits, including modulation techniques as well as radio transmitters and receivers. Antennas and microwave techniques are also discussed. In addition, every effort has been made to include the latest in high-tech communication components, circuits, and equipment. Data communication techniques (including modems and local area networks, fiber optics, and satellite communication) and advanced applications (such as cellular telephones, facsimile, and radar) are some of the topics covered.

My sincere hope is that this book will not only help stimulate interest in the exciting field of electronic communications but will also help students learn the latest techniques they will surely encounter on the job.

Special thanks go to Mr. Harold Reeb and Mr. William Coleman of McGraw-Hill's Continuing Education Center for permission to use excerpts from the home study course material I created for them in Chapters 8, 11, and 13 of this book.

Louis E. Frenzel

Safety

Electric and electronic circuits can be dangerous. Safe practices are necessary to prevent electrical shock, fires, explosions, mechanical damage, and injuries resulting from the improper use of tools.

Perhaps the greatest hazard is electrical shock. A current through the human body in excess of 10 milliamperes can paralyze the victim and make it impossible to let go of a "live" conductor or component. Ten milliamperes is a rather small amount of electrical flow: It is only *ten one-thousandths* of an ampere. An ordinary flashlight uses more than 100 times that amount of current!

Flashlight cells and batteries are safe to handle because the resistance of human skin is normally high enough to keep the current flow very small. For example, touching an ordinary 1.5-V cell produces a current flow in the microampere range (a microampere is one-millionth of an ampere). This amount of current is too small to be noticed.

High voltage, on the other hand, can force enough current through the skin to produce a shock. If the current approaches 100 milliamperes or more, the shock can be fatal. Thus, the danger of shock increases with voltage. Those who work with high voltage must be properly trained and equipped.

When human skin is moist or cut, its resistance to the flow of electricity can drop drastically. When this happens, even moderate voltages may cause a serious shock. Experienced technicians know this, and they also know that so-called low-voltage equipment may have a high-voltage section or two. In other words, they do not practice two methods of working with circuits: one for high voltage and one for low voltage. They follow safe procedures at all times. They do not assume protective devices are working. They do not assume a circuit is off even though the switch is in the OFF position. They know the switch could be defective.

As your knowledge and experience grow, you will learn many specific safe procedures for dealing with electricity and electronics. In the meantime:

1. Always follow procedures.
2. Use service manuals as often as possible. They often contain specific safety information.
3. Investigate before you act.
4. When in doubt, *do not act*. Ask your instructor or supervisor.

General Safety Rules for Electricity and Electronics

Safe practices will protect you and your fellow workers. Study the following rules. Discuss them with others, and ask your instructor about any you do not understand.

1. Do not work when you are tired or taking medicine that makes you drowsy.
2. Do not work in poor light.
3. Do not work in damp areas with wet shoes or clothing.
4. Use approved tools, equipment, and protective devices.
5. Avoid wearing rings, bracelets, and similar metal items when working around exposed electric circuits.
6. Never assume that a circuit is off. Double-check it with an instrument that you are sure is operational.
7. Some situations require a "buddy system" to guarantee that power will not be turned on while a technician is still working on a circuit.
8. Never tamper with or try to override safety devices such as an interlock (a type of switch that automatically removes power when a door is opened or a panel removed).
9. Keep tools and test equipment clean and in good working condition. Replace insulated probes and leads at the first sign of deterioration.
10. Some devices, such as capacitors, can store a *lethal* charge. They may store this charge for long periods of time. You must be certain these devices are discharged before working around them.

11. Do not remove grounds and do not use adaptors that defeat the equipment ground.
12. Use only an approved fire extinguisher for electrical and electronic equipment. Water can conduct electricity and may severely damage equipment. Carbon dioxide (CO_2) or halogenated-type extinguishers are usually preferred. Foam-type extinguishers may also be desired in some cases. Commercial fire extinguishers are rated for the type of fires for which they are effective. Use only those rated for the proper working conditions.
13. Follow directions when using solvents and other chemicals. They may be toxic, flammable, or may damage certain materials such as plastics.
14. A few materials used in electronic equipment are toxic. Examples include tantalum capacitors and beryllium oxide transistor cases. These devices should not be crushed or abraded, and you should wash your hands thoroughly after handling them. Other materials (such as heat shrink tubing) may produce irritating fumes if overheated.
15. Certain circuit components affect the safe performance of equipment and systems. Use only exact or approved replacement parts.
16. Use protective clothing and safety glasses when handling high-vacuum devices such as picture tubes and cathode-ray tubes.
17. Don't work on equipment before you know proper procedures and are aware of any potential safety hazards.
18. Many accidents have been caused by people rushing and cutting corners. Take the time required to protect yourself and others. Running, horseplay, and practical jokes are strictly forbidden in shops and laboratories.

Circuits and equipment must be treated with respect. Learn how they work and the proper way of working on them. Always practice safety; your health and life depend on it.

Electronic workers use specialized safety knowledge.

CHAPTER 1

Introduction to Electronic Communications

∎

CHAPTER OBJECTIVES

This chapter will help you to:

1. *Name* the four main elements of any communications system and *state* what each does.
2. *Define* modulation and *explain* why it is necessary and important.
3. *List* six examples of common simplex and duplex communications systems in wide use today.
4. *List* the twelve major segments of the electromagnetic spectrum and *give* the frequency ranges of each.
5. *Convert* between measures of frequency and wavelength.
6. *Define* bandwidth and *compute* the bandwidth of a given piece of spectrum given the upper and lower frequency ranges.

The field of electronics can be roughly divided into three major subfields: computers, communications, and control. Virtually every electronic application can be classified as part of one of these three divisions. The largest field in terms of the sales of equipment and services and number of employees is the computer field. It is also the youngest of the three. The communications industry is the second largest in size and is most certainly the oldest since electronics really started with radio communications. The smallest field is control. The computer industry is concerned with the development and servicing of computer hardware and software used by business, industry, and government for the processing, storage, and retrieval of data. The communications industry is concerned with electronic equipment used for the transfer of information between two or more points. That information may be voice, television pictures, computer data, or some other form of electronic information. The control field is concerned with electric power as well as various kinds of electronic components and circuits used to operate lights, heating elements, electric motors, and other devices. Although computers, communications, and control are three distinct fields, many applications involve the use of two or even all three disciplines. For example, computers are widely combined with control systems to automate factories and plants. Many computers are also part of networks that allow them to exchange data. In addition, the technologies from communications and control are often combined to perform remote operations. In this text, of course, we are concerned with communications. We will give you a broad look at all the components, circuits, techniques, and applications of electronic communications. When you complete this text, you will have a good overview of the many ways communications is used and how each way is accomplished.

∎

1-1 THE IMPORTANCE OF COMMUNICATIONS

Communication is the basic process of exchanging information. It is what humans do to convey their thoughts, ideas, and feelings to one another. Humans have been communicating with one another from the beginning of humankind. Most humans communicated through the spoken word, as they do today. Yet a considerable amount of communication was nonverbal. Body movements and facial expressions were effective communication tools. Later, written communication was developed. Humans wrote letters to one another and eventually invented newspapers and books. Although the bulk of human communication today is still oral, a huge volume of information is exchanged by means of the written word. If anything, there is a glut of information in printed form.

Two of the main barriers to human communication are language and distance. When humans of different tribes, nations, or races come together, they often find that they do not speak the same language. This language barrier continues today, but it can be overcome as humans learn the languages of others and can serve as interpreters.

Communicating over long distances is another problem. Most human communication in the beginning was limited to face-to-face conversations. However, long-distance communication attempts were made as tribesmen signaled one another with drums or smoke signals. Other early forms of long-distance communication were blowing a horn, lighting a signal fire, or waving a flag (semaphore). But despite these long-distance communication attempts, transmission distances were limited. If the signal could be launched from a hill, mountain, or chain of high towers, distances of several miles could usually be achieved. Beyond that, other forms of communication were necessary.

Long-distance communication was extended by the written word. Messages and letters were transported from one place to another. For many years, long-distance communication was limited to the sending of verbal or written messages by human runner, horseback, ship, and later by trains.

Human communication took a dramatic turn in the late nineteenth century when electricity was discovered and its many applications were explored. The telegraph was invented in 1844 and the telephone in 1876. In 1887 radio was discovered, and it was demonstrated in 1895. From that point on, the exchange of information took a great leap forward. Figure 1-1 is a time line of important milestones in the evolution of human communications.

1440	Gutenberg invents the printing press.
1844	Morse patents the telegraph.
1866	First successful use of a transatlantic telegraph cable.
1876	Bell invents and patents the telephone.
1879	Eastman develops photographic film.
1887	Hertz discovers radio waves.
1895	Marconi demonstrates wireless telegraphy.
1901	Marconi makes first transatlantic radio transmission.
1903	The Fleming "valve" is invented.
1906	De Forest invents the triode vacuum tube and the first radiotelephone broadcast.
1923	Television is invented.
1931	Radio astronomy is discovered.
1940–45	Radar is perfected and helps win World War II.
1948	The transistor is invented.
1954	Color television broadcasting begins.
1959	The integrated circuit is invented.
1962	First communications satellite.

Fig. 1-1 Time line of milestones in human and electronic communications.

Well-known forms of electronic communications, such as radio and television, have increased our ability to share information. Today, they are a major part of our lives. It is hard to imagine what our lives would be like without the knowledge and information that comes our way from around the world by the various means of electronic communications. It has made our society what it is today. In fact, the emphasis in our society has now shifted from that of manufacturing and mass production of goods to the accumulation, packaging, and exchange of information. Ours is an information society, and a key part of it is communications. Without electronic communications, we could not apply the available information in a timely way.

Self-Test

Supply the missing word(s) in each statement.

1. The three major fields of electronics are _____ . The largest is the _____ field.
2. Communication is defined as the process of _____ .
3. Most human communication is _____ even though there is a glut of _____ communication.

4. Two major barriers to human communication are _____ .
5. Electronic communications came into being in the _____ .

1-2 THE ELEMENTS OF A COMMUNICATIONS SYSTEM

All electronic communications systems have the basic form shown in Fig. 1-2. The basic components are a transmitter, a communications channel or medium, and a receiver. In most systems, a human generates a message that we call the information, or intelligence, signal. This signal is inputted to the transmitter which then transmits the message over the communications channel. The message is picked up by the receiver and is relayed to another human. Along the way, noise is added to the message in the communications channel. Noise is the general term applied to any interference that degrades the transmitted information. Let's take a closer look at each of these basic elements.

The *transmitter* is a collection of electronic components and circuits designed to convert the information into a signal suitable for transmission over a given communications medium. It may be as simple as a microphone or as complex as a microwave radio transmitter.

The *communications channel* is the *medium* by which the electronic signal is sent from one place to another. In its simplest form, the medium may simply be a pair of wires that carry a voice signal from a microphone to a headset. The communications medium may also be a fiber-optic cable or ''light pipe'' that carries the message on a light wave.

On the other hand, the medium may be radio. *Radio* is the broad general term applied to any form of wireless communication from one point to another. Radio makes use of the electromagnetic spectrum where signals are communicated from one point to another by converting them into electric and magnetic fields that propagate readily over long distances.

Although the medium supports the transmission of information, it also attenuates it. Any type of media degrades the signal and causes it to appear much lower in amplitude at the receiver. Considerable amplification of the signal, both at the transmitter and the receiver, is required for successful communication.

The *receiver* is another collection of electronic components and circuits that accept the transmitted message from the channel and convert it back into a form understandable by humans. Again it may be a simple earphone or a complex electronic receiver.

Noise is random, undesirable electric energy that enters the communications system via the communicating medium and interferes with the transmitted message. However, some noise is also produced in the receiver. Noise comes from the atmosphere (e.g., from lightning which produces static), from outer space where the sun and other stars emit various kinds of radiation that can interfere with communications, and from electrical interference created by manufactured equipment. The electric ignition systems of cars, electric motors, fluorescent lights, and other types of equipment generate signals that can also interfere with the transmission of the message.

Finally, many electronic components generate noise internally due to thermal agitation of the atoms. Although such noise signals are low level, they can often seriously interfere with the extremely low-level signals that appear greatly attenuated at the receiver after being transmitted over a long distance. In some cases, noise completely obliterates the message. At other times, it simply causes interference, which, in turn, means some of the message is completely missed or misinterpreted.

Noise is one of the more serious problems of electronic communications. For the most

Transmitter

Medium

Radio

Receiver

Noise

Fig. 1-2 **The basic elements of any communications system.**

part, it cannot be completely eliminated. However, there are ways to deal with it, as you will discover.

Self-Test

Supply the missing word(s) in each statement.

6. The three main elements of any communications system are _____ .
7. The three major types of communications paths are _____ .
8. The _____ converts the message into a form compatible with the selected medium.
9. The _____ converts the message from the medium into a form understandable by a human.
10. Undesirable interference in communications is _____ which is added to the signal in the _____ .
11. The communications media greatly _____ and _____ the information signal.
12. Three common sources of interference are _____ .

1-3 TYPES OF ELECTRONIC COMMUNICATIONS

There are three ways in which electronic communications is classified: one-way or two-way transmissions, analog versus digital signals, and baseband or modulated signals. Let's consider each of these categories in more detail.

There are two basic types of electronic communications. The simplest is one-way communications, normally referred to as *simplex*. In simplex communications, the information travels in one direction only. A common example of simplex communications is radio and TV broadcasting. Another example is the information transmitted by the telemetry system of a satellite to earth. The telemetry system transmits information about the physical status of the satellite including its position and temperature.

The bulk of electronic communications, however, is two-way. For example, when individuals communicate with one another over the telephone, each can transmit and hear simultaneously. Such two-way communications is referred to as *full duplex*.

Another form of two-way communications is where only one party transmits at a time. This is known as *half duplex*. The communication is two-way, but the direction alternates. Examples of half-duplex communication are

most radio communications such as those used in the military, fire, police, and other services. Citizens band (CB) and amateur radio are half duplex in operation.

Another way to categorize electronic communications is by the types of intelligence signals transmitted. There are two types of signals: analog and digital. An *analog* signal is a continuously varying voltage or current. A typical analog signal is a sine wave tone. Voice and video voltages are analog signals. The other type of transmitted signal falls under the broad general category of *digital*. The earliest forms of both wire and radio communications used a type of on/off digital code. The telegraph used Morse code, whereas radio telegraphy used an international code of dots and dashes. Data used in computers is also digital where binary codes representing numbers, letters, and special symbols are transmitted by wire or radio. The most commonly used digital code in communications is the American Standard Code for Information Interchange (ASCII, pronounced "ass key").

Although digital transmission can be made up of signals that originated in digital form, such as telegraphy messages or computer data, analog signals may also be transmitted in digital form. It is very common today to take voice or video analog signals and "digitize" them with an analog-to-digital converter. There are many benefits to transmitting data in digital form as you will learn later. Regardless of whether the original information signals are analog or digital, they are all referred to as *baseband signals*.

In a communications system, the information signals may be transmitted by themselves over the medium or may be used to modulate a carrier for transmission over the medium. Putting the original voice, video, or digital signals directly into the medium is referred to as *baseband transmission*. For example, in many telephone and intercom communications systems, it is the voice itself that is placed on the wires and transmitted over some distance to the receiver. In some computer networks, the digital signals are applied directly to coaxial cables for transmission to another computer.

On the other hand, there are many instances when the baseband signals are incompatible with the media. For example, voice signals cannot be transmitted directly by radio. Although, theoretically this is possible, realistically it is impractical. To transmit baseband signals by radio, modulation techniques must be used.

Modulation is the process of having a baseband voice, video, or digital signal modify another, higher-frequency signal called the *carrier*. The information to be sent is said to be impressed upon the carrier.

The carrier is usually a sine wave that is higher in frequency than the highest intelligence signal frequency. Three basic characteristics of the carrier can be changed by the information signal: amplitude, frequency, and phase. The information or intelligence signal is also called the modulating signal or wave.

The two most common methods of modulation are amplitude modulation (AM) and frequency modulation (FM). In AM, the baseband signal varies the amplitude of the higher-frequency carrier signal as shown in Fig. 1-3(*a*). In FM, the baseband signal varies the frequency of the carrier as Figure 1-3(*b*) shows.

Shifting the phase of the carrier in accordance with the intelligence signal produces phase modulation (PM). Phase modulation, in turn, produces FM, so a PM signal looks like the FM signal in Fig. 1-3(*b*).

The higher-frequency carrier signal is more compatible with the medium (such as free space) and propagates through it with greater efficiency. At the receiver the carrier is demodulated to extract the original baseband information signal and transmitted message. Another name for the demodulation process is *detection*.

The use of modulation also permits another technique known as multiplexing to be used. *Multiplexing* is the process of transmitting two or more signals simultaneously over the same channel. The greater part of this book is devoted to discussions of both modulation and multiplexing techniques.

Self-Test

Supply the missing word(s) in each statement.

13. One-way communications is called _____ . An example is _____ .
14. Simultaneous two-way communications is called _____ . An example is _____ .
15. Two-way communications where each party takes turns transmitting is referred to as _____ .
16. Voice and video signals are continuous _____ voltages.
17. On/off or coded signals are referred to as _____ signals.
18. Voice and video signals may be transmitted digitally if they are first passed through a(n) _____ .
19. An original voice, video, or data voltage is called the _____ signal.
20. To make the transmitted signal compatible with the medium, the process of _____ must be used where the _____ signal is impressed upon a higher-frequency signal called the _____ .
21. Recovering the originally transmitted signal is called _____ .
22. The process of transmitting two or more baseband signals simultaneously over a common medium is called _____ .

1-4 A SURVEY OF COMMUNICATIONS APPLICATIONS

The applications of electronic techniques to communications are so common and pervasive that you are already familiar with most of

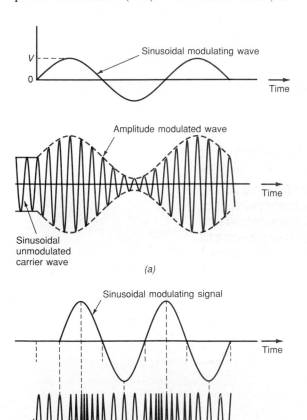

Fig. 1-3 Types of modulation. (*a*) Amplitude modulation and (*b*) frequency modulation.

them. You use the telephone, listen to the radio, and watch TV. You may also use other forms of communications such as cellular telephones, ham radios, CB radios, pagers, or remote-control garage door openers. But there are many more. The purpose of this section is to summarize all the various kinds of electronics communications, not only to show you the breadth of the field but also its importance.

Figure 1-4 lists all the various major applications of electronic communications. The list is divided into simplex and duplex categories.

SIMPLEX (One Way)

1. **AM and FM radio broadcasting.** Commercial stations broadcast music, news, weather reports, as well as other programs and information.
2. **TV broadcasting.** Commercial stations broadcast a wide variety of entertainment, informational, and educational programs.
3. **Cable television.** The distribution of movies, sports events, and other programs by local cable companies to subscribers by coax cable. Cable stations originate some programming, but they primarily distribute "packaged" programming received by satellite from services such as HBO, CNN, etc.
4. **Facsimile.** The transmission of printed visual material over the telephone lines. A facsimile, or fax, machine scans a photo or other document and converts it into electronic signals that are sent over the telephone system for reproduction in original printed form by another fax machine at the receiving end.
5. **Wireless remote control.** A mechanism that controls missiles, satellites, robots, and other vehicles or remote plants or stations. A garage door opener is a special form of remote control which uses a tiny battery-operated transmitter in the car to operate a receiver in the garage that activates a motor to open or close the door.
6. **Paging services.** A radio system for paging individuals, usually in connection with their work. Persons carry tiny battery-powered receivers or "beepers" that can pick up signals from a local paging station that receives telephone requests to locate and page individuals when they are needed.
7. **Navigation and direction-finding services.** Transmission by special stations of signals that can be picked up by receivers with highly directional antennas for the purpose of identifying exact location (latitude and longitude) or determining direction and/or distance from a station. Such systems employ both land-based and satellite stations. The services are used primarily by boats and ships or airplanes, although systems for cars and trucks are being developed.
8. **Telemetry.** The transmission of measurements over a long distance. Telemetry systems use sensors to determine physical conditions (temperature, pressure, flow rate, voltages, frequency, etc.) at a remote location. The sensors modulate a carrier signal that is sent by wire or radio to a remote receiver which stores and/or displays the data for analysis. Telemetry systems permit remote equipment and systems to be monitored for the purpose of determining their status and condition. Examples are satellites, rockets, pipelines, plants, and factories.
9. **Radio astronomy.** Radio signals, including infrared, are emitted by virtually all heavenly bodies such as stars and planets. By using large directional antennas and sensitive high-gain receivers, these signals may be picked up and used to plot star locations and study the universe. Radio astronomy is an alternative and supplement to traditional optical astronomy.
10. **Surveillance.** *Surveillance* means discreet monitoring or "spying." Electronic techniques are widely used by police forces, governments, the military, business and industry, and others to gather information for the purpose of gaining some competitive advantage. Techniques include phone taps, tiny wireless "bugs," clandestine listening stations, as well as reconnaissance airplanes and satellites.
11. **Teletext and Viewdata.** The transmission of text and graphical data to be displayed on a TV receiver with an appropriate adapter. Teletext data is transmitted by telephone lines, whereas Viewdata or Videotext is transmitted during the vertical picture blanking intervals on standard TV broadcast signals.
12. **Music services.** The transmission of continuous background music for doctors' offices, stores, elevators, and so on by local FM radio stations on special high-frequency subcarriers that cannot be picked up by conventional FM receivers.

DUPLEX (Two Way)

13. **Telephones.** One-on-one verbal communications over the vast worldwide telephone networks employing wire, radio relay stations and satellites. Cordless telephones provide short distance convenience communication without restrictive wires. Cellular radio systems provide telephone service in vehicles and for portable use.
14. **Two-way radio.** Commercial, industrial, and governmental communications between vehicles or between vehicles and a base station. *Examples:* police, fire, taxi, forestry service, trucking companies, etc. Other forms of two-way radios are used in aircraft, marine, military, and space applications. Government applications are broad and diverse and include embassy communications, Treasury, Secret Service, and CIA.
15. **Radar.** A special form of communications that makes use of reflected microwave signals for the purpose of detecting ships, planes, and missiles and for determining their range, direction, and speed. Most radar is used in military applications, but civilian aircraft and marine services also use it. Police use radar in speed detection and enforcement.
16. **Sonar.** Underwater communications in which audible baseband signals use water as the transmission medium. Submarines and ships use sonar to detect the presence of enemy submarines. Passive sonar uses audio receivers to pick up water, prop, and other noises. Active sonar is like an underwater radar where reflections from a transmitted audio pulse are used to determine direction, range, and speed of an underwater target.
17. **Amateur radio.** A hobby for individuals interested in radio communications. Individuals may become licensed "hams" to build and operate two-way radio equipment for personal communication with other "hams."
18. **Citizens radio.** Citizens or citizens band (CB) radio is a special service that any individual may use for personal communication with others. Most CB radios are used in trucks and cars for exchanging information about traffic conditions, speed traps, and for emergencies.
19. **Data communications.** The transmission of binary data between computers. Computers frequently use the telephone system as the medium. Devices, called modems, make the computers and telephone networks compatible. Data communications also take place over terrestrial microwave relay links and satellites.
20. **Local area networks (LANs).** Wired (or wireless) interconnections of personal computers (PCs) or PCs and mini or mainframe computers within an office or building for the purpose of sharing mass storage, peripherals, and data.

Fig. 1-4 Applications of electronic communications.

Read through the list to refresh your memory or possibly to discover some new applications you may not know about.

Self-Test

Answer the following questions.

23. Two methods of transmitting visual data over the telephone network are _____ .
24. A common household remote-control unit is the _____ .
25. The signaling of individuals at remote locations is called _____ .
26. Performing, recording, and analyzing measurements at a distance is done with _____ equipment.
27. Radio astronomy is based on the fact that stars and other heavenly bodies emit _____ .
28. List four ways radio is used in the telephone system.
29. Radar is based on the use of _____ radio signals.
30. Underwater radar is called _____ .
31. The two types of sonar are _____ .
32. The radio communications hobby is called _____ .
33. Computers exchange digital data over the telephone network by using devices called _____ .
34. Limited interconnections of PCs and other computers in offices or buildings are called _____ .

1-5 THE ELECTROMAGNETIC SPECTRUM

Before it can be transmitted, information must be converted into electronic signals compatible with the medium. For example, a microphone changes voice into a voltage of varying frequency and amplitude. This baseband signal is then passed over wires to a receiver or headphone. This is the way the telephone system works. A tremendous amount of information is transmitted in this way.

Instead of using wires, free space can be used. The information is converted into electronic signals which radiate into space. Such signals consist of both electric and magnetic fields. These so-called electromagnetic signals travel through space for long distances. Electromagnetic signals are also referred to as *radio-frequency (RF) waves*.

Electromagnetic waves are signals that oscillate; that is, the amplitudes of the electric and magnetic fields vary at a specific rate. The field intensities fluctuate up and down a given number of times per second. The electromagnetic waves vary sinusoidally. Their frequency is measured in cycles per second (cps) or hertz (Hz). These oscillations may occur at a very low frequency or at an extremely high frequency. This entire range of frequencies is referred to as the *electromagnetic spectrum*. It includes signals such as the 60-Hz power line frequency and audio (voice) signals at the low end. In the midrange are the most commonly used radio frequencies for two-way communications, television, and other applications. At the upper end of the spectrum are infrared and visible light. Figure 1-5 on the next page shows the entire electromagnetic spectrum. Both frequency and wavelength are given.

Remember the relationship between frequency f and wavelength λ (see page 8).

$$\lambda = \frac{300}{f}$$

where λ is in meters and f is in megahertz. For instance, if $f = 21$ MHz, then

$$\lambda = \frac{300}{21} = 14.29 \text{ m}$$

If you know the wavelength in meters, you can compute the corresponding frequency with the expression

$$f = \frac{300}{\lambda}$$

A wavelength of 2.4 m expressed as a frequency is:

$$f = \frac{300}{2.4} = 125 \text{ MHz}$$

Most information signals to be transmitted occur at the lower frequencies in the spectrum. They are used to modulate a carrier of a higher frequency. The high-frequency signal gives the lower-frequency information signals a ride through space. A receiver then picks up the modulated electromagnetic field, converts it back into a signal voltage, amplifies it, and demodulates it to recover the original information.

The inevitable question is, "Why can't the information signals be radiated directly?" Well, in theory they can be, but it is generally impractical. Let's take voice signals as an example. First, these occur in the 300- to 3000-Hz frequency range. After being amplified, they could be sent to a long wire antenna instead of a speaker. The resulting electromagnetic waves would be propagated through space to a receiver. In order for this to be done effectively,

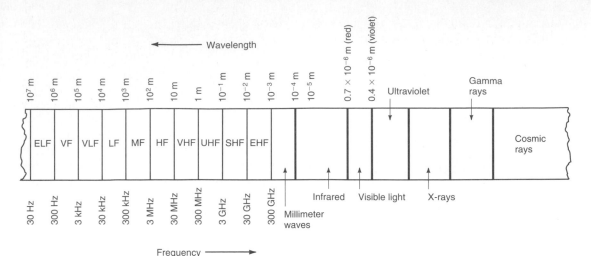

Fig. 1-5 The electromagnetic spectrum used in electronic communications.

FREQUENCY AND WAVELENGTH

Frequency is simply the number of times a particular phenomenon occurs in a given period of time. It may be the number of voltage polarity alternations or electromagnetic field oscillations that take place in one second. Each alternation or oscillation is called a cycle and the frequency is measured in cycles per second (cps). In electronics, the term *hertz* is used to express frequency in cps. For example, 60 cps = 60 Hz.

Figure A shows a sine wave variation of voltage. One positive and one negative alternation form a cycle. If 1000 of the cycles take place in one second, the frequency is 1000 Hz.

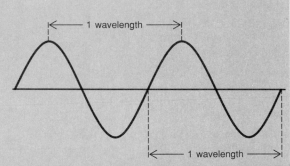

Fig. B

Wavelength is also the distance travelled by an electromagnetic wave during the time of one cycle. Electromagnetic waves travel at a speed of 300,000,000 meters per second or about 186,000 miles per second. The wavelength (λ) can be computed by dividing the speed by the frequency (f) in Hz.

$$\lambda = \frac{300,000,000}{f}$$

For example, the wavelength of a 4,000,000 Hz signal is

$$\lambda = \frac{300,000,000}{4,000,000} = 75 \text{ meters (m)}$$

If the frequency is expressed in megahertz, the formula can be simplified to

$$\lambda = \frac{300}{f}$$

A 150-MHz signal has a wavelength of

$$\lambda = \frac{300}{150} = 2 \text{ m}$$

A 430-MHz signal has a wavelength of

$$\lambda = \frac{300}{430} = 0.697 \text{ m}$$

or about 70 cm.

Fig. A

To express higher frequencies, the suffixes k, M, and G are used. The suffix k stands for kilo, M for mega, and G for giga. Also, k = 1000, M = 1,000,000, and G = 1,000,000,000. A frequency of 1000 Hz is usually expressed as 1 kHz or 1 kilohertz. A frequency of 7,000,000 Hz is expressed as 7 MHz or 7 megahertz. A very high frequency such as 4,000,000,000 Hz is given as 4 GHz or 4 gigahertz.

Wavelength is the distance between two points of similar cycles of a periodic wave. Figure B shows wavelength at several points on a sine wave.

the antenna would have to be enormous. Antenna length is usually one-quarter or one-half the length of the waves it is set up to transmit. For audio signals, the antenna would be many miles long. That, of course, would be impractical because of the size and cost.

Second, simultaneously transmitted audio signals would interfere with one another since they occupy the same frequency range. The audio portion of the spectrum would be nothing but a jumble of hundreds or thousands of simultaneous communications.

For these reasons, the information signal, be it audio, video, or data, modulates a high-frequency carrier. At higher frequencies antenna sizes are much smaller and signals propagate farther. There is also more room at the higher frequencies, so many channels can be formed to carry thousands of simultaneous communications without interference.

The frequency spectrum is divided into segments for the purpose of classifying the various portions and how they are used. These segments are designated in Fig. 1-6. The frequency and wavelength range for each segment is given. Let's summarize briefly the signal characteristics and applications of each segment.

Name	Frequency	Wavelength
Extremely low frequencies (ELF)	30–300 Hz	10^7–10^6 m
Voice frequencies (VF)	300–3000 Hz	10^6–10^5 m
Very low frequencies (VLF)	3–30 kHz	10^5–10^4 m
Low frequencies (LF)	30–300 kHz	10^4–10^3 m
Medium frequencies (MF)	300 kHz–3 MHz	10^3–10^2 m
High frequencies (HF)	3–30 MHz	10^2–10^1 m
Very high frequencies (VHF)	30–300 MHz	10^1–1 m
Ultra high frequencies (UHF)	300 MHz–3 GHz	1–10^{-1} m
Super high frequencies (SHF)	3–30 GHz	10^{-1}–10^{-2} m
Extremely high frequencies (EHF)	30–300 GHz	10^{-2}–10^{-3} m
Infrared	—	0.7–10 μm
The visible spectrum (light)	—	0.4×10^{-6}–0.8×10^{-6} m

Units of Measurement and Abbreviations:
kHz = 1000 Hz
MHz = 1000 kHz = 1×10^6 = 1,000,000 Hz
GHz = 1000 MHz = 1×10^6 = 1,000,000 kHz
= 1×10^9 = 1,000,000,000 Hz
m = meter
μm = micron = $\frac{1}{1,000,000}$ m = 1×10^{-6} m

Fig. 1-6 Segments of the electromagnetic spectrum.

EXTREMELY LOW FREQUENCIES

Extremely low frequencies (ELFs) are those in the 30- to 300-Hz range. These include ac power line frequencies (50 and 60 Hz are common) as well as those frequencies in the low end of the human hearing range.

VOICE FREQUENCIES

Voice frequencies (VFs) are those in the range of 300 to 3000 Hz. This is the normal range of human speech. Although human hearing extends from approximately 20 to 20,000 Hz, most intelligible sound occurs in the VF range.

VERY LOW FREQUENCIES

Very low frequencies (VLFs) include the higher end of the human hearing range up to about 15 or 20 kilohertz (kHz). Many musical instruments also make sounds in this range as well as in the ELF and VF ranges. The VLF range is also used in some government and military communications. For example, VLF radio transmission is used by the navy to communicate with submarines.

LOW FREQUENCIES

Low frequencies (LFs) are those in the 30- to 300-kHz range. The primary communications services in this range are those used in aeronautical and marine navigation. Frequencies in this range are also used as subcarriers. *Subcarriers* are signals which carry the baseband modulating information but which, in turn, modulate another higher-frequency carrier.

MEDIUM FREQUENCIES

Medium frequencies (MFs) are in the 300- to 3000-kHz [3-megahertz (MHz)] range. The major application of frequencies in this range is AM radio broadcasting (535 to 1605 kHz). Other services in this range include various marine and aeronautical communications applications.

HIGH FREQUENCIES

High frequencies (HFs) are those in the 3- to 30-MHz range. These are the frequencies generally known as short waves. All kinds of two-way radio communications take place in this range as well as some shortwave radio broadcasting. Voice of America and Radio Free Europe broadcasts occur in this range. Government and military services use these frequencies for two-way communications. Amateur radio and CB communications also occur in this part of the spectrum.

VERY HIGH FREQUENCIES

Very high frequencies (VHFs) cover the 30- to 300-MHz range. This is an extremely popular frequency range and is used by many services including mobile radio, marine and aeronautical communications, FM radio broadcasting (88 to 108 MHz), and television channels 2 through 13. Radio amateurs also have numerous bands in this frequency range.

ULTRAHIGH FREQUENCIES

Ultrahigh frequencies (UHF) cover the 300- to 3000-MHz range. This too is an extremely widely used portion of the frequency spectrum. It includes the UHF television channels 14 through 83. It is also widely used for land mobile communications and services such as cellular telephones. The military services widely use these frequencies for communications. In addition, some radar and navigation services occupy this portion of the frequency spectrum. Radio amateurs also have bands in this part of the spectrum. Incidentally, frequencies above the 1000-MHz [1 gigahertz (GHz)] range are called *microwaves*.

SUPERHIGH FREQUENCIES

The superhigh frequencies (SHFs) are those in the 3- to 30-GHz range. These are microwave frequencies that are widely used for satellite communications and radar. Some specialized forms of two-way radio communications also occupy this region.

EXTREMELY HIGH FREQUENCIES

Extremely high frequencies (EHFs) extend from 30 to 300 GHz. Equipment used to generate and receive signals in this range is extremely complex and expensive. Presently there is only a limited amount of activity in this range, but it does include satellite communications and some specialized radar. As technological developments permit equipment advances, this frequency range will be more widely used. Signals directly above this range are generally referred to as millimeter waves.

INFRARED

Those electromagnetic signals whose frequencies are higher than 300 GHz are not referred to as radio waves. Special names are given to the various bands in the spectrum beyond that point. The infrared region is sandwiched between the highest radio frequencies and the visible portion of the electromagnetic spectrum. It occupies the range between approximately 0.01 millimeter (mm) and 700 nanometers (nm) or 0.7 to 10 microns (μm). Infrared frequencies are often given in microns, where a micron is one millionth of a meter (m). Infrared is divided into two areas, long infrared (0.01 mm to 1000 nm) and short infrared (1000 to 700 nm).

Infrared refers to radiation generally associated with heat. Anything that produces heat generates infrared signals. Infrared is produced by light bulbs, our bodies, and any physical equipment that generates heat. Infrared signals can also be generated by special types of light-emitting diodes.

Infrared signals are used for various special kinds of communications. For example, infrared is used in astronomy to detect stars and other physical bodies in the heavens. Infrared is also used for guidance in weapons systems where the heat radiated from airplanes or missiles can be picked up by infrared detectors and used to guide missiles toward these targets. Infrared is also used in most new TV remote-control units where special coded signals are transmitted by infrared to the TV receiver for the purpose of changing channels, setting the volume, and other functions.

Infrared signals also have many of the same properties as light. Optical devices such as lenses and mirrors are often used to process and manipulate infrared signals.

THE VISIBLE SPECTRUM

Just above the infrared region is the visible spectrum we ordinarily refer to as light. Light is a special type of electromagnetic radiation that has a wavelength in the 0.4- to 0.8-μm range. Light wavelengths are usually expressed in terms of angstroms (Å). An angstrom is one ten-thousandth of a micron. The visible range is approximately 8000 Å (red) to 4000 Å (violet).

Light is widely used for various kinds of communications. Light waves can be modulated and transmitted through glass fibers just as electric signals can be transmitted over wires. Fiber optics is one of the fastest growing specialties of communications electronics. The great advantage of light wave signals is that their very high frequency gives them the ability to handle a tremendous amount of information. That is, the bandwidth of the baseband signals may be very wide.

Light signals can also be transmitted through free space. Various types of communications systems have been created using a laser that generates a light beam at a specific visible frequency. Lasers generate an extremely narrow beam of light which is easily modulated with voice, video, and data information.

Beyond the visible region are the x-rays, gamma rays, and cosmic rays. These are all forms of electromagnetic radiation, but they do not figure into communications systems and, as a result, we will not cover them here.

Self-Test

Answer the following questions.

35. Signals that travel through free space for long distances are called _____ .
36. Radio waves are made up of _____ fields.
37. A signal with a frequency of 18 MHz has a wavelength of _____ m.
38. Common power line frequencies of _____ and _____ Hz are in the _____ range.
39. Audio signals are not transmitted by electromagnetic waves because
    a. Antennas would be too long.
    b. Audio signals do not radiate.
    c. Simultaneous transmissions would interfere.
    d. The frequency is too low.
    (Choose all that apply.)
40. The human hearing range is approximately _____ to _____ Hz.
41. The frequency range of the human voice is _____ to _____ Hz.
42. True or false. Radio transmissions do not occur in the VLF and LF ranges.
43. AM broadcast stations are in the _____ range.
44. HF signals are also called _____ .
45. TV (channels 2 to 13) and FM broadcasting is in the _____ part of the spectrum.
46. List five major uses of the UHF band.
47. A frequency of 1 GHz is the same as _____ MHz.
48. Frequencies above 1 GHz are called _____ .
49. The SHF and EHF ranges are primarily used by _____ communications.
50. The frequencies just beyond the EHF range are called _____ waves.
51. One micron is the same as _____ m.
52. Infrared signals are usually derived from _____ sources.

53. The spectrum range of infrared signals is _____ to _____ μm.
54. One angstrom is equal to _____ μm.
55. The visible light range is from _____ to _____ Å.
56. Light signals use two mediums in electronic communications: _____ .

1-6 BANDWIDTH

Bandwidth is that portion of the electromagnetic spectrum occupied by a signal. It is also the frequency range over which an information signal is transmitted or over which a receiver or other electronic circuit operates. More specifically, bandwidth (BW) is the difference between the upper and lower frequency limits of the signal or the equipment operation range. Figure 1-7 shows the bandwidth of the voice frequency range from 300 to 3000 Hz. The upper frequency is f_2 and the lower frequency is f_1. The bandwidth then is

$$BW = f_2 - f_1$$
$$= 3000 - 300$$
$$= 2700 \text{ Hz}$$

When information is modulated onto a carrier somewhere in the electromagnetic spectrum, the resulting signal occupies a small portion of the spectrum surrounding the carrier frequency. For example, in AM broadcasting, audio signals up to 5 kHz may be transmitted. The modulation process causes other signals, called *sidebands*, to be generated at frequencies above and below the carrier frequency by an amount equal to the modulating frequency. If the carrier frequency is 1000 kHz, or 1 MHz, and the modulating frequency

Fig 1-7 Bandwidth is the frequency range over which equipment operates or that portion of the spectrum occupied by the signal. This is the voice frequency bandwidth.

is 5 kHz, then sidebands will be produced at 1000 − 5 = 995 kHz and at 1000 + 5 = 1005 kHz. In other words, the modulation process generates other signals which take up spectrum space. It is not just the carrier at 1000 kHz that is transmitted. The range of frequencies that contain the information is called the bandwidth. The term *channel bandwidth* is used to describe the range of frequencies required to transmit the desired information.

The bandwidth of the AM signal described above is the difference between the highest and lowest transmitting frequencies.

$$BW = 1005 \text{ kHz} - 995 \text{ kHz}$$
$$= 10 \text{ kHz}$$

In this case, the channel bandwidth is 10 kHz. An AM broadcast signal, therefore, takes up a 10-kHz chunk of spectrum space. Signals transmitting on the same frequency or on overlapping frequencies will, of course, interfere with one another. For this reason, you can see that only a limited number of signals can be transmitted in the frequency spectrum. As communications activities have grown over the years, there has been a continuous demand for more frequency channels over which communications can be conducted.

This has caused a push for the development of equipment that operates at the higher frequencies. Prior to World War II, frequencies above 1 GHz were virtually unused since there were no electronic components suitable for generating signals at those frequencies. But technological developments over the years have given us many microwave components such as klystrons, magnetrons, traveling-wave tubes, and today even transistors that work in the microwave range.

The benefit of using the higher frequencies is that a signal of a given bandwidth represents a smaller percentage of the spectrum at the higher frequencies than at the lower frequencies. For example, the 10-kHz-wide AM signal discussed earlier represents 1 percent of the spectrum at 1000 kHz.

$$\text{Percent} = \frac{10 \text{ kHz}}{1000 \text{ kHz}} \times 100$$
$$= 1\%$$

But at 1 GHz, or 1,000,000 kHz, it only represents a thousandth of 1 percent.

$$\text{Percent} = \frac{10 \text{ kHz}}{1,000,000 \text{ kHz}} \times 100$$
$$= 0.001\%$$

What this really means is that there are many more 10-kHz channels at the higher frequencies than at the lower frequencies. In other words, there is more spectrum space for information signals at the higher frequencies.

The higher frequencies also permit wider-bandwidth signals to be used. A TV signal, for example, occupies a bandwidth of 6 MHz. Such a signal cannot be used to modulate a carrier in the MF or HF ranges. Television signals are transmitted in the VHF and UHF portions of the spectrum where sufficient space is available.

Today, virtually all the frequency spectrum between approximately 30 kHz and 300 MHz has been spoken for. Granted, there are some open areas and portions of the spectrum that are not heavily used, but for the most part, the spectrum is filled with communications activities of all kinds. There is tremendous competition for these frequencies, not only between companies, individuals, and government services but also between the different nations of the world. The electromagnetic spectrum is one of our most precious natural resources. Because of this, communications engineering is devoted to making the best use of that finite spectrum. A considerable amount of effort goes into developing communications techniques that will minimize the bandwidth required to transmit given information. By minimizing the bandwidth, spectrum space is conserved and this gives other services or users an opportunity to take advantage of it. Many of the techniques we will discuss later in this book evolved in an effort to minimize transmission bandwidth.

The United States and other countries recognized early on that the frequency spectrum was indeed finite and a valuable resource. For that reason, governments set up agencies to control the use of the spectrum. In the United States, Congress created the Communications Act of 1934. Along with its various amendments, the act determined how the spectrum space was used. It also established the Federal Communications Commission (FCC) which is a regulatory body whose sole purpose is allocating spectrum space, issuing licenses, setting standards, and policing the airwaves. The FCC controls all telephone and radio communications in this country and, in general, regulates all electromagnetic emissions. The National Telecommunications and Information Administration (NTIA) performs a similar function for government and military services. Other countries have similar organizations.

There is also an international organization to

which all countries belong. This is the International Telecommunications Union (ITU), an agency of the United Nations. It has 154 member countries that meet at regular intervals to promote cooperation and negotiate national interests. Various committees of the ITU set standards for various areas within the communications field. The ITU brings together the various countries to discuss how the frequency spectrum is to be divided up and shared. Because many of the signals generated in the spectrum do not carry for long distances, countries may use these frequencies simultaneously without interference. On the other hand, there are some ranges of the frequency spectrum that can literally carry signals around the world. High-frequency shortwave signals are an example. As a result, countries must negotiate with one another for various portions of the high-frequency spectrum and coordinate usage.

Self-Test

Supply the missing information in each statement.

57. The spectrum space occupied by a signal is called the _____ .

58. The new signals above and below the carrier frequency produced by the modulation process are called _____ .

59. A signal occupies the frequency range from 1.050 to 1.175 MHz. Its bandwidth is _____ kHz.

60. Wide-bandwidth signals must be transmitted at _____ frequencies.

61. Percentagewise, there is less spectrum space at the _____ frequencies.

62. Many communications electronics techniques are designed in order to conserve _____.

63. Electronic communications in the United States is regulated by a set of laws called the _____ .

64. The regulatory body for electronic communications in the United States is the _____ .

65. Government and military communications are coordinated by the _____ .

66. The electromagnetic spectrum is managed worldwide by the _____ organization.

SUMMARY

1. The three major fields of electronics are computers, communications, and control. The computer segment is the largest; communications is the second largest.
2. Communication is the process of exchanging information.
3. Most human communication is oral, but a great deal of it is also in written or printed form.
4. The two main barriers to communication are language and distance.
5. Major electrical discoveries in the mid- and late nineteenth century made possible the development of electronic communications over long distances.
6. The telegraph (1844) and telephone (1876) were the first two long-distance communications systems.
7. Radio was discovered in 1887, and wireless telegraphy was demonstrated in 1895.
8. Electronic communications plays a vital role in all our lives and is essential to the success of our information society.
9. The major elements of a communications system are a transmitter to send a message, a communications medium, a receiver to pick up the message, and noise.
10. The three primary communications media are wires, free space, and fiber-optic cable.
11. Radio waves are signals made up of electric and magnetic fields that propagate over long distances.
12. Noise is any interference that disturbs the legible transmission of a signal. Noise is produced by the atmosphere, heavenly bodies, manufactured electrical equipment, and thermal agitation in electronic components.
13. The transmission medium greatly attenuates and degrades the transmitted signal.
14. Electronic communications may be either one-way or two-way. One-way transmission is called simplex or broadcasting.
15. Two-way communication is called duplex. In half-duplex communications, only one of the two parties can transmit at a time.

In full duplex, both parties may transmit and receive simultaneously.

16. Information signals may be either analog or digital. Analog signals are smooth, continuous voltage variations such as voice or video. Digital signals are binary pulses or codes.

17. The information signal, called the baseband signal, is often transmitted directly over the communications medium.

18. In most communications systems, the baseband signal is used to modulate a higher-frequency carrier signal than is transmitted by radio.

19. Modulation is the process of having an information signal modify a carrier signal in some way. Common examples are AM and FM.

20. The baseband signal cannot usually be transmitted through space by radio because the antennas required are too long and because multiple baseband signals transmitting simultaneously would interfere with one another.

21. Multiplexing is the process of transmitting two or more signals simultaneously over the same channel or medium.

22. Besides TV, there are several other methods of transmitting visual or graphical information; they are facsimile, videotex, teletext.

23. Simplex transmission of special signals from land-based or satellite stations is used by ship and airplanes for navigation.

24. Telemetry is measurement at a distance. Sensors convert physical characteristics to electric signals which modulate a carrier transmitted to a remote location.

25. Radio astronomy supplements optical astronomy by permitting the location and mapping of stars by the radio waves they emit.

26. Radar uses the reflection of radio waves from remote objects for the detection of their presence, direction, and speed.

27. Underwater radar is called active sonar. Passive sonar is simply listening underwater for the detection of objects of interest.

28. Two forms of personal communications services are CB radio and amateur or "ham" radio, which is a technical hobby as well as a communications service.

29. Data communications is the transmission of computer and other digital data via the telephone system, microwave links, or satellite.

30. Devices called modems permit digital data to be transmitted over the analog telephone networks.

31. Interconnections of PCs for the exchange of information are called local area networks.

32. The electromagnetic spectrum is that range of frequencies from approximately 30 Hz to visible light over which electronic communications takes place.

33. The greatest portion of the spectrum covers radio waves, which are oscillating electric and magnetic fields that radiate for long distances.

34. Wavelength (λ) is the distance (in meters) between corresponding points on successive cycles of a periodic wave: $\lambda = 300/f$ (f is in megahertz). It is also the distance that an electromagnetic wave travels in the time it takes for one cycle of oscillation.

35. The range of human hearing is approximately 20 to 20,000 Hz. The voice frequency range is 300 to 3000 Hz.

36. Amplitude-modulated broadcasting occurs in the MF range from 300 kHz to 3 MHz.

37. The high-frequency range (3 to 30 MHz), or shortwave, is used for worldwide two-way communications and broadcasting.

38. Television broadcasting occurs in the VHF and UHF ranges.

39. Frequencies above 1 GHz are called microwaves.

40. The SHF and EHF bands are used primarily for satellite communications and radar.

41. Those frequencies directly above 300 GHz are called millimeter waves.

42. Electromagnetic signals produced primarily by heat sources are called infrared. They cover the 0.7- to 100-μm range.

43. A micron is one millionth of a meter.

44. Visible light occupies the region above infrared. Its wavelength is 4000 to 8000 Å.

45. An angstrom is one ten-thousandth of a micron.

46. Bandwidth is the spectrum space occupied by a signal, the frequency range of a transmitted signal, or the range of frequencies accepted by a receiver. It is the difference between the upper and lower frequencies of the range in question.

47. There is more spectrum space available at the higher frequencies. For a given bandwidth signal, more channels can be accommodated at the higher frequencies.

48. Spectrum space is a precious natural resource.

49. In the United States, the FCC regulates the use of the spectrum and most forms of electronic communications according to the Communications Act of 1934.

50. Most countries belong to the ITU, an organization devoted to worldwide cooperation and negotiation on spectrum usage.

51. The NTIA coordinates government and military communications in the United States.

CHAPTER REVIEW QUESTIONS

Choose the letter that best answers each question.

1-1. Communication is the process of
  a. Keeping in touch.
  b. Broadcasting.
  c. Exchanging information.
  d. Entertainment by electronics.

1-2. Two key barriers to human communication are
  a. Distance.
  b. Cost.
  c. Ignorance.
  d. Language.

1-3. Electronic communications was discovered in which century?
  a. Sixteenth
  b. Eighteenth
  c. Nineteenth
  d. Twentieth

1-4. Which of the following is *not* a major communications medium?
  a. Free space
  b. Water
  c. Wires
  d. Fiber-optic cable

1-5. Random interference to transmitted signals is called
  a. Adjacent channel overlap.
  b. Cross talk.
  c. Garbage-in–garbage-out.
  d. Noise.

1-6. The communications medium causes the signal to be
  a. Amplified.
  b. Modulated.
  c. Attenuated.
  d. Interfered with.

1-7. Which of the following is *not* a source of noise?
  a. Another communications signal
  b. Atmospheric effects
  c. Manufactured electrical systems
  d. Thermal agitation in electronic components

1-8. One-way communications is called
  a. Half duplex.
  b. Full duplex.
  c. Monocomm.
  d. Simplex.

1-9. Simultaneous two-way communications is called
  a. Half duplex.
  b. Full duplex.
  c. Bicomm.
  d. Simplex.

1-10. The original electrical information signal to be transmitted is called the
  a. Modulating signal.
  b. Carrier.
  c. Baseband signal.
  d. Source signal.

1-11. The process of modifying a high-frequency carrier with the information to be transmitted is called
  a. Multiplexing.
  b. Telemetry.
  c. Detection.
  d. Modulation.

1-12. The process of transmitting two or more information signals simultaneously over the same channel is called
  a. Multiplexing.
  b. Telemetry.
  c. Mixing.
  d. Modulation.

1-13. Continuous voice or video signals are referred to as being
  a. Baseband.
  b. Analog.
  c. Digital.
  d. Continuous waves.

1-14. Recovering information from a carrier is known as

a. Demultiplexing.

b. Modulation.

c. Detection.

d. Carrier recovery.

1-15. Transmission of graphical information over the telephone network is accomplished by

a. Television.

b. CATV.

c. Videotext.

d. Facsimile.

1-16. Measuring physical conditions at some remote location and transmitting this data for analysis is the process of

a. Telemetry.

b. Instrumentation.

c. Modulation.

d. Multiplexing.

1-17. Receiving electromagnetic emissions from stars is called

a. Astrology.

b. Optical astronomy.

c. Radio astronomy.

d. Space surveillance.

1-18. A personal communications hobby for individuals is

a. Ham radio.

b. Electronic bulletin board.

c. CB radio.

d. Cellular radio.

1-19. Radar is based upon

a. Microwaves.

b. A water medium.

c. The directional nature of radio signals.

d. Reflected radio signals.

1-20. A frequency of 27 MHz has a wavelength of approximately

a. 11 m.

b. 27 m.

c. 30 m.

d. 81 m.

1-21. Radio signals are made up of

a. Voltages and currents.

b. Electric and magnetic fields.

c. Electrons and protons.

d. Noise and data.

1-22. The voice frequency range is

a. 30 to 300 Hz.

b. 300 to 3000 Hz.

c. 20 Hz to 20 kHz.

d. 0 Hz to 15 kHz.

1-23. Another name for signals in the HF range is

a. Microwaves.

b. RF waves.

c. Shortwaves.

d. Millimeter waves.

1-24. Television broadcasting occurs in which ranges?

a. HF

b. EHF

c. VHF

d. UHF

1-25. Electromagnetic waves produced primarily by heat are called

a. Infrared rays.

b. Microwaves.

c. Shortwaves.

d. X-rays.

1-26. A micron is

a. One-millionth of a foot.

b. One-millionth of a meter.

c. One-thousandth of a meter.

d. One ten-thousandth of an inch.

1-27. The frequency range of infrared rays is approximately

a. 30 to 300 GHz.

b. 4000 to 8000 Å.

c. 1000 to 10,000 Å.

d. 0.7 to 100 μm.

1-28. The approximate wavelength of red light is

a. 1000 μm.

b. 7000 Å.

c. 3500 Å.

d. 4000 Å.

1-29. Which of the following is *not* used for communications?

a. X-rays

b. Millimeter waves

c. Infrared

d. Microwaves

1-30. A signal occupies the spectrum space from 1.115 to 1.122 GHz. The bandwidth is

a. 0.007 MHz.

b. 7 MHz.

c. 237 MHz.

d. 700 MHz.

1-31. In the United States, the electromagnetic spectrum is regulated and managed by

    a. Business and industry.   *c.* FCC.

    b. ITU.   *d.* The United Nations.

1-32. For a given bandwidth signal, more channel space is available for signals in the range of

    a. VHF.   *c.* SHF.

    b. UHF.   *d.* EHF.

CRITICAL THINKING QUESTIONS

1-1. Considering the wide range of communications applications outlined in Fig. 1-4, which one of these interests you the most as a career? Why?

1-2. Could light be used as a carrier for an intelligence signal? Explain.

1-3. What if all the available spectrum space from ELF to microwave was used up or otherwise assigned? How could communications capability be added?

1-4. List all the electronic communications techniques and services that you use on a regular basis. Which additional services would you like to use?

1-5. Discuss how each of the following could be used as the transmission medium for intelligence signals: (a) the earth, (b) a rubber hose, and (c) a piece of string.

Answers to Self-Tests

1. computers, communications, control, computers
2. exchanging information
3. oral, written
4. distance, language
5. late nineteenth century
6. transmitter, receiver, channel or medium
7. wire, radio, fiber-optic cable
8. transmitter
9. receiver
10. noise, communications channel
11. degrades, attenuates
12. the atmosphere, manufactured equipment, thermal agitation in components
13. simplex, radio and TV broadcasting
14. full duplex, telephone communications
15. half duplex
16. analog
17. digital
18. analog-to-digital converter
19. baseband
20. modulation, baseband, carrier
21. demodulation or detection
22. multiplexing
23. facsimile, teletext
24. garage door opener
25. paging
26. telemetry
27. radio waves
28. microwave relay, satellites, cordless phones, cellular phones
29. reflected
30. sonar (active)
31. active, passive
32. amateur or "ham" radio
33. modems
34. local area networks
35. electromagnetic waves or radio-frequency (RF) waves
36. electric, magnetic
37. 16.67 ($^{300}/_{18} = 16.67$)
38. 50, 60, ELF
39. *a, c*
40. 20, 20,000
41. 300, 3000
42. false
43. MF
44. shortwaves
45. VHF
46. land mobile, cellular telephones, military, radar
   and navigation, amateur radio
47. 1000
48. microwaves
49. radar, satellite
50. millimeter
51. $^1/_{1,000,000}$
52. heat
53. 0.7, 10
54. $^1/_{10,000}$
55. 4000, 8000
56. fiber-optic cables, free space
57. bandwidth
58. sidebands
59. 125 (1.175 − 1.050 = 0.125 MHz = 125 kHz)
60. higher
61. lower
62. spectrum space
63. Communications Act of 1934
64. Federal Communications Commission
65. National Telecommunications and Information Administration
66. International Telecommunications Union

CHAPTER 2
Amplitude Modulation and Single-Sideband Modulation

∎

CHAPTER OBJECTIVES

This chapter will help you to:

1. *Recognize* an AM signal in the time domain (oscilloscope display), the frequency domain (spectral display), or in trigonometric equation form.
2. *Calculate* the percentage of modulation of an AM signal given waveform measurements.
3. *Calculate* the upper and lower sidebands of an AM signal given the carrier and modulating signal frequencies.
4. *Calculate* the sideband power in an AM wave given the carrier power and the percentage of modulation.
5. *Define* the terms DSB and SSB and *state* the benefits of SSB over an AM signal.

One of the principal techniques used in electronic communications is modulation. Modulation is the process of having the information to be transmitted alter a higher-frequency signal for the purpose of transmitting the information somewhere in the electromagnetic spectrum via radio, wire, or fiber-optic cable. Without modulation, electronic communications would not exist as we know it today. Communications electronics is largely the study of various modulation techniques and of the modulator and demodulator circuits that make modulation possible. The three principal types of electronic communications are amplitude modulation (AM), frequency modulation (FM), and phase modulation (PM). The oldest and simplest form of modulation is AM. We will cover AM along with a derivation known as single-sideband modulation in this chapter. Chapter 3 will cover amplitude modulator and demodulator circuits.

∎

2-1 AMPLITUDE MODULATION PRINCIPLES

Information signals such as voice, video, or binary data are sometimes transmitted directly from one point to another over some communications medium. For example, voice signals are transmitted by way of wires in the telephone system. Coaxial cables carry video signals between two points, and twisted-pair cable is often used to carry binary data from one point to another. However, when transmission distances are far, cables are sometimes impractical. In such cases, radio communications is used. To carry out reliable long-distance radio communication, a high-frequency signal must be used. It is simply impractical to convert the information signal directly to electromagnetic radiation. Excessively long antennas and interference between signals would result if information signals were transmitted directly. For this reason, it is desirable to translate the information signal to a point higher in the electromagnetic frequency spectrum. It is the process of modulation that creates a higher-frequency signal containing the original information.

Modulation can also be defined as the process of modifying the characteristic of one signal in accordance with some characteristic of another signal. In most cases, the information signal, be it voice, video, binary data, or some other information, is normally used to modify a higher-frequency signal known as the carrier. The information signal is usually called the *modulating signal,* and the higher-frequency signal which is being modulated is called the *carrier* or *modulated wave.* The carrier is usually a sine wave, while the information signal can be of any shape permitting both analog and digital signals to be transmitted. In most cases, the carrier frequency is considerably higher than the highest information frequency to be transmitted.

In AM, the information signal varies the amplitude of the carrier sine wave. In other words, the instantaneous value of the carrier amplitude changes in accordance with the amplitude and frequency variations of the modulating signal. Figure 2-1 shows a single-frequency sine wave modulating a higher-frequency carrier signal. Note that the carrier frequency remains constant during the modulation process but that its amplitude varies in accordance with the modulating signal. An increase in the modulating signal amplitude causes the amplitude of the carrier to increase. Both the positive and negative peaks of the carrier wave vary with the modulating signal. An increase or decrease in the amplitude of the modulating signal causes a corresponding increase or decrease in both the positive and negative peaks of the carrier amplitude.

If you interconnect the positive and negative peaks of the carrier waveform with an imaginary line (shown dashed in Fig. 2-1), then you recreate the exact shape of the modulating information signal. This imaginary line on the carrier waveform is known as the *envelope,* and it is the same as the modulating signal.

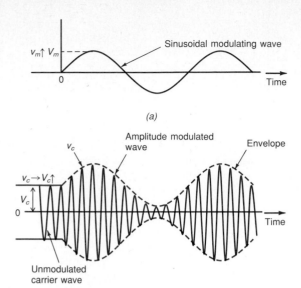

Fig. 2-1 Amplitude modulation. (*a*) The modulating or information signal. (*b*) The modulated carrier.

Because complex waveforms like that shown in Fig. 2-1 are difficult to draw, they are usually simplified by representing the high-frequency carrier wave as simply many equally spaced vertical lines whose amplitudes vary in accordance with a modulating signal. Figure 2-2 shows a sine wave tone modulating a higher-frequency carrier. We will use this method of representation throughout this book.

The signals illustrated in Figs. 2-1 and 2-2 show the variation of the carrier signal with respect to time. Such signals are said to be in

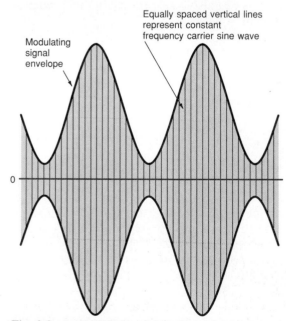

Fig. 2-2 A simplified method of representing an AM high-frequency sine wave.

Time domain

DEGREE VERSUS RADIAN MEASURE OF ANGLES

You may recall that a sine wave reaches 70.7 percent of its maximum value at a phase angle of 45°. In general, the instantaneous value of a sine wave can be found by

$$v = V_m \times \sin \theta$$

where v = the instantaneous value
V_m = the maximum value
θ = the phase angle

Example: A 1-MHz sine wave has a peak or maximum value of 18 V. What is its instantaneous value at a phase angle of 45°?

$$\sin 45° = 0.707$$

$$v = 18\text{ V} \times 0.707 = 12.7\text{ V}$$

In communications, the phase angle may be stated in an equivalent way using the frequency of the signal and some time of interest. This is known as *radian* measure. For example, the instantaneous value of a signal can be found with:

$$v = V_m \times \sin(2\pi ft)$$

where f = the frequency of the signal
t = the time of interest

A 1-MHz signal has a period of 1 μs. One period equals one cycle with 360°. A phase angle of 45° corresponds to 45/360 or ⅛ cycle, which is a time of 1 μs divided by 8(0.125 μs).

Example: Use the second equation for finding the instantaneous value of a 1-MHz sine signal with a peak value of 18 volts (V) at a time of 0.125 μs. (Your calculator must be in the *radian* mode.)

$$v = 18\text{ V} \times \sin (6.28 \times 1\text{ MHz} \times 0.125\text{ μs})$$
$$= 12.7\text{ V}$$

Conclusion: The two equations for finding v are equivalent. The first is based on angular measure, and the second is based on radian measure. Keep in mind that 1 rad = 57.3°. In our example, 45° = 45/57.3 = 0.785 rad.

$$\sin 0.785 = 0.707$$

---

the time domain. Time-domain signals are the actual variation of voltage over time. They are what you would see displayed on the screen of an oscilloscope. We will show the time-domain signals created by the various types of modulation. Later you will see that modulated signals can also be expressed in the frequency domain.

As you know, sinusoidal alternating current (ac) signals can be represented mathematically by trigonometric functions. For example, we can express the sine wave carrier with the simple expression

$$v_c = V_c \sin 2\pi f_c t$$

In this expression, v_c represents the instantaneous value of the sine wave voltage at some specific time in the cycle. The V_c represents the peak value of the sine wave as measured between zero and the maximum amplitude of either the positive or negative-going alternations. See Fig. 2-1. The term f_c is the frequency of the carrier sine wave. Finally, t represents some particular point in time during the ac cycle.

In the same way, a sine wave modulating signal can also be expressed with a similar formula:

$$v_m = V_m \sin 2\pi f_m t$$

where f_m = the frequency of the modulating signal

Referring back to Fig. 2-1, you can see that the modulating signal uses the peak value of the carrier rather than zero as its reference point. The envelope of the modulating signal varies above and below the peak carrier amplitude. That is, the zero reference line of the modulating signal coincides with the peak value of the unmodulated carrier. Because of this, the relative amplitudes of the carrier and modulating signal are important. In general, the amplitude of the modulating signal should be less than the amplitude of the carrier. If the amplitude of the modulating signal is greater than the amplitude of the carrier, distortion will occur. Distortion causes incorrect information to be transmitted. It is important in AM that the peak value of the modulating signal be less than the peak value of the carrier.

Using the mathematical expressions for the carrier and the modulating signal, we can create a new mathematical expression for the complete modulated wave. First, keep in mind that the peak value of the carrier is the reference point for the modulating signal. The modulating signal value adds to or subtracts from the peak value of the carrier. This instantaneous value of either the top or bottom voltage envelope can be computed by the simple expression

$$v_1 = V_c + v_m$$

Substituting the trigonometric expression for v_m, we get

$$v_1 = V_c + V_m \sin 2\pi f_m t$$

All this expression says is that the instantaneous value of the modulating signal algebraically adds to the peak value of the carrier. As you can see, the value of v_1 is really the envelope of the carrier wave. For that reason, we can write the instantaneous value of the complete modulated wave v_2 as

$$v_2 = v_1 \sin 2\pi f_c t$$

In this expression, the peak value of carrier wave V_c from the first equation given is replaced by v_1. Now, substituting the previously derived expression for v_1 and expanding, we get

$$v_2 = (V_c + V_m \sin 2\pi f_m t) \sin 2\pi f_c t$$

$$= \underbrace{V_c \sin 2\pi f_c t}_{Carrier} + \underbrace{(V_m \sin 2\pi f_m t)(\sin 2\pi f_c t)}_{+\ modulation\ \times\ carrier}$$

This expression consists of two parts: the first part is simply the carrier waveform, and the second part is the carrier waveform *multiplied* by the modulating signal waveform. It is this second part of the expression that is characteristic of AM. A circuit must be able to produce mathematical multiplication of analog signals in order for AM to occur.

The circuit used for producing AM is called a *modulator*. Its two inputs, the carrier and the modulating signal, and the resulting output are shown in Fig. 2-3. Amplitude modulators compute the product of the carrier and modulating signals.

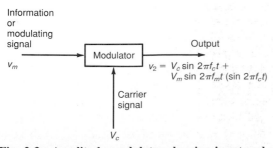

Fig. 2-3 Amplitude modulator showing input and output signals.

Information or modulating signal v_m → Modulator → Output $v_2 = V_c \sin 2\pi f_c t + V_m \sin 2\pi f_m t (\sin 2\pi f_c t)$; Carrier signal V_c

Self-Test

Answer the following questions.

1. Modulation causes the information signal to be _____ to a higher frequency for more efficient transmission.
2. During modulation, the information signal _____ the amplitude of a high-frequency signal called the _____ .
3. The circuit used to produce modulation is called a _____ . Its two inputs are _____ .
4. In AM, the instantaneous _____ of the carrier varies in accordance with the information signal.
5. True or false. The carrier frequency is usually lower than the modulating frequency.
6. The outline of the peaks of the carrier signal is called the _____ , and it has the same shape as the _____ .
7. Voltages varying over time are said to be _____ signals.
8. The trigonometric expression for the carrier is $v_c = $ _____ .
9. True or false. The carrier frequency remains constant during AM.
10. An amplitude modulator performs the mathematical operation of
    a. Addition.   c. Multiplication.
    b. Subtraction.   d. Division.

2-2 MODULATION INDEX AND PERCENTAGE OF MODULATION

In order for proper AM to occur, the modulating signal voltage V_m must be less than the carrier voltage V_c. Therefore, the relationship between the amplitudes of the modulating signal and carrier is important. This relationship is expressed in terms of a ratio known as the modulation index m (also called modulation factor, modulation coefficient, or the degree of modulation). The m is simply the ratio of the modulating signal voltage to the carrier voltage:

$$m = \frac{V_m}{V_c}$$

The modulation index should be a number between 0 and 1. If the amplitude of the modulating voltage is higher than the carrier voltage, m will be greater than 1. This will cause severe distortion of the modulated waveform. This is illustrated in Fig. 2-4. Here a sine wave information signal modulates a sine wave carrier, but the modulating voltage is much greater

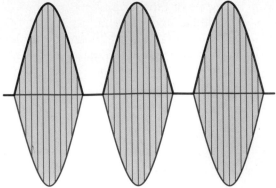

Fig. 2-4 Distortion of the envelope caused by overmodulation where the modulating signal amplitude V_m is greater than the carrier signal V_c.

than the carrier voltage. This condition is called *overmodulation*. As you can see, the waveform is flattened near the zero line. The received signal will produce an output waveform in the shape of the envelope, which in this case is a sine wave whose negative peaks have been clipped off. By keeping the amplitude of the modulating signal less than the carrier amplitude, no distortion will occur. The ideal condition for AM is where $V_m = V_c$ or $m = 1$, since this will produce the greatest output at the receiver with no distortion.

The modulation index can be determined by measuring the actual values of the modulation voltage and the carrier voltage and computing the ratio. However, it is more common to compute the modulation index from measurements taken on the composite modulated wave itself. Whenever the AM signal is displayed on an oscilloscope, the modulation index can be computed from V_{max} and V_{min} as shown in Fig. 2-5.

The peak value of the modulating signal V_m is one-half the difference of the peak and trough values and is computed with the expression

$$V_m = \frac{V_{max} - V_{min}}{2}$$

By observing Fig. 2-5, you can see V_{max} is the peak value of the signal during modulation, while V_{min} is the lowest value, or trough, of the modulated wave. The V_{max} is one-half the peak-to-peak value of the AM signal or $V_{max(p-p)}/2$. Subtracting V_{min} from V_{max} produces the peak-to-peak value of the modulating signal. One-half of that, of course, is simply the peak value.

The peak value of the carrier signal V_c is the average of the V_{max} and V_{min} values and is computed with the expression

$$V_c = \frac{V_{max} + V_{min}}{2}$$

Substituting these values in our original formula for the modulation index produces the result

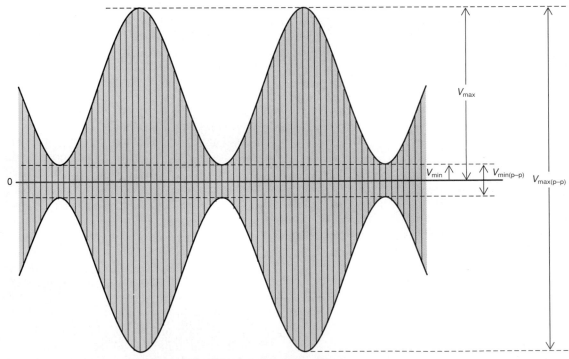

Fig. 2-5 An AM wave showing peaks (V_{max}) and troughs (V_{min}).

$$m = \frac{V_{max} - V_{min}}{V_{max} + V_{min}}$$

The values for V_{max} and V_{min} can be read directly from an oscilloscope screen and plugged into the formula to compute the modulation index.

For example, suppose the V_{max} value read from the graticule on the oscilloscope screen is 4.6 divisions and V_{min} is 0.7 divisions. The modulation index is then

$$m = \frac{4.6 - 0.7}{4.6 + 0.7}$$

$$= \frac{3.9}{5.3}$$

$$= 0.736$$

Whenever the modulation index is multiplied by 100, the degree of modulation is expressed as a percentage. The amount or depth of AM is more commonly expressed as percent modulation rather than as a fractional value. In the example above, the percentage of modulation is 100% × m or 73.6 percent. The maximum amount of modulation without signal distortion, of course, is 100 percent where V_c and V_m are equal. At this time, $V_{min} = 0$ and $V_{max} = 2V_m$, where V_m is the peak value of the modulating signal.

In practice, it is desirable to operate with as close to 100 percent modulation as possible. In this way, the maximum information signal amplitude is transmitted. More information signal power is transmitted, thereby producing a stronger, more intelligible signal. When the modulating signal amplitude varies randomly over a wide range, it is impossible to maintain 100 percent modulation. A voice signal, for example, changes amplitude as a person speaks. Only the peaks of the signal produce 100 percent modulation.

Self-Test

Choose the letter that best answers each question.

11. Which of the following is the most correct?
    a. V_m should be greater than V_c.
    b. V_c should be greater than V_m.
    c. V_m should be equal to or less than V_c.
    d. V_c must always equal V_m.
12. Which of the following is *not* another name for modulation index?
    a. Modulation reciprocal
    b. Modulation factor

    c. Degree of modulation
    d. Modulation coefficient
13. The degree or depth of modulation occurring expressed as a percentage is computed using the expression
    a. $2V_m$.
    b. $100/m$.
    c. $m/100$.
    d. 100% × m.

Supply the missing information in each statement.

14. The modulation index is the ratio of the peak voltage of the _____ to the _____ .
15. An AM wave displayed on an oscilloscope has values of $V_{max} = 3.8$ and $V_{min} = 1.5$ as read from the graticule. The percentage of modulation is _____ percent.
16. The ideal percentage of modulation for maximum amplitude of information transmission is _____ percent.
17. To achieve 85 percent modulation of a carrier of $V_c = 40$ volts (V), a modulating signal of $V_m = $ _____ is needed.
18. The peak-to-peak value of an AM signal is 30 V. The peak-to-peak value of the modulating signal is 12 V. The percentage of modulation is _____ percent.

2-3 SIDEBANDS AND THE FREQUENCY DOMAIN

Whenever a carrier is modulated by an information signal, new signals at different frequencies are generated as part of the process. These new frequencies, called *side frequencies* or *sidebands,* occur in the frequency spectrum directly above and directly below the carrier frequency. More specifically, the sidebands occur at frequencies that are the sum and difference of the carrier and modulating frequencies. Assuming a carrier frequency of f_c and a modulating frequency of f_m, the upper sideband f_{USB} and lower sideband f_{LSB} are computed as follows:

$$f_{USB} = f_c + f_m$$

$$f_{LSB} = f_c - f_m$$

The existence of these additional new signals that result from the process of modulation can also be proven mathematically. This can be done by starting with the equation for an AM signal v_2 described previously.

$$v_2 = V_c \sin 2\pi f_c t + (V_m \sin 2\pi f_m t)(\sin 2\pi f_c t)$$

There is a trigonometric identity that says that the product of two sine waves is

$$\sin A \sin B = \frac{\cos (A - B)}{2} - \frac{\cos (A + B)}{2}$$

By substituting this identity into the expression for our modulated wave, the instantaneous amplitude of the signal becomes

$$e_2 = \overbrace{V_c \sin 2\pi f_c t}^{\text{Carrier}} + \overbrace{\frac{V_m}{2} \cos 2\pi t(f_c - f_m)}^{\text{LSB}}$$

$$\underbrace{- \frac{V_m}{2} \cos 2\pi t(f_c + f_m)}_{\text{USB}}$$

As you can see, the second and third terms of this expression contain the sum $f_c + f_m$ and difference $f_c - f_m$ of the carrier and modulating signal frequencies. The first element in the expression is simply the carrier wave to which

is added the difference frequency and the sum frequency.

By algebraically adding the carrier and the two sideband signals together, the standard AM waveform described earlier is obtained. This is illustrated in Fig. 2-6. This is solid proof that an AM wave contains not only the carrier but also the sideband frequencies. Observing an AM signal on an oscilloscope, you can see the amplitude variations of the carrier with respect to time. This is called a time-domain display. It gives no indication of the existence of the sidebands, although the modulation process does indeed produce them.

Another method of showing the sideband signals is to plot the carrier and sideband amplitudes with respect to frequency. This is illustrated in Fig. 2-7. Here the horizontal axis represents frequency, and the vertical axis represents the amplitudes of the signals. A plot of signal amplitude versus frequency is

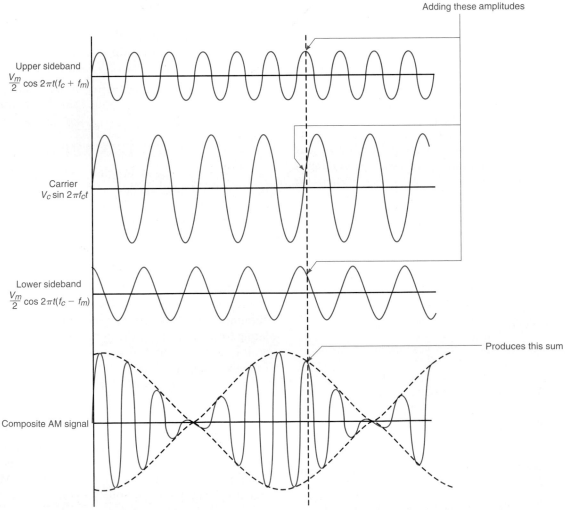

Fig. 2-6 The algebraic sum of the carrier and sideband signals is the AM signal.

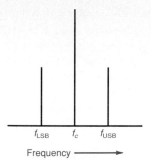

Fig. 2-7 A frequency-domain display of an AM signal.

referred to as a frequency-domain display. A test instrument known as a spectrum analyzer will display the frequency domain of a signal.

Whenever the modulating signal is more complex than a single sine wave tone, multiple upper and lower side frequencies will be produced. For example, a voice signal consists of many different-frequency sine wave components mixed together. Recall that voice frequencies occur in the 300- to 3000-Hz range. Therefore, voice signals will produce a range of frequencies above and below the carrier frequency as shown in Fig. 2-8. These sidebands take up spectrum space. You can compute the total bandwidth of the AM signal by computing the maximum and minimum sideband frequencies. This is done by finding the sum and difference of the carrier frequency and maximum modulating frequency, 3000 Hz, or 3 kHz, in the audio example above. If the carrier frequency is 2.8 MHz, or 2800 kHz, then the maximum and minimum sideband frequencies are

$$f_{USB} = 2800 + 3 = 2803 \text{ kHz}$$
$$f_{LSB} = 2800 - 3 = 2797 \text{ kHz}$$

The total bandwidth (BW) then is simply the difference between the upper and lower sideband frequencies or

Fig. 2-8 The upper and lower sidebands of a voice modulator AM signal.

$$BW = f_{USB} - f_{LSB}$$
$$= 2803 - 2797$$
$$= 6 \text{ kHz}$$

As it turns out, the bandwidth of the AM signal is simply twice the highest frequency in the modulating signal. With a voice signal whose maximum frequency is 3 kHz, the total bandwidth would simply be twice this, or 6 kHz.

When other complex signals such as pulses or rectangular waves modulate a carrier, again a broad spectrum of sidebands will be produced. According to the Fourier theory, complex signals such as square waves, triangular waves, sawtooth waves, or distorted sine waves are simply made up of a fundamental sine wave and numerous harmonic signals at different amplitudes. The classical example is that of a square wave which is made up of a fundamental sine wave and all odd harmonics. A modulating square wave will produce sidebands at frequencies of the fundamental square wave as well as at the third, fifth, seventh, etc., harmonics. The resulting frequency-domain plot would appear like that shown in Fig. 2-9(*a*) on the next page. As you can see, pulses generate extremely wide bandwidth signals. In order for the square wave to be transmitted and received without distortion or degradation, all the sidebands must be passed by the antennas and the transmitting and receiving circuits.

Figure 2-9(*b*) shows the relationship between the time and frequency domain presentations of the modulating square wave. The time domain shows the individual sine wave harmonics that, when added together, produce the square wave. The frequency domain shows the signal amplitudes of the harmonics that modulate the carrier and produce sidebands.

Self-Test

Answer the following questions.

19. New signals generated by the modulation process that appear directly above and below the carrier frequency are called _____ .

20. An AM radio station operating at 630 kHz is permitted to broadcast audio frequencies up to 5 kHz. The upper and lower side frequencies are
$$f_{USB} = \text{_____ kHz}$$
$$f_{LSB} = \text{_____ kHz}$$

Sidebands produced by the fundamental and its harmonics

Fundamental

Seventh harmonic

Fifth harmonic

Third harmonic

f_c

(a)

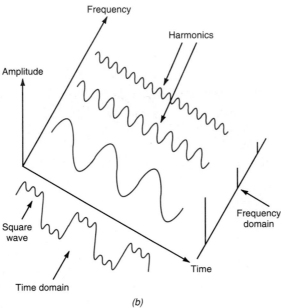

Frequency

Harmonics

Amplitude

Square wave

Frequency domain

Time

Time domain

(b)

Fig. 2-9 (a) Frequency spectrum of an AM signal modulated by a square wave. (b) Relationship between time and frequency domain displays of the modulating signal.

21. The total bandwidth of the signal in the above example is _____ kHz.
22. A signal whose amplitude is displayed with respect to time is called a _____-domain display. The test instrument used to present such a display is the _____ .
23. A signal whose amplitude is displayed with respect to frequency is called a _____-domain display. The test instrument used to present such a display is the _____ .
24. Complex modulating signals containing many frequencies produce multiple _____ thus occupying more spectrum space.

25. The AM signal that occupies the greatest bandwidth is the one modulated by a
   a. 1-kHz sine wave.
   b. 5-kHz sine wave.
   c. 1-kHz square wave.
   d. 5-kHz square wave.
26. The composite AM signal can be recreated by algebraically adding which three signals?

2-4 AMPLITUDE MODULATION POWER DISTRIBUTION

To communicate by radio, the AM signal is amplified by a power amplifier and fed to the antenna with a characteristic impedance R. The total transmitted power divides itself between the carrier and the upper and lower sidebands. The total transmitted power P_T is simply the sum of the carrier power P_c and the power in the two sidebands P_{USB}, and P_{LSB}. This is expressed by the simple equation below

$$P_T = P_c + P_{LSB} + P_{USB}$$

The power in the sidebands depends upon the value of the modulation index. The greater the percentage of modulation, the higher the sideband power. Of course, maximum power appears in the sidebands when the carrier is 100 percent modulated. The power in each sideband P_s is given by the expression

$$P_s = P_{LSB} = P_{USB} = \frac{P_c \, (m^2)}{4}$$

Assuming 100 percent modulation where the modulation factor $m = 1$, the power in each sideband is one-fourth, or 25 percent, of the carrier power. Since there are two sidebands, their power together represents 50 percent of the carrier power. For example, if the carrier power is 100 watts (W), then at 100 percent modulation, 50 W will appear in the sidebands, 25 W in each. The total transmitted power then is the sum of the carrier and sideband powers or 150 W.

As you can see, the carrier power represents two-thirds of the total transmitted power assuming 100 percent modulation. With a carrier power of 100 W and a total power of 150 W, the carrier power percentage can be computed.

$$Carrier \ power \ percentage = \frac{100}{150}$$

$$= 0.667 \ (or \ 66.7\%)$$

The percentage of power in the sidebands can be computed in a similar way:

$$Sideband\ power\ percentage = \frac{50}{150}$$

$$= 0.333\ (or\ 33.3\%)$$

The carrier itself conveys no information. The carrier can be transmitted and received, but unless modulation occurs, no information will be transmitted. When modulation occurs, sidebands are produced. It is easy to conclude, therefore, that all the transmitted information is contained within the sidebands. Only one-third of the total transmitted power is allotted to the sidebands, while the remaining two-thirds of it is literally wasted on the carrier. Obviously, although it is quite effective and still widely used, AM is a very inefficient method of modulation.

At lower percentages of modulation, the power in the sidebands is even less. You can compute the amount of power in a sideband with the previously given expression. Assume a carrier power of 500 W and a modulation of 70 percent. The power in each sideband then is

$$P_S = \frac{P_c(m^2)}{4}$$

$$= \frac{500(0.7)^2}{4}$$

$$= \frac{500(0.49)}{4}$$

$$= 61.25\ W$$

At 70 percent modulation, only 61.25 W appears in each sideband for a total sideband power of 122.5 W. The carrier power, of course, remains unchanged at 500 W.

One way to calculate the total AM power is to use the formula

$$P_T = P_c(1 + m^2/2)$$

where P_c = unmodulated carrier power
m = modulation index

For example, if the carrier power is 1200 W and the percentage of modulation is 90 percent, the total power is

$$P_T = 1200(1 + 0.9^2/2) = 1200(1.405) = 1686\ W$$

If you subtract the carrier power, this will leave the power in both sidebands.

$$P_T = P_c + P_{LSB} + P_{USB}$$

$$P_{LSB} + P_{USB} = P_T - P_c$$
$$= 1686 - 1200 = 486\ W$$

Since the sideband powers are equal, the power in each sideband is 486/2 = 243 W.

In many cases, the power output from an AM transmitter is not measured directly. Instead, the output current in the antenna is measured and the power is calculated. The total modulated current I_T is determined by the unmodulated carrier current I_c and the modulation index m.

$$I_T = I_c\sqrt{(1 + m^2/2)}$$

If the unmodulated carrier current is 4 A and the percentage of modulation is 70 percent, the total output current is

$$I_T = 4\sqrt{(1 + 0.7^2/2)} = 4\sqrt{1.245}$$
$$= 4(1.116) = 4.46\ A$$

In practice, 100 percent modulation is difficult to maintain. The reason for this is that typical information signals, such as voice and video, do not have constant amplitudes. The complex voice and video signals vary over a wide amplitude and frequency range, so 100 percent modulation only occurs on the peaks of the modulating signal. For this reason, the average sideband power is considerably less than the ideal 50 percent produced by full 100 percent modulation. With less sideband power transmitted, the received signal is weaker and communication is less reliable.

Despite its inefficiency, AM is still widely used because it is simple and effective. It is used in AM radio broadcasting, CB radio, TV broadcasting, and in some aircraft and marine communications.

▐▐▶ **YOU MAY RECALL** that there are three basic ways to calculate the power dissipated in a load. These are:

$$P = V(I)$$
$$P = V^2/R$$
$$P = I^2R$$

Simply select the formula for which you have the values of current, voltage, or resistance. In an AM radio transmitting station, R is the load resistance, which is an antenna. To a transmitter, the antenna looks like a resistance. Although an antenna is not actually a physical resistor, it does appear to be one. This resistance is referred to as the *characteristic resistance of the antenna*. You will learn more about it in a later chapter.

Measurements of voltage at an antenna are difficult and rarely made. As a result, the first two formulas are seldom used in transmitter measurements. Current in an antenna is easier to measure because accurate radio-frequency current meters are available. Therefore, current is the quantity normally available to determine power. For example, if you know that the unmodulated carrier produces a current of 2.5 A in an antenna with a characteristic resistance of 73 Ω, the power is:

$$P = I^2R = (2.5)^2(73) = 6.25(73) = 456.25 \text{ W}$$

If the carrier is modulated, the antenna current will be higher because of the additional power in the sidebands. The total antenna current I_T is

$$I_T = I_c\sqrt{(1 + m^2/2)}$$

where I_c = unmodulated carrier current
m = modulation index

The total AM power then is

$$P_T = (I_T)^2R$$

To determine the total power, monitor the total modulated antenna current and make the calculation above, given the antenna resistance.

If you measure both the modulated and the unmodulated carrier antenna currents, you can compute the percentage of modulation by rearranging the formula above.

$$m = \sqrt{2[(I_T/I_c)^2 - 1]}$$

Assume that you measured the unmodulated carrier current and found it to be 1.8 A. With modulation, the total current was 2 A. The percentage of modulation is:

$$m = \sqrt{2[(2/1.8)^2 - 1]} = \sqrt{2[1.234 - 1]}$$
$$= \sqrt{2(0.234)} = \sqrt{0.468} = 0.684 \text{ or } 68.4\%$$

Self-Test

Choose the letter which best answers each question.

27. The total sideband power is what percentage of the carrier power for 100 percent modulation?
    a. 25 percent
    b. 50 percent
    c. 100 percent
    d. 150 percent
28. Information in an AM signal is conveyed in the
    a. Carrier.
    b. Sidebands.
    c. Both together.

Supply the missing information in each statement.

29. The load into which the AM signal power is dissipated is a(n) _____ .
30. The total transmitted power is the sum of the _____ and _____ powers.
31. A 5-kW carrier with 60 percent modulation produces _____ kW in each sideband.
32. In an AM signal with a carrier of 18 W and a modulation percentage of 75 percent, the total power in the sidebands is _____ W.
33. An AM signal with a carrier of 1 kW has 100 W in each sideband. The percentage of modulation is _____ percent.
34. An AM transmitter has a carrier power of 200 W. The percentage of modulation is 60 percent. The total signal power is _____ W.
35. The total AM signal power is 2800 W. The carrier power is 2000 W. The power in one sideband is _____ W. The percentage of modulation is _____ .
36. The unmodulated carrier current in an antenna is 1.5 A. When the carrier is modulated by 95 percent, the total antenna current is _____ A.

2-5 SINGLE-SIDEBAND COMMUNICATIONS

It is obvious from the previous discussion that AM is an inefficient and, therefore, wasteful method of communications. Two-thirds of the transmitted power appears in the carrier which itself conveys no information. The real information is contained within the sidebands. One way to overcome this problem is simply to suppress the carrier. Since the carrier does not provide any useful information, there is no reason why it has to be transmitted. By suppressing the carrier, the resulting signal is simply the upper and lower sidebands. Such a signal is referred to as a *double-sideband suppressed carrier (DSSC or DSB)* signal. The benefit, of course, is that no power is wasted on the carrier and that the power saved can be put into the sidebands. Double-sideband suppressed carrier modulation is simply a special case of AM with no carrier.

A typical DSB signal is shown in Fig. 2-10. This signal is simply the algebraic sum of the two sinusoidal sidebands. This is the signal produced when a carrier is modulated by a single-tone sine wave information signal. During the modulation process, the carrier is suppressed, but the two sideband signals remain. Even though the carrier is suppressed, the time-domain DSB signal is a sine wave at the carrier frequency varying in amplitude as shown. Note that the envelope of this waveform is not the same as the modulating signal as it is in a pure AM signal with carrier.

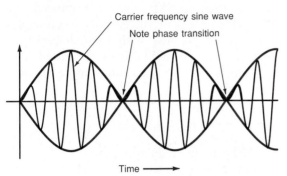

Fig. 2-10   A time-domain display of a DSB AM signal.

A frequency-domain display of a DSB signal is given in Fig. 2-11. Note that the spectrum space occupied by a DSB signal is the same as that for a conventional AM signal.

Double-sideband suppressed carrier signals are generated by a circuit called a *balanced modulator*. The purpose of the balanced modulator is to produce the sum and difference frequencies but to cancel or balance out the carrier. You will learn more about balanced modulators in Chap. 3.

By using DSB AM, considerable power is saved by eliminating the carrier. This power can be put into the sidebands for stronger signals over longer distances. But considering the

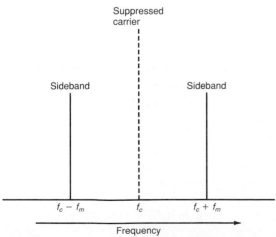

Fig. 2-11   A frequency-domain display of a DSB signal.

DSB signal further, we find that it is redundant. That is, the basic information is transmitted twice, once in each sideband. Since the sidebands are the sum and difference of the carrier and modulating signals, the information must be contained in both of them. As it turns out, there is absolutely no reason to transmit both sidebands in order to convey the information. One sideband may be suppressed. The remaining sideband is called a *single-sideband suppressed carrier (SSSC or SSB)* signal.

The SSB signal offers four major benefits. First, the spectrum space occupied by the SSB signal is only half that of AM and DSB signals. This greatly conserves spectrum space and allows more signals to be transmitted in the same frequency range. It also means there should be less interference between signals.

The second benefit is that all the power previously devoted to the carrier and other sideband can be channeled into the single sideband, thereby producing a stronger signal that should carry farther and be more reliably received at greater distances.

The third benefit is that there is less noise on the signal. Noise gets added to all signals in the communications medium or in the receiver itself. Noise is a random voltage made up of an almost infinite number of frequencies. If the signal bandwidth is restricted and the receiver circuits are made with a narrower bandwidth, a great deal of the noise is filtered out. Since the SSB signal has less bandwidth than an AM or a DSB signal, logically there will be less noise on it. This is a major advantage in weak signal long-distance communications.

The fourth advantage of SSB signals is that they experience less fading than an AM signal. Fading means that a signal alternately increases and decreases in strength as it is picked up by the receiver. It occurs because the carrier and sidebands may reach the receiver shifted in time and phase with respect to one another. The carrier and sidebands, which are on separate frequencies, are affected by the ionosphere in different ways. The ionosphere bends the carrier and sideband signals back to earth at slightly different angles so that sometimes they reach the receiver in such a way that they cancel out one another rather than adding up to the desired AM wave. The result is fading. With SSB there is only one sideband, so this kind of fading does not occur.

An SSB signal has some unusual characteristics. First, when no information or modulating signal is present, no RF signal is transmitted. In a standard AM transmitter, the carrier is still transmitted even though it may not be modulated. This is the condition that might occur during a voice pause on an AM broadcast station. But since there is no carrier transmitted in an SSB system, no signals are present if the information signal is zero. Sidebands are generated only during the modulation process, such as when someone speaks into a microphone.

Figure 2-12 shows the frequency- and time-domain displays of an SSB signal produced when a steady 2-kHz sine wave tone modulates a 14.3-MHz carrier. Amplitude modulation would produce sidebands of 14.298 and 14.302 MHz. In SSB, only one sideband is used. Figure 2-11(*a*) shows that only the upper sideband is generated. The RF signal is simply a constant-power 14.302-MHz sine wave. A time-domain display of the SSB signal appears in Fig. 2-12(*b*).

(a) Frequency domain

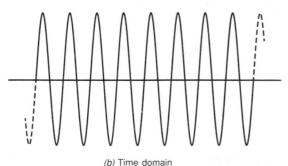

(b) Time domain

Fig. 2-12 An SSB signal produced by a 2-kHz sine wave modulating a 14.3-MHz sine wave carrier.

Of course, most information signals transmitted by SSB are not pure sine waves. A more common modulation signal is voice with its varying frequency and amplitude content. The voice signal will create a complex RF SSB signal which varies in frequency and amplitude over the narrow spectrum defined by the voice signal bandwidth.

When the voice or other modulating signal is zero, then no SSB signal is produced. An SSB RF signal is produced only when modulation occurs. In AM, with no modulating signal, the carrier would still be transmitted. This explains why SSB is so much more efficient than AM.

In conventional AM, the transmitted power is distributed among the carrier and two sidebands. If we assume a carrier power of 100 W and 100 percent modulation, each sideband will contain 25 W of power. The total transmitted power will be 150 W. The communication effectiveness of this conventional AM transmitter is established by the combined power in the sidebands, or 50 W in this example.

An SSB transmitter sends no carrier, so the carrier power is zero. Such a transmitter will have the same communication effectiveness as a conventional AM unit running much more power. A 50-W SSB transmitter will equal the performance of an AM transmitter running a total of 150 W, since they both show 50 W of total sideband power. The power advantage of SSB over AM is 3:1.

In SSB, the transmitter output is expressed in terms of peak envelope power (PEP). This is the maximum power produced on voice amplitude peaks. The PEP is computed by the familiar expression

$$P = \frac{V^2}{R}$$

where P = output power
V = root mean square (rms) output voltage
R = load resistance (usually antenna characteristic impedance)

As an example, assume that a voice signal produces a 120-volt (V) peak-to-peak signal across a 50-ohm (Ω) load. The rms voltage is 0.707 times the peak value. The peak value is one-half the peak-to-peak voltage. In this example, the rms voltage is

$$0.707\left(\frac{120}{2}\right) = 42.42 \text{ V}$$

The peak envelope power then is

$$PEP = \frac{V^2}{R}$$

$$= \frac{(42.42)^2}{50}$$

$$= 36 \text{ W}$$

The PEP input power is simply the direct-current (dc) input power of the transmitter final amplifier stage at the instant of the voice envelope peak. It is the final amplifier stage dc supply voltage multiplied by the maximum amplifier current that occurs at the peak or

$$PEP = V_s I_{max}$$

where V_s = the amplifier supply voltage
I_{max} = the current peak

A 300-V supply with a peak current of 0.6 ampere (A) produces a PEP of

$$PEP = 300(0.6) = 180 \text{ W}$$

It is important to point out that the PEP occurs only occasionally.

Voice amplitude peaks are produced only when very loud sounds are generated during certain speech patterns or when some word or sound is emphasized. During normal speech levels, the input and output power levels are much less than the PEP level. The average power is typically only one-fourth to one-third of the PEP value with typical human speech.

$$P_{avg} = \frac{PEP}{3} \quad or \quad P_{avg} = \frac{PEP}{4}$$

With a PEP of 180 W, the average power is only 45 to 60 W. Typical SSB transmitters are designed to handle only the average power level on a continuous basis, not the PEP.

The transmitted sideband will, of course, change in frequency and amplitude as a complex voice signal is applied. This sideband will occupy the same bandwidth as one sideband in a fully modulated AM signal with carrier.

Incidentally, it does not matter whether the upper or lower sideband is used since the information is contained in either. A filter is typically used to remove the unwanted sideband.

Both DSB and SSB techniques are widely used in communications. Pure SSB signals are used in telephone systems as well as in two-way radio. Two-way SSB communications is used in marine applications, in the military, and by hobbyists known as radio amateurs.

The DSB signals are used in FM and TV broadcasting to transmit two-channel stereo signals. They are also used in some types of

phase-shift keying which is used for transmitting binary data.

An unusual form of AM is that used in television broadcasting. A TV signal consists of the picture (video) signal and the audio signal which have different carrier frequencies. The audio carrier is frequency-modulated, but the video information amplitude-modulates the picture carrier. The picture carrier is transmitted, but one sideband is partially suppressed.

Video information typically contains frequencies as high as 4.2 MHz. A fully amplitude-modulated television signal would then occupy $2(4.2) = 8.4$ MHz. This is an excessive amount of bandwidth that is wasteful of spectrum space because not all of it is required to reliably transmit a TV signal. To reduce the bandwidth to the 6-MHz maximum allowed by the FCC for TV signals, a portion of the lower sideband of the TV signal is suppressed leaving only a small vestige of the lower sideband. Such an arrangement is known as a *vestigial sideband signal*. It is illustrated in Fig. 2-13. Video signals above 0.75 MHz (750 kHz) are suppressed in the lower sideband, while all video frequencies are transmitted in the upper vestigial sideband.

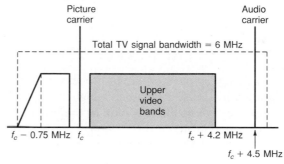

Fig. 2-13 **Vestigial sideband transmission of a TV picture signal.**

Self-Test

Answer the following questions.

37. An AM signal without a carrier is called a(n) _____ signal.
38. True or false. Two sidebands must be transmitted to retain all the information.
39. The acronym SSB means
    a. Single sideband with carrier.
    b. Single sideband with suppressed carrier.
    c. Double sideband with no carrier.
    d. Double sideband with carrier.

40. A major benefit of DSB and SSB is
    a. Higher power can be put into the sideband(s).
    b. Greater power consumption.
    c. More carrier power.
    d. Double the sideband power.
41. List four benefits of SSB over AM and DSB.
42. List two common uses of SSB.
43. A common use of DSB is
    a. Two-way communication.
    b. Telephone systems.
    c. FM/TV stereo.
    d. Satellite communications.
44. True or false. In SSB, no signal is transmitted unless the information signal is present.
45. The type of AM signal that is used in TV broadcasting is called _____ transmission.
46. An SSB signal produces a peak-to-peak voltage of 720 V on voice peaks across a 75-Ω antenna. The PEP output is _____ W.
47. An SSB transmitter has a 150-V supply. Voice peaks produce a current of 2.3 A. The PEP input is _____ W.
48. The average output power of an SSB transmitter rated at 12 W PEP is in the _____ to _____ range.

2-6 FREQUENCY CONVERSION

Frequency conversion is the process of translating a modulated signal to a higher or lower frequency while still retaining all the originally transmitted information. Although modulation itself is a form of frequency translation, frequency conversion is often used before or after transmission or reception to provide some benefit. There are many occasions where it is desirable to convert a signal to a higher or lower frequency for more convenient processing. In radio receivers, high-frequency radio signals are regularly translated to a lower frequency where improved gain and selectivity can be obtained. In satellite communications, the original signal is more conveniently generated at a lower frequency but is then translated to a higher frequency for transmission. Any modulation on the original signal is retained.

Frequency conversion is a form of AM. It is carried out by a mixer circuit. In some applications, the mixer is referred to as a converter.

The function performed by the mixer is called *heterodyning*.

Figure 2-14 shows the simplified symbol used to represent a mixer. The mixer, like a modulator, accepts two inputs. The signal f_s which is to be translated to another frequency is applied to one input, whereas the sine wave from an oscillator f_o is applied to the other input. The signal to be translated may be a simple sine wave or any complex modulated signal containing sidebands.

The mixer, like an amplitude modulator, essentially performs a mathematical multiplication of its two input signals. The oscillator is the carrier, while the signal to be translated is the modulating signal. The output will contain not only the carrier signal but also the sidebands formed when the local oscillator and input signal are mixed. The output of the mixer, therefore, consists of these signals:

$$f_s$$
$$f_o$$
$$f_o + f_s$$
$$f_o - f_s \quad \text{or} \quad f_s - f_o$$

Only one of these signals is the desired one. For example, to translate the input signal to a lower frequency, the lower sideband or difference signal $f_o - f_s$ will be chosen. The local oscillator frequency will be chosen such that when the information signal is subtracted from it, a signal with the desired lower frequency is obtained. When translating upward in frequency, the upper sideband or sum signal $f_o + f_s$ is chosen. Again the local oscillator frequency will determine what the new, higher frequency will be. A tuned circuit or filter is normally used at the output of the mixer to select the desired signal.

For example, an FM signal at 107.1 MHz is translated by an FM radio receiver to an intermediate frequency of 10.7 MHz for amplification

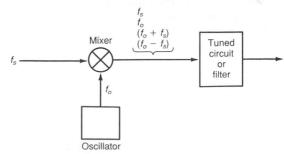

Fig. 2-14   A mixer used for frequency translation.

and detection. A local oscillator frequency of 96.4 MHz is used. The mixer output signals are

$$f_s = 107.1 \text{ MHz}$$

$$f_o = 96.4 \text{ MHz}$$

$$f_o + f_s = 96.4 + 107.1$$
$$= 203.5 \text{ MHz}$$

$$f_s - f_o = 107.1 - 96.4$$
$$= 10.7 \text{ MHz}$$

A tuned circuit selects the 10.7-MHz signal and rejects the others.

Keep in mind that although the original signal is translated in frequency, any modulation regardless of type will continue to appear on this signal. In the above example, the 10.7-MHz output signal contains the original frequency modulation. The result is as if the carrier frequency of the input signal were changed as well as all the sideband frequencies. Such frequency conversion makes it possible to shift a signal from one part of the spectrum to another as required by the application.

Self-Test

Answer the following questions.

49. The process of translating a signal to a higher or lower frequency for more convenient processing is called _____ .
50. The circuit used for translating the frequency of a signal is referred to as a(n) _____ or _____ .
51. The input signals to a frequency translation circuit are $f_s = 3.7$ MHz and $f_o = 4.155$ MHz. The output signals are _____ .
52. In the example given in the text, what other local oscillator frequency could be used to produce the 10.7-MHz output with a 107.1 = MHz input?
53. Unwanted mixer output signals are eliminated by a _____ .
54. True or false. The local oscillator is modulated.
55. True or false. Any modulation on the input signal to be translated is retained.
56. The operation carried out by a mixer is known as _____ .

SUMMARY

1. Modulation is the process of having the information to be communicated modify a higher-frequency signal called a carrier.
2. Amplitude modulation (AM) is the oldest and simplest form of modulation.
3. In AM, the amplitude of the carrier is changed in accordance with the amplitude and frequency or the characteristics of the modulating signal. The carrier frequency remains constant.
4. The amplitude variation of the carrier peaks has the shape of the modulating signal and is referred to as the envelope.
5. A time-domain display shows amplitude versus time variation of AM and other signals.
6. Amplitude modulation is produced by a circuit called a modulator which has two inputs and an output.
7. The modulator performs a mathematical multiplication of the carrier and information signals. The output is their analog product.
8. The ratio of the peak voltage value of the modulating signal V_m to the peak value of the carrier V_c is called the modulation index m ($m = V_m/V_c$). It is also referred to as the modulation coefficient or factor and the degree of modulation.
9. The ideal value for m is 1. Typically m is less than 1. The condition where m is greater than 1 should be avoided as it introduces serious distortion of the modulating signal. This is called overmodulation.
10. When the modulation index is multiplied by 100, it is called the percentage of modulation.
11. The percentage of modulation can be computed from AM waveforms displayed on an oscilloscope by using the expression

$$\% \text{ mod.} = \frac{100(V_{\max} - V_{\min})}{V_{\max} + V_{\min}}$$

where $V_{\max}$ is the maximum peak carrier amplitude and $V_{\min}$ is the minimum peak carrier amplitude.
12. The new signals generated by the modulation process are called sidebands and occur at frequencies above and below the carrier frequency.
13. The upper f_{USB} and lower f_{LSB} sideband

frequencies are the sum and difference of the carrier frequency f_c and the modulating frequency f_m and are computed with the expressions

$$f_{USB} = f_c + f_m$$

$$f_{LSB} = f_c - f_m$$

14. A display of signal amplitudes with respect to frequency is called a frequency-domain display.
15. An AM signal can be viewed as the carrier signal added to the sideband signals produced by AM.
16. The total transmitted power in an AM signal is the sum of the carrier and sideband powers $P_T = P_c + P_{USB} + P_{LSB}$) and is distributed among the carrier and sidebands. This power distribution varies with the percentage of modulation. The total power is

$$P_T = P_c(1 + m^2/2)$$

The power in each sideband is

$$P_s = \frac{P_c(m^2)}{4}$$

17. The higher the percentage of modulation, the greater the sideband power and the stronger and more intelligible the transmitted and received signal.
18. Despite its simplicity and effectiveness, AM is a highly inefficient method of modulation.
19. In an AM signal, the carrier contains no information. Any transmitted information lies solely in the sideband. For that reason, the carrier may be suppressed and not transmitted.
20. An AM signal with suppressed carrier is called a double-sideband (DSB) signal.
21. Since the same transmitted information is contained in both upper and lower sidebands, one is redundant. Full information can be transmitted using only one sideband.
22. An AM signal with no carrier and one sideband is called a single-sideband (SSB) signal. The upper and lower sidebands contain the same information, and one is not preferred over the other.
23. The main advantage of an SSB AM signal over an AM or DSB signal is that it occupies one-half the spectrum space.

24. Both DSB and SSB signals are more efficient in terms of power usage. The power wasted in the useless carrier is saved, thereby allowing more power to be put into the sidebands.
25. Power in an SSB transmitter is rated in terms of peak envelope power (PEP), the power that is produced on voice peaks. PEP output is computed using the expression

$$PEP = \frac{V^2}{R}$$

where PEP is in watts and V is the rms voltage across the antenna load impedance R. The PEP input is computed using the expression

$$PEP = V_s \times I_{max}$$

where V_s is the dc supply voltage of the final amplifier stage and I_{max} is the amplifier current on voice peaks.
26. The average output of an SSB transmitter is one-fourth to one-third of the PEP value.
27. Frequency conversion is a form of AM used to translate signals to higher or lower frequencies for improved processing.
28. Frequency conversion is carried out by a circuit called a mixer or converter.
29. The mixer performs analog multiplication of the input signal and a local oscillator signal.
30. The frequency conversion process faithfully retains any modulation (AM, FM, etc.) on the input. The translated signal contains the same modulation.
31. The output of a mixer consists of the local oscillator signal f_o and the sum and difference frequencies of the local oscillator and input frequencies:

$$f_s$$
$$f_o$$
$$f_o + f_s$$
$$f_o - f_s \quad \text{or} \quad f_s - f_o$$

32. Either the sum or the difference frequency is selected with a filter, while the others are rejected.
33. The process of mixing is also known as heterodyning.

Choose the letter that best answers each question.

2-1. Having an information signal change some characteristic of a carrier signal is called
  a. Multiplexing.
  b. Modulation.
  c. Duplexing.
  d. Linear mixing.

2-2. Which of the following is *not* true about AM?
  a. The carrier amplitude varies.
  b. The carrier frequency remains constant.
  c. The carrier frequency changes.
  d. The information signal amplitude changes the carrier amplitude.

2-3. The opposite of modulation is
  a. Reverse modulation.
  b. Downward modulation.
  c. Unmodulation.
  d. Demodulation.

2-4. The circuit used to produce modulation is called a
  a. Modulator.
  b. Demodulator.
  c. Variable gain amplifier.
  d. Multiplexer.

2-5. A modulator circuit performs what mathematical operation on its two inputs?
  a. Addition
  b. Multiplication
  c. Division
  d. Square root

2-6. The ratio of the peak modulating signal voltage to the peak carrier voltage is referred to as
  a. The voltage ratio.
  b. Decibels.
  c. The modulation index.
  d. The mix factor.

2-7. If *m* is greater than 1, what happens?
  a. Normal operation.
  b. Carrier drops to zero.
  c. Carrier frequency shifts.
  d. Information signal is distorted.

2-8. For ideal AM, which of the following is true?
  a. $m = 0$
  b. $m = 1$
  c. $m < 1$
  d. $m > 1$

2-9. The outline of the peaks of a carrier has the shape of the modulating signal and is called the
  a. Trace.
  b. Waveshape.
  c. Envelope.
  d. Carrier variation.

2-10. Overmodulation occurs when
  a. $V_m > V_c$
  b. $V_m < V_c$
  c. $V_m = V_c$
  d. $V_m = V_c = 0$

2-11. The values of V_{max} and V_{min} as read from an AM wave on an oscilloscope are 2.8 and 0.3. The percentage of modulation is
  a. 10.7 percent.
  b. 41.4 percent.
  c. 80.6 percent.
  d. 93.3 percent.

2-12. The new signals produced by modulation are called
  a. Spurious emissions.
  b. Harmonics.
  c. Intermodulation products.
  d. Sidebands.

2-13. A carrier of 880 kHz is modulated by a 3.5-kHz sine wave. The LSB and USB are, respectively,
  a. 873 and 887 kHz.
  b. 876.5 and 883.5 kHz.
  c. 883.5 and 876.5 kHz.
  d. 887 and 873 kHz.

2-14. A display of signal amplitude versus frequency is called the
  a. Time domain.
  b. Frequency spectrum.
  c. Amplitude spectrum.
  d. Frequency domain.

2-15. Most of the power in an AM signal is in the
a. Carrier.   c. Lower sideband.
b. Upper sideband.   d. Modulating signal.

2-16. An AM signal has a carrier power of 5 W. The percentage of modulation is 80 percent. The total sideband power is
a. 0.8 W.   c. 2.5 W.
b. 1.6 W.   d. 4.0 W.

2-17. For 100 percent modulation, what percentage of power is in each sideband?
a. 25 percent   c. 50 percent
b. 33.3 percent   d. 100 percent

2-18. An AM transmitter has a percentage of modulation of 88. The carrier power is 440 W. The power in one sideband is
a. 85 W.   c. 170 W.
b. 110 W.   d. 610 W.

2-19. An AM transmitter antenna current is measured with no modulation and found to be 2.6 amperes. With modulation, the current rises to 2.9 amperes. The percentage of modulation is
a. 35 percent.   c. 42 percent.
b. 70 percent.   d. 89 percent.

2-20. What is the carrier power in the problem above if the antenna resistance is 75 ohms?
a. 195 W.   c. 507 W.
b. 631 W.   d. 792 W.

2-21. In an AM signal, the transmitted information is contained within the
a. Carrier.   c. Sidebands.
b. Modulating signal.   d. Envelope.

2-22. An AM signal without the carrier is called a(n)
a. SSB.   c. FM signal.
b. Vestigial sideband.   d. DSB.

2-23. What is the minimum AM signal needed to transmit information?
a. Carrier plus sidebands   c. One sideband
b. Carrier only   d. Both sidebands

2-24. The main advantage of SSB over standard AM or DSB is
a. Less spectrum space is used.   c. Less power is consumed.
b. Simpler equipment is used.   d. A higher modulation percentage.

2-25. In SSB, which sideband is the best to use?
a. Upper   c. Neither
b. Lower   d. Depends upon the use

2-26. The typical audio modulating frequency range used in radio and telephone communications is
a. 50 Hz to 5 kHz.   c. 100 Hz to 10 kHz.
b. 50 Hz to 15 kHz.   d. 300 Hz to 3 kHz.

2-27. An AM signal with a maximum modulating signal frequency of 4.5 kHz has a total bandwidth of
a. 4.5 kHz.   c. 9 kHz.
b. 6.75 kHz.   d. 18 kHz.

2-28. Distortion of the modulating signal produces harmonics which cause an increase in the signal
a. Carrier power.   c. Sideband power.
b. Bandwidth.   d. Envelope voltage.

2-29. The process of translating a signal, with or without modulation, to a higher or lower frequency for processing is called
a. Frequency multiplication.   c. Frequency shift.
b. Frequency division.   d. Frequency conversion.

2-30. Frequency translation is carried out by a circuit called a
a. Translator.   c. Balanced modulator.
b. Converter.   d. Local oscillator.

2-31. An input signal of 1.8 MHz is mixed with a local oscillator of 5 MHz. A filter selects the difference signal. The output is
a. 1.8 MHz.   c. 5 MHz.
b. 3.2 MHz.   d. 6.8 MHz.

2-32. The output of an SSB transmitter with a 3.85-MHz carrier and a 1.5-kHz sine wave modulating tone is
a. a 3.8485-MHz sine wave.
b. a 3.85-MHz sine wave.
c. 3.85-, 3.8485-, and 3.8515-MHz sine waves.
d. 3848.5- and 3851.5-MHz sine waves.

2-33. An SSB transmitter produces a 400-V peak-to-peak signal across a 52-Ω antenna load. The PEP output is
a. 192.2 W.   c. 769.2 W.
b. 384.5 W.   d. 3077 W.

2-34. The output power of an SSB transmitter is usually expressed in terms of
a. Average power.   c. Peak-to-peak power.
b. RMS power.   d. Peak envelope power.

2-35. An SSB transmitter has a PEP rating of 1 kW. The average output power is in the range of
a. 150 to 450 W.   c. 250 to 333 W.
b. 100 to 300 W.   d. 3 to 4 kW.

CRITICAL THINKING QUESTIONS

2-1. Draw the waveforms of a sine wave carrier modulated by a square wave at 100 percent and 50 percent.

2-2. Would it be possible to transmit one intelligence signal in the upper sideband and a different intelligence signal in the lower sideband of an AM or DSB signal? Explain.

Answers to Self-Tests

1. translated
2. varies, carrier
3. modulator, carrier, modulating signal
4. amplitude
5. false
6. envelope, modulating signal
7. time-domain
8. $V_c \sin 2\pi f_c t$
9. true
10. c
11. c
12. a
13. d
14. modulating signal, carrier
15. 43.4
16. 100
17. 34 V
18. 66.67
19. sidebands
20. 635, 625
21. 10

22. time, oscilloscope
23. frequency, spectrum analyzer
24. sidebands
25. d
26. Carrier, upper sideband, lower sideband
27. b
28. b
29. antenna
30. carrier, sideband
31. 0.45
32. 5.06
33. 63.25
34. 236 W
35. 400 W, 89.44 percent
36. 1.8 amperes
37. DSB
38. false
39. b
40. a
41. less spectrum space, more

power in the sidebands with greater efficiency, less noise, little or no fading
42. telephone systems, two-way radio
43. c
44. true
45. vestigial sideband
46. 863.7
47. 345
48. 3-, 4-W
49. frequency conversion
50. mixer, converter
51. 4.155 MHz, 7.855 MHz, 455 kHz, 3.7 MHz
52. 117.8 MHz
53. filter or tuned circuit
54. false
55. true
56. heterodyning

CHAPTER 3

Amplitude Modulation Circuits

■

CHAPTER OBJECTIVES

This chapter will help you to:

1. *Explain* the operation of a low-level diode modulator and a high-level collector modulator.
2. *Explain* the operation of a diode detector circuit.
3. *State* the function of a balanced modulator and *explain* the operation of diode and IC balanced modulators.
4. *Draw* a block diagram of a filter-type SSB generator and *name* three types of filters used to eliminate one sideband.
5. *Recognize* and *explain* the operation of diode, bipolar, and FET mixer circuits.

──────

Over the years, hundreds of circuits have been developed to produce AM. These modulator circuits cause the amplitude of the carrier to be varied in accordance with the modulating information signal. There are circuits to produce AM, DSB, and SSB. In this chapter, you will examine some of the more common and widely used discrete-component and integrated-circuit (IC) amplitude modulators. This chapter also covers demodulators. A demodulator is the circuit that recovers the original information signal from the modulated wave. Demodulators for AM, DSB, and SSB will also be discussed.

■

3-1 AMPLITUDE MODULATORS

There are two basic ways to produce amplitude modulation. The first is to multiply the carrier by a gain or attenuation factor that varies with the modulating signal. The second is to linearly mix or algebraically add the carrier and modulating signals and then apply the composite signal to a nonlinear device or circuit. All amplitude modulators are based upon one of these two methods.

You can see how the first method works by referring to the basic AM equation in Chap. 2. It is:

$$V_{am} = V_c \sin 2\pi f_c t + (V_m \sin 2\pi f_m t)(\sin 2\pi f_c)$$

We know that the modulation index is

$$m = V_m/V_c$$

Therefore

$$V_m = mV_c$$

Substituting this in the equation above and factoring gives:

$$V_{am} = \sin 2\pi f_c t(V_c + mV_c \sin 2\pi f_m t)$$
$$= V_c \sin 2\pi f_c t(1 + m \sin 2\pi f_m t)$$

You can see in this equation that AM is accomplished by multiplying the carrier by a factor equal to 1 plus the modulating sine wave. If we can create a circuit whose gain or attenuation can be varied in accordance with the modulating signal, AM is produced by passing the carrier through it. Certain types of amplifiers and voltage dividers can be created to do this.

In the second method, we linearly mix the carrier and modulating signals. Then we use this composite voltage to vary the current in a nonlinear device. A nonlinear device is one whose current is proportional to but does not vary linearly with the applied voltage. One nonlinear device is the field-effect transistor (FET) whose response curve is approximately a parabola, indicating that the current in the device is proportional to the square of the input voltage (see Fig. 3-1). We say that such a device or circuit has a *square law response*. Squaring the sum of the carrier and modulating signals produces the classic AM equation described in Chap. 3. Other unwanted signals such as the second harmonics (also called second-order products) are produced by this circuit. But a filter or tuned circuit used on the output will reject these unwanted products and leave only the carrier and sidebands. This exponential relationship is the basis for all amplitude modulation, mixing, and heterodyning.

Diodes and bipolar transistors are also nonlinear devices. Their response is not square law, but they also are capable of producing amplitude modulation. The advantage of using a true square-law device (such as an FET) is that only second-order products are produced. Other nonlinear devices, such as diodes and BJTs, produce third-order products as well. Since filters can remove the unwanted high-order products, this is often not a significant problem. So diodes and BJTs are useful as amplitude modulators.

One of the oldest and simplest amplitude modulators is shown in Fig. 3-2. It consists of a resistive mixing network, a diode rectifier, and an *LC* tuned circuit. The carrier is applied to one input resistor and the modulating signal to the other. The mixed signals appear across

Fig. 3-2 Amplitude modulation with a diode.

R_3. This network causes the two signals to be linearly mixed, that is, algebraically added. If both the carrier and modulating signal are sine waves, the waveform resulting at the junction of the two resistors is shown in Fig. 3-3(c) on the next page. The result is the carrier wave riding on the modulating signal. The important thing to remember about this signal is that it is not AM. The two signals have simply been added together or linearly mixed. Modulation is a multiplication process and not an addition process.

The composite waveform is applied to a diode rectifier which has an exponential response. The diode is connected so that it is forward-biased by the positive-going half cycles of the input wave. During the negative portions of the wave, the diode is cut off and no signal passes. The current through the diode is a series of positive-going pulses whose amplitude varies in proportion with the amplitude of the modulating signal. See Fig. 3-3(d).

These positive-going pulses are applied to the parallel tuned circuit made up of L and C. Both L and C resonate at the carrier frequency. Each time the diode conducts, a pulse of current flows through the tuned circuit. The coil and capacitor repeatedly exchange energy causing an oscillation or "ringing" at the resonant frequency. What happens is that the ringing or flywheel action of the tuned circuit creates a negative half cycle for every positive input pulse. High-amplitude positive pulses cause the tuned circuit to produce high-amplitude negative pulses. Low-amplitude positive pulses produce corresponding low-amplitude negative pulses. The resulting waveform across the tuned circuit is AM as Fig. 3-3(e) illustrates. Of course, the Q of the tuned circuit should be high enough to produce a clean sine wave but also low enough so that its bandwidth will accommodate the sidebands generated. Another way to look at this is to view the LC circuit as a bandpass filter that passes the carrier and

Fig. 3-1 A square law response curve produced by a diode or appropriately biased transistor.

From page 38:

Multiplication

On this page:

Square law response

Multiplication

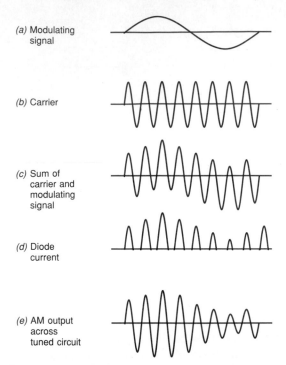

(a) Modulating signal

(b) Carrier

(c) Sum of carrier and modulating signal

(d) Diode current

(e) AM output across tuned circuit

Fig. 3-3   Waveforms in the diode modulator.

sidebands but rejects the unwanted products of the exponential response.

Another popular diode modulator is the type used to connect video games and personal computers to a TV set. The video output of the game or computer must be fed to the TV set. It is done through the TV set's antenna terminals. An oscillator operating on VHF channel 3 or 4 furnishes the carrier. The complex signal from the game or computer is used to amplitude modulate this carrier using a single diode. The TV set believes that it is receiving a legitimate TV signal and will, of course, display the information contained in the modulating signal. A typical circuit is shown in Fig. 3-4.

An oscillator Q_1 generates the carrier. It operates on VHF channels 3 or 4 as determined by the position of SW_1. The oscillator frequency is determined by the values of L_1, C_1, C_2, C_3, and C_4. A transistor switch Q_2 connects or shorts C_2 to change the frequency. With SW_1 in the position shown, a positive base bias is applied to Q_2 through R_1. This turns on Q_2. It acts as a very low resistance, effectively shorting out C_2. Then L_1 and C_1 resonate on channel 3.

With SW_1 in the channel 4 position, the bias is removed from Q_2. This cuts off Q_2, so it has no effect on the circuit. Since C_2 appears in series with C_1 to lower the total capacitance across L_1, the circuit resonates at the higher, channel 4, frequency. Both L_1 and C_2 are ad-

justable so that the oscillator can be accurately tuned to each channel; L_1 is adjusted for channel 3, and then C_2 is adjusted for channel 4.

Transistor switch Q_2 could be replaced by a simple manual switch like SW_1 to simplify the circuit. To do this, SW_1 has to be mounted very close to C_2. Any long wires connecting the switch to the tuned circuit will introduce stray capacitance that will detune the circuit. With the circuit in Fig. 3-4, Q_2 is mounted directly at C_2 and long wires may be used from the base of Q_2 to SW_1 since only dc rather than RF is switched.

Capacitors C_3 and C_4 partly determine the frequency of oscillation and also form a voltage divider. The voltage across C_4 is applied to the diode modulator D_1. Voltage divider R_2-R_3 biases D_1 on. The modulating signal is applied to the cathode of the diode through the radio frequency choke (RFC). The RFC keeps the carrier signal out of the video circuits. With this arrangement, D_1 is used as a variable resistor. The variation of the modulating signal causes the current through D_1 to vary linearly. Then D_1 and R_4 form a variable voltage divider or attenuator. The resistance of D_1 is varied by the video signal thereby varying the carrier current in the primary winding of T_1. The result is AM. The R_4 is adjustable to vary the percentage of modulation and the output amplitude over a narrow range.

The AM signal is coupled through T_1 and voltage divider C_5-C_6 to the TV set. In some units, a balun transformer is used to convert the 75-Ω output impedance to 300 Ω which is the standard antenna input impedance of a TV set on the VHF channels.

Another simple amplitude modulator is shown in Fig. 3-5. It consists of an operational amplifier (op amp) and a field-effect transistor (FET) used as a variable resistor. The op amp is connected as a noninverting amplifier for the carrier signal. The gain A of the circuit for the oscillator signal is given by the expression

$$\text{Gain A} = 1 + \frac{R_f}{R_i}$$

The feedback resistance R_f is a fixed value, while R_i is the resistance of an N-channel junction FET. A negative dc bias keeps the gate-source junction reverse-biased. A modulating signal is applied to the gate through capacitor C_1.

The carrier signal is applied to the noninverting (+) input of the op amp. By changing the gain of the amplifier in accordance with the modulating signal, AM is produced. With

Fig. 3-4 An RF modulator commonly used to connect video games and personal computers to a TV set.

zero modulating signal, the FET resistance is a fixed value and, therefore, the carrier output amplitude will be constant. As a sine wave modulating signal is applied, the resistance of the FET will vary. A positive-going modulating input signal will cause the FET resistance to decrease, whereas a negative-going modulating signal will cause it to increase. Increas-

$$\text{Gain } A = 1 + \frac{R_f}{R_i}$$

Fig. 3-5 Using a JFET to vary the gain of an op amp to produce AM.

ing the FET resistance causes the op-amp circuit gain to decrease and vice versa. The result is an AM signal at the op-amp output.

The critical component in this circuit is the FET. The FET should be biased such that its source-to-drain resistance is as linear as possible over a wide range. The best point of linearity can be predicted from the operating curves of the FET. The bias on the gate can then be set for the center of this linear range.

Other circuits for producing AM using the variable resistance principle are shown in Fig. 3-6. These circuits use PIN diodes to produce AM at VHF, UHF, and microwave frequencies. The PIN diodes are a special type of silicon junction diode designed for use at frequencies above approximately 100 MHz. When forward-biased, these diodes act as variable resistors. The resistance of the diode varies linearly with the amount of current flowing through it. A high current produces a low resistance, whereas a low current produces a high resistance. By using the modulating signal to vary the forward-bias current through the PIN diode, AM is produced.

In Fig. 3-6(a), two PIN diodes are connected back to back and are forward-biased by a fixed negative dc voltage. The modulating signal is applied to the diodes through capacitor C_1. This ac modulating signal rides on the dc bias and, therefore, adds to and subtracts from it. In doing so, it varies the resistance of the PIN diodes. These diodes appear in series with the carrier oscillator and the load. A positive-going modulating signal will reduce the bias on the PIN diodes causing their resistance to go up. This causes the amplitude of the carrier to be reduced across the load. A negative-going modulating signal will add to the forward bias, causing the resistance of the diodes to go down, thereby increasing the carrier amplitude.

A variation of the PIN diode modulator circuit is shown in Fig. 3-6(b). The diodes are arranged in a pi network. This configuration is used when it is necessary to maintain a constant circuit impedance even under modulation.

In both circuits of Fig. 3-6, the PIN diodes form a variable attenuator circuit whose attenuation varies with the amplitude of the modulating signal. Such modulator circuits introduce a considerable amount of attenuation and, therefore, must be followed by amplifiers to increase the AM signal to a usable level. Despite this disadvantage, PIN modulators are widely

(a)

(b)

Fig. 3-6   High-frequency amplitude modulators using PIN diodes.

used because they are one of the only methods available to produce AM at microwave frequencies.

The modulator circuits we have been discussing are known as low-level modulator circuits. "Low level" refers to the fact that the signals are generated at relatively low voltage and power amplitudes. Before an AM signal is transmitted, its power level must be increased. In systems using low-level modulation, the AM signal is applied to one or more linear amplifiers as shown in Fig. 3-7. These may be class A, class AB, or class B linear amplifier circuits. These raise the output level of the circuit to the desired power level before the AM

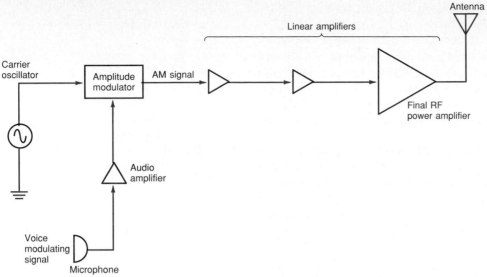

Fig. 3-7 Low-level modulation systems use linear power amplifiers to increase the AM signal level before transmission.

signal is fed to the antenna. The key point here is that linear amplifiers must be used so as not to distort the AM signal.

High-level modulation is also possible. In high-level modulation, the modulator varies the voltage and power in the final RF amplifier stage of the transmitter. One example of a high-level modulator circuit is the collector modulator shown in Fig. 3-8. The output stage of the transmitter is a high-power class C amplifier. Class C amplifiers conduct for only a portion of the positive half cycle of their input signal. The collector current pulses cause the tuned circuit to oscillate or ring at the desired output frequency. The tuned circuit, therefore, reproduces the negative portion of the carrier signal.

The modulator is a linear power amplifier that takes the low-level modulating signal and amplifies it to a high power level. The modulating output signal is coupled through modulation transformer T_1 to the class C amplifier. The secondary winding of the modulation transformer is connected in series with the collector supply voltage V_{cc} of the class C amplifier.

With a zero modulation input signal, there will be zero modulation voltage across the secondary of T_1. Therefore, the collector supply voltage will be applied directly to the class C amplifier, and the output carrier will be a steady sine wave.

When the modulation signal occurs, the ac voltage across the secondary of the modulation transformer will be added to and subtracted from the collector supply voltage. This varying supply voltage is then applied to the class

C amplifier. Naturally, the amplitude of the current pulses through transistor Q_1 will vary. As a result, the amplitude of the carrier sine wave varies in accordance with the modulated signal. For example, when the modulation signal goes positive, it adds to the collector supply voltage, thereby increasing its value and causing higher current pulses and a higher amplitude carrier. When the modulating signal goes negative, it subtracts from the collector supply voltage making it less. For that reason, the class C amplifier current pulses are smaller, thereby causing a lower amplitude carrier output.

For 100 percent modulation, the peak of the modulating signal across T_1 must be equal to

Fig. 3-8 A high-level collector modulator.

the supply voltage. When the positive peak occurs, the voltage applied to the collector is twice the collector supply voltage. When the modulating signal goes negative, it subtracts from the collector supply voltage. When the negative peak is equal to the supply voltage, the effective voltage applied to the collector of Q_1 is zero, producing zero carrier output. This is illustrated in Fig. 3-9.

In practice, 100 percent modulation cannot be achieved with the high-level collector modulator circuit shown in Fig. 3-8. To overcome this problem, the driver amplifier stage driving the final class C amplifier is also collector-modulated simultaneously. This process is shown in Fig. 3-10. The output of the modulation transformer is connected in series with the collector supply voltage both to the driver transistor Q_1 and to the final class C amplifier Q_2. With this arrangement, solid 100 percent modulation is possible. This technique is widely used in low-power CB transmitters.

High-level modulation produces the best type of AM, but it requires an extremely high-power modulator circuit. In fact, the power supplied by the modulator must be equal to one-half the total class C power amplifier rating for 100 percent modulation. If the class C amplifier has an input power of 1000 W, the modulator must be able to deliver one-half this amount, or 500 W.

Self-Test

Choose the letter which best answers each statement.

1. In the modulator circuit of Fig. 3-2, the carrier and modulating signals are
   a. Added.
   b. Subtracted.
   c. Multiplied.
   d. Divided.

2. In Fig. 3-2, D_1 acts as a(n)
   a. Capacitor.
   b. Rectifier.
   c. Variable resistor.
   d. Adder.
3. In Fig. 3-4, D_1 acts as a(n)
   a. Capacitor.
   b. Rectifier.
   c. Variable resistor.
   d. Adder.

Supply the missing information in each statement.

4. AM can be produced by passing the carrier through a circuit whose _____ or _____ can be varied in accordance with the modulating signal.
5. The name of the nonlinear response of a device that produces AM with only second-order products is _____ .
6. A component that has an exponential response ideal for producing AM is the _____ .
7. In Fig. 3-2, the negative peaks of the AM signal are supplied by the _____ .
8. In Fig. 3-5, the FET acts like a(n) _____ .
9. In Fig. 3-5, AM is produced by varying the _____ of the op-amp circuit.
10. In Fig. 3-5, a positive-going modulating signal causes the current in the FET to _____ , its resistance to _____ , and the gain of the circuit to _____ .
11. When forward-biased, a PIN diode acts like a(n) _____ .
12. PIN diode modulators are used only at frequencies above about _____ .
13. A PIN diode modulator is a variable _____ .
14. High current in a PIN diode means that its resistance is _____ .
15. The AM signals generated by low-level modulating circuits must have their power level increased by a(n) _____ before being transmitted.

Modulating signal across the secondary of T_1 and the composite supply voltage applied to Q_1

$+V_{cc}$

0V

Fig. 3-9 For 100 percent modulation the peak of the modulating signal must be equal to V_{cc}.

Fig. 3-10 **For 100 percent modulation both the driver and the final class C amplifier are modulated.**

16. In a high-level AM transmitter, the output stage is usually a class _____ amplifier.

17. A high-level modulator like that in Fig. 3-8 is referred to as a _____ modulator.

18. The output of a high-level modulator causes the _____ applied to the final RF amplifier to vary with the amplitude of the modulating signal.

19. The final amplifier of a high-level modulation CB transmitter has an input power of 5 W. The modulator must be able to supply a power of _____ W for 100 percent modulation.

20. The final RF power amplifier has a supply voltage of 12 V. For 100 percent AM using a high-level modulator, the peak ac output of the modulation transformer must be _____ V.

21. To achieve 100 percent high-level modulation of an RF power amplifier, its _____ must also be modulated.

3-2 AMPLITUDE DEMODULATORS

A *demodulator* is a circuit that accepts a modulated signal and recovers the original modulating information. Also known as a *detector,* a demodulator circuit is the key circuit in any radio receiver. In fact, the demodulator circuit may be used alone as the simplest form of radio receiver.

The simplest and most widely used amplitude demodulator is the diode detector shown in Fig. 3-11 on the next page. The AM signal is usually transformer-coupled as indicated. It is applied to a basic half-wave rectifier circuit consisting of D_1 and R_1. The diode conducts when the positive half cycles of the AM signals occur. During the negative half cycles, the diode is reverse-biased and no current flows through it. As a result, the voltage across R_1 is a series of positive pulses whose amplitude varies with the modulating signal.

To recover the original modulating signal, a capacitor is connected across resistor R_1. Its

Fig. 3-11   The diode detector AM modulator.

value is critical to good performance. The value of this capacitor is carefully chosen so that it has a very low impedance at the carrier frequency. At the frequency of the modulating signal, it has a much higher impedance. The result is that the capacitor effectively shorts or filters out the carrier, thereby leaving the original modulating signal.

Another way to look at the operation of the diode detector is to assume that the capacitor charges quickly to the peak value of the pulses passed by the diode. When the pulse drops to zero, the capacitor retains the charge but discharges into resistor R_1. The time constant of C and R_1 is chosen to be long compared to the period of the carrier. As a result, the capacitor discharges only slightly during the time that the diode is not conducting. When the next pulse comes along, the capacitor again charges to its peak value. When the diode cuts off, the capacitor will again discharge a small amount into the resistor. The resulting waveform across the capacitor is a close approximation to the original modulating signal. Because of the capacitor charging and discharging, the recovered signal will have a small amount of ripple on it. This causes distortion of the modulating signal. However, because the carrier frequency is usually many times higher than the modulating frequency, these ripple variations are barely noticeable. In Fig. 3-11, the

ripple is quite pronounced because the carrier frequency is low.

The output of the detector is the original modulating signal. Because the diode detector recovers the envelope of the AM signal which is the modulating signal, the circuit is sometimes referred to as an *envelope detector.*

The basic diode detector circuit is really a complete radio receiver in its own right. In fact, this circuit is the same as that used in the crystal radio receivers of the past. The crystal refers to the diode. In Fig. 3-12, the diode detector circuit is redrawn, showing an antenna connection and headphones. A long wire antenna picks up the radio signal which is inductively coupled to the tuned circuit. The variable capacitor C_1 is used to select a station. The diode detector D_1 recovers the original modulating information which causes current flow in the headphones. The headphones serve as the load resistance, whereas capacitor C_2 removes the carrier. The result is a simple radio receiver with very weak reception because no amplification is provided. Typically a germanium diode is used because its voltage threshold is lower than that of a silicon diode and permits reception of weaker signals. Such a receiver can easily be built to receive standard AM broadcasting stations.

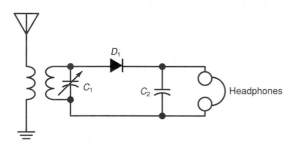

Fig. 3-12   A crystal radio receiver.

The performance of the basic diode detector can be improved by using a full-wave rectifier circuit as shown in Fig. 3-13. Here, two diodes and a center-tapped secondary on the RF transformer are used to form a standard full-wave rectifier circuit. With this arrangement, diode D_1 will conduct on the positive half cycle, while D_2 will conduct on the negative half cycle. This diode detector produces a higher average output voltage which is much easier to filter. The capacitor value necessary to remove the carrier can be half the size of the capacitor value used in a half-wave diode detector. The primary benefit of this circuit is

Fig. 3-13 A full-wave diode detector for AM signals.

that the higher modulating frequencies will not be distorted by ripple or attenuated as much as in the half-wave detector circuit.

Although more sophisticated transistorized amplitude demodulators have been developed over the years, in most cases they are not used. The basic diode detector is so simple and effective and its performance so good that rarely is any other circuit considered.

Self-Test

Answer the following questions.

22. The purpose of a _____ is to recover the original modulating signal from an AM wave.
23. The most widely used amplitude demodulator is called a _____ .
24. The most critical component in the circuit of Fig. 3-11 is _____ .
25. The charging and discharging of C_1 in Fig. 3-11 produces _____ which causes _____ of the modulating signal.
26. Another name for the demodulator in Fig. 3-11 is _____ detector.
27. List the two main benefits of the full-wave amplitude demodulator over the half-wave circuit.
28. True or false. An amplitude demodulator is a complete radio receiver.

3-3 BALANCED MODULATORS

A balanced modulator generates a DSB signal. The inputs to a balanced modulator are the carrier and a modulating signal. The output of a balanced modulator is the upper and lower sidebands. The balanced modulator suppresses the carrier, leaving only the sum and difference

frequencies at the output. The output of a balanced modulator can be further processed by filters or phase-shifting circuitry to eliminate one of the sidebands, thereby resulting in an SSB signal.

One of the most popular and widely used balanced modulators is the diode ring or lattice modulator illustrated in Fig. 3-14(*a*) on the next page. It consists of an input transformer T_1, an output transformer T_2, and four diodes connected in a bridge circuit. The carrier signal is applied to the center taps of the input and output transformers. The modulating signal is applied to the input transformer T_1. The output appears across the secondary of the output transformer T_2.

Sometimes you will see the lattice modulator drawn as shown in Fig. 3-14(*b*). Physically the connections are the same, but the operation of the circuit can be more easily visualized with this circuit.

The operation of the lattice modulator is relatively simple. The carrier sine wave, which is usually considerably higher in frequency and amplitude than the modulating signal, is used as a source of forward and reverse bias for the diodes. The carrier turns the diodes off and on at a high rate of speed. The diodes act like switches which connect the modulating signal at the secondary of T_1 to the primary of T_2.

To see how the circuit works, refer to Fig. 3-15 on page 49. Assume that the modulating input is zero. When the polarity of the carrier is as illustrated in Fig. 3-15(*a*), diodes D_1 and D_2 are forward-biased. At this time, D_3 and D_4 are reverse-biased and act like open circuits. As you can see, current divides equally in the upper and lower portions of the primary winding of T_2. The current in the upper part of the winding produces a magnetic field that is equal and opposite to the magnetic field produced by the current in the lower half of the secondary. Therefore, these magnetic fields cancel each other out and no output is induced in the secondary. Thus, the carrier is effectively suppressed.

When the polarity of the carrier reverses as shown in Fig. 3-15(*b*), diodes D_1 and D_2 are reverse-biased and diodes D_3 and D_4 conduct. Again, the current flows in the secondary winding of T_1 and the primary winding of T_2. The equal and opposite magnetic fields produced in T_2 cancel each other out and thus result in zero carrier output. The carrier is effectively bal-

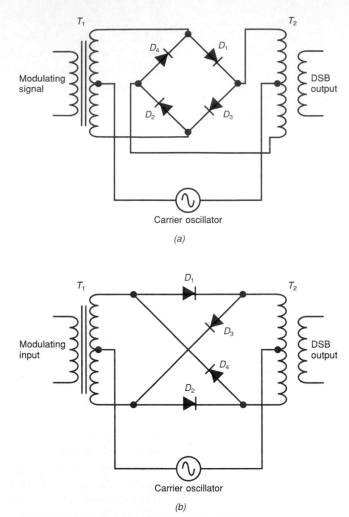

anced out. The degree of carrier suppression depends upon the degree of precision with which the transformers are made and the placement of the center tap to ensure perfectly equal upper and lower currents and magnetic field cancellation. The degree of the carrier attenuation also depends upon the diodes. The greatest carrier suppression will occur when the diode characteristics are perfectly matched. A carrier suppression of 40 decibels (dB) is achievable with well-balanced components.

Now assume that a low-frequency sine wave is applied to the primary of T_1 as the modulating signal. The modulating signal will appear across the secondary of T_1. The diode switches will connect the secondary of T_1 to the primary of T_2 at different times depending upon the carrier polarity. Refer to Fig. 3-15. When the carrier polarity is as shown in Fig. 3-15(a), diodes D_1 and D_2 conduct and act as closed switches.

At this time, D_3 and D_4 are reverse-biased and are effectively not in the circuit. As a result, the modulating signal at the secondary of T_1 is applied to the primary of T_2 through D_1 and D_2.

When the carrier polarity reverses, D_1 and D_2 cut off and D_3 and D_4 conduct. Again a portion of the modulating signal at the secondary of T_1 will be applied to the primary of T_2, but this time the leads have been effectively reversed because of the connections of D_3 and D_4. The result is a 180° phase reversal. If the modulating signal is positive, the output will be negative with this connection and vice versa.

The carrier is operating at a considerably higher frequency than the modulating signal (Fig. 3-16 on page 50). Therefore, the diodes will switch off and on at a high rate of speed, causing portions of the modulating signal to be

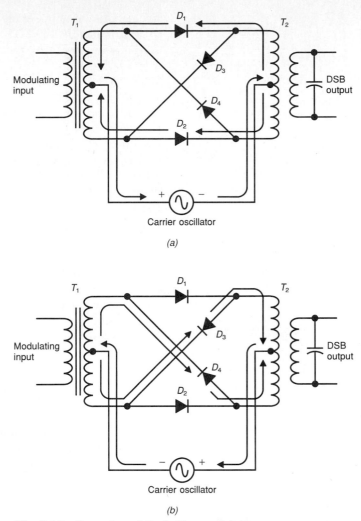

(a)

(b)

Fig. 3-15   Operation of the lattice modulator.

passed through the diodes at different times. The DSB signal appearing across the primary of T_2 is illustrated in Fig. 3-16(*c*). The steep rise and fall of the waveform is caused by the rapid switching of the diodes. The waveform contains harmonics of the carrier because of the switching action. Ordinarily, the secondary of T_2 is a resonant circuit as shown, and, therefore, the high-frequency harmonic content is filtered out, leaving a signal that appears like that shown in Fig. 3-16(*d*). This is a DSB signal.

There are several important things to notice about this signal. First, the output waveform is occurring at the carrier frequency. This is true even though the carrier has been removed. If you take two sine waves occurring at the sideband frequencies and algebraically add them together, the result is a sine wave signal at the carrier frequency but with the amplitude

variation shown in Fig. 3-16(*c*) or (*d*). Observe that the envelope of the output signal is not the shape of the modulating signal. Note also the phase reversal of the signal in the very center of the waveform. This is one way you can tell if the signal being observed is a true DSB signal.

Although diode ring lattice modulators can be constructed of discrete components, they are usually available in a single module containing the transformers and diodes in a sealed package. The unit can be used as an individual component. The transformers are carefully balanced, and matched hot-carrier diodes are used to provide a wide operating frequency range and superior carrier suppression.

The diode lattice modulator shown in Fig. 3-14 uses one low-frequency iron-core transformer for the modulating signal and an air-core transformer for the RF output. This is an inconvenient arrangement because the low-

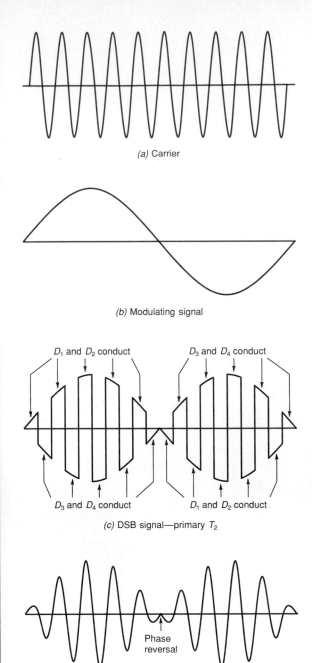

(a) Carrier

(b) Modulating signal

D_1 and D_2 conduct D_3 and D_4 conduct

D_3 and D_4 conduct D_1 and D_2 conduct

(c) DSB signal—primary T_2

Phase
reversal

(d) DSB output

**Fig. 3-16   Waveforms in the lattice-type balanced
modulator.**

frequency transformer is large and expensive.
More commonly, two RF transformers are
used in the configuration shown in Fig. 3-17.
Here the modulating signal is applied to the
center taps of the RF transformers. The operation of the circuit is similar to that of the previously discussed circuit.

Another widely used balanced modulator circuit uses differential amplifiers. A typical circuit is shown in Fig. 3-18. This is the circuit of
the popular 1496/1596 IC balanced modulator.

This circuit can work at carrier frequencies up
to approximately 100 MHz and can achieve a
carrier suppression of 50 to 65 dB.

In Fig. 3-18, transistors Q_7 and Q_8 are constant current sources. The constant current
sources are biased with a single external resistor and the negative supply. These current
sources supply equal values of current to the
two differential amplifiers. One differential amplifier is made up of Q_1, Q_2, and Q_5, while the
other is made up of Q_3, Q_4, and Q_6. The modulating signal is applied to the bases of Q_5 and
Q_6. These transistors are connected in the current paths to the differential transistors and,
therefore, will vary the amplitude of the current in accordance with the modulating signal.
The currents in Q_5 and Q_6 will be 180° out of
phase with each other. As the current in Q_5 increases, the current through Q_6 decreases, and
vice versa.

The differential transistors Q_1 through Q_4 operate as switches. These transistors are controlled by the carrier. When the carrier input
is such that the lower input terminal is positive with respect to the upper input terminal,
transistors Q_1 and Q_4 will conduct and act as
closed switches and Q_2 and Q_3 will be cut off.
When the polarity of the carrier signal reverses, Q_1 and Q_4 will be cut off and Q_2 and
Q_3 will conduct and act as closed switches.
These differential transistors, therefore, serve
the same switching purpose as the diodes in
the lattice modulator circuit discussed previously. That is, they switch the modulating signal off and on at the carrier rate.

To see how the circuit works, assume that a
high-frequency carrier wave is applied to the
switching transistors Q_1 through Q_4 and that a
low-frequency sine wave is applied to the modulating signal input at Q_5 and Q_6. Assume that
the modulating signal is positive-going so that
the current through Q_5 is increasing while the
current through Q_6 is decreasing.

When the carrier polarity is positive, Q_1 and
Q_4 conduct. Since the current through Q_5 is increasing, the current through Q_1 and R_2 will
increase proportionately; therefore, the output
voltage at the collector of Q_1 will go in a negative direction. The current through Q_6 is decreasing; therefore, the current through Q_4 and
R_1 is decreasing. The output voltage at the collector of Q_4 is hence increasing.

When the carrier polarity reverses, Q_2 and
Q_3 conduct. Now the increasing current of Q_5
is passed through Q_2 and R_1. Therefore, the
output voltage begins to decrease. The decreas-

Fig. 3-17 A modified version of the lattice modulator not requiring an iron-core transformer for the low-frequency modulating signal.

Fig. 3-18 Integrated-circuit balanced modulator.

ing current through Q_6 is now passed through Q_3 and R_2. This decreasing current causes an increasing output voltage. The result of the carrier switching off and on and the modulating signal varying as indicated produces the classical DSB output signal described before. See Fig. 3-16(c). The signal at R_1 is the same as the signal at R_2, but the two are 180° out of phase.

Another type of IC that can be used as a balanced modulator is an analog multiplier. This device produces the true product of two analog input signals. The product of two sine wave inputs contains the sum and difference frequencies but not the carrier. Therefore, the analog multiplier is often used to generate DSB signals. The primary difference between IC balanced modulators and analog multipliers is that the balanced modulator is a switching circuit. The carrier causes the differential-amplifier transistors to turn off and on to switch the modulating signal. The carrier may be a rectangular wave. The analog multiplier uses differential amplifiers, but they operate in the linear mode. It produces the true product of two analog inputs.

Self-Test

Choose the letter which best answers each statement.

29. A balanced modulator eliminates which of the following from its output?
    a. Upper sideband
    b. Lower sideband
    c. Carrier
    d. Both sidebands
30. The output signal of a balanced modulator is
    a. AM.
    b. SSB.
    c. FM.
    d. DSB.
31. Which has better carrier suppression?
    a. Lattice modulator
    b. IC balanced modulator

Supply the missing information in each statement.

32. A balanced modulator using a diode bridge is called a(n) _____ .
33. In the balanced modulator of Fig. 3-14, the diodes are used as _____ .
34. A balanced modulator has a carrier frequency of 1.9 MHz and a modulating sine wave of 2.6 kHz. The output signals are _____ and _____ kHz.

35. In Fig. 3-18, transistors _____ supply a constant current.
36. In Fig. 3-18, transistors Q_1 to Q_4 operate as _____ .

Determine whether the statement is true or false.

37. An IC analog multiplier may be used for DSB signal generation.

3-4 SSB CIRCUITS

There are two primary methods of generating SSB signals. These are the filter method and the phasing method. The filter method is by far the simplest and most widely used, but we will discuss both types here.

Figure 3-19 shows a general block diagram of an SSB transmitter using the filter method. The modulating signal, usually voice from a microphone, is applied to the audio amplifier whose output is fed to one input of a balanced modulator. A crystal oscillator provides the carrier signal which is also applied to the balanced modulator. The output of the balanced modulator is a DSB signal. An SSB signal is produced by passing the DSB signal through a highly selective band-pass filter. This filter selects either the upper or lower sideband.

The filter, of course, is the critical component in the filter method SSB generator. Its primary requirement is that it have high selectivity so that it passes only the desired sideband and rejects the other. The filters are usually designed with a bandwidth of approximately 2.5 to 3 kHz, making them only wide enough to pass standard voice frequencies. The sides of the filter response curve are extremely steep providing excellent rejection of the other sideband.

The filter is a fixed tuned device, meaning that the frequencies that it can pass cannot be changed. Therefore, the carrier oscillator frequency must be chosen so that the sidebands fall within the filter bandpass. Usually, the filter is tuned to a frequency in the 455-kHz, 3.35-MHz, or 9-MHz range. Other frequencies are also used, but many commercially available filters are in these frequency ranges.

It is also necessary to select either the upper or lower sideband. Since the same information is contained in both sidebands, it generally makes no difference which one is selected. However, various conventions in different communications services have chosen

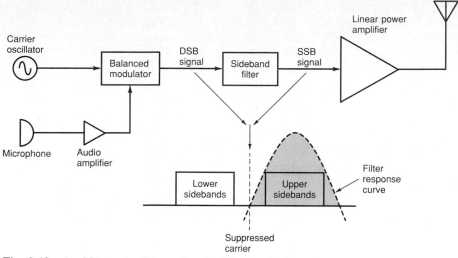

Fig. 3-19 An SSB transmitter using the filter method.

either the upper or lower sideband as a standard. These vary from service to service, and it is necessary to know whether it is an upper or lower sideband to properly receive an SSB signal.

There are two methods of selecting the sideband. Many transmitters simply contain two filters, one that will pass the upper sideband and the other that will pass the lower sideband. A switch is used to select the desired sideband. See Fig. 3-20(*a*).

The other method of selecting the sideband is to provide two carrier oscillator frequencies. Two crystals change the carrier oscillator frequency to force either the upper sideband or the lower sideband to appear in the filter bandpass.

Fig. 3-20 Methods of selecting the upper or lower sideband. (*a*) Two filters and (*b*) two carrier frequencies.

Assume a simple example where the bandpass filter is fixed at 1000 kHz. The modulating signal f_m is 2 kHz. The balanced modulator generates the sum and difference frequencies. Therefore, the carrier frequency f_c must be chosen so that the USB or LSB is at 1000 kHz.

The balanced modulator outputs are USB $= f_c + f_m$ and LSB $= f_c - f_m$. To put the USB at 1000 kHz, the carrier must be

$$f_c + f_m = 1000$$
$$f_c + 2 = 1000$$
$$f_c = 1000 - 2$$
$$= 998 \text{ kHz}$$

To set the LSB at 1000 kHz, the carrier must be

$$f_c - f_m = 1000$$
$$f_c - 2 = 1000$$
$$f_c = 1000 + 2$$
$$= 1002 \text{ kHz}$$

Crystal filters are by far the most commonly used filters in SSB transmitters. They are low in cost and relatively simple to design. Their very high Q provides extremely good selectivity.

Crystal filters are made from the same type of quartz crystals normally used in crystal oscillators. When a voltage is applied across a crystal, it will vibrate at a specific resonant frequency. This resonant frequency is a function of the size, thickness, and direction of cut of the crystal. Crystals can be cut and ground for almost any frequency in the 100-kHz to 100-MHz range. Its frequency of vibration is extremely stable, and, therefore, crystals are widely used to supply signals on exact frequencies with good stability.

The schematic symbol and the equivalent circuit of a quartz crystal are shown in Fig. 3-21. The crystal acts as a resonant LC circuit. The series LCR part of the equivalent circuit represents the crystal itself, whereas the parallel capacitance C_p is the capacitance of the metal mounting plates with the crystal as the dielectric.

Fig. 3-22 shows the impedance variations of the crystal as a function of frequency. At frequencies below the crystal's resonant frequency, the circuit appears capacitive and has a high impedance. However, at some frequency, the reactances of the equivalent inductance L and the series capacitance C_s are

(a)

(b)

Fig. 3-21 (*a*) **Quartz-crystal–equivalent electric circuit, and** (*b*) **schematic symbol.**

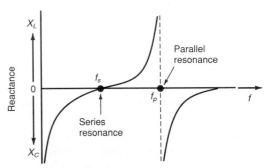

Fig. 3-22 **Impedance variation of a quartz crystal as a function of frequency.**

equal, and the circuit will resonate. You should remember that a series circuit is at resonance when $X_L = X_C$. At this series resonant frequency f_s, the circuit is resistive. The resistance of the crystal is extremely low, thereby giving the circuit an extremely high Q. Values of Q in the 10,000 to 1,000,000 range are common. This makes the crystal a highly selective series resonant circuit.

If the frequency of the signal applied to the crystal is above f_s, the crystal appears inductive. At some frequency, the reactance of the parallel capacitance C_p equals the reactance of the net inductance. When this occurs, a parallel resonant circuit is formed. At this parallel resonant frequency f_p, the impedance of the circuit is resistive but extremely high.

Because the crystal has both series and parallel resonant frequencies which are close together, it makes an ideal component for use in filters. By combining crystals with selected series and parallel resonant points, highly selective filters with any desired bandpass can be constructed.

The most commonly used crystal filter is the full crystal lattice shown in Fig. 3-23. Crystals

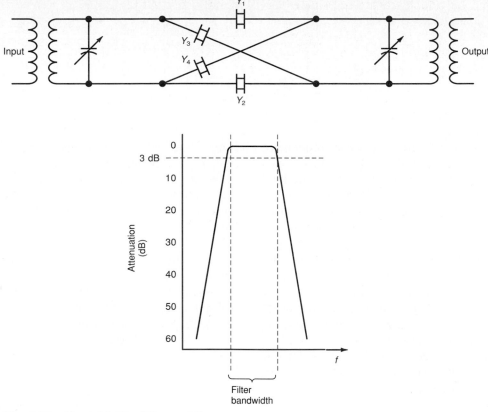

Fig. 3-23　Crystal lattice filter and its response curve.

Y_1 and Y_2 resonate at one frequency, while crystals Y_3 and Y_4 resonate at another frequency. The difference between the two crystal frequencies determines the bandwidth of the filter. The 3-dB down bandwidth will be approximately 1.5 times the crystal frequency spacing. For example, if the Y_1-Y_2 frequency is 9 MHz and the Y_3-Y_4 frequency is 9.002 MHz, the difference is $9.002 - 9.000 = 0.002$ MHz $= 2$ kHz. The 3-dB bandwidth then is 1.5×2 kHz $= 3$ kHz.

The crystals are also chosen so that the parallel resonant frequency of Y_3-Y_4 equals the series resonant frequency of Y_1-Y_2. The series resonant frequency of Y_3-Y_4 is equal to the parallel resonant frequency of Y_1-Y_2. The result is a passband with extremely steep attenuation. Signals outside the passband are rejected as much as 50 to 60 dB below those inside the passband. See the response curve in Fig. 3-23. Such a filter can easily separate one sideband from another.

A popular variation of the crystal lattice filter is shown in Fig. 3-24. It uses only a center-tapped inductor rather than two transformers as in Fig. 3-23. Its primary advantage is that it is easily inserted between transistor amplifier stages.

Fig. 3-24　A popular variation of the crystal lattice filter.

Another type of crystal filter is the ladder filter shown in Fig. 3-25 on the next page. All the crystals in this filter are cut for exactly the same frequency. The number of crystals used and the values of the shunt capacitors set the bandwidth. Crystals with frequencies near 9 MHz are commonly used. Because of the low cost and wide availability of crystals for CB radios with frequencies near 27 MHz, very

Fig. 3-25   A crystal ladder filter. All crystals are ground for the same frequency.

simple low-cost filters with these frequencies are easily constructed. At least six crystals must usually be cascaded to achieve the kind of selectivity needed in SSB applications.

Other types of filters are also used to remove the unwanted sideband. These include both ceramic and mechanical filters. Ceramic is a manufactured crystal-like compound. It has the same piezoelectric qualities as quartz. Ceramic disks can be made so that they vibrate at a fixed frequency, thereby providing filtering actions. Ceramic filters are very small and inexpensive and are, therefore, widely used in transmitters and receivers. Typical center frequencies are 455 kHz and 10.7 MHz. These are available in different bandwidths up to 350 kHz.

Mechanical filters are also used in SSB-generating equipment. These mechanical filters consist of small metal disks coupled together with rods to form an assembly that vibrates or resonates over a narrow frequency range. The diameter and thickness of the disks determine the resonant frequency, whereas the number of disks and their spacing and method of coupling determine the bandwidth. The ac signal to be filtered is applied to a coil which creates

a magnetic field. This magnetic field works against a permanent magnet to produce mechanical motion in the disks. If the input signal is within the bandpass resonant frequency range of the disks, they will vibrate freely. This vibration is mechanically coupled to a coil. The moving coil cuts the field of a permanent magnet inducing a voltage in the coil. This is the output signal. If the input signal is outside of the resonant frequency range of the disks, they will not vibrate and little or no output will be produced. Such mechanical assemblies are extremely effective bandpass filters. Most are designed to operate over the 200- to 500-kHz range. A 455-kHz mechanical filter is commonly used.

The phasing method of SSB generation uses a phase-shift technique that causes one of the sidebands to be canceled out. A block diagram of a phasing-type SSB generator is shown in Fig. 3-26. It uses two balanced modulators instead of one. The balanced modulators effectively eliminate the carrier. The carrier oscillator is applied directly to the upper balanced modulator along with the audio modulating signal. Then both the carrier and the modulating signal are shifted in phase by 90°

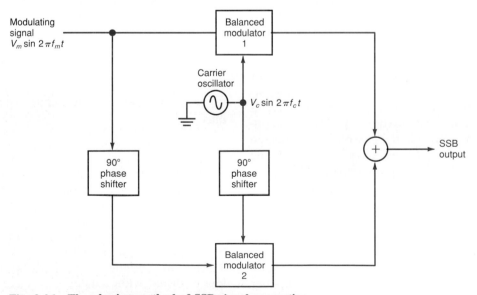

Fig. 3-26   The phasing method of SSB signal generation.

and applied to the second, lower, balanced modulator. The two balanced modulator outputs are then added together algebraically. The phase-shifting action causes one sideband to be canceled out when the two balanced modulator outputs are combined.

The carrier signal is $V_c \sin 2\pi f_c t$. The modulating signal is $V_m \sin 2\pi f_m t$. Balanced modulator 1 produces the product of these two signals:

$$(V_m \sin 2\pi f_m t)(V_c \sin 2\pi f_c t)$$

Applying a trigonometric identity,

$$(V_m \sin 2\pi f_m t)(V_c \sin 2\pi f_c t) =$$

$$0.5[\cos (2\pi f_c - 2\pi f_m)t - \cos (2\pi f_c + 2\pi f_m)t]$$

Note that these are the sum and difference frequencies or the upper and lower sidebands.

It is important to remember that a cosine wave is simply a sine wave shifted by 90°. A cosine wave has exactly the same shape as a sine wave, but it occurs 90° earlier in time. The cosine wave leads a sine wave by 90° or the sine wave lags a cosine wave by 90°.

The 90° p[hase] [s]hifters in Fig. 3-26 create cosine wave[s] nals which a[re] tor 2 to pro[duce]

$(V_$
Another co[lates this t[

$(V_m$
$0.5[\cos(2$

Now, if [
gether, th[
difference[
lower sid[

A pha[
causes t[
put by [
have be[
shift. A[
is show[
section[
section[
put tha[
up of[
leads [
then b[
put g[
other[

Fig. 3-27　A fixed-frequency, 90° *RC* phase shifter.

The most difficult part of creating a phasing-type SSB generator is designing a circuit that maintains a constant 90° phase shift over a wide range of modulating frequencies. Keep in mind that the definition of phase shift is a time shift between sine waves of the same frequency. An *RC* network produces a specific amount [of] phase shift at only one frequency [because th]e capacitive reactance varies with [frequency.] In the carrier phase shifter, this is [no prob]lem since the carrier is maintained [at a const]ant frequency. But the modulating [signal is u]sually a band of frequencies, typi[cally th]e audio range from 300 to 3000 Hz.

[One of] the circuits commonly used to pro[duce a 90]° phase shift over a wide bandwidth [is shown] in Fig. 3-28 on the next page. The [phase sh]ift difference between output 1 and [output 2] is 90° ±1.5° over the 300- to 3000-Hz [range. T]he resistor and capacitor values are [carefull]y selected to ensure this phase accu[racy. P]hase-shift inaccuracies will cause in[comple]te cancellation of the undesired side[band.]

[The] phasing method can be used to select [either] the upper or lower sideband. This is [done b]y changing the phase shift of either the [carrier] or carrier signals to the balanced mod[ulator] inputs. For example, applying the direct [carrier] signal to balanced modulator 2 in Fig. [3-26] and the 90° phase-shifted signal to bal[ance]d modulator 1 will cause the upper [side]band to be selected instead of the lower [side]band. The phase relationship of the carrier [woul]d also be switched to make this change.

[T]he output of the phasing generator is a low-[leve]l SSB signal. The degree of suppression of [the] carrier depends upon the quality of the bal-

Product detector

Fig. 3-28 A phase shifter that produces a 90° shift over the 300- to 3000-Hz range.

anced modulators. The precision of the phase shifting determines the degree of suppression of the unwanted sideband. The design and adjustment of phasing-type SSB generators is critical in order to ensure complete suppression of the undesired sideband. The SSB output is then applied to linear RF amplifiers where its power level is increased before being applied to the transmitting antenna.

To demodulate an SSB signal, you must reinsert the carrier at the receiver. Assume that you generate an SSB signal by modulating a 9-MHz carrier with a 2-kHz sine wave intelligence signal. A balanced modulator suppresses the 9-MHz carrier but generates the upper and lower sideband frequencies of 9.002 and 8.998 MHz respectively. Assume that a sharp bandpass filter selects the upper sideband of 9.002 MHz and suppresses the lower sideband. At the receiver you will get only the 9.002-MHz signal. But what you want is the 0.002-MHz or 2-kHz intelligence signal.

Recovering the original modulating signal is a matter of mixing the received signal with a locally generated carrier. Inside the receiver is an oscillator that is set to 9 MHz. The oscillator signal is applied to a mixer circuit along with the incoming signal. The mixer forms the sum and difference frequencies of 9.002 + 9.000 = 18.002 MHz and 9.002 − 9.000 = 0.002 MHz or 2 kHz. The 18.002-MHz signal is filtered out, leaving the desired 2 kHz intelligence signal.

The demodulator for SSB signals is therefore a mixer. Typically, a balanced modulator is used for this purpose. Any of the previously described circuits will work. A balanced modulator used for this purpose is generally referred to as a *product detector*.

Because it is difficult to set the internal local oscillator to the exact frequency of the original carrier, the frequency of the recovered intelligence signal may be slightly higher or lower than the original 2 kHz. For voice transmissions, this means that the recovered voice may be higher or lower in pitch. To correct for this effect, the internal oscillator is usually made variable so that the operator of the receiver can adjust it and tune for most intelligible and natural sounding output.

Self-Test

Supply the missing information in each statement.

38. The most common way of generating an SSB signal is the _____ method.
39. A filter capable of passing the desired sideband while rejecting the other sideband must have good _____ .
40. A balanced modulator has a 3-MHz carrier input and a modulating signal input of 1.5 kHz. To pass the lower sideband, a filter must have a center frequency of _____ MHz.
41. The most popular filter used to select the desired sideband in an SSB generator uses _____ for selectivity.
42. Name the two ways of generating either the upper or lower sideband in a filter-type SSB generator.
43. A quartz crystal acts like a highly selective _____ circuit.
44. In a filter-type SSB generator, a crystal lattice filter is used. The two crystal frequencies are 3.0 and 3.0012 MHz.

58 CHAPTER 3 AMPLITUDE MODULATION CIRCUITS

The filter bandwidth is approximately _____ kHz.

45. Mechanical filters provide selectivity because they _____ at a specific frequency.

46. A ceramic filter is similar in operation to a(n) _____ filter.

47. The operating frequency range of a mechanical filter is _____ to _____ kHz.

48. In the phasing method of SSB generation, _____ is used to cancel the undesired sideband.

49. A _____ circuit is commonly used to demodulate or recover an SSB signal.

50. The circuit used to demodulate an SSB signal is typically called a(n) _____ .

3-5 MIXERS AND CONVERTERS

As we indicated in the previous chapter, mixing is a special kind of AM used for frequency translation purposes. It is often necessary or desirable to convert a signal along with any modulation that may be on it to another higher or lower frequency for further processing. The circuits used to perform frequency translation or conversion are called mixers or converters rather than modulators. The function they perform is referred to as heterodyning and is identical to that of an amplitude modulator.

There are two basic types of frequency conversion, up conversion and down conversion. *Up conversion* means that the input signal along with any modulation is translated to a higher frequency. In *down conversion,* the input signal is translated to some lower frequency.

Figure 3-29 shows a simplified diagram of a mixer circuit. The two inputs to the mixer are the input signal along with any modulation and the input from a local oscillator (LO). The output will consist of the sum and difference frequencies of the two inputs. In some mixers, the LO and input signals may also appear in the output. Of course, the sum and difference frequencies are what we are interested in. A tuned circuit or filter at the output of the mixer selects the desired output signal while rejecting the others.

The mixer can be one of many different components or circuits. Its primary characteristic is nonlinearity. Any device or circuit whose output does not vary linearly with the input can be used as a mixer. A diode is a good example.

The mixer, because it is an amplitude modulator, basically performs the function of analog multiplication. It is this multiplication process that generates the sum and difference frequencies. For up conversion, the sum of the two inputs is the selected output, while for down conversion, the difference between the two inputs is selected.

A wide variety of circuits have been developed to perform mixing operations. Many of those circuits used for AM are perfectly suitable in frequency translation applications. For example, one of the most widely used types of mixer is the simple diode modulator described in Sec. 3-1. The circuit is simple and quite effective. Such diode mixers are the most common type found in microwave applications.

A simple single-diode mixer circuit is shown in Fig. 3-30. The input signal is applied to the primary winding of transformer T_1. The signal is coupled to the secondary winding and applied to the diode mixer. The LO signal is coupled to the diode by way of capacitor C_1. The output signals, including both inputs, are de-

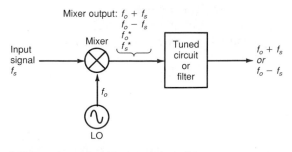

* May or may not be in the output depending upon the type of mixer.

Fig. 3-29   Basic mixer circuit.

Fig. 3-30   A simple diode mixer.

veloped across the tuned circuit, which selects either the sum or difference frequency and eliminates the others.

Another popular mixer circuit using two diodes is a singly balanced mixer illustrated in Fig. 3-31. The input signal is applied to the primary of the transformer, whereas the LO is applied to the center tap of the secondary. The output signal is developed across an RFC and then applied to a filter or tuned circuit which selects the sum or difference frequency. In the one- or two-diode mixer, the LO signal is very much larger than the input signal. For this reason, the diodes are turned off and on like switches by the LO signal.

In small-signal applications, germanium diodes, because of their relatively low turn-on voltage, are used in mixers. Silicon diodes also make excellent RF mixers. The best diode mixers at VHF, UHF, and microwave frequencies are hot-carrier or Schottky barrier diodes.

Balanced modulators are also widely used as mixers. These circuits offer the advantage of having the carrier eliminated from the output, thereby making the job of filtering much easier. Any balanced modulator described previously can be used in mixing applications. Both the diode lattice balanced modulator and the integrated differential amplifier-type balanced modulator are quite effective in mixing applications. A version of the balanced modulator shown in Fig. 3-17, known as a doubly balanced mixer and illustrated in Fig. 3-32, is one of the most popular mixers at VHF and UHF. It may be the single best mixer available.

One of the most common applications for a mixer is in radio receivers. The mixer is used to convert the incoming signal to a lower frequency where it is easier to obtain the high gain and selectivity required. Although any mixer

Fig. 3-32 A doubly balanced mixer very popular at high frequencies.

circuit previously described can be used in this application, special types of mixers have also been developed. Typically these mixers are made up of a single transistor biased into the nonlinear range so that it produces analog multiplication. The primary benefit of transistor mixers over the diode mixers is that gain is obtained with a transistor stage. An example of such a mixer circuit is illustrated in Fig. 3-33. This bipolar transistor is typically biased as if it were a linear amplifier, but the bias is adjusted so that the collector current does not vary linearly with variations in the base current. The class of operation is AB rather than A. This results in analog multiplication which

Fig. 3-31 A singly balanced mixer.

Fig. 3-33 A bipolar transistor mixer.

produces the sum and difference frequencies. In this circuit, both the incoming signal and LO are applied to the base of the transistor. The tuned circuit in the collector usually selects the difference frequency.

Improved performance can be obtained by using an FET as shown in Fig. 3-34. Like the bipolar transistor, the FET is biased so that it operates in a nonlinear portion of its characteristic. The input signal is applied to the gate, while the LO signal is coupled to the source. Again the tuned circuit in the drain selects the difference frequency.

Another popular FET mixer is shown in Fig. 3-35. Here a dual-gate metal-oxide-semiconductor FET (MOSFET) is used. The input signal is applied to one gate, while the LO is coupled to the other gate. Dual-gate MOSFETs provide superior performance in mixing applications. Today, bipolar transistors are rarely used in mixing applications except at very low radio frequencies. In receivers built for VHF, UHF, and microwave applications, junction FETs and dual-gate MOSFETs are widely used as mixers because of their high gain and low noise. Gallium arsenide (GaAs) rather than silicon FETs are preferred at the higher frequencies because of their lower noise contribution and higher gain. Diode mixers are also used in many high-frequency receivers.

It is possible to combine the functions of the mixer and LO in one circuit. An example of

Fig. 3-35 A dual-gate MOSFET mixer.

such a circuit is shown in Fig. 3-36 on the next page. Known as an autodyne converter, the transistor operates as a class C LO and as a mixer. Transformer T_1 and the transistor form a Hartley oscillator.

The lower winding of T_1 and C_1 form a tuned circuit that resonates at the desired LO frequency. Tuning is accomplished with C_1. Transformer T_2 and the transistor comprise the mixer. Transformer T_2 resonates at the difference frequency. The LO signal and input signal combine inside the transistor to produce the four standard mixer outputs. The difference frequency is selected by T_2.

When both mixing and oscillator functions are performed by a single transistor, the circuit is usually referred to as a converter. Converters are only used on the lower RF frequencies and are common in AM radios and inexpensive communications receivers operating below 30 MHz.

A typical integrated circuit mixer is shown in Fig. 3-37(a) on page 63. Known as the NE602, it consists of a double-balanced mixer circuit made up of two cross-connected differential amplifiers. Also on the chip is an NPN transistor that can be connected as a stable oscillator circuit and a dc voltage regulator. The device is housed in an 8-pin dual in-line IC package. It operates from a single dc power supply voltage of 4.5 to 8 V. The circuit can be used at frequencies up to 500 MHz, which makes it useful in HF, VHF, and low-frequency UHF applications. The oscillator will operate up to about 200 MHz. An external LC-tuned circuit or a crystal is required to set

Fig. 3-34 A JFET mixer.

Fig. 3-36 An autodyne converter that combines a mixer and LO.

the operating frequency. The oscillator is internally connected to one input of the mixer.

Figure 3-37(b) shows the circuit details of the mixer itself. Q_1 and Q_2 form a differential amplifier with current source Q_3, and Q_4 and Q_5 form another differential amplifier with current source Q_6. Note that the inputs are connected in parallel. The collectors are cross-connected; that is, the collector of Q_1 is connected to the collector of Q_4 instead of Q_3, as would be the case for a parallel connection. In the same way, the collector of Q_2 is connected to the collector of Q_5. This connection results in a balanced modulator-like circuit in which the internal oscillator signal is suppressed, leaving only the sum and difference signals in the output. The output may be balanced or single-ended as the application requires. A filter or tuned circuit must be connected to the output to select the desired sum or difference signal.

A typical circuit using the NE602 IC mixer is shown in Fig. 3-38 on page 64. It is powered by 6 V dc. R_1 and C_1 are used for decoupling. A resonant transformer T_1 couples the 49-MHz input signal to the mixer. C_2 resonates with the transformer secondary at the input frequency. C_3 is an ac bypass connecting pin 2 to ground. External components C_4 and L_1 form a tuned circuit that sets the oscillator to 59 MHz. C_5 and C_6 form a capacitive voltage divider that connects the on-chip NPN transis-

tor as a Colpitts oscillator circuit. C_7 is an ac coupling and blocking capacitor. The output is taken from pin 5 and connected to a ceramic bandpass filter. Note the special symbol for the ceramic filter. The output, in this case the difference signal, or $59 - 49 = 10$ MHz, appears across R_2. The balanced mixer circuit suppresses the 59-MHz oscillator signal. The sum signal of 108 MHz is eliminated by the ceramic filter.

Self-Test

Choose the letter which best answers each statement.

51. The mixing process is
   a. Linear.
   b. Nonlinear.
52. In up conversion, which signal is selected?
   a. $f_1 - f_2$
   b. $f_1 + f_2$
53. Which transistor mixer is preferred?
   a. Bipolar
   b. FET

Supply the missing information in each statement.

54. Another name for the mixing process is
_____ .

55. The purpose of a mixer is to perform
_____ .

Top view

(a) Block diagram and pinout

(b) Simplified schematic

Fig. 3-37   NE602 IC mixer.

56. The output signals produced by a mixer with inputs f_1 and f_2 are _____ , _____ , _____ , and _____ .

57. The type of mixer most used at microwave frequencies is a(n) _____ .

58. The two mixers preferred at VHF and UHF are _____ .

59. Transistor mixers provide _____ , while diode mixers do not.

60. A mixer producing down conversion has an input of 1390 kHz and an LO frequency of 1845 kHz. The output is _____ kHz.

61. A single-transistor circuit that combines the mixer and LO functions is known as a(n) _____ .

62. The mixer circuit in the popular NE602 IC is _____ .

Fig. 3-38 **NE602 mixer used for frequency translation.**

63. In addition to a mixer and a voltage regulator, the NE602 mixer IC contains a(n) _____ .

64. The output of a NE602 mixer IC is 5.5 MHz. The local oscillator is set to 26.8 MHz. The input frequency is _____ .

Determine whether each statement is true or false.

65. Mixing is the same as AM.

66. Any balanced modulator may be used as a mixer.

SUMMARY

1. Amplitude modulation can be accomplished by multiplying the carrier sine wave by a gain or attenuation factor that varies in accordance with the intelligence signal.

2. Amplitude modulation can be carried out by linearly combining the carrier and intelligence signals then applying the result to a nonlinear component or circuit. A diode is an example.

3. The simplest AM circuit uses resistors to linearly mix the carrier and information signal, a diode to rectify the result, and a tuned circuit to complete the waveform.

4. Amplitude modulation can be produced by feeding the carrier to a voltage divider or other network and then varying one of the resistive elements in accordance with the modulating signal. Common variable resistive components are diodes and FETs. PIN diodes are used at UHF and microwave frequencies.

5. Low-level modulation is the process of generating the AM signal at low voltage and/or power levels and then using linear amplifiers to increase the power level.

6. High-level modulation is the process of amplitude modulating the final power amplifier of a transmitter.

7. High-level modulation is accomplished with a collector (plate in vacuum tubes) modulator that varies the collector supply voltage in accordance with the modulating signal.

8. For 100 percent high-level modulation, the modulation amplifier must produce an output wave whose peak-to-peak is 2 times the collector supply voltage.

9. For 100 percent high-level modulation, the modulation amplifier must generate an output power that is one-half of the final RF power amplifier input power ($P_i = V_{cc} \times I_c$).

10. The simplest and best amplitude demodulator is the diode detector. The AM signal

is rectified by a diode and then filtered by a capacitor to recover the envelope which is the original modulating information.

11. Balanced modulators are AM circuits that cancel or suppress the carrier but generate a DSB output signal that contains the upper (sum) and lower (difference) sideband frequencies.

12. A popular balanced modulator is the lattice modulator that uses a diode bridge circuit as a switch. The carrier turns the diodes off and on letting segments of the modulating signal through to produce a DSB output signal. A carrier suppression of 40 dB is possible.

13. Another widely used balanced modulator is an integrated circuit (IC) using differential amplifiers as switches to switch the modulating signal at the carrier frequency. A popular device is the 1496 or 1596. Carrier suppression can be as high as 50 to 65 dB.

14. The most common way of generating an SSB signal is to use the filter method which incorporates a balanced modulator followed by a highly selective filter that passes either the upper or lower sideband.

15. To make both sidebands available, SSB generators use two filters, one for each sideband, or switch the carrier frequency to put the desired sideband into the fixed filter bandpass.

16. Most SSB filters are made with quartz crystals.

17. A quartz crystal is a frequency-determining component that acts like an *LC* circuit with a very high Q.

18. Crystals have series and parallel resonant modes. These can be combined into a lattice (bridge) circuit that provides extremely sharp selectivity over a desired bandwidth.

19. Ceramic filters use ceramic resonators that act like crystals but are smaller and lower in cost.

20. Mechanical bandpass filters are also used in SSB generators. These devices use multiple resonant disks that vibrate at some frequency in the 200- to 500-kHz range.

21. The phasing method of SSB generation uses two balanced modulators and 90° phase shifters for the carrier and modulating signal to produce two DSB signals that when added cause one sideband to be canceled out.

22. In phasing-type SSB generators, the accuracy of the phase shifters determines the degrees of unwanted sideband suppression.

23. Precision *RC* networks are normally used to produce the desired 90° phase shifts.

24. A demodulator for SSB is a mixer such as a balanced modulator that is called a product detector. The carrier is reinserted in the receiver with a local oscillator.

25. Frequency translation or conversion is an AM process that converts a signal with any modulation to a higher (up conversion) or lower (down conversion) frequency.

26. Frequency conversion is produced by a circuit called a mixer.

27. Another name for frequency conversion is heterodyning.

28. Almost any low-level AM circuit can be used for mixing.

29. One of the most commonly used mixers is a single diode. Germanium or hot-carrier diodes are used for mixing at very high radio frequencies.

30. The inputs to a mixer are the signal to be translated f_s and a sine wave f_o from a local oscillator (LO). The outputs are f_o, f_s, $f_o + f_s$, and $f_o - f_s$. A tuned circuit at the output selects either the sum or difference frequency while suppressing the others. Any modulation on the input appears on the output signal.

31. Bipolar and field-effect transistors can be used as mixers by operating them in the nonlinear region of their characteristics.

32. Transistor mixers offer the benefit of gain over diode mixers.

33. Balanced modualtors are widely used as mixers.

34. Doubly balanced modulators and GaAs FETs are the best mixers at VHF, UHF, and microwave frequencies.

35. A single transistor can be connected to perform both the mixer and LO functions. Such a mixer is called a converter and is used primarily in AM receivers at frequencies below 30 MHz.

36. A popular IC mixer is the NE602, which contains a cross-connected differential amplifier mixer circuit and an on-chip local oscillator.

Choose the letter that best answers each question.

3-1. Amplitude modulation is the same as
  a. Linear mixing.
  b. Analog multiplication.
  c. Signal summation.
  d. Multiplexing.

3-2. In a diode modulator, the negative half of the AM wave is supplied by a(n)
  a. Tuned circuit.
  b. Transformer.
  c. Capacitor.
  d. Inductor.

3-3. Amplitude modulation can be produced by
  a. Having the carrier vary a resistance.
  b. Having the modulating signal vary a capacitance.
  c. Varying the carrier frequency.
  d. Varying the gain of an amplifier.

3-4. Amplitude modulators that vary the carrier amplitude with the modulating signal by passing it through an attenuator work on the principle of
  a. Rectification.
  b. Resonance.
  c. Variable resistance.
  d. Absorption.

3-5. In Fig. 3-4, D_1 is a
  a. Variable resistor.
  b. Mixer.
  c. Clipper.
  d. Rectifier.

3-6. The component used to produce AM at very high frequencies is a
  a. Varactor.
  b. Thermistor.
  c. Cavity resonator.
  d. PIN diode.

3-7. Amplitude modulation generated at a very low voltage or power amplitude is known as
  a. High-level modulation.
  b. Low-level modulation.
  c. Collector modulation.
  d. Minimum modulation.

3-8. A collector modulator has a supply voltage of 48 V. The peak-to-peak amplitude of the modulating signal for 100 percent modulation is
  a. 24 V.
  b. 48 V.
  c. 96 V.
  d. 120 V.

3-9. A collector-modulated transmitter has a supply voltage of 24 V and a collector current of 0.5 A. The modulator power for 100 percent modulation is
  a. 6 W.
  b. 12 W.
  c. 18 W.
  d. 24 W.

3-10. The circuit that recovers the original modulating information from an AM signal is known as a
  a. Modulator.
  b. Demodulator.
  c. Mixer.
  d. Crystal set.

3-11. The most commonly used amplitude demodulator is the
  a. Diode mixer.
  b. Balanced modulator.
  c. Envelope detector.
  d. Crystal filter.

3-12. A circuit that generates the upper and lower sidebands but no carrier is called a(n)
  a. Amplitude modulator.
  b. Diode detector.
  c. Class C amplifier.
  d. Balanced modulator.

3-13. The inputs to a balanced modulator are 1 MHz and a carrier of 1.5 MHz. The outputs are
  a. 500 kHz.
  b. 2.5 MHz.
  c. 1.5 MHz.
  d. All the above.
  e. *a* and *b*

3-14. A widely used balanced modulator is called the
  a. Diode bridge circuit.
  b. Full-wave bridge rectifier.
  c. Lattice modulator.
  d. Balanced bridge modulator.

3-15. In a diode ring modulator, the diodes act like
a. Variable resistors.   c. Rectifiers.
b. Switches.   d. Variable capacitors.

3-16. The output of a balanced modulator is
a. AM.   c. SSB.
b. FM.   d. DSB.

3-17. The principal circuit in the popular 1496/1596 IC balanced modulator is a
a. Differential amplifier.   c. Bridge.
b. Rectifier.   d. Constant current source.

3-18. The most commonly used filter in SSB generators uses
a. LC networks.   c. Crystals.
b. Mechanical resonators.   d. RC networks and op amps.

3-19. The equivalent circuit of a quartz crystal is a
a. Series resonant circuit.
b. Parallel resonant circuit.
c. Neither a nor b
d. Both a and b

3-20. A crystal lattice filter has crystal frequencies of 27.5 and 27.502 MHz. The bandwidth is approximately
a. 2 kHz.   c. 27.501 MHz.
b. 3 kHz.   d. 55.502 MHz.

3-21. An SSB generator has a sideband filter centered at 3.0 MHz. The modulating signal is 3 kHz. To produce both upper and lower sidebands, the following carrier frequencies must be produced:
a. 2.7 and 3.3 MHz.   c. 2997 and 3003 kHz.
b. 3.3 and 3.6 MHz.   d. 3000 and 3003 kHz.

3-22. In the phasing method of SSB generation, one sideband is canceled out due to
a. Phase shift.   c. Carrier suppression.
b. Sharp selectivity.   d. Phase inversion.

3-23. A balanced modulator used to demodulate a SSB signal is called a(n)
a. Transponder.   c. Converter.
b. Product detector.   d. Modulator.

3-24. Frequency translation is done with a circuit called a
a. Summer.   c. Filter.
b. Multiplier.   d. Mixer.

3-25. The inputs to a mixer are f_o and f_m. In down conversion, which of the following mixer output signals is selected?
a. f_o c. $f_o - f_m$
b. f_m d. $f_o + f_m$

3-26. Mixing for frequency conversion is the same as
a. Rectification.   c. Linear summing.
b. AM.   d. Filtering.

3-27. Which of the following can be used as a mixer?
a. Balanced modulator   c. Diode modulator
b. FET   d. All the above

3-28. The desired output from a mixer is usually selected with a
a. Phase-shift circuit.   c. Resonant circuit.
b. Crystal filter.   d. Transformer.

3-29. The two inputs to a mixer are the signal to be translated and a signal from a(n)
a. Modulator.   c. Antenna.
b. Filter.   d. LO.

3-30. An NE602 mixer IC has a difference output of 10.7 MHz. The input is 146.8 MHz. The local oscillator frequency is
a. 101.9 MHz.   c. 131.6 MHz.
b. 125.4 MHz.   d. 157.5 MHz.

CRITICAL THINKING QUESTIONS

3-1. If AM can be achieved by varying the gain or attenuation of the carrier, name one or more devices or circuits that can be used for this purpose.

3-2. Could the 1496 IC balanced modulator in Fig. 3-18 be used as a mixer? Why?

3-3. Is it possible for one AM signal to amplitude-modulate a carrier on another frequency? If so, describe what would happen. What would the output spectrum look like?

Answers to Self-Tests

1. *a*
2. *b*
3. *c*
4. gain, attenuation
5. square law
6. diode
7. tuned circuit
8. variable resistor
9. gain
10. increase, decrease, increase
11. resistor
12. 100 MHz
13. attenuator
14. low
15. linear amplifier
16. C
17. collector
18. supply voltage
19. 2.5
20. 12
21. driver
22. demodulator
23. diode detector
24. C_1
25. ripple, distortion
26. envelope

27. smaller filter capacitor, less ripple and distortion
28. true
29. *c*
30. *d*
31. *b*
32. lattice modulator or diode ring
33. switches
34. 1897.4, 1902.6
35. Q_7, Q_8
36. switches
37. true
38. filter
39. selectivity
40. 2.9985
41. crystals
42. Use one filter for each sideband; select the carrier frequency so that the desired sideband is in the filter passband.
43. tune, resonant, or *LC*
44. 1.8 (3.0012 − 3.0 = 0.0012 MHz = 1.2 kHz; 1.2 × 1.5 = 1.8 kHz)
45. vibrate or resonate

46. crystal
47. 200 to 500
48. phase shift
49. mixer or balanced modulator
50. product detector
51. *b*
52. *b*
53. *b*
54. heterodyning
55. frequency conversion or translation
56. f_1, f_2, $f_1 + f_2$, $f_1 - f_2$
57. diode
58. doubly balanced mixer, GaAs FET
59. gain
60. 455 (1845 − 1390 = 455)
61. converter
62. cross-connected differential amplifier
63. local oscillator
64. 21.3 or 32.3 MHz
65. true
66. true

CHAPTER 4

Frequency Modulation

∎

CHAPTER OBJECTIVES

This chapter will help you to:

1. *Define and explain* the processes of frequency modulation (FM) and phase modulation (PM) and state their differences.
2. *Calculate* the modulation index given the maximum deviation and maximum modulating frequency, *determine* the significant number of sidebands in an FM signal, and *calculate* the bandwidth of an FM signal.
3. *Define* pre-emphasis and de-emphasis, *state* their benefits, and *show* how they are accomplished.
4. *Name* the advantages and disadvantages of FM and PM compared to AM.

Modulation is the process of modifying a carrier wave in accordance with an information signal to be transmitted. A carrier wave is usually a high-frequency sine wave. In examining a carrier sine wave, you can see that there are two obvious ways to modify it. First, its amplitude may be varied. Changing the amplitude of the carrier produces AM. The other carrier characteristic that can be changed is the frequency. It is also possible to impress an information signal on a carrier by changing its frequency. Although not immediately obvious, another characteristic of a carrier that can be changed is its phase shift. By varying the amount of phase shift that a carrier experiences, information can be impressed upon the carrier. This is known as phase modulation (PM). As it turns out, varying the phase shift of a carrier produces FM. Therefore, both FM and PM are closely related to one another. Together they are collectively referred to as types of angle modulation. Since FM is generally superior in performance to AM, it is widely used in many areas of communications electronics. In this chapter, we will introduce you to the fundamentals of FM and PM. Circuits for producing FM and PM and frequency demodulators are covered in Chap. 5.

∎

4-1 FREQUENCY MODULATION PRINCIPLES

In FM, the carrier amplitude remains constant, while the carrier frequency is changed by the modulating signal. As the amplitude of the information signal varies, the carrier frequency will shift in proportion. As the modulating signal amplitude increases, the carrier frequency increases. If the amplitude of the modulating signal decreases, the carrier frequency decreases. The reverse relationship can also be implemented. A decreasing modulating signal will increase the carrier frequency above its center value, whereas an increasing modulating signal will decrease the carrier frequency below its center value. As the modulating signal amplitude varies, the carrier frequency varies above and below its normal center frequency with no modulation. The amount of change in carrier frequency produced by the modulating signal is known as the *frequency deviation*. Maximum frequency deviation occurs at the maximum amplitude of the modulating signal.

The frequency of the modulating signal determines how many times per second the carrier frequency deviates above and below its nominal center frequency. If the modulating signal is a 100-Hz sine wave, then the carrier frequency will shift above and below the center frequency 100 times per second. This is called the frequency deviation rate.

An FM signal is illustrated in Fig. 4-1(c).

Normally the carrier [Fig. 4-1(a)] is a sine wave, but it is shown as a triangular wave here to simplify the illustration. With no modulating signal applied, the carrier frequency is a constant-amplitude sine wave at its normal constant center frequency.

The modulating information signal [Fig. 4-1(b)] is a low-frequency sine wave. As the sine wave goes positive, the frequency of the carrier increases proportionately. The highest frequency occurs at the peak amplitude of the modulating signal. As the modulating signal amplitude decreases, the carrier frequency decreases. When the modulating signal is at zero amplitude, the carrier will be at its center frequency point.

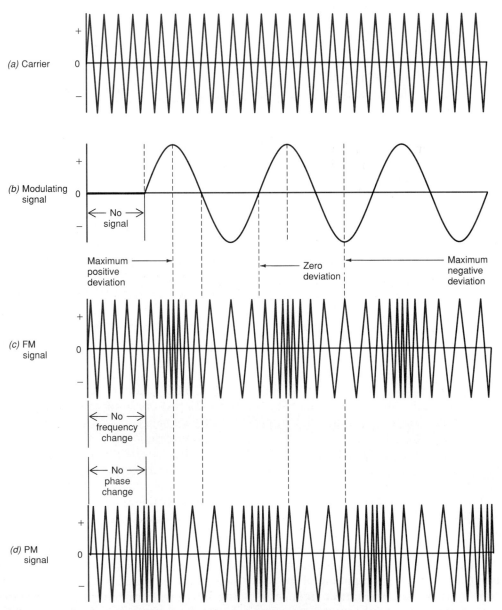

Fig. 4-1 Frequency modulation and phase modulation signals. The carrier is drawn as a triangular wave for simplicity, but in practice it is a sine wave.

Now when the modulating signal goes negative, the carrier frequency will decrease. The carrier frequency will continue to decrease until the peak of the negative half cycle of the modulating sine wave is reached. Then, as the modulating signal increases toward zero, the frequency will again increase. Note in Fig. 4-1(c) how the carrier sine waves seem to be first "compressed" and then "stretched" by the modulating signal.

Assume a carrier frequency of 50 MHz. If the peak amplitude of the modulating signal causes a maximum frequency shift of 200 kHz, the carrier frequency will deviate up to 50.2 MHz and down to 49.8 MHz. The total frequency deviation is $50.2 - 49.8 = 0.4\ MHz = 400\ kHz$. In practice, however, the frequency deviation is expressed as the amount of frequency shift of the carrier above or below the center frequency. Therefore, the frequency deviation in the example above is said to be ± 200 kHz. This means that the modulating signal varies the carrier above and below its center frequency by 200 kHz. The frequency of the modulating signal determines the rate of frequency deviation but has no effect on the amount of deviation which is strictly a function of the amplitude of the modulating signal.

Self-Test

Answer the following questions.

1. The general name given to both FM and PM is _____ modulation.
2. True or false. In FM, the carrier amplitude remains constant with modulation.
3. The amount of frequency shift during modulation is called the _____ .
4. The amount of frequency shift in FM is directly proportional to the _____ of the modulating signal.
5. As the modulating signal amplitude goes positive, the carrier frequency _____ .
   As the modulating signal amplitude goes negative, the carrier frequency _____ .

4-2 PHASE MODULATION

Another way to produce angle modulation is to vary the amount of phase shift of a constant-frequency carrier in accordance with a modulating signal. The resulting output is a PM signal. Imagine a modulator circuit whose basic function is to produce a phase shift. Remember that a phase shift refers to a time separation between two sine waves of the same frequency. Assume that we can build a phase shifter that causes the amount of phase shift to vary with the amplitude of the modulating signal. The greater the amplitude of the modulating signal, the greater the phase shift. Assume further that positive alternations of the modulating signal produce a lagging phase shift and negative signals produce a leading phase shift.

If a constant-amplitude–constant-frequency carrier sine wave is applied to the phase shifter, the output of the phase shifter will be a PM wave. As the modulating signal goes positive, the amount of phase lag increases with the amplitude of the modulating signal. This means that the carrier output is delayed. That delay increases with the amplitude of the modulating signal. The result at the output is as if the constant-frequency carrier signal had been stretched out or its frequency lowered.

When the modulating signal goes negative, the phase shift becomes leading. This causes the carrier sine wave to be effectively speeded up or compressed. The result is as if the carrier frequency had been increased.

Phase modulation produces frequency modulation. Since the amount of phase shift is varying, the effect is as if the carrier frequency is changed. Since FM is produced by PM, the latter is often referred to as indirect FM.

It is important to point out that it is the dynamic nature of the modulating signal that causes the frequency variation at the output of the phase shifter. In other words, FM is only produced as long as the phase shift is being varied. One way to understand this better is to assume a modulating signal like that shown in Fig. 4-2(a) on the next page. It is a triangular wave whose positive and negative peaks have been clipped off at a fixed amplitude. During time t_o, the signal is zero so the carrier is at its center frequency.

Applying this modulating signal to a frequency modulator will produce the signal shown in Fig. 4-2(b). During the time the waveform is rising (t_1), the frequency is increasing. During the time that the positive amplitude is constant (t_2), the FM output frequency is constant. During the time the amplitude decreases and goes negative (t_3), the frequency will decrease. Then, during the constant-amplitude negative alternation (t_4), the frequency remains constant at a lower frequency. During t_5, the frequency increases.

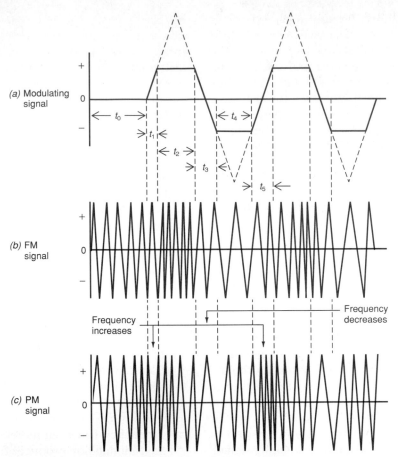

Fig. 4-2 A frequency shift occurs in PM only when the modulating signal amplitude varies.

When the modulating signal is applied to a phase shifter, the output frequency will change only during the time that the amplitude of the modulating signal is varying. Refer to the PM signal in Fig. 4-2(*c*). During increases or decreases in amplitude (t_1, t_3, and t_5), a varying frequency will be produced. However, during the constant-amplitude positive and negative peaks no frequency change takes place. The output of the phase shifter will simply be the carrier frequency which has been shifted in phase. This clearly illustrates that frequency variations take place only if the modulating signal amplitude is varying.

As it turns out, the maximum frequency deviation produced by a phase modulator occurs during the time that the modulating signal is changing at its most rapid rate. For a sine wave modulating signal, the rate of change of the modulating signal is greatest when the modulating wave changes from plus to minus or from minus to plus. The maximum rate of change of modulating voltage occurs exactly at the zero crossing points. You can see this in

Fig. 4-2(*c*). In contrast, note that in an FM wave the maximum deviation occurs at the peak positive and negative amplitude of the modulating voltage. So although a phase modulator does indeed produce FM, maximum deviation occurs at different points of a modulating signal. Of course, this is irrelevant since both the FM and PM waves contain exactly the same information and when demodulated will reproduce the original modulating signal.

In FM, maximum deviation occurs at the peak positive and negative amplitudes of the modulating signal. In PM, the maximum amount of leading or lagging phase shift occurs at the peak amplitudes of the modulating signal. Recall that we said that the frequency deviation at the output of the phase shifter depends upon the rate of change of the modulating signal. The faster the modulating signal voltage varies, the greater the frequency deviation produced. Because of this, the frequency deviation produced in PM increases with the frequency of the modulating signal. The higher the modulating signal frequency, naturally the shorter its period and the faster the voltage changes. Higher modulating voltages produce greater phase shift which, in turn, produces greater frequency deviation. However, higher modulating frequencies produce a faster rate of change of the modulating voltage and, therefore, also produce greater frequency deviation. In PM then, the carrier frequency deviation is proportional to both the modulating frequency and amplitude. In FM, frequency deviation is proportional only to the amplitude of the modulating signal regardless of its frequency. The relationship between carrier deviation and modulating signal characteristics is summarized in Fig. 4-3.

To make PM compatible with FM, we must compensate for the deviation produced by the frequency changes in the modulating signal. This is illustrated in Fig. 4-4. This low-pass filter causes the higher modulating frequencies to be attenuated in amplitude. Although the higher modulating frequencies will produce a greater rate of change and thus a greater frequency deviation, this is offset by the lower amplitude of the modulating signal which will produce less phase shift and less frequency deviation. This network compensates for the excess frequency deviation caused by higher modulating frequencies. The result is an output that is the same as an FM signal. The FM produced by a phase modulator is called indirect FM.

Although both FM and PM are widely used in communications systems, most angle modulation is PM. The reason for this is that a crystal oscillator with higher frequency accuracy and stability can be used to produce the carrier. In FM, crystal oscillators cannot, in general, be frequency-modulated over a very wide range. However, the crystal oscillator can drive

Fig. 4-4   A low-pass filter compensates for higher phase shift and frequency deviation at the higher modulating frequencies to produce indirect FM.

a phase modulator which can produce the desired FM. Further, most phase modulators are simpler to implement than frequency modulators. You will see this when practical FM and PM circuits are discussed in the next chapter.

Self-Test

Answer the following questions.

6. In PM, the carrier _____ is varied in proportion to the amplitude of the modulating signal.
7. A varying phase shift produces a(n) _____ .
8. A phase modulator produces a frequency deviation only when the modulating signal is _____ .
9. In PM, the frequency deviation is proportional to both the modulating signal _____ and _____ .
10. When the modulating signal amplitude crosses zero, the phase shift and frequency deviation in a phase modulator are
    a. At a maximum.
    b. At a minimum.
    c. Zero.
11. A phase modulator may use a low-pass filter to offset the effect of increasing carrier

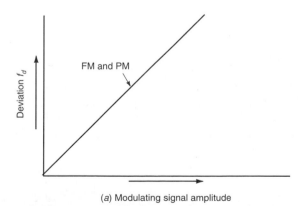

Fig. 4-3   Relationship between frequency deviation and modulating signal amplitude and frequency for FM and PM.

frequency deviation for increasing modulating

   a. Amplitude.
   b. Frequency.
   c. Phase shift.

12. The FM produced by a phase modulator is known as _____ .

4-3 SIDEBANDS AND THE MODULATION INDEX

Any modulation process produces sidebands. As you saw in AM, when a constant-frequency sine wave modulates a carrier, two side frequencies are produced. The side frequencies are the sum and difference of the carrier and the modulating frequency. In FM and PM too, sum and difference sideband frequencies are produced. In addition, a theoretically infinite number of pairs of upper and lower sidebands are also generated. As a result, the spectrum of an FM/PM signal is usually wider than an equivalent AM signal. A special narrowband FM signal whose bandwidth is only slightly wider than that of an AM signal can also be generated.

Figure 4-5 shows an example of the spectrum of a typical FM signal produced by modulating a carrier with a single-frequency sine wave. Note that the sidebands are spaced from the carrier f_c and are spaced from one another

by a frequency equal to the modulating frequency f_m. If the modulating frequency is 500 Hz, the first pair of sidebands are above and below the carrier by 500 Hz. The second pair of sidebands are above and below the carrier by 2×500 Hz = 1000 Hz, or 1 kHz, and so on. Note also that the amplitudes of the sidebands vary. If each sideband is assumed to be a sine wave with a frequency and amplitude as indicated in Fig. 4-5 and all these sine waves were added together, then the FM signal producing them would be created.

As the amplitude of the modulating signal varies, of course, the frequency deviation will change. The number of sidebands produced, their amplitude, and their spacing depend upon the frequency deviation and modulating frequency. Keep in mind that an FM signal has a constant amplitude. If that FM signal is a summation of the sideband frequencies, then you can see that the sideband amplitudes must vary with frequency deviation and modulating frequency if their sum is to produce a fixed-amplitude FM signal.

Although the FM process produces an infinite number of upper and lower sidebands, only those with the largest amplitudes are significant in carrying the information. Typically any sideband whose amplitude is less than 1 percent of the unmodulated carrier is considered insignificant. As a result, this markedly narrows the bandwidth of an FM signal.

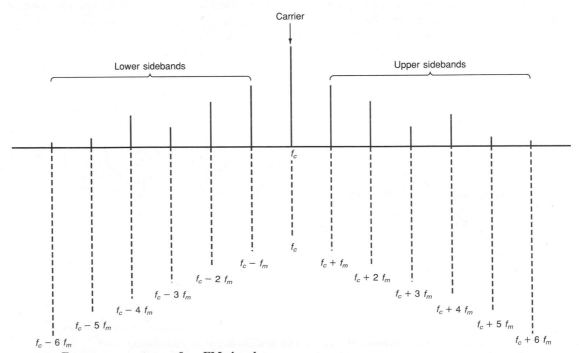

Fig. 4-5 Frequency spectrum of an FM signal.

As indicated earlier, the number of significant sidebands and their amplitudes are dependent upon the amount of frequency deviation and the modulating frequency. The ratio of the frequency deviation to the modulating frequency is known as the modulation index m.

$$m = \frac{f_d}{f_m}$$

where f_d is the frequency deviation and f_m is the modulating frequency.

For example, assume that the maximum frequency deviation of the carrier is ± 25 kHz while the maximum modulating frequency is 10 kHz. The modulation index, therefore, is

$$m = \frac{25}{10} = 2.5$$

In most communications systems using FM, maximum limits are put on both the frequency deviation and the modulating frequency. For example, in standard FM broadcasting, the maximum permitted frequency deviation is 75 kHz, while the maximum permitted modulating frequency is 15 kHz. This produces a modulation index of

$$m = \frac{75}{15} = 5$$

Whenever the maximum allowable frequency deviation and the maximum modulating frequency are used in computing the modulation index, m is known as the *deviation ratio*.

Knowing the modulation index, you can compute the number and amplitudes of the significant sidebands. This is done through a complex mathematical process known as the Bessel functions. This mathematical computation is beyond the scope of this text. In general, it is not necessary to know how to make the calculation as the Bessel functions have been computed and tabulated for a wide range of modulation indexes. An example is given in Fig. 4-6. The left-hand column gives the modulation index. The remaining columns indicate the relative amplitudes of the carrier and the various pairs of sidebands. Any sideband with a relative carrier amplitude of less than 1 percent (0.01) has been eliminated. Note that some of the carrier and sideband amplitudes have negative signs. This means that the signal represented by that amplitude is simply shifted in phase 180° (phase inversion).

As you can see, the spectrum of an FM signal varies considerably in bandwidth depending upon the modulation index. The higher the modulation index, the wider the bandwidth of the FM signal. When spectrum conservation is necessary, the bandwidth of an FM signal may be deliberately restricted by putting an upper limit on the modulation index.

Figure 4-7 on the next page shows several examples of an FM signal spectrum with different modulation indexes. Compare the examples to the entries in the table of Fig. 4-6. The unmodulated carrier has a relative amplitude of 1.0. With modulation, the carrier amplitude decreases while the amplitudes of the various sidebands increase. With some values of modulation index, the carrier can disappear completely.

The total bandwidth of an FM signal can be determined by knowing the modulation index and using the table in Fig. 4-6. For example,

Modulation Index	Carrier	Sidebands (Pairs)															
		1st	2d	3d	4th	5th	6th	7th	8th	9th	10th	11th	12th	13th	14th	15th	16th
0.00	1.00	—	—	—	—	—	—	—	—	—	—	—	—	—	—	—	—
0.25	0.98	0.12	—	—	—	—	—	—	—	—	—	—	—	—	—	—	—
0.5	0.94	0.24	0.03	—	—	—	—	—	—	—	—	—	—	—	—	—	—
1.0	0.77	0.44	0.11	0.02	—	—	—	—	—	—	—	—	—	—	—	—	—
1.5	0.51	0.56	0.23	0.06	0.01	—	—	—	—	—	—	—	—	—	—	—	—
2.0	0.22	0.58	0.35	0.13	0.03	—	—	—	—	—	—	—	—	—	—	—	—
2.5	−0.05	0.50	0.45	0.22	0.07	0.02	—	—	—	—	—	—	—	—	—	—	—
3.0	−0.26	0.34	0.49	0.31	0.13	0.04	0.01	—	—	—	—	—	—	—	—	—	—
4.0	−0.40	−0.07	0.36	0.43	0.28	0.13	0.05	0.02	—	—	—	—	—	—	—	—	—
5.0	−0.18	−0.33	0.05	0.36	0.39	0.26	0.13	0.05	0.02	—	—	—	—	—	—	—	—
6.0	0.15	−0.28	−0.24	0.11	0.36	0.36	0.25	0.13	0.06	0.02	—	—	—	—	—	—	—
7.0	0.30	0.00	−0.30	−0.17	0.16	0.35	0.34	0.23	0.13	0.06	0.02	—	—	—	—	—	—
8.0	0.17	0.23	−0.11	−0.29	−0.10	0.19	0.34	0.32	0.22	0.13	0.06	0.03	—	—	—	—	—
9.0	−0.09	0.24	0.14	−0.18	−0.27	−0.06	0.20	0.33	0.30	0.21	0.12	0.06	0.03	0.01	—	—	—
10.0	−0.25	0.04	0.25	0.06	−0.22	−0.23	−0.01	0.22	0.31	0.29	0.20	0.12	0.06	0.03	0.01	—	—
12.0	−0.05	−0.22	−0.08	0.20	0.18	−0.07	−0.24	−0.17	0.05	0.23	0.30	0.27	0.20	0.12	0.07	0.03	0.01
15.0	−0.01	0.21	0.04	0.19	−0.12	0.13	0.21	0.03	−0.17	−0.22	−0.09	0.10	0.24	0.28	0.25	0.18	0.12

Fig. 4-6   A table showing carrier and sideband amplitudes for different modulation indexes of FM signals. Based on the Bessel functions.

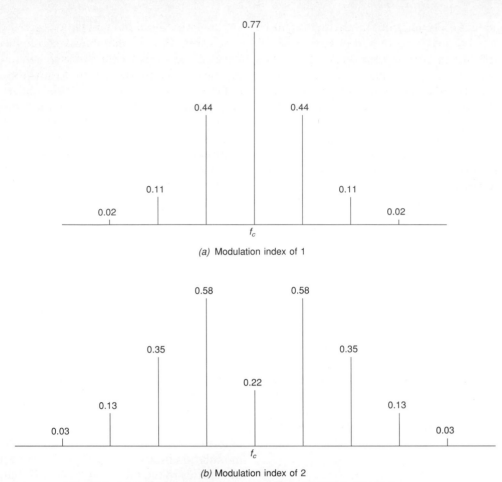

(a) Modulation index of 1

(b) Modulation index of 2

Fig. 4-7 Significant numbers of sidebands for different modulation indexes.

assume the modulation index is 2. Referring to the table, you can see that this produces four significant pairs of sidebands. The bandwidth can then be determined with the simple formula

$$BW = 2Nf_{m \, max}$$

where N is the number of significant sidebands

Using the example above and assuming a highest modulating frequency of 2.5 kHz, the bandwidth of the FM signal is

$$BW = 2(2.5)(4)$$
$$= 20 \text{ kHz}$$

An FM signal with a modulation index of 2 and a highest modulating frequency of 2.5 kHz will then occupy a bandwidth of 20 kHz.

An alternative way to calculate the bandwidth of an FM signal is to use Carson's rule. This rule takes into consideration only the power in the most significant sidebands whose amplitudes are greater than 2 percent of the carrier. These are the sidebands whose values are 0.02 or more in Fig. 4-6. Carson's rule is given by the expression

$$BW = 2(f_{d \, max} + f_{m \, max})$$

In this expression, $f_{d \, max}$ is the maximum frequency deviation, and $f_{m \, max}$ is the maximum modulating frequency.

Assuming a maximum frequency deviation of 5 kHz and a maximum modulating frequency of 2.5 kHz, the bandwidth would be

$$BW = 2(5 \text{ kHz} + 2.5 \text{ kHz})$$
$$= 2(7.5 \text{ kHz}) = 15 \text{ kHz}$$

Comparing the bandwidth with that computed in the preceding example, you can see that Carson's rule gives a smaller bandwidth. It has been determined that, if a circuit or system has that bandwidth (per Carson's rule), sufficient sideband power will be passed to ensure full intelligibility of the information signal.

In AM, the amount or degree of modulation is usually given as a percentage of modulation. The percentage of modulation is the ratio of the amplitude of the modulating signal to the amplitude of the carrier. When the two factors are equal, the ratio is 1 and we say that 100 percent modulation occurs. Should the modulating signal amplitude become greater than the carrier amplitude, then overmodulation and distortion occur.

Such conditions do not exist with FM or PM. Since the carrier amplitude remains constant during modulation with FM and PM, the percentage-of-modulation indicator used in AM has no meaning. Further, increasing the amplitude or the frequency of the modulating signal will not cause overmodulation or distortion. Increasing the modulating signal amplitude simply increases the frequency deviation. This, in turn, increases the modulation index which simply produces more significant sidebands and a wider bandwidth. For practical reasons of spectrum conservation and receiver performance, there is usually some limit put on the upper frequency deviation and the upper modulating frequency. As indicated earlier, the ratio of the maximum frequency deviation permitted to the maximum modulating frequency is referred to as the deviation ratio.

The audio in broadcast TV is transmitted by FM. The maximum deviation permitted is 25 kHz, and the maximum modulating frequency is 15 kHz. This produces a deviation ratio of

$$d = \frac{25}{15}$$
$$= 1.6666$$

In standard two-way mobile radio communications using FM, the maximum permitted deviation is usually 5 kHz. The upper modulating frequency is usually limited to 2.5 kHz which is high enough for intelligible voice transmission. This produces a deviation ratio of

$$d = \frac{5}{2.5}$$
$$= 2$$

The maximum deviation permitted can be used in a ratio with the actual carrier deviation to produce a percentage of modulation for FM. Remember, in commercial FM broadcasting the maximum allowed deviation is 75 kHz. If the modulating signal is producing only a maximum deviation of 60 kHz, then the FM percentage of modulation is

FM percent modulation

$$= \frac{\text{actual carrier deviation}}{\text{maximum carrier deviation}} \times 100$$
$$= \frac{60}{75} \times 100$$
$$= 80\%$$

When maximum deviations are specified, it is important that the percentage of modulation be held to less than 100 percent. The reason for this is that FM stations operate in assigned frequency channels. These are adjacent to other channels containing other stations. If the deviation is allowed to exceed the maximum, a greater number of pairs of sidebands will be produced and the signal bandwidth may be excessive. This can cause undesirable adjacent channel interference.

Self-Test

Answer the following questions.

13. True or false. An FM signal produces more sidebands than an AM signal.
14. The bandwidth of an FM signal is proportional to the _____ .
15. The maximum frequency deviation of an FM signal is 10 kHz. The maximum modulating frequency is 3.33 kHz. The deviation ratio is _____ .
16. An FM signal has a modulation index of 2.5. How many significant pairs of sidebands are produced? (See Fig. 4-6.)
17. In an FM signal, the modulating frequency is a 1.5-kHz sine wave. The carrier frequency is 1000 kHz. The frequencies of the third significant sidebands are _____ and _____ kHz.
18. Refer to Fig. 4-6. What is the relative amplitude of the fourth significant pair of sidebands in an FM signal with a deviation ratio of 4?
19. The amplitudes of the sidebands in an FM signal are dependent upon a mathematical process known as _____ .
20. An FM signal has a deviation ratio of 3. The maximum modulating signal is 5 kHz. The bandwidth of the signal is _____ kHz.
21. If the maximum allowed deviation is 5 kHz but the actual deviation is 3.75 kHz, the percentage of modulation is _____ percent.
22. A negative sign on the carrier and sideband amplitudes in Fig. 4-6 means a(n) _____ .
23. True or false. The carrier in an FM signal can never drop to zero amplitude.
24. Calculate the bandwidth of an FM signal with a maximum deviation of 10 kHz and a maximum modulating signal frequency of 4 kHz. Use the two methods given in

the text, significant sidebands, and Carson's rule, and compare your answers. Explain the difference.

4-4 FREQUENCY MODULATION VERSUS AMPLITUDE MODULATION

In general, FM is considered to be superior to AM. Although both modulation signals are suitable for transmitting information from one place to another and both are capable of equivalent fidelity and intelligibility, FM typically offers some significant benefits over AM. The advantages and disadvantages of FM when compared to AM are summarized in the table of Fig. 4-8. Let's consider each of them in more detail.

Advantages	Disadvantages
1. Better noise immunity 2. Rejection of interfering signals because of "capture effect" 3. Better transmitter efficiency	1. Excessive use of spectrum space 2. More complex and costly circuits

Fig. 4-8 Advantages and disadvantages of FM compared to AM.

The primary benefit of FM over AM is its superior noise immunity. Noise is interference to a signal generated by lightning, motors, automotive ignition systems, and any power line switching that produces transients. Such noise is typically narrow spikes of voltage with very high frequency content. They add to a signal and interfere with it. If the noise signals are strong enough, they can completely obliterate the information signal.

Noise is essentially amplitude variations. An FM signal, on the other hand, has a constant carrier amplitude. Because of this, FM receivers contain limiter circuits that deliberately restrict the amplitude of the received signal. Any amplitude variations occurring on the FM signal are effectively clipped off. This does not hurt the information content of the FM signal since it is contained solely within the frequency variations of the carrier. Because of the clipping action of the limiter circuits, noise is almost completely eliminated.

Another major benefit of FM is that interfering signals on the same frequency will be effectively rejected. Because of the limiters built into FM receivers, a peculiar effect takes place when two or more FM signals occur simultaneously on the same frequency. If the signal of one is more than twice the amplitude of the other, the stronger signal will "capture" the channel and will totally eliminate the weaker, interfering signal. This is known as the *capture effect* in FM. When two AM signals occupy the same frequency, both signals will generally be heard regardless of their relative signal strengths. When one AM signal is significantly stronger than the other, naturally the stronger signal will be intelligible; however, although the weaker signal will be unintelligible, it will still be heard in the background. When the signal strengths of the AM signals are nearly the same, they will interfere with one another making both of them nearly unintelligible. In FM, the capture effect allows the stronger signal to dominate while the weaker signal is eliminated.

However, when the strengths of the two FM signals begin to be nearly the same, the capture effect may cause the signals to alternate in their domination of the frequency. At some time one signal will be stronger than the other, and it will capture the channel. At other times, the signal strengths will reverse, and the other signal will capture the channel. You may have experienced this effect yourself when listening to the FM radio in your car while driving on the highway. You may be listening to a strong station on a particular frequency, but as you drive, you move away from that station. At some point, you may begin to pick up the signal from another station on the same frequency. When the two signals are approximately the same amplitude, you will hear one station dominate and then the other as the signal amplitudes vary during your driving. However, at some point, the stronger signal will eventually dominate. In any case, once the strong signal dominates, the weaker is not heard at all on the channel.

Despite the fact that FM has superior noise rejection qualities, noise still interferes with an FM signal. This is particularly true for the high-frequency components in the modulating signal. Since noise is primarily sharp spikes of energy, it contains a considerable number of harmonics and other high-frequency components. These high frequencies can at times be larger in amplitude than the high-frequency content of the modulating signal. This causes a form of frequency distortion that can make the signal unintelligible.

Most of the content of a modulating signal, particularly voice, is at lower frequencies. In voice communications systems, the bandwidth of the modulating signal is deliberately limited to a maximum of approximately 3 kHz. The voice is still intelligible despite the bandwidth limitations. After all, telephones cut off at 3 kHz and give good voice quality. However, music would be severely distorted by such a narrow bandwidth because it contains high-frequency components necessary to high fidelity. Typically, however, these high-frequency components are of a lower amplitude. For example, musical instruments typically generate their signals at low frequencies but contain many lower level harmonics that give them their unique sound. If their unique sound is to be preserved, then the high-frequency components must be passed. This is the reason for such a wide bandwidth in high-fidelity sound systems. Since these high-frequency components are at a very low level, noise can obliterate them.

To overcome this problem, most FM systems use a technique known as *pre-emphasis*

that helps offset high-frequency noise interference. At the transmitter, the modulating signal is passed through a simple network which amplifies the high-frequency components more than the low-frequency components. The simplest form of such a circuit is a simple high-pass filter of the type shown in Fig. 4-9(a). Specifications dictate a time constant t of 75 microseconds (μs) where $t = R_1C$. Any combination of resistor and capacitor (or resistor and inductor) giving this time constant will be satisfactory. Such a circuit has a cutoff frequency f_{co} of 2122 Hz. This means that frequencies higher than 2122 Hz will be linearly enchanced. The output amplitude increases with frequency at a rate of 6 dB per octave. The pre-emphasis curve is shown in Fig. 4-9(b). This pre-emphasis circuit increases the

(a) Pre-emphasis circuit

(b) Pre-emphasis curve

(c) De-emphasis circuit

(d) De-emphasis curve

(e) Combined frequency response

Fig. 4-9 **Pre-emphasis and de-emphasis.**

energy content of the higher-frequency signals so that they will tend to become stronger than the high-frequency noise components. This improves the signal-to-noise ratio and increases intelligibility and fidelity.

The pre-emphasis circuit also has an upper break frequency f_u where the signal enhancement flattens out. See Fig. 4-9(b). This upper break frequency is computed with the expression

$$f_u = R_1 + \frac{R_2}{2\pi R_1 R_2 C}$$

It is usually set at some very high value beyond the audio range. An f_u of greater than 30 kHz is typical.

To return the frequency response to its normal level, a de-emphasis circuit is used at the receiver. This is a simple low-pass filter with a time constant of 75 μs. See Fig. 4-9(c). It features a cutoff of 2122 Hz and causes signals above this frequency to be attenuated at the rate of 6 dB per octave. The response curve is shown in Fig. 4-9(d). As a result, the pre-emphasis at the transmitter is exactly offset by the de-emphasis circuit in the receiver, providing a normal frequency response. The combined effect of pre-emphasis and de-emphasis is to increase the high-frequency components during transmission so that they will be stronger and not masked by noise.

A third advantage of FM over AM is in transmitting efficiency. Recall that AM can be produced by both low-level and high-level modulation techniques. The most effective is high-level modulation where a class C amplifier is used as the final RF power stage and is modulated by a high-power modulation amplifier. The AM transmitter must produce both very high RF and modulating signal power. Additionally, at very high power levels large modulation amplifiers are impractical. Under such conditions, low-level modulation must be used. The AM signal is generated at a lower level and then amplified with linear amplifiers to produce the final RF signal. Because linear amplifiers operate class A or class B, they are far less efficient than class C amplifiers. However, linear amplifiers must be used if the AM information is to be preserved.

An FM signal has a constant amplitude, and, therefore, it is not necessary to use linear amplifiers to increase its power level. In fact, FM signals are always generated at a lower level and then amplified by a series of class C amplifiers to increase their power. The result is greater use of available power because class C amplifiers are far more efficient.

Refer again to Fig. 4-8. Although the advan-

tages of FM generally outweigh its disadvantages, these disadvantages can weigh against FM in some applications. Perhaps the greatest disadvantage of FM is that it simply uses too much spectrum space. The bandwidth of an FM signal is considerably wider than that of an AM signal transmitting similar information. Although the modulation index can be kept low to minimize the bandwidth used, still the bandwidth is typically larger than that of an AM signal. Further, reducing the modulation index also reduces the noise immunity of the FM signal. In commercial two-way FM radio systems, the maximum allowed deviation is 5 kHz with a maximum modulating frequency of 3 kHz. This produces a deviation ratio of $\frac{5}{3} = 1.67$. This is usually referred to as *narrowband FM (NBFM)*.

Since FM occupies so much bandwidth, it typically has been used only at the very high frequencies. In fact, it is rarely used in communications below frequencies of 30 MHz. Most FM communications work is done at the VHF, UHF, and microwave frequencies. It is only in these portions of the spectrum where adequate bandwidth is available for FM signals and where line-of-sight transmission is prevalent. This means that the range of communication is more limited.

The other major disadvantage of FM is that the circuits used for modulation and demodulation are usually far more complex than those for AM. Typically the circuits are more critical and are more difficult to design and adjust than the simple circuits used for amplitude modulation and demodulation. This also makes the cost of FM circuitry higher than that of AM circuitry. The advantages usually offset the increases in cost and complexity.

Self-Test

Answer the following questions.

25. The main advantage of FM over AM is its immunity from _____ .
26. Noise is primarily a variation in
    a. Amplitude.
    b. Frequency.
    c. Phase.
27. FM receivers reject noise because of built-in _____ circuits.
28. The _____ in an FM receiver causes a stronger signal to dominate a weaker signal on the same frequency.
29. Typically FM transmitters are more efficient than AM transmitters because they use class _____ amplifiers.

30. The biggest disadvantage of FM is its excessive use of _____ .
31. True or false. An AM circuit is usually more complex and expensive than an FM circuit.
32. Noise interferes primarily with _____ modulating frequencies.
33. The method used to offset the effect of noise in FM transmissions by boosting high frequencies is known as _____ .
34. To boost high frequencies a(n) _____ circuit is used.
35. To correct for the high-frequency boost, a(n) _____ circuit is used at the receiver.
36. The time constant of a pre-emphasis circuit is _____ μs.
37. In an FM receiver, frequencies above _____ Hz are attenuated 6 dB per octave.

SUMMARY

1. In FM, the information signal varies the frequency of the carrier.
2. The amount of frequency change from the carrier center frequency is called the frequency deviation.
3. In FM, the deviation is proportional to the amplitude of the modulating signal.
4. During FM, the carrier amplitude remains constant.
5. Both FM and PM are types of angle modulations.
6. In PM, the phase shift of the carrier is varied by the amplitude of the modulating signal.
7. Phase modulation produces frequency modulation.
8. The FM produced by PM is called indirect FM.
9. Maximum frequency deviation in a PM signal occurs where the rate of change of the modulating signal amplitude is greatest, which is at its zero-crossing points.
10. Frequency deviation does not occur at the output of a phase modulator unless the modulating signal amplitude varies.
11. The amount of frequency deviation produced by a phase modulator increases with the modulating frequency.
12. To produce true FM from a PM signal, the amplitude of the modulating signal must be decreased with frequency so that frequency deviation does not change with modulating frequency.
13. In PM, a low-pass filter on the modulating signal compensates for increased frequency deviation at the higher modulating frequencies.
14. Frequency modulation produces pairs of sidebands spaced from the carrier in multiples of the modulating frequency.
15. The modulation index m of an FM signal is the ratio of the frequency deviation f_d to the modulating frequency f_m ($m = f_d/f_m$).
16. The deviation ratio is the maximum frequency deviation divided by the maximum modulating frequency.
17. The modulation index determines the number of significant pairs of sidebands in an FM signal.
18. The amplitudes of the carrier and sidebands vary with the modulation index and can be calculated with a mathematical procedure known as the Bessel functions.
19. The carrier or sideband amplitudes are zero at some modulation indexes.
20. The bandwidth of an FM signal is proportional to the modulation index.
21. There are two ways to calculate the bandwidth of an FM signal.

$$BW = 2Nf_{m\ max}$$
$$BW = 2(f_{d\ max} + f_{m\ max})$$

22. For FM, the percentage of modulation is the ratio of the actual frequency deviation and the maximum allowed frequency deviation multiplied by 100.
23. The primary advantage of FM over AM is its immunity to noise.
24. Noise is short-duration amplitude variations caused by lightning, motors, auto ignitions, power transients, and other sources.
25. Limiter circuits in FM receivers clip off noise signals.
26. Another benefit of FM over AM is the capture effect that allows the strongest signal on a frequency to dominate without interference from the other signal.
27. A third benefit of FM over AM is greater transmitter efficiency since class C amplifiers may be used.
28. A major disadvantage of FM is that its bandwidth is wider than the bandwidth of AM.

29. The spectrum space taken up by an FM signal may be limited by carefully controlling the deviation ratio.

30. Another disadvantage of FM is that the circuits to produce and demodulate it are usually more complex and expensive than AM circuits.

31. Noise occurs primarily at high frequencies; therefore, noise interferes more with high modulating frequencies.

32. Interference from high-frequency noise can be minimized by boosting the amplitude of high-frequency modulating signals prior to modulation. This is called pre-emphasis.

33. Pre-emphasis is accomplished by passing the modulating signal through an *RC* network that linearly boosts the amplitude of frequencies above 2122 Hz in proportion to frequency. This increases the signal-to-noise ratio at the higher frequencies.

34. The effect of pre-emphasis is corrected for in an FM receiver by de-emphasizing the higher frequencies by passing them through an *RC* low-pass filter.

35. The pre-emphasis and de-emphasis networks have a time constant of 75 μs and a cutoff frequency of 2122 Hz.

CHAPTER REVIEW QUESTIONS

Choose the letter that best answers each question.

4-1. The amount of frequency deviation from the carrier center frequency in an FM transmitter is proportional to what characteristic of the modulating signal?
   a. Amplitude
   b. Frequency
   c. Phase
   d. Shape

4-2. Both FM and PM are types of what kind of modulation?
   a. Amplitude
   b. Phase
   c. Angle
   d. Duty cycle

4-3. If the amplitude of the modulating signal decreases, the carrier deviation
   a. Increases.
   b. Decreases.
   c. Remains constant.
   d. Goes to zero.

4-4. On an FM signal, maximum deviation occurs at what point on the modulating signal?
   a. Zero-crossing points
   b. Peak positive amplitude
   c. Peak negative amplitude
   d. Both *b* and *c*

4-5. In PM, a frequency shift occurs while what characteristic of the modulating signal is changing?
   a. Shape
   b. Phase
   c. Frequency
   d. Amplitude

4-6. Maximum frequency deviation of a PM signal occurs at
   a. Zero crossing points.
   b. Peak positive amplitude.
   c. Peak negative amplitude.
   d. Peak positive or negative amplitudes.

4-7. In PM, carrier frequency deviation is *not* proportional to:
   a. Modulating signal amplitude.
   b. Carrier amplitude and frequency.
   c. Modulating signal frequency.
   d. Modulator phase shift.

4-8. To compensate for increases in carrier frequency deviation with an increase in modulating signal frequency, what circuit is used between the modulating signal and the phase modulator?
   a. Low-pass filter
   b. High-pass filter
   c. Phase shifter
   d. Bandpass filter

4-9. The FM produced by PM is called
   a. FM.
   b. PM.
   c. Indirect FM.
   d. Indirect PM.

4-10. If the amplitude of the modulating signal applied to a phase modulator is constant, the output signal will be
   a. Zero.
   b. The carrier frequency.
   c. Above the carrier frequency.
   d. Below the carrier frequency.

4-11. A 100-MHz carrier is deviated 50 kHz by a 4-kHz signal. The modulation index is

 a. 5.   c. 12.5.
 b. 8.   d. 20.

4-12. The maximum deviation of an FM carrier is 2 kHz by a maximum modulating signal of 400 Hz. The deviation ratio is

 a. 0.2   c. 8.
 b. 5.   d. 40.

4-13. A 70-kHz carrier has a frequency deviation of 4 kHz with a 1000-Hz signal. How many significant sideband pairs are produced?

 a. 4   c. 6
 b. 5   d. 7

4-14. What is the bandwidth of the FM signal described in question 4-13 above?

 a. 4 kHz   c. 14 kHz
 b. 7 kHz   d. 28 kHz

4-15. What is the relative amplitude of the third pair of sidebands of an FM signal with $m = 6$?

 a. 0.11   c. 0.24
 b. 0.17   d. 0.36

4-16. A 200-kHz carrier is modulated by a 2.5-kHz signal. The fourth pair of sidebands are spaced from the carrier by

 a. 2.5 kHz.   c. 10 kHz.
 b. 5 kHz.   d. 15 kHz.

4-17. An FM transmitter has a maximum deviation of 12 kHz and a maximum modulating frequency of 12 kHz. The bandwidth by Carson's rule is

 a. 24 kHz.   c. 36.8 kHz.
 b. 33.6 kHz.   d. 48 kHz.

4-18. The maximum allowed deviation of the FM sound signal in TV is 25 kHz. If the actual deviation is 18 kHz, the percent modulation is

 a. 43 percent.   c. 96 percent.
 b. 72 percent.   d. 139 percent.

4-19. Which of the following is *not* a major benefit of FM over AM?

 a. Greater efficiency   c. Capture effect
 b. Noise immunity   d. Lower complexity and cost

4-20. The primary disadvantage of FM is its

 a. Higher cost and complexity.   c. Noise susceptibility.
 b. Excessive use of spectrum space. d. Lower efficiency.

4-21. Noise is primarily

 a. High-frequency spikes.   c. Random level shifts.
 b. Low-frequency variations.   d. Random frequency variations.

4-22. The receiver circuit that rids FM of noise is the

 a. Modulator.   c. Limiter.
 b. Demodulator.   d. Low-pass filter.

4-23. The phenomenon of a strong FM signal dominating a weaker signal on a common frequency is referred to as the

 a. Capture effect.   c. Quieting factor.
 b. Blot out.   d. Domination syndrome.

4-24. The AM signals generated at a low level may only be amplified by what type of amplifier?

 a. Op amp   c. Class C
 b. Linear   d. Push-pull

4-25. Frequency modulation transmitters are more efficient because their power is increased by what type of amplifier?

 a. Class A   c. Class C
 b. Class B   d. All the above.

4-26. Noise interferes mainly with modulating signals that are
    a. Sinusoidal.   *c.* Low frequencies.
    b. Nonsinusoidal.   *d.* High frequencies.

4-27. Pre-emphasis circuits boost what modulating frequencies before modulation?
    a. High frequencies   *c.* Low frequencies
    b. Mid-range frequencies   *d.* All the above

4-28. A pre-emphasis circuit is a
    a. Low-pass filter.   *c.* Phase shifter.
    b. High-pass filter.   *d.* Bandpass filter.

4-29. Pre-emphasis is compensated for at the receiver by a
    a. Phase inverter.   *c.* High-pass filter.
    b. Bandpass filter.   *d.* Low-pass filter.

4-30. The cutoff frequency of pre-emphasis and de-emphasis circuits is
    a. 1 kHz.   *c.* 5 kHz.
    b. 2.122 kHz.   *d.* 75 kHz.

CRITICAL THINKING QUESTIONS

4-1. Discuss the effects and implications of using FM on the standard AM broadcast bands that use intelligence signal modulating frequencies up to 5 kHz and a 10 kHz standard station spacing. Assume a maximum deviation of 10 kHz.

4-2. The standard FM broadcast band has a frequency range of 88 to 108 MHz. Stations are spaced every 200 kHz beginning at 88.1 MHz and up to 107.9 MHz. The maximum permitted deviation is 75 kHz, with modulating frequencies up to 15 kHz. Compute the bandwidth of an FM station using two methods and then discuss how that bandwidth compares with the channel spacing.

4-3. TV sound is FM. The maximum deviation is 25 kHz for monaural and 50 kHz for stereo. The maximum modulating frequency is 15 kHz. What is the maximum bandwidth of monaural and stereo broadcasts using Carson's rule?

4-4. Explain how the bandwidth of an FM signal is affected by transmitting a square-wave modulating signal.

Answers to Self-Tests

1. angle
2. true
3. deviation
4. amplitude
5. increases, decreases (the reverse could also be true)
6. phase shift
7. frequency shift or deviation
8. changing or varying
9. amplitude, frequency
10. *a*
11. *b*
12. indirect FM
13. true
14. modulation index
15. 3 ($m = {}^{10}\!/_{3.33} = 3$)
16. 5
17. 995.5, 1004.5 [3(1.5) = 4.5 KHz; 1000 ± 4.5 = 995.5 and 1004.5 kHz]
18. 0.28
19. Bessel functions
20. 60 [2(5)(6) = 60 kHz]
21. 75 [(3.75/5)100 = 75%]
22. phase inversion or 180° shift
23. false
24. BW = 40 kHz; BW = 28 kHz using Carson's rule; Carson's rule gives narrower bandwidths because sidebands of less than 2 percent amplitude are not considered
25. noise
26. *a*
27. limiter (or clipper)
28. capture effect
29. C
30. spectrum space
31. false
32. high
33. pre-emphasis
34. high-pass filter
35. de-emphasis
36. 75
37. 2122

CHAPTER 5

Frequency Modulation Circuits

■

CHAPTER OBJECTIVES

This chapter will help you to:

1. *Describe* the operation of a voltage-variable capacitor and *show* how it is used to make frequency and phase modulator circuits in FM transmitters.
2. *Explain* the operation of a reactance modulator.
3. *Explain* the operation of varactor and transistor phase modulators.
4. *Describe* the operation of the Foster-Seeley discriminator, ratio detector, pulse-averaging discriminator, quadrature detector, and differential peak detector FM demodulator circuits.
5. *Explain* the operation and characteristics of a phase-locked loop and *show* how it is used for FM demodulation.

An incredible variety of circuits have been devised to produce FM and PM signals. There are direct FM circuits and circuits that produce FM indirectly by PM techniques. Direct FM circuits make use of techniques for actually varying the frequency of the carrier oscillator in accordance with the modulating signal. The PM techniques use a phase shifter after the carrier oscillator. In this chapter, some of the more commonly used FM and PM circuits will be discussed. Frequency demodulator circuits convert the FM signal back into the original modulating signal. Dozens of different circuits have been devised. The most popular and widely used demodulator circuits will be covered in detail here. Special emphasis will be given to the phase-locked-loop demodulator which provides the best overall frequency demodulation.

■

5-1 FREQUENCY MODULATORS

The basic concept of FM is to vary the carrier frequency in accordance with the modulating signal. The carrier is generated by either an *LC* or crystal oscillator circuit. The object then is to find a way to change the frequency of oscillation. In an *LC* oscillator, the carrier frequency is fixed by the values of the inductance and capacitance in a tuned circuit. The carrier frequency, therefore, may be changed by varying either this inductance or the capacitance. The idea is to find a circuit or component that converts a modulating voltage into a corresponding change in capacitance or inductance.

When the carrier is generated by a crystal oscillator, the frequency is fixed by the crystal. However, keep in mind that the equivalent circuit of a crystal is an *LCR* circuit with both series and parallel resonant points. By connecting an external capacitor to the crystal, minor variations in operating frequency can be obtained. Again the objective is to find a circuit or component whose capacitance will change in response to the modulating signal.

The component most frequently used in this application is a *varactor* or *voltage-variable capacitor (VVC)*. Also known as a variable capacitance diode or varicap, this component is basically a semiconductor junction diode that is operated in a reverse-bias mode.

Refer to Fig. 5-1 on the next page. When a

From page 85:

Voltage-variable capacitor

On this page:

Depletion region

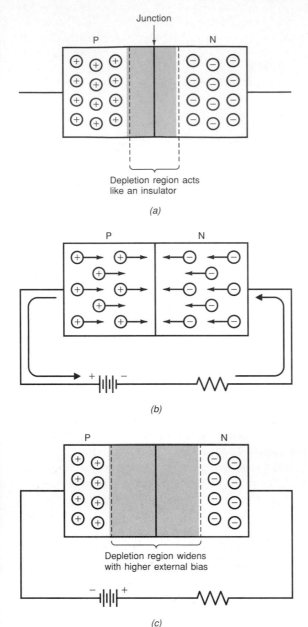

Junction

P   N

Depletion region acts like an insulator

(a)

P   N

(b)

P   N

Depletion region widens with higher external bias

(c)

Fig. 5-1   Depletion region in a junction diode.

junction diode is formed, P- and N-type semiconductors are joined to form a junction. Some electrons in the N-type material drift over into the P-type material and neutralize the holes there. Thus a thin region where there are no free carriers, holes, or electrons is formed. This is called the *depletion region*. It acts like a thin insulator that prevents current from flowing through the device. See Fig. 5-1(*a*).

If you apply a forward bias to the diode, it will conduct. The external potential forces the holes and electrons toward the junction where they combine and cause a continuous current inside the diode as well as externally. The depletion layer simply disappears. See Fig. 5-1*b*.

If an external reverse bias is applied to the diode as in Figure 5-1(*c*), no current will flow. The bias actually increases the width of the depletion layer. The width of this depletion mode depends upon the amount of the reverse bias. The higher the reverse bias, the wider the depletion layer and the less chance for current flow.

A reverse-biased junction diode appears to be a small capacitor. The P- and N-type materials act as the two plates of the capacitor, while the depletion region acts as the dielectric. The width of the depletion layer determines the width of the dielectric and, therefore, the amount of capacitance. If the reverse bias is high, the depletion region will be wide and the dielectric will cause the plates of the capacitor to be widely spaced, producing a low value of capacitance. Decreasing the amount of reverse bias narrows the depletion region, and, therefore, the plates of the capacitor will be effectively closer together and produce a higher capacitance.

All junction diodes exhibit this characteristic of variable capacitance as the reverse bias is changed. However, varactors or VVCs have been designed to optimize this particular characteristic. Such diodes are made so that the capacitance variations are as wide and linear as possible.

Voltage-variable capacitors are made with a wide range of capacitance values. Most units have a nominal capacitance in the range of 1 to 200 picofarads (pF). The capacitance variation range can be as high as 12 to 1. Figure 5-2 shows the curve for a typical diode. A maximum capacitance of 80 pF is obtained at 1 V. With 60 V applied, the capacitance drops to 20 pF, a 4 to 1 range. The operating range is usually restricted to the linear center portion of the curve.

Fig. 5-2   Capacitance vesus reverse junction voltage for a typical varactor.

Figure 5-3 shows the basic concept of a varactor frequency modulator. The L_1 and C_1 represent the tuned circuit of the carrier oscillator. Varactor diode D_1 is connected in series with capacitor C_2 across the tuned circuit. The value of C_2 is made very large at the operating frequency so that its reactance is very low. As a result, when C_2 is connected in series with the lower capacitance of D_1, the effect is as if D_1 were connected directly across the tuned circuit. The total effective circuit capacitance then is the capacitance of D_1 in parallel with C_1. This fixes the center carrier frequency.

The capacitance of D_1, of course, is controlled by two factors: a fixed dc bias and the modulating signal. In Fig. 5-3, the bias on D_1 is set by the voltage divider which is made up of R_1 and R_2. Usually either R_1 or R_2 is made variable so that the center carrier frequency can be adjusted over a narrow range. The modulating signal is applied through C_3 and the RFC. The C_3 is a blocking capacitor that keeps the dc bias out of the modulating signal circuits. The RFC is a radio frequency choke whose reactance is high at the carrier frequency to prevent the carrier signal from getting into the modulating signal circuits.

The modulating signal derived from the microphone is amplified and applied to the modulator. As the modulating signal varies, it adds to and subtracts from the fixed-bias voltage. Thus the effective voltage applied to D_1 causes its capacitance to vary. This, in turn, produces a deviation of the carrier frequency as desired. A positive-going signal at point A adds to the reverse bias, decreasing the capacitance and increasing the carrier frequency. A negative-going signal at A subtracts from the bias, increasing the capacitance and decreasing the carrier frequency.

The main problem with the circuit in Fig. 5-3 is that most LC oscillators are simply not stable enough to provide a carrier signal. Despite the quality of the components and the ex-cellence of design, the LC oscillator frequency will vary due to temperature changes, circuit voltage variations, and other factors. Such instabilities cannot be tolerated in most modern electronic communications systems. As a result, crystal oscillators are normally used to set the carrier frequency. Crystal oscillators not only provide a highly accurate carrier frequency, but also their frequency stability is superior over a wide temperature range.

It is possible to vary the frequency of a crystal oscillator by changing the value of capacitance in series or in parallel with the crystal. Figure 5-4 on the next page shows a typical crystal oscillator. When a small value of capacitance is connected in series with the crystal, the crystal frequency can be "pulled" slightly from its natural resonant frequency. By making the series capacitor a varactor diode, frequency modulation of the crystal oscillator can be achieved. The modulating signal is applied to the varactor diode D_1 which changes the oscillator frequency.

The important thing to note about an FM crystal oscillator is that only a very small frequency deviation is possible. Greater deviations can be achieved with LC oscillators. Rarely can the frequency of a crystal oscillator be changed more than several hundred hertz from the nominal crystal value. The resulting deviation may be less than the total deviation desired. To achieve a total frequency shift of 75 kHz as in commercial FM broadcasting, other techniques must be used. In two-way (narrowband) FM communications systems, the narrower deviations are acceptable.

Although a deviation of only several hundred cycles may be possible at the crystal oscillator frequency, the total deviation can be increased by using frequency multiplier circuits after the carrier oscillator. When the FM signal is applied to a frequency multiplier, both the frequency of operation and the amount of deviation is increased. Typical frequency mul-

Fig. 5-3 Frequency modulation with a VVC.

Fig. 5-4 Frequency modulation of a crystal oscillator with a VVC.

tipliers may increase the base (oscillator) frequency by 24 to 32 times.

For example, assume that the desired output frequency from an FM transmitter is 168 MHz and that the carrier is generated by a 7-MHz crystal oscillator. This is followed by frequency multiplier circuits that increase the frequency by a factor of 24 (7 MHz × 24 = 168 MHz).

Assume further that the desired maximum frequency deviation is 5 kHz. Frequency modulation of the crystal oscillator may only produce a maximum deviation of 200 Hz. When multiplied by the factor of 24 in the frequency multiplier circuits, this deviation, of course, is increased to 200 × 24 = 4800 Hz, or 4.8 kHz. Frequency multiplier circuits will be discussed in more detail in the chapter on transmitters.

Another way to produce direct frequency modulation is to use a reactance modulator. This circuit uses a transistor amplifier to act like either a variable capacitor or an inductor. When the circuit is connected across the tuned circuit of an oscillator, the oscillator frequency can be varied by applying the modulating signal to the amplifier.

A reactance modulator is illustrated in Fig. 5-5. It is basically a standard common-emitter class A amplifier. Resistors R_1 and R_2 form a voltage divider to bias the transistor into the linear region. R_3 is an emitter bias resistor which is bypassed with capacitor C_3. Instead of a collector resistor, a radio frequency choke (RFC$_2$) is used to provide a high impedance load at the operating frequency.

Now, note that the collector of the transistor is connected to the top of the tuned circuit in the oscillator. Capacitor C_4 has a very low impedance at the oscillator frequency. Its main purpose is to keep the direct current from the collector of Q_1 from being shorted to ground through the oscillator coil L_o. As you can see, the reactance modulator circuit is connected directly across the parallel tuned circuit that sets the oscillator frequency.

The oscillator signal from the tuned circuit V_o is connected back to an RC phase-shift circuit made up of C_s and R_s. Capacitor C_2 in series with R_s has a very low impedance at the operating frequency, so it does not affect the phase shift. However, it does prevent R_s from disturbing the dc bias on Q_1. The value of C_s is chosen so that its reactance at the oscillator frequency is about 10 or more times the value of R_s. If the reactance is much greater than the resistance, the circuit will appear predominantly capacitive; therefore the current through the capacitor and R_s will lead the applied voltage by about 90°. This means that the voltage across R_s that is applied to the base of Q_1 leads the voltage from the oscillator.

Since the collector current is in phase with the base current, which in turn is in phase with the base voltage, the collector current in Q_1 leads the oscillator voltage V_o by 90°. Of course, any circuit whose current leads its applied voltage by 90° looks capacitive to the source voltage. This means that the reactance modulator looks like a capacitor to the oscillator-tuned circuit.

The modulating signal is applied to the modulator circuit through C_1 and RFC$_1$. The RFC helps keep the RF signal from the oscillator out of the audio circuits from which the modulating signal usually comes. The audio modulating signal will vary the base voltage and current of Q_1 according to the intelligence to be transmitted. The collector current will also vary in proportion. As the collector current amplitude varies, the phase-shift angle changes with respect to the oscillator voltage, which is interpreted by the oscillator as a change in the capacitance. So, as the modulating signal changes, the effective capacitance of the circuit varies and the oscillator frequency is varied accordingly. An increase in capacitance lowers the frequency, whereas a lower capacitance increases the frequency. The circuit produces direct frequency modulations.

If you reverse the positions of R_s and C_s in the circuit of Fig. 5-5, the current in the phase

Fig. 5-5   A reactance modulator.

shifter will still lead the oscillator voltage by 90°. However, it is voltage from across the capacitor that is now applied to the base of the transistor. This voltage lags the oscillator voltage by 90°. With this arrangement, the reactance modulator acts like an inductor. The equivalent inductance changes as the modulating signal is applied. Again, the oscillator frequency is varied in proportion to the amplitude of the intelligence signal.

The reactance modulator is one of the best FM circuits because it can produce frequency deviation over a wide frequency range. It is also highly linear; that is, distortion is minimal. The circuit can also be implemented with a field-effect transistor (FET) in place of the NPN bipolar shown in Fig. 5-5.

Oscillators whose frequencies are controlled by an external input voltage are generally referred to as *voltage-controlled oscillators (VCOs)*. A voltage-controlled crystal oscillator is generally referred to as a VXO. Although VCOs are used primarily for FM, there are other applications where some form of voltage-to-frequency conversion is required.

In high-frequency communications circuits, VCOs are ordinarily implemented with discrete-component transistor and varactor diode circuits. However, there are many different types of lower-frequency VCOs in common use. These include IC VCOs using *RC* multivibrator-type oscillators whose frequency can be controlled over a wide range by an ac or dc input voltage. These VCOs typically have an operating range of less than 1 Hz to approximately 1 MHz. The output is either a square or triangular wave rather than a sine wave. A typical

IC VCO is shown in Fig. 5-6(*a*) on the next page. This is a general block diagram of the popular NE566. External resistor R_1 at pin 6 sets the value of current produced by the internal current sources. The current sources linearly charge and discharge external capacitor C_1 at pin 7. An external voltage V_C applied at pin 5 may also be used to vary the amount of current produced by the current sources.

The Schmitt trigger circuit is a level detector which switches when the capacitor charges or discharges to a specific voltage level. The Schmitt trigger controls the current source by switching the current sources between charging and discharging. A linear sawtooth of voltage is developed across the capacitor by the current source. This is buffered and made available at pin 4. The Schmitt trigger output is a square wave at the same frequency available at pin 3. If a sine wave output is desired, usually the triangle wave is filtered with a tuned circuit resonant to the desired carrier frequency.

A complete FM circuit is shown in Fig. 5-6(*b*). The current sources are biased with a voltage divider made up of R_2 and R_3. The modulating signal is applied through C_2 to the voltage divider at pin 5. The 0.001-μF capacitor between pins 5 and 6 is used to prevent unwanted spurious oscillations.

The center carrier frequency of the circuit is set by the values of R_1 and C_1. Carrier frequencies up to 1 MHz may be used with this IC. The outputs may be filtered or used to drive other circuits such as a frequency multiplier if higher frequencies and deviations are necessary. The modulating signal can vary the

(a) Block diagram of NE566 IC VCO   (b) Basic frequency modulator using the NE566 VCO

Fig. 5-6 Frequency modulation with an IC VCO.

carrier frequency over a nearly 10:1 range, making very large deviations possible. The deviation is linear with respect to the input amplitude over the entire range.

Self-Test

Choose the letter which best answers each statement.

1. Increasing the reverse bias on a voltage-variable capacitor causes its capacitance to
   a. Increase.
   b. Decrease.
2. Connecting a VVC across a parallel *LC* circuit causes the resonant frequency to
   a. Increase.
   b. Decrease.
3. In the circuit of Fig. 5-3, a negative-going modulating signal causes the carrier frequency to
   a. Increase.
   b. Decrease.
4. A crystal is operating in its series resonant mode. A VVC is connected in series with it. The crystal frequency
   a. Increases.
   b. Decreases.
5. Which is capable of greater frequency deviation?
   a. *LC* oscillator
   b. Crystal oscillator

Supply the missing information in each statement.

6. Another name for voltage-variable capacitor is _____ .
7. Most VVCs have a nominal capacitance in the _____ to _____ pF range.

8. A crystal oscillator has superior _____ over an *LC* oscillator.
9. The acronym VCO means _____ .
10. A voltage-variable crystal oscillator is referred to as a(n) _____ .
11. Carrier frequency and frequency deviation may be increased by using a(n) _____ after the carrier oscillator.
12. A reactance modulator is set up to act like an inductive reactance. If the modulating signal increases in amplitude, the effective inductance decreases. This causes the oscillator frequency to _____ .
13. An IC VCO normally uses a combination of _____ and _____ to set the center operating frequency.

Determine whether each statement is true or false.

14. Voltage-variable capacitors should not be forward-biased.
15. For highly stable carrier generators, *LC* oscillators are preferred over crystal oscillators.
16. A reactance modulator is used with crystal oscillators.
17. IC VCOs operate primarily at frequencies below 1 MHz.

5-2 PHASE MODULATORS

Most modern FM transmitters use some form of PM to produce indirect FM. The reason for using PM instead of direct FM is that the carrier oscillator can be optimized for frequency accuracy and stability. Crystal oscillators or crystal-controlled frequency synthesizers can be used to set the carrier frequency accurately and maintain solid stability.

The output of the carrier oscillator is fed to a phase modulator where the phase shift is made to vary in accordance with the modulating signal. Since phase variations produce frequency variations, indirect FM is the result.

The simplest phase shifters are RC networks like those shown in Fig. 5-7(a) and (b). Depending upon the values of R and C, the output of the phase shifter can be set to any phase angle between 0° and 90°. In Fig. 5-7(a) the output leads the input by some angle between 0° and 90°. For example, when X_C equals R, the phase shift is 45°.

A low-pass version of the same RC filter can also be used as shown in Fig. 5-7(b). Here the output is taken from across the capacitor, so it lags the input voltage by some angle between 0° and 90°.

One of these simple phase-shift circuits can be used as a phase modulator if the resistance or capacitance can be made to vary with the modulating signal. One way to do this is to substitute a transistor for the resistor in the circuit in Fig. 5-7(a). The resulting phase-shift circuit is shown in Fig. 5-7(c). The transistor simply acts as a variable resistor that varies in response to the modulating signal. If the modulating signal increases, the transistor base current and collector current increase. Therefore, the effective transistor resistance decreases. Lowering the resistance increases the amount of phase shift. This causes a corresponding frequency increase. If the amplitude of the modulated signal decreases, the base and collector currents decrease and the effective transistor resistance increases. This decreases the amount of phase shift, and, as a result, the amount of frequency shift decreases. An FET can be substituted for the bipolar transistor in Fig. 5-7(c) with comparable results.

Fig. 5-7(d) shows how a varactor can be used to implement a simple low-pass phase-shift modulator. Here the modulating signal causes the capacitance of the varactor to change. If the modulating signal amplitude increases, it adds to the varactor bias from R_1 and R_2, thereby causing the capacitance to decrease. This causes the reactance to increase; thus, the circuit produces less phase shift. A decreasing modulating signal subtracts from the reverse bias on the varactor diode, thereby increasing the capacitance or decreasing the capacitive reactance. This increases the amount of phase shift.

(a) Output leads input: 0°–90°

(b) Output lags input: 0°–90°

(c) Transistor modulator

(d) Varactor modulator

Fig. 5-7 Simple PM circuits.

An improved phase modulator is shown in Fig. 5-8. It uses a phase shifter made up of a capacitor and the variable resistance of a field-effect transistor Q_1. The carrier signal from a

Fig. 5-8 An improved phase modulator.

crystal oscillator or a phase-locked loop frequency synthesizer is applied directly to the output through C_1 and C_2. The carrier signal is also applied to the gate of the FET through C_1. The series capacitance of C_1 and C_2 and the FET source to drain resistance produce a leading phase shift of current in the FET and a leading voltage at the output. The carrier signal applied to the gate of the FET also varies the FET current. C_1 and R_1 produce a leading phase shift less than 90°. The leading voltage across R_1 also controls the current in Q_1. With two signals controlling the FET current, the result is a phasor sum of the two currents.

The modulating signal is applied to the gate of the FET. RFC_1 keeps the carrier RF out of the audio circuits. The audio signal now also controls the FET current. This changes the amplitude relationships of the other two controlling inputs, thereby producing a phase shift that is directly proportional to the amplitude of the modulating signal. The carrier output at the FET drain varies in phase and amplitude. The signal is then usually passed on to a class C amplifier or frequency multiplier which removes the amplitude variations but preserves the phase and frequency variations.

The problem with such simple phase modulators is that they are capable of producing only a small amount of phase shift because of the narrow range of linearity of the transistor or varactor. The total amount of phase shift is essentially limited to ±20°. Such a limited amount of phase shift produces, in turn, only a limited frequency shift.

An improved method of PM is to use a parallel tuned circuit to produce the phase shift. At resonance, the parallel resonant circuit will have a very high resistance. Off resonance, the circuit will act inductively or capacitively and, as a result, will produce a phase shift between its current and applied voltage.

Figure 5-9 shows the basic impedance Z response curve of a parallel resonant circuit. Also shown is the phase variation A. At the resonant frequency f_r, the inductive and capacitive reactances are equal, and, therefore, their effects cancel one another. The result is an extremely high impedance at f_r. The circuit acts resistive at this point, and, therefore, the phase angle between the current and the applied voltage is zero.

At frequencies below resonance, X_L decreases and X_C increases. This causes the circuit to act like an inductor. Therefore, the current will lag the applied voltage. Above resonance, X_L increases while X_C decreases. This causes the circuit to act capacitively, and the current leads the applied voltage. If the Q of the resonant circuit is relatively high, the phase shift will be quite pronounced as shown in Fig. 5-9. The same effect is achieved if f is constant and either L or C is varied. For a relatively small change in L or C, a significant phase shift

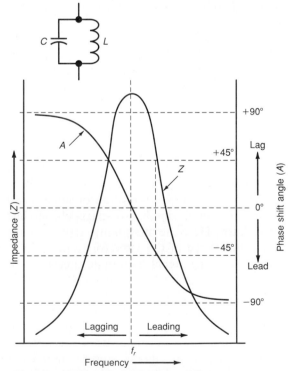

Fig. 5-9 Impedance and phase shift versus frequency of a parallel resonant circuit.

can be produced. The idea then is to cause the inductance or capacitance to vary with the modulating voltage and thus produce a phase shift.

A variety of circuits have been developed based on this technique, but one illustrating its operation is shown in Fig. 5-10. The parallel tuned circuit made up of L, C_1, and C_2 is part of the output circuit of an RF amplifier driven by the carrier oscillator. Capacitor C_1 is large so that its reactance at the carrier frequency is low. Therefore, the resonant frequency is controlled by C_2. A varactor diode D_1 is connected in parallel with C_2 in the tuned circuit and, therefore, will provide a capacitance change with the modulating signal. The voltage divider made up of R_1 and R_2 sets the reverse bias on D_1. The value of C_3 is very large and simply acts as a dc blocking capacitor, preventing bias from being applied to the tuned circuit. Its value is very large, so it is essentially an ac short at the carrier frequency. Therefore, it is the capacitance of D_1 and C_2 that controls the resonant frequency.

The modulating signal is first passed through a low-pass network R_3-C_5 that provides the amplitude compensation necessary to produce FM. The modulating signal appears across potentiometer R_4. In this way, the desired amount of modulating signal can be tapped off and applied to the phase-shift circuit. The potentiometer acts as a deviation control. The higher the modulating voltage, the greater the frequency deviation. The modulating signal is applied to the varactor diode through capacitor C_4. The RFC has a high impedance at the carrier frequency to minimize the loading of the tuned circuit which reduces Q. With zero modulating voltage, the value of the capacitance of D_1 along with capacitor C_2 and the inductor L set

the resonant frequency of the tuned circuit. The PM output across L is inductively coupled to the output.

When the modulating signal goes negative, it subtracts from the reverse bias of D_1. This increases the capacitance of the circuit and lowers the reactance, making the circuit appear capacitive. Thus a leading phase shift is produced. The parallel LC circuit looks like a capacitor to output resistance of the RF amplifier, so the output lags the input. A positive-going modulating voltage will decrease the capacitance, and thus the tuned circuit will become inductive and produce a lagging phase shift. The LC circuit looks like an inductor to output resistance of the RF amplifier, so the output leads the input. The result at the output is a relatively wide phase shift which, in turn, produces excellent linear frequency deviation.

Although phase modulators are relatively easy to implement, they have two main disadvantages. First, the amount of phase shift they produce and the resulting frequency deviation is relatively low. For that reason, the carrier is usually generated at a lower frequency and frequency multipliers are used to increase the carrier frequency and the amount of frequency deviation. Second, all the phase-shift circuits described above produce amplitude variations as well as phase changes. In the simple phase-shift circuits of Fig. 5-7, the phase shifters are all voltage dividers. When the value of one of the components is changed, the phase shifts but the output amplitude changes as well. This is also true of the tuned-circuit phase shifter. As a result, some means must be used to remove the amplitude variations.

Both these problems are solved by feeding the output of the phase modulator to class C

Fig. 5-10 One form of phase modulator.

amplifiers used as frequency multipliers. The class C amplifiers eliminate any amplitude variations, while at the same time they increase the carrier frequency and the deviation to the desired final values. Class C frequency multipliers will be discussed in the next chapter.

Self-Test

Answer the following questions.

18. Phase modulation is called _____ FM.

19. True or false. Phase modulation produces frequency variation as well as amplitude variation of the carrier.

20. In a simple RC network, the phase shift is between _____ and _____ degrees.

21. In the circuits of Fig. 5-7, phase is varied by changing the _____ or _____ .

22. Larger linear phase shifts are obtained with a(n) _____ circuit.

23. The control element in a phase modulator is usually a(n) _____ .

24. In Fig. 5-10, the component which adjusts the deviation is _____ .

25. If a parallel LC circuit is at resonance, increasing C will cause the current to _____ (lead, lag) the applied voltage.

26. In Fig. 5-7(b), if C is decreased, will the phase shift increase or decrease?

27. In the phase modulator of Fig. 5-8, the FET current is the phasor sum of the currents produced by _____ and _____ .

28. The small phase shifts produced by indirect FM are increased by sending the PM signal to a(n) _____ circuit.

5-3 FREQUENCY DEMODULATORS

There are literally dozens of circuits used to demodulate or detect FM and PM signals. The well-known Foster-Seeley discriminator and the ratio detector were among the most widely used frequency demodulators at one time, but today these circuits have been replaced with more sophisticated IC demodulators. Of course, they are still found in older equipment. The most widely used detectors today include the pulse-averaging discriminator, the quadrature detector, and the phase-locked loop. In this section, we will take a look at all these widely used demodulator circuits.

One of the best frequency demodulators is the Foster-Seeley discriminator shown in Fig. 5-11. The FM signal is applied to the primary of the RF transformer T_1. The primary and secondary windings are resonated at the carrier frequency with C_1 and C_2. The parallel tuned circuit in the primary of T_1 is connected in the collector of a limiter amplifier Q_1 that removes amplitude variations from the FM signal.

The signal across the primary of T_1 is also passed through capacitor C_3 and appears directly across an RFC. The voltage appearing across the RFC is exactly the same as that appearing across the primary winding simply because C_3 and C_5 are essentially short circuits at the carrier frequency. The voltage across this RFC is designated V_3.

The current flowing in the primary winding of T_1 induces a voltage in the secondary winding. Because the secondary winding is center-tapped, the voltage across the upper portion V_1 will be 180° out of phase with the voltage across the lower portion V_2. The voltage induced into the secondary winding is 90° out of phase with the voltage across the primary winding. When both the primary and secondary windings of an air-core transformer are tuned resonant circuits, the phase relationship between the voltages across the primary and secondary will be 90°. This means that the voltages V_1 and V_2 will also be 90° out of phase with V_3, the voltage across the RFC. This phase relationship is shown in the vector diagrams in Fig. 5-11. In Fig. 5-11(a), the input is the unmodulated carrier frequency.

The remainder of the circuit consists of two diode detector circuits similar to those used for AM detection. The voltage $V_{1\text{-}3}$ applied to D_1, R_1, and C_4 is the sum of voltages V_1 and V_3. The voltage $V_{2\text{-}3}$ applied to D_2, R_2, and C_5 is the sum of voltages V_2 and V_3. Since voltages V_1 and V_2 are out of phase with voltage V_3, their respective sums $V_{1\text{-}3}$ and $V_{2\text{-}3}$, are vector sums as illustrated in Fig. 5-11(a).

On one half cycle of the primary voltage, D_1 conducts and current flows through R_1 and charges C_4. On the next half cycle, D_2 conducts and current flows through R_2 and charges C_5. The voltages across R_1 and R_2, designated V_A and V_B, are identical because V_1 and V_2 are the same. Since these two voltages are equal but of the opposite polarity, the voltage between point A and ground is zero. At the carrier center frequency with no modulation, the modulator output is therefore zero.

(a) Carrier frequency   (b) Above carrier frequency   (c) Below carrier frequency

Fig. 5-11 The Foster-Seeley discriminator.

The secondary of T_1 and capacitor C_2 form a series resonant circuit. The reason for this is that the voltage induced into the secondary appears in series with that winding. At resonance, the inductive reactance of the secondary winding equals the capacitive reactance of C_2. At that time, the current flowing in the circuit is exactly in phase with the voltage induced into the secondary. The output is derived from a portion of the secondary winding which is an inductor. The voltage across the secondary winding, therefore, is 90° out of phase with the current in the circuit.

Should the input frequency change as would be the case with FM, the secondary winding will no longer be at resonance. For example, if the input frequency increases, the inductive reactance will be higher than the capacitive reactance, making this circuit inductive. This causes the phase relationship between V_1 and V_2 to change with respect to V_3. If the input is above the resonant frequency, then V_1 will lead V_3 by a phase angle less than 90°. Since V_1 and V_2 remain 180° out of phase, V_2 will then lag V_3 by an angle of more than 90°. This change in phase relationship is shown in Fig. 5-8 (b). When the new vector sums of V_1 and V_3 and V_2 and V_3 are computed, it is found that the voltage $V_{1\text{-}3}$ applied to D_1 is greater than the voltage $V_{2\text{-}3}$ applied to D_2. Therefore, the voltage across R_1 will be greater than the voltage across R_2 and the net output voltage will be positive with respect to ground.

If the frequency deviation is lower than the center frequency, voltage V_1 will lead by an angle of more than 90° while voltage V_2 will lag by an angle less than 90°. The resulting vector additions of V_1 and V_3 and V_2 and V_3 are shown in Fig. 5-11(c). This time, $V_{2\text{-}3}$ is greater than $V_{1\text{-}3}$. As a result, the voltage across R_2 will be greater than the voltage across R_1 and the net output voltage will be negative.

As you can see, as the frequency deviates above and below the center frequency, the output at point A increases or decreases, and, therefore, the original modulating signal is recovered. This signal is passed through the de-emphasis network consisting of R_3 and C_6. The output is then applied to an amplifier or to other circuits as required. Figure 5-12 on the next page shows the output voltage occurring across point A and ground with respect to frequency deviation. It is the linear range in the center that produces the accurate reproduction of the original modulating signal.

The discriminator circuit is sensitive to in-

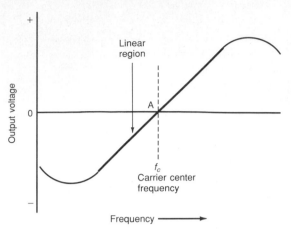

Fig. 5-12   Output voltage of the discriminator.

put amplitude variations. Higher inputs will produce a greater amount of signal at the output, whereas lower inputs create less output. These output variations will be interpreted as frequency changes, and the output will be an incorrect reproduction of the modulating signal. For that reason, all amplitude variations must be removed from the FM signal before being applied to the discriminator. This is usually handled by limiter circuits that will be discussed later.

Another widely used demodulator is the ratio detector. It is similar in appearance to the discriminator but has several important differences as is shown in Fig. 5-13. The FM signal is applied to the RF transformer T_1 with its center tapped secondary. The FM signal is also passed through capacitor C_3 and applied across the RFC as in the discriminator. The circuit uses two diodes, but note that the direction of D_2 is reversed from that in the discriminator. The voltages V_{1-3} and V_{2-3} applied to D_1 and D_2, respectively, are again a composite of V_1 and V_3 (V_{1-3}) and V_2 and V_3 (V_{2-3}) as before.

Another major difference in the ratio detector is the use of a very large capacitor C_6 connected across the output. The load resistors R_1 and R_2 are equal in value, and their common connection is at ground. The output is taken from between points C and ground in the circuit. Capacitors C_4 and C_5 and resistors R_1 and R_2 form a bridge circuit. The voltage across capacitors C_4 and C_5 is the bridge input voltage, while the output is taken between points C and D.

With no modulation on the carrier, the voltage V_{1-3} applied to D_1 is the same as the voltage V_{2-3} applied to D_2. Therefore, capacitors C_4 and C_5 charge to the same voltage with the polarity shown. Since C_6 is connected across these two capacitors, it will charge to the sum of their voltages. Because C_6 is a very large capacitor, usually tantalum or electrolytic, it takes several cycles of the input signal for the capacitor to charge fully. However, once it charges, it will maintain a relatively constant voltage. Since R_1 and R_2 are equal, their voltage drops will be equal. Also, the voltage drops across C_4 and C_5 are equal. The bridge circuit, therefore, is balanced. Looking between points C and D, you will see 0 V because the potential is the same.

Assume that at the center carrier frequency, the voltage drops across C_4 and C_5 are each 2 V. This means the charge on C_6 is 4 V. Then the voltages across R_1 and R_2 are each 2 V.

If the frequency increases, the phase relationship in the circuit will change as described previously for the discriminator circuit [Fig. 5-11(a)]. This will cause the voltage across C_4 to be greater than the voltage across C_5. Assume that the voltage across C_4 is 3 V and the voltage across C_5 is 1 V. The voltages across R_1 and R_2 remain the same at 2 V each because the charge on C_6 does not change. The bridge

Fig. 5-13   A ratio detector.

is now unbalanced, and an output voltage will appear between points C and D in the circuit. Using point B as a reference, the voltage at point C is 1 V positive, and the voltage across R_2 is 2 V positive. Therefore, the voltage difference at C is -1 V.

If the frequency decreases, then the phase relationship will be such that the charge on C_5 will be greater than the charge on C_4. If the voltage across C_5 is $+3$ V with respect to B and the voltage across R_2 remains 2 V, then point C is $+1$ V. The bridge is unbalanced, but in the opposite direction, and the output voltage is of the opposite polarity.

The primary advantage of the ratio detector over the discriminator is that it is essentially insensitive to noise and amplitude variations. The reason for this is the very large capacitor C_6. Since it takes a long time for this capacitor to charge or discharge, short noise pulses or minor amplitude variations are totally smoothed out. However, the average dc voltage across C_6 is the same as the average signal amplitude. This voltage can, therefore, be used in automatic gain control applications.

A simplified block diagram of a pulse-averaging discriminator is shown in Fig. 5-14. The FM signal is applied to a zero-crossing detector or a clipper/limiter which generates a binary voltage level change each time the FM signal varies from minus to plus, or plus to minus. The result is a rectangular wave containing all the frequency variations of the original signal but without amplitude variations.

The FM square wave is then applied to a one-shot multivibrator. The one-shot or monostable multivibrator generates a fixed-amplitude, fixed-width dc pulse on the leading edge of each FM cycle. The duration of the one shot is set so that it is less than the period of the highest frequency expected during maximum deviation.

The one-shot output pulses are then fed to a simple RC low-pass filter which averages the dc pulses to recover the original modulating signal.

The waveforms for the pulse-averaging discriminator are shown in Fig. 5-15 on the next page. Note how at the low frequencies, the one-shot pulses are widely spaced. At the higher frequencies, the one-shot pulses occur very close together. When these pulses are applied to the averaging filter, a dc output voltage is developed. The amplitude of this dc voltage is directly proportional to the frequency deviation.

When a one-shot pulse occurs, the capacitor in the filter charges to the amplitude of the pulse. When the pulse turns off, the capacitor discharges into the load. If the RC time constant is high, the charge on the capacitor will not decrease much. If the time interval between pulses is long, however, the capacitor will lose some of its charge into the load, so the average dc output will be low. When the pulses occur rapidly, the capacitor has little time to discharge between pulses, and, therefore, the average voltage across it remains higher. As you can see, the filter output voltage varies in amplitude with the frequency deviation. Note in Fig. 5-15 how the voltage rises linearly as the frequency increases. The original modulating signal is developed across the filter output. The filter components are carefully selected to minimize the ripple caused by the charging and discharging of the capacitor while at the same time providing the necessary high-frequency response for the original modulating signal.

The pulse-averaging discriminator is a very high quality frequency demodulator. Prior to the availability of low-cost ICs, its use was limited to expensive telemetry and industrial control applications. Today, the pulse-averaging discriminator is easily implemented with low-cost ICs, and has, therefore, found its way into many electronic products.

Another popular frequency demodulator is the quadrature detector. Its primary application is in TV audio demodulation, although it is also used in some FM radio systems. Most IC frequency demodulators are of the quadrature type.

Fig. 5-14 A pulse-averaging discriminator for frequency demodulation.

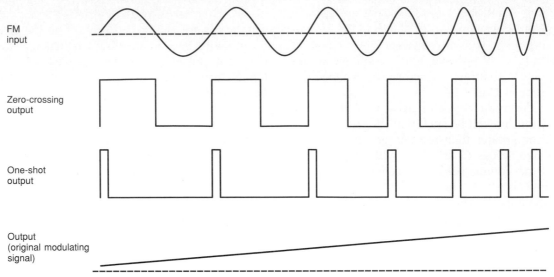

FM input

Zero-crossing output

One-shot output

Output (original modulating signal)

Fig. 5-15 **Waveforms for the pulse-averaging discriminator.**

The term *quadrature* refers to a 90° phase shift between two signals. The quadrature detector uses a phase-shift circuit to produce a phase shift of 90° at the unmodulated carrier frequency. The most commonly used phase-shift arrangement is shown in Fig. 5-16. The FM signal is applied through a very small capacitor C_1 to the parallel tuned circuit which is adjusted to resonance at the center carrier frequency. The tuned circuit appears as a high value of pure resistance at resonance. The small capacitor has a very high reactance compared to the tuned circuit impedance. The output across the tuned circuit then at the carrier frequency is very close to 90° leading the input. Now, when FM occurs, the carrier frequency will deviate above and below the resonant frequency of the tuned circuit. The result will be an increasing or decreasing amount of phase shift between the input and the output.

The two quadrature signals are then fed to a phase detector circuit. The phase detector is nothing more than a circuit whose output is a function of the amount of phase shift between two input signals. The most commonly used phase detector is a balanced modulator using differential amplifiers like those discussed in Chap. 3. The output of the phase detector is a series of pulses whose width varies with the amount of phase shift between the two signals. These signals are averaged in an *RC* low-pass filter to recreate the original modulating signal.

Normally the sinusoidal FM input signals to the phase detector are at a very high level. Therefore, they will drive the differential amplifiers in the phase detector into cutoff and saturation. The differential transistors will act as switches, so the output will be as a series of pulses. No limiter is needed if the input signal is large enough. The output pulse duration is determined by the amount of phase shift. You might think of the phase detector as simply an AND gate whose output is on only when the two input pulses are on and is off if either one or both of the inputs is off.

Figure 5-17 shows the typical waveforms involved in a quadrature detector. When there

Fig. 5-16 **A quadrature FM detector.**

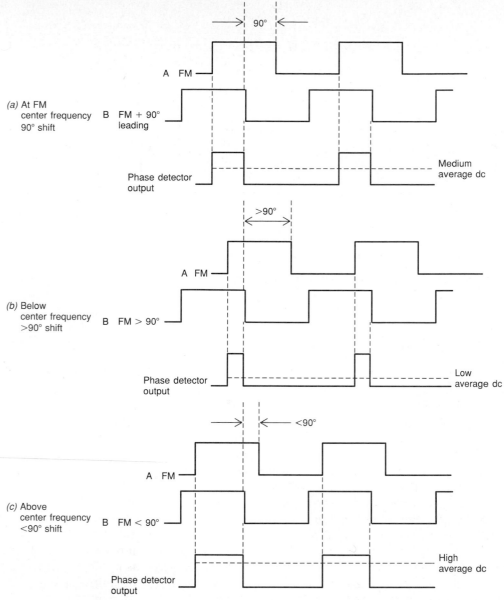

Fig. 5-17   Waveforms in the quadrature detector.

is no modulation, the two input signals are exactly 90° out of phase and, therefore, provide an output pulse width as indicated. When the frequency increases, the amount of phase shift decreases and, therefore, causes the output pulse width to be greater. The wider pulses averaged by the *RC* filter produce a higher average output voltage. This, of course, corresponds to the higher amplitude required to produce the higher carrier frequency.

When the FM signal frequency decreases, there is more phase shift and, as a result, the output pulses will be narrower. When averaged, the narrower pulses will produce a lower average output voltage, thus corresponding to the original lower modulating signal.

Another integrated circuit FM demodulator is the differential peak detector. It is found in TV sets and other consumer electronic products. Usually the demodulator is only one of many other circuits on the chip. A general schematic diagram of the circuit is shown in Fig. 5-18(*a*) on the next page.

The circuit is an enhanced differential amplifier. Transistors Q_3 and Q_4 form the differential amplifier, and the other transistors are emitter followers. Transistor Q_7 is a current source. Note that a tuned circuit is connected between the bases of Q_1 and Q_6. L_1, C_1, and C_2 are discrete components that are external to the chip. The FM input is applied to the base of Q_1. A typical input is a 4.5-MHz FM sound carrier in a TV set.

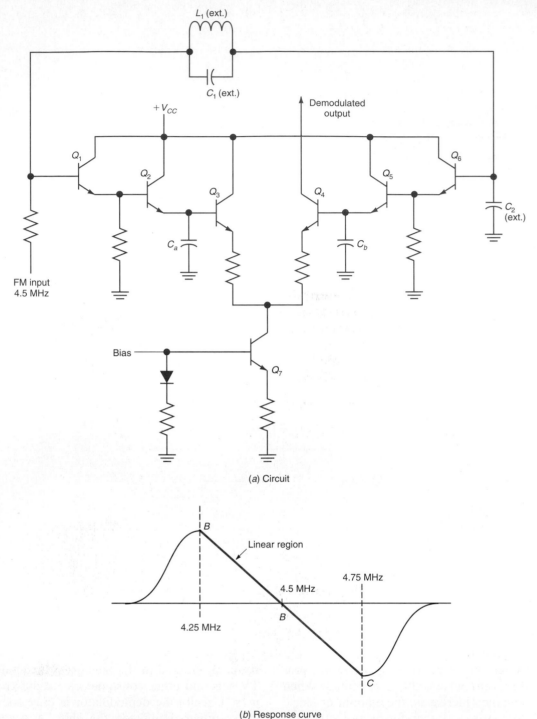

(a) Circuit

(b) Response curve

Fig. 5-18 A differential peak detector FM demodulator.

Q_1 and Q_6 are emitter followers that provide high input impedance and power amplification to drive two other emitter followers, Q_2 and Q_5. Q_2 and Q_5, along with on-chip capacitors C_a and C_b, are peak detectors. C_a and C_b charge and discharge as the voltages at the two inputs vary. For example, assume that the input is a carrier sine wave. When the input to Q_1 goes positive, the voltage at the emitter of Q_2 also goes positive, and C_a charges to the

peak ac value that appears across the tuned circuit. When the peak of the sine wave is past, the voltage declines to zero and then reverses polarity for the negative half-cycle of the sine wave. As the voltage drops, capacitor C_a retains the positive peak value as a charge. The input impedance to Q_3 is relatively high and so does not significantly discharge C_a. C_a simply acts as a temporary storage cell for the peak voltage that cycle.

Capacitor C_b on Q_5 also forms a peak detector circuit. It too will charge up to the peak of the input from the opposite end of the tuned circuit. If the 4.5-MHz carrier is unmodulated, the two capacitors will charge up to the same value. These equal voltages appear at the inputs to the differential amplifier. Remember that the output of a differential amplifier is proportional to the difference of the two inputs. If the inputs are equal, the difference is zero, and so is the output. The output current from Q_4 when this condition exists is used as the zero reference. A lower or higher current will represent a varying frequency.

The external tuned circuit is designed so that L_1 forms a series-resonant circuit with C_2 at a frequency lower than the center frequency, usually 4.5 MHz. A typical series-resonant frequency is 4.25 MHz. L_1 forms a parallel resonant circuit, with C_1 at a frequency above the center frequency. This frequency is usually 4.75 MHz.

To see how the circuit works, assume that an input is gradually increasing in frequency from below 4.25 MHz to above 4.75 MHz. At very low deviation frequencies, the effect on the tuned circuit is minimal and the inputs to the differential amplifiers are about equal; therefore the output is near zero. Now, as the input frequency increases, at some point it reaches the series-resonant point of L_1 and C_2. When this occurs, the total impedance of the series circuit is resistive and very low. This effectively shorts the input of Q_1 to ground, making it near zero. At the same time, the voltage across C_2 is at its peak value because of resonance, so the input to Q_6 is maximum. The result is a very high output current from Q_4. This condition is illustrated at point A in Fig. 5-18(b).

As the input frequency continues to rise, at some point the center frequency of 4.5 MHz is reached, at which time the output will be zero as described earlier. This is shown at point B in Fig. 5-18(b). The frequency continues to increase. At some point, L_1 will be parallel resonant with C_1. This produces a very high

impedance at the base of Q_1, so the voltage is maximum at this time. The lower reactance of C_1 at this higher frequency makes the input to Q_6 very low. As a result, the difference between the two inputs is very high and the output current at Q_4 becomes very high in the opposite (negative) direction. This is point C in Fig. 5-18(b). As you can see, the variation in output current from point A to point C is very linear as the frequency increases. This change in output amplitude with a change in frequency results in excellent demodulation of FM signals.

The best frequency demodulator is the phase-locked loop (PLL). A PLL is a frequency- or phase-sensitive feedback control circuit. It is used not only in frequency demodulation but also in frequency synthesizers, as you will see in a later chapter, and in various filtering and signal detection applications.

All PLLs have the three basic elements illustrated in Fig. 5-19. A phase detector or mixer is used to compare the input or reference signal to the output of a VCO. The VCO frequency is varied by the dc output voltage from a low-pass filter. It is the output of the phase detector that the low-pass filter uses to produce the dc control voltage.

The primary job of the phase detector is to compare the two input signals and generate an output signal that when filtered will control the VCO. If there is a phase or frequency difference between the input and VCO signals, the phase detector output will vary in proportion. The filtered output will adjust the VCO frequency in an attempt to correct for the original frequency or phase change. This dc control voltage is called the *error signal* and is also the feedback in this circuit.

To examine the operation of the PLL, assume initially that no input signal is applied. At this time, the phase detector and low-pass filter outputs are zero. The VCO then operates at what is called its *free-running frequency*. This is the normal operating frequency of the VCO as determined by its internal frequency-determining components.

PLL demodulator

Error signal

Free-running frequency

Fig. 5-19 **The elements of a phase-locked loop (PLL).**

Now assume that an input signal close to the frequency of the VCO is applied. The phase detector will compare the VCO free-running frequency to the input frequency and produce an output voltage proportional to the frequency difference. The resulting dc voltage is applied to the VCO. This dc voltage is such that it forces the VCO frequency to move in a direction that reduces the dc error voltage. The error voltage forces the VCO frequency to change in the direction that reduces the amount of phase or frequency difference between the VCO and the input. At some point, the error voltage will eventually cause the VCO frequency to equal the input frequency. When this happens, the PLL is said to be in a "locked" condition. Although the input and VCO frequencies are equal, there will be a phase difference between them that produces the dc output voltage which causes the VCO to produce the frequency that will keep the circuit locked.

If the input frequency should change, the phase detector and low-pass filter will produce a new value of dc control voltage that will force the VCO output frequency to change until it is equal to the new input frequency. Because of this action, the PLL is said to *track* the input signal. Any input frequency variation will be matched by a VCO frequency change so that the circuit remains locked.

The VCO in a PLL, therefore, is capable of tracking the input frequency over a wide range. The range of frequencies over which the PLL will track the input signal and remain locked is known as the lock range. The *lock range* is usually a band of frequencies above and below the free-running frequency of the VCO. If the input signal frequency is out of the lock range, the PLL will not lock. When this occurs, the VCO output frequency jumps to its free-running frequency.

If an input frequency within the lock range is applied to the PLL, the circuit will immediately adjust itself and remain in a locked condition. The phase detector will determine the phase difference between the VCO free-running and input frequencies and generate the error signal that will force the VCO to equal the input frequency. Once the input signal is captured, the PLL remains locked and will track any changes in the input signal as long as it remains within the lock range.

The range of frequencies over which the PLL will capture an input signal is known as the *capture range*. It is much narrower than

the lock range, but it is generally centered around the free-running frequency of the VCO as is the lock range. See Fig. 5-20.

f_o = VCO free running frequency

Fig. 5-20   Capture and lock ranges of a PLL.

The characteristic that causes the PLL to capture signals within a certain frequency range causes it to act like a bandpass filter. Phase-locked loops are often used in signal conditioning applications where it is desirable to pass signals only in a certain range and reject signals outside of that range. The PLL is highly effective in eliminating the noise and interference on a signal.

Since the PLL responds to input frequency variations, naturally you can understand why it is useful in FM applications. If an FM signal is applied to the input, the VCO will track it. The VCO is, in effect, a frequency modulator that produces exactly the same FM signal as the input. In order for this to happen, however, the VCO input must be identical to the original modulating signal.

The VCO output follows the FM input signal because the error voltage produced by the phase detector and low-pass filter forces the VCO to track it. For that reason, the VCO output must be identical to the input signal if the PLL is to remain locked, and the error signal must be identical to the original modulating signal of the FM input. The low-pass filter cutoff frequency is designed such that it is capable of passing the original modulating signal.

The PLL is the best of all frequency demodulators in use because of the characteristics described earlier. Its ability to provide frequency selectivity and filtering give it a signal-to-noise ratio superior to any other type of FM detector. The linearity of the VCO ensures a highly accurate reproduction of the original modulating signal. Although PLLs are complex, they are easy to apply because they are readily available in low-cost IC form.

Self-Test

Supply the missing information in each statement.

29. In the Foster-Seeley discriminator and ratio detector circuits, an input frequency variation produces a(n) _____ that causes an output voltage variation.
30. In the pulse-averaging discriminator, a low-pass filter averages the fixed-width, fixed-amplitude pulses from a(n) _____ .
31. Should the input frequency increase or decrease for the average voltage output of the low-pass filter in a pulse-averaging discriminator to increase?
32. Quadrature refers to a _____ ° phase difference.
33. In Fig. 5-13, does input B lag or lead input A?
34. The phase detector in a quadrature detector is usually a(n) _____ .
35. The quadrature detector input circuit produces a varying _____ as the input frequency changes.
36. The varying-width pulses produced by the quadrature detector are converted into the original modulating signal by a(n) _____ .
37. The circuit in a differential peak detector IC that temporarily stores the peak value of an input sine wave is called a(n) _____ .
38. In the differential peak detector, the components _____ convert frequency variations into the amplitude variations that ultimately become the output.

39. The three main elements of a PLL are _____ .
40. The _____ circuit in a PLL recognizes _____ or _____ changes between the input and VCO signals.
41. The _____ part of a PLL is a simple frequency modulator.
42. It is the _____ signal in the PLL that varies the VCO frequency.
43. The range of frequencies over which a PLL will cause the input and VCO signals to remain synchronized is known as the _____ range.
44. If the PLL input is zero, the VCO will operate at its _____ frequency.
45. The range of frequencies over which a PLL will latch onto or recognize an input signal is called the _____ range.
46. Since a PLL will only respond to signals over a narrow frequency range, it acts like a(n) _____ .
47. In a PLL frequency demodulator, the error signal is the _____ .
48. In a PLL demodulating an FM signal, the VCO output is an exact reproduction of the _____ .

Determine whether each statement is true or false.

49. The Foster-Seeley discriminator is sensitive to input amplitude as well as frequency variations.
50. The ratio detector requires a limiter.
51. The lock range of a PLL is narrower than the capture range.

SUMMARY

1. The component most widely used for FM or PM is the varactor diode or voltage-variable capacitor (VVC).
2. A VVC is a specially designed silicon junction diode optimized for large capacitive variations.
3. A reverse-biased junction diode will act as a small capacitor where the depletion region is the dielectric.
4. The capacitance of a varactor is inversely proportional to the reverse-biased voltage amplitude.
5. The most common frequency modulators use a varactor to vary the frequency of an *LC* circuit or crystal in accordance with the modulating signal.
6. A reactance modulator is an amplifier that is made to appear inductive or capacitive by phase shift. It is used to produce wide-deviation direct FM.
7. Crystal oscillators are preferred for their frequency stability over *LC* oscillators, but only very small frequency deviation is possible with crystal oscillators.
8. An IC VCO produces excellent deviation FM at frequencies below 1 MHz.

9. In a phase modulator, the carrier is shifted in phase in accordance with the modulating signal. This produces indirect FM.
10. One of the best phase modulators is a parallel tuned circuit controlled by a varactor.
11. Most phase modulators produce very small amounts of frequency deviation.
12. Frequency deviation and carrier frequency can be increased by passing them through a frequency multiplier.
13. One of the oldest and best frequency demodulators is the Foster-Seeley discriminator that is a phase detector whose output voltage increases or decreases with phase changes produced by input frequency deviation.
14. A Foster-Seeley discriminator is sensitive to input amplitude variations and, therefore, must be preceded by a limiter.
15. A variation of the Foster-Seeley discriminator is the ratio detector widely used in older TV receiver designs. A primary advantage of the ratio detector is that no limiter is needed.
16. A pulse-averaging discriminator converts an FM signal into a square wave of identical frequency variation using a zero-crossing detector, comparator, or limiter circuit. This circuit triggers a one shot that produces pulses that are averaged in a low-pass filter to reproduce the original modulating signal.
17. A quadrature detector uses a unique phase-shift circuit to provide quadrature (90°) FM input signals to a phase detector. The phase detector produces a different pulse width for different phase shifts. These pulse width variations are averaged in a low-pass filter to recover the modulating signal.
18. Quadrature detectors are available in IC form and are one of the most widely used TV audio demodulators.
19. A differential peak detector is an IC FM demodulator that uses a differential amplifier and capacitive storage peak detectors plus tuned circuits to translate frequency variations into voltage variations.
20. A phase-locked loop (PLL) is a feedback control circuit made up of a phase detector, voltage-controlled oscillator (VCO), and low-pass filter. The phase detector compares an input signal to the VCO signal and produces an output that is filtered by a low-pass filter into an error signal that controls the VCO frequency.
21. The PLL is synchronized or locked when the input and VCO frequencies are equal. Input frequency changes cause a phase or frequency shift which, in turn, produces an error signal that forces the VCO to track the input and reduce their difference to zero.
22. The range of frequencies over which a PLL will track an input is called the lock range. If the input strays outside the lock range, the PLL will go out of lock and the VCO will operate at its free-running frequency.
23. The capture range of a PLL is that narrow band of frequencies over which a PLL will recognize and lock onto an input signal. The capture range is narrower than the lock range and it makes the PLL look like a bandpass filter.
24. The PLL is the best frequency demodulator because its filtering action removes noise and interference and its highly linear output faithfully reproduces the original modulating signal.

CHAPTER REVIEW QUESTIONS

Choose the letter that best answers each question.

5-1. Another name for a VVC is
   a. PIN diode.
   b. Varactor diode.
   c. Snap diode.
   d. Hot-carrier diode.
5-2. The depletion region in a junction diode forms what part of a capacitor?
   a. Plates   *c.* Package
   b. Leads   *d.* Dielectric

5-3. Increasing the reverse bias on a varactor diode will cause its capacitance to
  a. Decrease.
  b. Increase.
  c. Remain the same.
  d. Drop to zero.

5-4. The capacitance of a varactor diode is in what general range?
  a. pF
  b. nF
  c. μF
  d. F

5-5. In Fig. 5-3, the varactor diode is biased by which components?
  a. R_1, R_2
  b. R_1, C_2
  c. L_1, C_1
  d. RFC, C_3

5-6. In Fig. 5-3, if the reverse bias on D_1 is reduced, the resonant frequency of C_1
  a. Increases.
  b. Decreases.
  c. Remains the same.
  d. Cannot be determined.

5-7. The frequency change of a crystal oscillator produced by a varactor diode is
  a. Zero.
  b. Small.
  c. Medium.
  d. Large.

5-8. A phase modulator varies the phase shift of the
  a. Carrier.
  b. Modulating signal.
  c. Both *a* and *b*.
  d. Neither *a* or *b*.

5-9. The widest phase variation is obtained with a(n)
  a. *RC* low-pass filter.
  b. *RC* high-pass filter.
  c. *LR* low-pass filter.
  d. *LC* resonant circuit.

5-10. In Fig. 5-7, R_4 is the
  a. Pre-emphasis circuit.
  b. De-emphasis circuit.
  c. Deviation control.
  d. Frequency determining component in the tuned circuit.

5-11. The small frequency change produced by a phase modulator can be increased by using a(n)
  a. Amplifier.
  b. Mixer.
  c. Frequency multiplier.
  d. Frequency divider.

5-12. A crystal oscillator whose frequency can be changed by an input voltage is called a(n)
  a. VCO.
  b. VXO.
  c. VFO.
  d. VHF.

5-13. Which oscillators are preferred for carrier generators because of their good frequency stability?

a. LC

b. RC

c. LR

d. Crystal

5-14. Which of the following frequency demodulators requires an input limiter?

a. Foster-Seeley discriminator

b. Pulse-averaging discriminator

c. Quadrature detector

d. PLL

5-15. Which discriminator averages pulses in a low-pass filter?

a. Ratio detector

b. PLL

c. Quadrature detector

d. Foster-Seeley discriminator

5-16. Which frequency demodulator is considered the best overall?

a. Ratio detector

b. PLL

c. Quadrature

d. Pulse-averaging discriminator

5-17. In Fig. 5-8, the voltage at point A when the input frequency is below the FM center frequency is

a. Negative.

b. Positive.

c. Zero.

d. Indeterminant.

5-18. In Fig. 5-8, R_3 and C_6 form which kind of circuit?

a. Carrier filter

b. Pulse-averaging filter

c. Pre-emphasis

d. De-emphasis

5-19. In Fig. 5-10, the voltage across C_6 is

a. Inversely proportional to signal amplitude.

b. Directly proportional to signal amplitude.

c. Directly proportional to frequency deviation.

d. Constant.

5-20. In a pulse-averaging discriminator, the pulses are produced by a(n)

a. Astable multivibrator.

b. Zero-crossing detector.

c. One shot.

d. Low-pass filter.

5-21. A reactance modulator looks like a capacitance of 35 pF in parallel with the oscillator-tuned circuit whose inductance is 50 μH and capacitance is 40 pF. What is the center frequency of the oscillator prior to FM?

a. 1.43 MHz

b. 2.6 MHz

c. 3.56 MHz

d. 3.8 MHz

5-22. Which of the following is true about the NE566 IC?

a. It is a VCO.

b. Its output is sinusoidal.

c. It is an FM demodulator.

d. It uses LC-tuned circuits.

5-23. An FM demodulator that uses a differential amplifier and tuned circuits to convert frequency variations into voltage variations is the

a. Quadrature detector.

b. Foster-Seeley discriminator.

c. Differential peak detector.

d. Phase-locked loop.

5-24. The output amplitude of the phase detector in a quadrature detector is proportional to
  a. Pulse width.
  b. Pulse frequency.
  c. Input amplitude.
  d. The phase shift value at center frequency.
5-25. The input to a PLL is 2 MHz. In order for the PLL to be locked, the VCO output must be
  a. 0 MHz.
  b. 1 MHz.
  c. 2 MHz.
  d. 4 MHz.
5-26. Decreasing the input frequency to a locked PLL will cause the VCO output to
  a. Decrease.
  b. Increase.
  c. Remain constant.
  d. Jump to the free-running frequency.
5-27. The range of frequencies over which a PLL will track input signal variations is known as the
  a. Circuit bandwidth.
  b. Capture range.
  c. Band of acceptance.
  d. Lock range.
5-28. The band of frequencies over which a PLL will acquire or recognize an input signal is called the
  a. Circuit bandwidth.
  b. Capture range.
  c. Band of acceptance.
  d. Lock range.
5-29. Over a narrow range of frequencies, the PLL acts like a
  a. Low-pass filter.
  b. Bandpass filter.
  c. Tunable oscillator.
  d. Frequency modulator.
5-30. The output of a PLL frequency demodulator is taken from the
  a. Low-pass filter.
  b. VCO.
  c. Phase detector.
  d. None of the above.

CRITICAL THINKING QUESTIONS

5-1. A standard diode detector AM demodulator like that in Fig. 3-11 can be used to demodulate frequency modulation. Explain how this is possible. (Hint: Consider how the FM signal is affected by the tuned circuit before the diode.)
5-2. Describe what you think might be the result of demodulating an FM signal that had been passed by a tuned circuit whose bandwidth was not wide enough to pass all the FM sidebands.

Answers to Self-Tests

1. *b*
2. *b*
3. *b*
4. *a*
5. *a*
6. varactor diode (or varicap)
7. 1,200
8. frequency stability
9. voltage-controlled oscillator
10. VXO
11. frequency multiplier
12. increase
13. resistance, capacitance
14. true
15. false
16. false
17. true
18. indirect

19. true
20. 0, 90
21. resistance, capacitance
22. parallel resonant (or tuned)
23. varactor
24. $R0_4$
25. lead
26. decrease
27. C_1, C_2, Q_1, and C_1, R_1, Q_1
28. frequency multiplier
29. phase shift
30. one-shot or monostable multivibrator
31. increase
32. 90
33. lead
34. differential amplifier
35. phase shift

36. low-pass filter
37. peak detector
38. C_1, C_2, L_1
39. phase detector, VCO, low-pass filter
40. phase detector, frequency, phase
41. VCO
42. error
43. lock
44. free-running
45. capture
46. bandpass filter
47. modulating signal or information signal
48. FM input
49. true
50. false
51. false

CHAPTER 6

Radio Transmitters

∎

CHAPTER OBJECTIVES

This chapter will help you to:

1. *Draw* a block diagram of a CW, an AM, an FM, or an SSB transmitter and *explain* the operation of each major circuit.
2. *State* the four major classes of RF power amplifiers and the advantages and disadvantages of each.
3. *Explain* the biasing and operation of a class C power amplifier.
4. *Explain* the operation of frequency multiplier.
5. *Calculate* the component values in an LC impedance-matching network to match an output impedance to an input impedance.
6. *Show* how transformers and baluns are used in impedance matching.
7. *State* the benefits of speech processing and how it is achieved.

A radio transmitter takes the information to be communicated and converts it into an electronic signal compatible with the communications medium. Typically this process involves carrier generation, modulation, and amplification. The signal is then fed by wire, coax, or waveguide to an antenna that launches it into free space. In this chapter you will learn about the circuits commonly used in radio transmitters. These include amplifiers, frequency multipliers, impedance-matching networks, and speech-processing circuits.

∎

6-1 INTRODUCTION TO TRANSMITTERS

The *transmitter* is the electronic unit that accepts the information signal to be transmitted and converts it into an RF signal capable of being transmitted over long distances. Every transmitter has three basic functions. First, the transmitter must generate a signal of the correct frequency at a desired point in the spectrum. Second, it must provide some form of modulation that causes the information signal to modify the carrier signal. Third, it must provide sufficient power amplification to ensure that the signal level is high enough so that it will carry over the desired distance.

The simplest form of transmitter is an oscillator as shown in Fig. 6-1 on the next page.

The oscillator generates a carrier signal of the desired frequency. The frequency here is determined by a crystal. Information to be transmitted is expressed in a special form of code using dots and dashes to represent letters of the alphabet and numbers. Information transmitted in this way is referred to as *continuous-wave (CW) transmission*. A key which is simply a convenient hand-operated switch is used in the emitter to turn the oscillator off and on to produce the dots and dashes. The oscillator will produce a short burst of RF energy for a dot and a longer RF burst for a dash. Although such a simple transmitter may have a power of 1 W or less, at the right frequency and with a good antenna it is capable of sending signals halfway around the world.

The basic CW transmitter just described can

Fig. 6-1   The simplest transmitter, a CW oscillator.

be greatly improved by simply adding a power amplifier to it. The result is shown in Fig. 6-2. The oscillator is still keyed off and on to produce dots and dashes, while the amplifier increases the power level of the signal. The result is a stronger signal that will carry farther and produce more reliable communications.

The basic oscillator-amplifier combination shown in Fig. 6-2 is the basis for virtually all radio transmitters. Many other circuits are added depending upon the type of modulation

used, the power level, and other considerations. Now let's take a look at some of these transmitters which use various types of modulation.

Figure 6-3 shows an AM transmitter. An oscillator generates the final carrier frequency. In most applications, this will be a crystal oscillator. Usually, transmitters operate on assigned frequencies or channels. Crystals provide the best way to obtain the desired frequency with good stability. In general, *LC* oscillators do not have the frequency stability required to stay on frequency. Temperature variations and other conditions will cause the frequency to drift outside of the limits imposed by the FCC.

The carrier signal is then fed to a buffer amplifier whose primary purpose is to isolate the oscillator from the remaining power amplifier stages. The buffer amplifier usually operates class A and provides a modest increase in power output. The main purpose of the buffer amplifier is to prevent load changes from causing frequency variations in the oscillator.

The signal from the buffer amplifier is applied to a driver amplifier. This is a class C am-

Fig. 6-2   A more powerful CW transmitter.

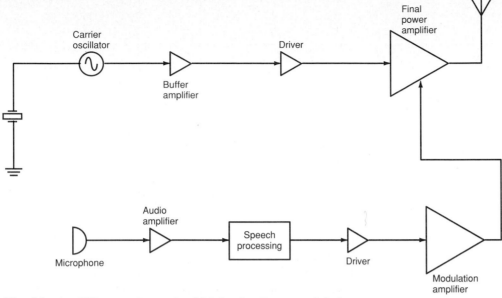

Fig. 6-3   An AM transmitter using high-level collector modulation.

plifier designed to provide an intermediate level of power amplification. The purpose of this circuit is to generate sufficient output power to drive the final power amplifier stage.

The final power amplifier, normally just referred to as the *final,* also operates class C at very high power. The actual amount of power depends upon the application. For example, in a CB transmitter, the power input is only 5 W. However, AM radio stations operate at the much higher powers of 250, 500, 1000, 5000, or 50,000 W.

All the RF circuits in the transmitter are usually solid-state, meaning they are implemented with either bipolar or field-effect transistors. Although bipolar transistors are by far the most common, FETs are being more widely used as they are now capable of handling high power. Transistors are also typically used in the final as long as the power level does not exceed several hundred watts. Individual RF power transistors can handle up to about 100 W. Several of these may be connected in parallel or in push-pull configurations to increase the power-handling capability further. However, transistors are rarely used in final amplifiers where the power output exceeds 500 W. For higher power levels, vacuum tubes are still used.

Refer again to Fig. 6-3. Assume that this is a voice transmitter. The input from the microphone is applied to a low-level class A audio amplifier. This amplifier boosts the small signal from the microphone to a higher voltage level. One or more stages of amplification may be used.

The voice signal is then fed to some form of

speech-processing circuit. "Speech processing" refers to filtering and amplitude control. The filtering ensures that only voice frequencies in a certain range are passed. This helps to minimize the bandwidth occupied by the signal. Most communications transmitters limit the voice frequency to the 300- to 3000-Hz range. This is adequate for intelligible communications. However, AM broadcast stations offer higher fidelity and allow frequencies up to 5 kHz to be used.

The speech processor also contains some kind of a circuit used to hold the amplitude to a particular level. High-level amplitude signals are compressed and often lower-level signals are given more amplification. The result is that overmodulation is prevented. This reduces the possibility of signal distortion and harmonics which produce wider sidebands that can cause adjacent channel interference.

After the speech processor, a driver amplifier is used. This driver amplifier increases the power level of the signal so that it is capable of driving the high-power modulation amplifier. In the AM transmitter of Fig. 6-3, high-level or collector modulation is used. As you saw earlier, the power output of the modulation amplifier must be one-half the input power of the RF amplifier. The high-power modulation amplifier usually operates class AB or class B push-pull to achieve such power levels.

Fig. 6-4 on the next page shows the typical configuration for an FM or PM transmitter. The indirect method of FM generation is used. A stable crystal oscillator is used to generate

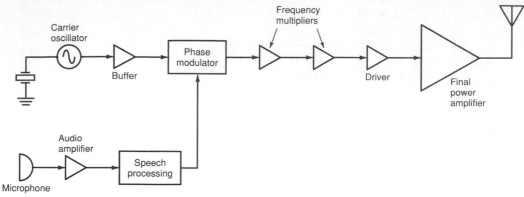

Fig. 6-4 A typical FM transmitter using indirect FM with a phase modulator.

the carrier signal, and a buffer amplifier is used to isolate it from the remainder of the circuitry. The carrier signal is then applied to a phase modulator similar to those previously discussed. The voice input is amplified and processed to limit the frequency range and prevent overdeviation. The output of the modulator is the desired FM signal.

Most FM transmitters are used in the VHF and UHF range, and crystals are not available to generate those frequencies directly. As a result, the carrier is usually generated at a frequency considerably lower than the final output frequency. To achieve the desired output frequency, one or more frequency multiplier stages are used. A *frequency multiplier* is a class C amplifier whose output frequency is some integer multiple of the input frequency. Most frequency multipliers increase the frequency by a factor of 2, 3, 4, or 5, and because they are class C amplifiers, also provide a modest amount of power amplification.

The frequency multiplier not only increases the carrier frequency to the desired output frequency, but it also multiplies the frequency deviation produced by the modulator. Many frequency and phase modulators generate only a small frequency shift, much lower than the desired final deviation. The design of the transmitter is such that the frequency multipliers will provide the correct amount of multiplication not only for the carrier frequency but also for the modulation deviation.

After the frequency multipliers, a class C driver amplifier is used to increase the power level sufficiently to operate the final power amplifier. The final power amplifier also operates class C.

Most FM communications transmitters operate at relatively low power levels, typically less than 100 W. All the circuits use transis-

tors even in the VHF and UHF range. For power levels beyond several hundred watts, vacuum tubes must be used. The final amplifier stages in FM broadcast transmitters typically use large vacuum tube class C amplifiers. In FM transmitters operating in the microwave range, klystrons, magnetrons, and traveling wave tubes are used to provide the final power amplification.

A typical SSB transmitter is shown in Fig. 6-5. An oscillator signal generates the carrier which is then fed to the buffer amplifier. The buffer amplifier supplies the carrier input signal to the balanced modulator. The audio amplifier and speech-processing circuits described previously provide the other input to the balanced modulator. The balanced modulator output is then fed to a sideband filter which selects either the upper or lower sideband.

The SSB signal is then fed to a mixer circuit which is used to translate the signal to its final operating frequency. Typically, the SSB signal is generated at a much lower frequency. This makes the balanced modulator and filter circuits simpler and easier to design. The mixer translates the SSB signal to a higher desired frequency. The other input to the mixer is derived from an LO set at a frequency that, when mixed with the SSB signal, will produce the desired operating frequency. The mixer may be set up so that the tuned circuit at its output selects either the sum or difference frequency. The oscillator frequency must be set to provide the desired output frequency. This LO may use crystals for fixed-channel operation. However, in some equipment such as that used by amateur radio operators, a variable-frequency oscillator (VFO) is used to provide continuous tuning over a desired range. In most modern communications, a frequency synthesizer is used to set the final output frequency.

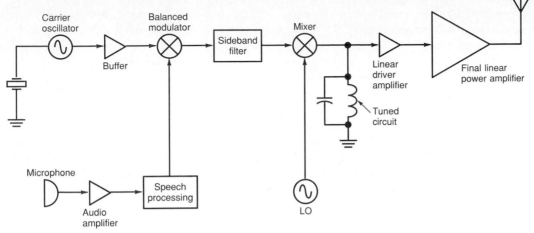

Fig. 6-5   An SSB transmitter.

The output of the mixer is the desired final carrier frequency containing the SSB modulation. It is then fed to linear driver and power amplifiers to increase the power level as required.

Self-Test

Supply the missing word(s) in each statement.

1. The simplest transmitter is a(n) _____ circuit.
2. Information sent in the form of coded dots and dashes is called _____ transmission and is abbreviated _____ .
3. Fixed-frequency or channel operation of a transmitter is obtained by using a(n) _____ .
4. An amplifier that isolates the carrier oscillator from other transmitter amplifiers is known as a _____ amplifier.
5. Intermediate power amplifier stages in a transmitter are commonly referred to as _____ .
6. The output RF power amplifier in a transmitter is sometimes called the _____ .
7. Audio circuits that prevent excessive signal bandwidth and overmodulation are called _____ circuits.
8. In an FM transmitter, special amplifiers called _____ are used to increase the carrier frequency and deviation to the desired output values.
9. In an SSB transmitter, the final output frequency is usually produced by a(n) _____ circuit before being amplified.
10. The power level of SSB signals must be

increased by _____ amplifiers to prevent distortion.
11. Frequency-modulated signals use _____ amplifiers for power amplification.

6-2 LINEAR AMPLIFIERS, CLASS C AMPLIFIERS, AND FREQUENCY MULTIPLIERS

There are two basic types of power amplifiers used in transmitters: linear and class C. Linear amplifiers provide an output signal that is an identical, enlarged replica of the input. Their output is directly proportional to their input; therefore, they faithfully reproduce an input but at a higher power level. All audio amplifiers are linear. Linear RF amplifiers must be used to increase the power level of varying-amplitude RF signals such as low-level AM or SSB signals. Frequency-modulated signals do not vary in amplitude and, therefore, may be amplified with more efficient, nonlinear class C amplifiers.

Linear amplifiers operate class A, AB, or B. The class of an amplifier indicates how it will be biased. A class A amplifier is biased so that it conducts continuously. The bias is set so that the input varies the collector (or drain) current over a linear region of the transistor's characteristic. In this way, its output is an amplified linear reproduction of the input. Usually we say that the class A amplifier conducts for 360° of an input sine wave.

A class B amplifier is biased at cutoff so that no collector current flows with zero input. The transistor conducts on only one-half of the sine wave input. In other words, it conducts for 180° of a sine wave input. This means that only one-half of the sine wave is amplified. Normally, two class B amplifiers are connected in a push-pull arrangement so that both the positive and negative alternations of the input are amplified simultaneously.

A class AB amplifier is biased near cutoff with some continuous collector current flow. It will conduct for more than 180° but for less than 360° of the input. It too is used primarily in push-pull amplifiers and provides better linearity than a class B amplifier but with less efficiency.

Class A amplifiers are linear but not very efficient. For that reason, they make poor power amplifiers. As a result, they are used primarily as small-signal voltage amplifiers or for low-power amplifiers. The buffer amplifiers described previously operate class A.

Class B and class C amplifiers are more efficient because current flows for only a portion of the input signal. They make good power amplifiers, the class C being the most efficient. Since both class B and class C amplifiers distort an input signal, special techniques are used to eliminate or compensate for the distortion. For example, class B amplifiers are operated in a push-pull configuration, while class C amplifiers use a resonant LC load to eliminate the distortion.

A simple class A buffer amplifier is shown in Fig. 6-6. The carrier oscillator signal is capacitively coupled to the input. The bias is derived from R_1, R_2, and R_3. The collector is

tuned with a resonant LC circuit at the operating frequency, and an inductively coupled secondary loop transfers power to the next stage. Buffers like this usually operate at a power level of well less than 1 W.

A high-power class A linear amplifier is shown in Fig. 6-7. Base bias is supplied by a complex constant-current circuit that is temperature-compensated. The RF input from a 50-Ω source is connected to the bias via an impedance-matching circuit made up of C_1, C_2, and L_1. The output is matched to a 50-Ω load by the impedance-matching network made up of L_2, L_3, C_3, and C_4. When connected to a proper heat sink, the transistor can generate up to 100 W of power up to about 30 MHz. The amplifier is designed for a specific frequency which is set by the input and output tuned circuits.

Keep in mind that a class A amplifier can achieve a maximum efficiency of 50 percent. That means only 50 percent of the dc power is converted to RF. The remaining power is dissipated in the transistor. For 100-W RF output, the transistor dissipates 100 W.

One problem that all RF amplifiers have is self-oscillation. When some of the output voltage finds its way back to the input of the amplifier with the correct amplitude and phase, the amplifier will oscillate. The amplifier may oscillate at its tuned frequency or, in some cases, at a much higher frequency. When the circuit oscillates at a higher frequency unrelated to the tuned frequency, the oscillation is referred to as *parasitic oscillation*. In either case, the oscillation is undesirable and prevents amplification from taking place or, in the case of parasitic oscillation, reduces the power amplification and introduces distortion of the signal.

Self-oscillation of an amplifier is usually the result of positive feedback that occurs because of the interelement capacitance of the amplifying device, whether it is a bipolar transistor, an FET, or a vacuum tube. In a bipolar transistor this is the collector-to-base capacitance C_{bc} as shown in Fig. 6-8(a). Transistor amplifiers are biased so that the emitter-base junction is forward-biased and the base-collector junction is reverse-biased. As you saw in a previous chapter, a reverse-biased diode or transistor junction acts like a capacitor. This small capacitance permits output from the collector to be fed back to the base. Depending upon the frequency of the signal, the value of the capacitance, and the values of stray inductances and capacitances in the circuit, the sig-

Fig. 6-6 Class A RF buffer amplifier.

Fig. 6-7 **High-power class A linear RF amplifier.**

(a) Canceling the effect of C_{bc} with an equivalent inductance, L_n

(b) Neutralization using a tapped collector coil and a neutralization capacitor, C_n

$V_f = 0$ when $C_n = C_{bc}$

(c) Equivalent circuit of part b

(d) Neutralization with a tapped input inductor

Fig. 6-8 **Neutralization circuits.**

nal fed back may be in phase with the input signal and high enough in amplitude to cause oscillation.

This interelement capacitance cannot be eliminated; therefore its effect must be compensated for. This process is called *neutralization*. The concept of neutralization is to feed back another signal equal in amplitude to the signal fed back through C_{bc} and 180° out of phase with it. The result is that the two signals will cancel one another.

Several methods of neutralization are shown in Fig. 6-8. In Fig. 6-8(*a*), a signal of equal and opposite phase is provided by the inductor L_n. Capacitor C_1 is a high-value capacitor used strictly for dc blocking to prevent collector voltage from being applied to the base. L_n is made adjustable so that its value can be set to make its reactance equal to the reactance of C_{bc} at the oscillation frequency. As a result, C_{bc} and L_n form a parallel resonant circuit that acts like a very high resistance at the resonant frequency. The result is effective cancellation of the positive feedback.

Figure 6-8(*b*) shows another type of neutralization using a tapped collector inductor and a neutralization capacitor C_n. The two equal halves of the collector inductance, the junction capacitance C_{bc}, and C_n form a bridge circuit as illustrated in Fig. 6-8(*c*). When C_n is adjusted to equal to C_{bc}, the bridge is balanced and no feedback signal V_f occurs. A variation of this is shown in Fig. 6-8(*d*), where a center-tapped base input inductor is used.

Most RF power transistors have an upper power limit of several hundred watts. To produce more power, two or more devices may be connected in parallel. A class B linear power amplifier using push-pull is shown in Fig. 6-9. The RF driving signal is applied to Q_1 and Q_2 through input transformer T_1. It provides impedance matching and base-drive signals to Q_1 and Q_2 that are 180° out of phase. An output transformer T_2 couples the power to the antenna or load. Bias is provided by R_1 and D_1.

For class B operation, Q_1 and Q_2 must be biased right at the cutoff point. The emitter-base junction of a transistor will not conduct until about 0.6 to 0.8 V of forward bias is applied because of the built-in potential barrier. This effect causes the transistors to be biased beyond cutoff, not right at it. A forward-biased silicon diode D_1 has about 0.7 V across it. This is used to put Q_1 and Q_2 right on the conduction threshold.

Now, when the positive half cycle of input RF occurs, the base of Q_1 will be positive and

Fig. 6-9   Broadband class B linear RF power amplifier.

the base of Q_2 will be negative. The Q_2 will be cut off, but Q_1 will conduct and amplify the positive half cycle. Collector current flows in the upper half of T_2 which induces an output voltage in the secondary.

On the negative half cycle of the RF input, the base of Q_1 is negative, so it is cut off. The base of Q_2 is positive, so Q_2 amplifies the negative half cycle. Current flows in Q_2 and the lower half of T_2, thereby completing a full cycle. The power is split between the two transistors.

The circuit in Fig. 6-9 is a broadband circuit, that is, untuned. It will amplify signals over a broad frequency range. A typical range may be 2 to 30 MHz. A low-power AM or SSB signal would be generated at the desired frequency, and then applied to this power amplifier prior to being sent to the antenna. With push-pull circuits, power levels of up to several hundred watts are possible.

Figure 6-10 shows another push-pull RF power amplifier. It uses two power MOSFETs and can produce an output up to 1 kilowatt (kW) over the 10- to 90-MHz range. It has a 12-dB power gain. The RF input driving power must be 63 W to produce the full 1-kW output. Toroidal transformers are used at the input and output for impedance matching. They provide broadband operation over the 10- to 90-MHz range without tuning. The 20-nanohenry (nH) chokes and 20-Ω resistors form neutralization circuits that provide out-of-phase feed-

back from output to input to prevent self-oscillation.

The key circuit in most AM and FM transmitters is the class C amplifier. It is used for power amplification in the form of drivers, frequency multipliers, and final amplifiers.

A class C amplifier is biased so that it conducts for less than 180° of the input. It will typically have a conduction angle of 90° to 150°. This means that current flows through it in short pulses. The question then is, "How is a complete signal amplified?" As you will see, a resonant tuned circuit is used for that purpose.

Figure 6-11(a) on the next page shows one way of biasing a class C amplifier. The base of the transistor is simply connected to ground through a resistor. No external bias voltage is applied. An RF signal to be amplified is applied directly to the base. The transistor will conduct on the positive half cycles of the input wave and will be cut off on the negative half cycles. Ordinarily, you would think that this would be a class B amplifier. However, this is not the case. Recall that the emitter-base junction of a bipolar transistor has a forward voltage threshold of approximately 0.7 V. In other words, the emitter-base junction will not really conduct until the base is more positive than the emitter by +0.7 V. Because of this, the transistor has an inherent built-in bias. When the input signal is applied, the collector current will not flow until the base is positive by 0.7 V. See Fig. 6-11(b). The result is that

Fig. 6-10   A 1-kW push-pull RF power amplifier using MOSFETs.

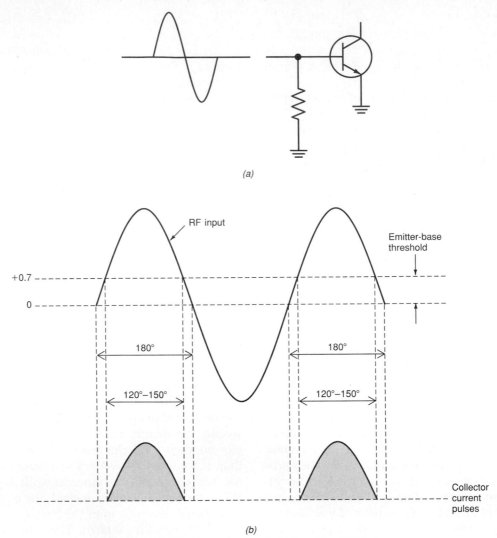

(a)

(b)

Fig. 6-11 **Using the internal emitter-base threshold for class C biasing.**

collector current flows through the transistor in positive pulses for less than the full 180° of the positive ac alternation.

In many low-power driver and multiplier stages, no special biasing provisions other than the inherent emitter-base junction voltage are required. The resistor between base and ground simply provides a load for the driving circuit.

In some cases, a narrower conduction angle than that provided by the circuit in Fig. 6-11(a) may be required. In this case, some form of bias must be applied. A simple way of supplying bias is with the *RC* network shown in Fig. 6-12(a). Here the signal to be amplified is applied through capacitor C_1. When the emitter-base junction conducts on the positive half cycle, C_1 will charge to the peak of the applied voltage less the forward drop across the emitter-base junction. On the negative half cycle of the input, the emitter-base junction, of

course, will be reverse-biased, so the transistor does not conduct. During this period of time, however, capacitor C_1 will discharge through R_1. This produces a negative voltage across R_1 which serves as a reverse bias on the transistor. By properly adjusting the time constant of R_1 and C_1, an average dc reverse-bias voltage will be established. The applied voltage will cause the transistor to conduct but only on the peaks. The higher the average dc bias voltage, the narrower the conduction angle and the shorter the duration of the collector current pulses. This method is referred to as *signal bias*.

Of course, negative bias can also be supplied to a class C amplifier from a fixed dc supply voltage as shown in Fig. 6-12(b). After the desired conduction angle is determined, the value of the reverse voltage can be determined. It is applied to the base through the RFC. The in-

(a) Signal bias

(b) External bias

(c) Self-bias

Fig. 6-12   Methods of biasing a class C amplifier.

Fig. 6-13   Class C amplifier operation.

coming signal is then coupled to the base and causes the transistor to conduct only on the peaks of the positive input alternations. This is called *external bias* but requires a separate negative dc supply.

Another biasing method is shown in Fig. 6-12(*c*). The bias is also derived from the signal as in Fig. 6-12(*a*). This arrangement is known as the *self-bias method*. When current flows in the transistor, a voltage is developed across R_1. Capacitor C_1 is charged and holds the voltage constant. This makes the emitter more positive than the base, which is the same as a negative voltage on the base. A strong input signal is required for proper operation.

All class C amplifiers have some form of tuned circuit connected in the collector as shown in Fig. 6-13. The primary purpose of this tuned circuit is to form the complete ac sine wave output. A parallel tuned circuit will ring or oscillate at its resonant frequency whenever it receives a dc pulse. The pulse will charge the capacitor which, in turn, will discharge into the inductor. The magnetic field in the inductor will increase and then collapse, causing a voltage to be induced. This induced voltage then recharges the capacitor in the opposite direction. This exchange of energy between the inductor and the capacitor is called the *flywheel effect* and produces a damped sine wave at the resonant frequency. If the resonant circuit receives a pulse of current every half cycle, the voltage across the tuned circuit will be a constant-amplitude sine wave at the resonant frequency. Even though the current flows through the transistor in short pulses, the class C amplifier output will be a continuous sine wave.

The tuned circuit in the collector also has another purpose which is to filter out unwanted harmonics. Any nonsinusoidal signal, such as a square wave or the short pulses that flow through the class C amplifier, consists of a fundamental sine wave and multiple harmonics. The short pulses in a class C amplifier are made up of second, third, fourth, fifth, etc., harmonics. In a high-power transmitter, signals will be radiated at these harmonic frequencies as well as at the fundamental resonant frequency. Such harmonic radiation can cause out-of-band interference. The purpose of the tuned circuit is to act as a selective filter that will eliminate these higher-order harmonics. If the Q of the tuned circuit is made high enough, the harmonics will be adequately suppressed.

It is important to point out that the Q of the tuned circuit is an important consideration in a class C amplifier. Remember that the bandwidth (BW) of a tuned circuit and its Q are related by the expression

$$\text{BW} = \frac{f_r}{Q} \qquad Q = \frac{f_r}{\text{BW}}$$

The Q of the tuned circuit in the class C amplifier should be selected so that it provides adequate attenuation of the harmonics but also has adequate bandwidth to pass the sidebands produced by the modulation process. If the Q of the tuned circuit is too high, the bandwidth will be very narrow and some of the higher-frequency sidebands will be eliminated. This will cause a form of frequency distortion called *sideband clipping* and may make some signals unintelligible or will at least limit the fidelity of reproduction.

One of the main reasons why class C amplifiers are preferred for RF power amplification over class A and class B amplifiers is that class C amplifiers have high efficiency. Because the current flows for less than 180° of the ac input cycle, the average current in the transistor is fairly low, meaning that the power dissipated by the device is low. A class C amplifier functions almost as a transistor switch that is off for over 180° of the input cycle. The switch conducts for approximately 90° to 150° of the input cycle. During the time that it conducts, its emitter-to-collector impedance is very low. Even though the peak current may be high, the total power dissipation is much less than that

in class A and class B circuits. For this reason, more of the dc power is converted to RF energy and passed on to the load, usually an antenna. The efficiency of most class C amplifiers is in the 60 percent to 85 percent range.

In discussing power in a class C amplifier, there is input power and output power. The input power is the average power consumed by the circuit. This is simply the product of the supply voltage and the average collector current.

$$P_{\text{in}} = V_{cc} I_C$$

For example, if the supply voltage is 13.5 V and the average dc collector current is 0.7 A, the input power is

$$P_{\text{in}} = 13.5(0.7) = 9.45 \text{ W}$$

The output power, of course, is the power actually transmitted to the load. The actual amount of power depends upon the efficiency of the amplifier. The output power can be computed with the familiar power expression

$$P_{\text{out}} = \frac{V^2}{R_L}$$

Here V is the RF output voltage at the collector of the amplifier, and R_L is the load impedance. When a class C amplifier is set up and operating properly, the peak-to-peak RF output voltage is 2 times the supply voltage, or $2V_{cc}$.

A special form of class C amplifier is the frequency multiplier. Any class C amplifier is capable of performing frequency multiplication if the tuned circuit in the collector resonates at some integer multiple of the input frequency.

For example, a frequency doubler can be constructed by simply connecting a parallel tuned circuit in the collector of a class C amplifier that resonates at twice the input frequency. When the collector current pulse occurs, it excites or rings the tuned circuit at twice the input frequency. A current pulse flows for every other cycle of the input.

A tripler circuit is constructed in the same way except that the tuned circuit resonates at 3 times the input frequency. In this way, the tuned circuit receives one input pulse for every three cycles of oscillation it produces. Multipliers can be constructed to increase the input frequency by any integer factor up to approximately 10. As the multiplication factor gets higher, the power output of the multiplier decreases. For most practical applications, the best result is obtained with multipliers of 2 and 3.

Another way to look at the operation of class

Fig. 6-14 **Frequency multiplication with class C amplifiers.**

C frequency multipliers is to remember that the nonsinusoidal current pulse is rich in harmonics. Each time the pulse occurs, the second, third, fourth, fifth, and higher harmonics are generated. The purpose of the tuned circuit in the collector is to act as a filter to select the desired harmonic.

In many applications, a multiplication factor greater than that achievable with a single multiplier stage is required. In such cases, two or more multipliers are cascaded. Figure 6-14 shows two multiplier examples. In the first case, multipliers of 2 and 3 are cascaded to produce an overall multiplication of 6. In the second example, three multipliers provide an overall multiplication of 30. The total multiplication factor is simply the product of the individual stage multiplication factors.

Self-Test

Answer the following questions.

12. Linear power amplifiers are used to raise the power level of _____ and _____ signals.
13. A _____ power amplifier is used to increase the power level of an FM signal.
14. Linear power amplifiers operate class _____ and _____ .
15. A class A transistor power amplifier has

an efficiency of 50 percent. The output power is 27 W. The power dissipated in the transistor is _____ W.
16. Class A amplifiers conduct for _____ degrees of a sine wave input.
17. The cause of self-oscillation in a transistor RF amplifier is _____ .
18. Self-oscillation in an RF amplifier that does not occur at the tuned frequency of the amplifier is called _____ .
19. Self-oscillation of an RF amplifier can be eliminated by the process of _____ .
20. True or false. Elimination of self-oscillation can be accomplished with either a feedback inductor or a capacitor.
21. Combining a feedback capacitor with the internal capacitance of the transistor forms a(n) _____ circuit which is used to cancel the self-oscillation.
22. True or false. With no input, a class B amplifier does not conduct.
23. Class B RF power amplifiers normally used a(n) _____ configuration.
24. A class C amplifier conducts for approximately _____ degrees to _____ degrees of the input signal.
25. In a class C amplifier, collector current flows in the form of _____ .
26. In a class C amplifier, a complete sinusoidal output signal is produced by a(n) _____ .

27. The efficiency of a class C amplifier is in the range of _____ to _____ percent.
28. The tuned circuit in the collector of a class C amplifier acts as a filter to eliminate _____ .
29. A class C amplifier whose output tuned circuit resonates at some integer multiple of the input frequency is called a(n) _____ .
30. Frequency multipliers with factors of 2, 3, 4, and 5 are cascaded. The input is 1.5 MHz. The output is _____ MHz.
31. A class C amplifier has a dc supply voltage of 28 V and an average collector current of 1.8 A. The power input is _____ W.

6-3 IMPEDANCE-MATCHING NETWORKS

One of the most important parts of any transmitter is the matching networks that connect one stage to another. In a typical transmitter, the oscillator generates the basic carrier signal which is then amplified usually by multiple stages before reaching the antenna. Since the idea is to increase the power of the signal, the interstage coupling circuits must permit an efficient transfer of power from one stage to the next. Finally, some means must be provided to connect the final amplifier stage to the antenna, again for the purpose of transferring the maximum possible amount of power.

The circuits used to connect one stage to another are known as *impedance-matching networks*. In most cases, they are either *LC* circuits, transformers, or some combination of the two. The basic function of a matching network is to provide for an optimum transfer of power through impedance-matching techniques.

Another important function of the matching network is to provide filtering and selectivity. Transmitters are designed to operate on a single frequency or on selectable narrow ranges of frequencies. The various amplifier stages in the transmitter must confine the RF generated to these frequencies. In class C amplifiers, a considerable number of high-amplitude harmonics are generated. These must be eliminated to prevent spurious radiation from the transmitter. The impedance-matching networks used for interstage coupling also provide this desired filtering and selectivity.

The basic problem of coupling is illustrated in Fig. 6-15(*a*). The driving stage appears as a signal source with an internal impedance of Z_i.

(a) Ideal case: $Z_i = Z_l$.

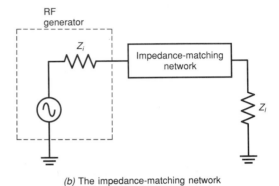

(b) The impedance-matching network makes Z_l "look like" Z_i.

Fig. 6-15   Impedance matching in RF circuits.

The stage being driven represents a load to the generator with its internal resistance of Z_l. Ideally, Z_i and Z_l are resistive. As you recall, maximum power transfer takes place when Z_i equals Z_l. Although this basic relationship is essentially true in RF as in dc circuits, it is a much more complex relationship. In RF circuits, Z_i and Z_l are seldom purely resistive and, in fact, usually include a reactive component of some type. Further, it is not always necessary to transfer maximum power from one stage to the next. The goal is to transfer a sufficient amount of power to the next stage so that it can provide the maximum amount of output of which it is capable.

In most cases, the two impedances involved are considerably different from one another, and, therefore, a very inefficient transfer of power takes place. To overcome this problem, an impedance-matching network is introduced between the two as illustrated in Fig. 6-15(*b*). This impedance-matching network is usually an *LC* circuit or transformer as indicated before.

One of the simplest forms of impedance-matching network is the L network. It consists of an inductor and a capacitor connected in various L-shaped configurations as shown in Fig. 6-16. The circuits shown in Fig. 6-16(*a*) and (*b*) are low-pass filters, while those shown in Fig. 6-16(*c*) and (*d*) are high-pass filters.

(a) $Z_l < Z_i$

(b) $Z_l > Z_i$

(c) $Z_l < Z_i$

(d) $Z_l > Z_i$

Fig. 6-16   Four L-type impedance-matching networks.

Typically, the low-pass networks are used so that harmonic frequencies are filtered out.

By proper design of the L-matching network, the load impedance can be "matched" to the source impedance. For example, the network in Fig. 6-16(a) causes the load resistance

to appear larger than it actually is. The load resistance Z_l appears in series with the inductor of the L network. The inductor and the capacitor are chosen to resonate at the transmitter frequency. When the circuit is at resonance, X_L equals X_C. To the generator impedance (Z_i), the complete circuit simply appears as a parallel resonant circuit. At resonance, the impedance represented by the circuit is very high. The actual value of the impedance depends upon the L and C values and the Q of the circuit. The higher the Q, the higher the impedance. The Q in this circuit is basically determined by the value of the load impedance. By proper selection of the circuit values, the load impedance can be made to appear as any desired value to the source impedance as long as Z_i is greater than Z_l.

By using the L network shown in Fig. 6-16(b), the impedance can be stepped down. In other words, the load impedance can be made to appear as if it were much smaller than it actually is. With this arrangement, the capacitor is connected in parallel to the load impedance. The parallel combination of C and Z_l can be transformed to appear as an equivalent series RL combination. The result is that the overall network appears as a series resonant circuit. Recall that a series resonant circuit has a very low impedance at resonance. The impedance is, in fact, the equivalent load impedance.

The design equations for L networks are given in Fig. 6-17 on the next page. We assume that the internal source and load impedances are resistive, so $Z_i = R_i$ and $Z_l = R_l$. The network of Fig. 6-17(a) assumes that $R_l < R_i$, while the network of Fig. 6-17(b) assumes that $R_i < R_l$.

Suppose we wish to match a 5-Ω transistor amplifier impedance to a 50-Ω antenna load at 120 MHz. In this case, $R_i < R_l$, so we use the formulas in Fig. 6-17(b).

$$X_L = \sqrt{R_i R_l - R_i^2}$$
$$= \sqrt{5(50) - 5^2}$$
$$= \sqrt{225}$$
$$= 15 \ \Omega$$

$$Q = \sqrt{\frac{R_l}{R_i} - 1}$$
$$= \sqrt{10 - 1} = \sqrt{9}$$
$$= 3$$

$$X_c = \frac{R_i R_l}{X_L} = \frac{5(50)}{15} = \frac{250}{15} = 16.67 \ \Omega$$

$$X_L = \sqrt{R_i R_l - (R_i)^2}$$

$$Q = \sqrt{\frac{R_l}{R_i} - 1}$$

$$X_c = \frac{R_i R_l}{X_L}$$

(a) $R_i < R_l$

$$X_L = \sqrt{R_i R_l - (R_i)^2}$$

$$Q = \sqrt{\frac{R_l}{R_i} - 1}$$

$$X_c = \frac{R_i R_l}{X_L}$$

(b) $R_i < R_l$

Fig. 6-17 The L network design equations.

To find the values of L and C at 120 MHz, we rearrange the basic reactance formulas.

$$X_L = 2\pi f L$$

$$L = \frac{X_L}{2\pi f} = \frac{15}{6.28 \times 120 \times 10^6} = 20 \text{ nH}$$

$$X_c = \frac{1}{2\pi f C}$$

$$C = \frac{1}{2\pi f X_c} = \frac{1}{6.28 \times 120 \times 10^6 \times 16.67}$$

$$= 80 \text{ pF}$$

In most cases, the internal impedance and load impedances are not purely resistive. In-ternal and stray reactances make the imped-ances complex. Figure 6-18 shows an example. Here the internal resistance is 5 Ω, but it in-cludes an internal inductance L_i of 8 nH. There is a stray capacitance C_l of 12 pF across the load. The way to deal with these reactances is simply to combine them with the L network values. In the example above, the calculation calls for an inductance of 20 nH. Since the stray inductance is in series with the L network inductance in Figure 6-18, naturally the values will add. As a result, the L network inductance can be less than the computed value by an amount equal to the stray inductance of 8 nH.

$$L = 20 - 8$$

$$= 12 \text{ nH}$$

Fig. 6-18 Incorporating the internal and stray reactances into the matching network.

By making the L network inductance 12 nH, the total circuit inductance will be correct when it adds to the stray inductance.

A similar thing occurs with the capacitance. The circuit calculations above call for a total of 80 pF. The L network capacitance and the stray capacitance will add as they are in parallel. Therefore, the L network capacitance can be less than the calculated value by the amount of the stray capacitance or

$$C = 80 - 12$$
$$= 68 \text{ pF}$$

Making the L network capacitance 68 pF gives the total correct capacitance when it is added to the stray capacitance.

Although L networks are widely used in impedance matching, they are inflexible with regard to their selectivity. When designing L networks, there is very little control over the Q of the circuit. The Q of the circuit is defined by the values of the internal and load impedances. The resulting Q is what you get but may not always be what you need to achieve the desired selectivity.

To overcome this problem, matching networks using three reactive elements can be used. The three most widely used impedance-matching networks containing three reactive components are illustrated in Fig. 6-19 on the next page. The network in Fig. 6-19(a) is known as a pi network because its configuration resembles the Greek symbol pi. The circuit in Fig. 6-19(b) is known as a T network, again because the circuit elements essentially resemble the letter T. The circuit in Fig. 6-19(c) is a T network using two capacitors. Note that all are low-pass filters which provide maximum harmonic attenuation. The pi and T networks can be designed to either step up or step down the impedances as required by the circuit. The capacitors are usually made variable so that the circuit can be tuned to resonance and adjusted for maximum power output.

The most popular circuit is the T network of Fig. 6-19(c). Often called an *LCC* network, it is widely used to match the low output impedance of a transistor power amplifier to the higher impedance of another amplifier or an antenna. The design formulas are given in Fig. 6-20 on page 127. Using our previous example of matching a 5-Ω source R_i to a 50-Ω load R_l at 120 MHz, let's calculate the *LCC* network. Assume a Q of 10. First calculate the inductance.

$$X_L = QR_i = 10(5) = 50 \ \Omega$$
$$L = \frac{X_L}{2\pi f} = \frac{50}{6.28(120 \times 10^6)} = 66.3 \text{ nH}$$

Next, calculate C_1.

$$X_{C_1} = R_l \sqrt{\frac{R_i(Q^2 + 1)}{R_l} - 1}$$
$$= 50 \sqrt{\frac{5(10^2 + 1)}{50} - 1}$$
$$= 50 \sqrt{\frac{505}{50} - 1} = 50\sqrt{9.1} = 150 \ \Omega$$
$$C_1 = \frac{1}{2\pi f X_c} \frac{1}{6.28(120 \times 10^6)(150)}$$
$$= 8.8 \times 10^{-12} = 8.8 \text{ pF}$$

Finally, calculate C_2.

$$X_{C_2} = \frac{R_i(Q^2 + 1)}{Q} \times \frac{1}{1 - X_{C_1}/QR_l}$$
$$= \frac{5(10^2 + 1)}{10} \times \frac{1}{1 - 150/(10 \times 50)}$$
$$= 50.5 \frac{1}{1 - 0.3} = 50.5(1.43) = 72 \ \Omega$$
$$C_2 = \frac{1}{2\pi f X_c} = \frac{1}{6.28(120 \times 10^6)(72)}$$
$$= 18.4 \text{ pF}$$

One of the best impedance-matching components is the transformer. Recall that iron-core transformers are widely used at lower frequencies to match one impedance to another. Any load impedance can be made to look like a desired load impedance if the correct value of the transformer turns ratio is selected.

Refer to Fig. 6-21 on page 127. The relationship between the turns ratio and the input and output impedances is

$$\frac{Z_i}{Z_l} = \left(\frac{N_p}{N_s}\right)^2$$

or

$$\frac{N_p}{N_s} = \sqrt{\frac{Z_i}{Z_l}}$$

What this formula says is that the ratio of the input impedance Z_i to the load impedance Z_l is equal to the square of the ratio of the number of turns on the primary N_p to the number of turns on the secondary N_s. As an example, to match a 5-Ω generator impedance to a 50-Ω load impedance, the turns ratio would be

$$\frac{N_p}{N_s} = \sqrt{\frac{Z_i}{Z_l}} = \sqrt{\frac{5}{50}} = \sqrt{0.1}$$

$$\frac{N_p}{N_s} = 0.316 \quad or \quad \frac{N_s}{N_p} = \frac{1}{3.16} = 0.316$$

This means that there are 3.16 times as many turns on the secondary as on the primary.

This relationship holds true only on iron-core transformers. When air-core transformers are used, the coupling between primary and secondary windings is not complete and, therefore, the impedance ratio is not as indicated. Although air-core transformers are widely used at RF frequencies and can be used for impedance matching, they are less efficient than iron-core transformers.

Special types of core materials have been created for use at very high frequencies so that

(a) Pi (π) network

(b) T network

(c) Two-capacitor T network

Fig. 6-19 Three-element matching networks.

T network

$Z_i = R_i$ X_L X_{C_1}

X_{C_2}

$Z_l = R_l$

$R_i < R_l$

Design Procedure:
1. Select a desired circuit Q
2. Calculate $X_L = QR_i$
3. Calculate X_{C_1}:

$$X_{C_1} = R_i \sqrt{\frac{R_l(Q^2 + 1)}{R_i} - 1}$$

4. Calculate X_{C_2}:

$$X_{C_2} = \frac{R_l(Q^2 + 1)}{Q} \times \frac{1}{\left(1 - \frac{X_{C_1}}{QR_l}\right)}$$

5. Compute final L and C values:

$$L = \frac{X_L}{2\pi f}$$

$$C = \frac{1}{2\pi f X_C}$$

Fig. 6-20   Design equations for an *LCC* T network.

RF generator

Z_i

Primary   Secondary Z_l

Turns ratio $= \dfrac{N_p}{N_s}$

Fig. 6-21   Impedance matching with an iron-core transformer.

iron-core transformers can be made. The core material is a ferrite or powdered iron. Both the primary and secondary windings are wound on a core of this material.

The most widely used core for RF transformers is the popular toroid. A *toroid* is a circular, doughnut-shaped core usually made of a special type of powdered iron. Copper wire is wound on the toroid to create the primary and secondary windings. A typical arrangement may appear like that shown in Fig. 6-22.

Powdered iron toroid

Primary   Secondary

Fig. 6-22   A toroid transformer.

Single-winding tapped coils called *autotransformers* are also used for impedance matching between RF stages. Figure 6-23 shows both impedance step-down and step-up arrangements. Toroids are commonly used.

Unlike air-core transformers, toroid transformers cause the magnetic field produced by

Z_i

N_p N_s Z_l

$$\frac{N_p}{N_s} = \sqrt{\frac{Z_i}{Z_l}}$$

(a) Step-down

Z_i

N_s Z_l

N_p

$$\frac{N_p}{N_s} = \sqrt{\frac{Z_i}{Z_l}}$$

(b) Step-up

Fig. 6-23   Impedance matching in an autotransformer.

CHAPTER 6   RADIO TRANSMITTERS   **127**

the primary to be completely contained within the core itself. This has several important advantages. First, a toroid will not radiate RF energy. Air-core coils radiate because the magnetic field produced around the primary is not contained. Transmitter and receiver circuits using air-core coils must be contained by magnetic shields to prevent them from interfering with other circuits. The toroid, on the other hand, confines the entire magnetic field, and, therefore, no shields are necessary.

Another benefit is that most of the magnetic field produced by the primary cuts the turns of the secondary winding. For this reason, the basic turns ratio, input-output voltage, and impedance formulas for standard low-frequency transformers also apply to high-frequency toroid transformers.

In most new RF designs, toroid transformers are used for RF impedance matching between stages. Further, the primary and secondary windings are sometimes used as inductors in tuned circuits. Alternately, toroid inductors can also be built. Toroid inductors have an advantage over air-core inductors for RF applications; that is, the high permeability of the core causes the inductance to be high. Recall that whenever an iron core is inserted into a coil of wire, the inductance will increase dramatically. For RF applications, what this means is that the desired values of inductance can be created by using fewer turns of wire. This results in smaller inductors. Further, fewer turns have less resistance, thereby giving the coil a higher Q than that obtainable with air-core coils.

Powdered-iron toroids are so effective that they have virtually replaced air-core coils in most modern transmitter designs. They are available in sizes from a fraction of an inch to several inches in diameter. In most applica-

tions, a minimum number of turns are required to create the desired inductance.

Figure 6-24 shows a toroid transformer used for interstage coupling between two class C amplifiers. The primary of the driver transformer is tuned to resonance by capacitor C_1. The capacitor is adjustable so that the exact frequency of operation can be set. The relatively high output impedance of the transistor is coupled with the low input impedance of the next class C stage by a step-down transformer which provides the desired impedance-matching effects. Usually, the secondary winding is only a few turns of wire and is not resonated. The circuit in Fig. 6-24 also shows a similar transformer used for output coupling to the antenna.

Transformers can be connected in unique ways to provide fixed impedance-matching characteristics over a wide range of frequencies. One of the most widely used configurations is shown in Fig. 6-25. This is a transformer usually wound on a toroid where the number of primary and secondary turns are equal giving the transformer a 1:1 turns ratio and a 1:1 impedance-matching ratio. The dots indicate the phasing of the windings. Note the unusual way in which the windings are connected. A transformer connected in this way is generally known as a *balun*. The name balun is derived from the first letters of the words balanced and unbalanced since such transformers are normally used to connect a balanced source to an unbalanced load or vice versa. In the circuit of Fig. 6-25(*a*), a balanced generator is connected to an unbalanced (grounded) load. In Fig. 6-25(*b*), an unbalanced (grounded) generator can be connected to a balanced load.

Figure 6-26 shows how a 1:1 turns ratio balun can be used for impedance matching. With the arrangement shown in Fig. 6-26(*a*),

Fig. 6-24 Using toroid transformers for coupling and impedance matching in class C amplifier stages.

$Z_i = Z_l$

(a) Balanced to unbalanced

$Z_i = Z_l$

(b) Unbalanced to balanced

Fig. 6-25 Balun transformers used for connecting balanced and unbalanced loads or generators.

an impedance step-up is obtained. A load impedance of 4 times the source impedance Z_i provides a correct match. The balun makes the load of 4 Z_i "look like" Z_i. In Fig. 6-26(*b*), an impedance step down is obtained. The balun makes the load Z_l "look like" $4Z_i$.

Many other balun configurations are possible with different impedance ratios. Several common 1:1 baluns can be interconnected for both 9:1 and 16:1 impedance transformation ratios. In addition, baluns may also be cascaded so that the output of one appears as the input to the other, and so on. By cascading baluns, impedances may be stepped up or stepped down by wider ratios.

An important point to note here is that the windings in a balun are not resonated with capacitors to a particular frequency. As a result, they operate over a wide frequency range. The winding inductances are made such that the coil reactances are 4 or more times that of the highest impedance being matched. In this way, the transformer will provide the designated impedance matching over a tremendous range of frequencies.

This broadband characteristic of balun transformers allows designers to construct broad-

band RF power amplifiers. Such amplifiers provide a specific amount of power amplification over a wide bandwidth. These amplifiers are preferred particularly in communications equipment that must operate in more than one frequency range. Rather than having one transmitter for each desired band, a single transmitter can be used.

When using conventional tuned amplifiers, some method of switching the correct tuned circuit into the circuit must be provided. Such switching networks are complex and expensive. Further, they introduce problems, particularly at the high frequencies. In order for them to perform effectively, the switches must be located very close to the tuned circuits so that stray inductances and capacitances are not introduced by the switch and the interconnecting leads.

One way to overcome the switching problem is simply to use a broadband amplifier. No switching or tuning is needed. The broadband amplifier provides the necessary amplification as well as impedance matching.

$Z_l = 4Z_i$

(a) Impedance step-up

$Z_l = \dfrac{Z_i}{4}$

$Z_l = \dfrac{Z_i}{4}$
or
$Z_i = 4Z_l$

(b) Impedance step-down

Fig. 6-26 Using a balun for impedance matching.

The primary problem with the broadband amplifier is that it does not provide the filtering necessary to get rid of harmonics. One way to overcome this problem is to generate the desired frequency at a lower power level allowing tuned circuits to filter out the harmonics and then to provide final power amplification with the broadband circuit. The broadband power amplifier operates either a class A or class B push-pull so that the inherent harmonic content of the output is very low.

Figure 6-27 shows a typical broadband linear amplifier. Note that two 4:1 balun transformers are cascaded at the input so that the low base input impedance is made to look like an impedance 16 times higher than the input. The output uses a 1:4 balun that steps up the very low output impedance of the final amplifier to an impedance 4 times higher to equal the antenna load impedance. In some transmitters, broadband amplifiers may be followed by low-pass filters which are used to eliminate undesirable harmonics in the output.

Self-Test

Answer the following questions.

32. Maximum power transfer occurs when the generator impedance _____ the load impedance.
33. A commonly used two-element (*LC*) circuit called a(n) _____ is widely used for impedance matching.
34. List the two popular types of three-element (*LC*) matching networks.

35. Most impedance-matching networks are _____ filters, so they eliminate harmonics.
36. The main advantage of pi and T networks over the L network is that the circuit _____ can be chosen.
37. True or false. Most impedance-matching networks are made adjustable (tunable).
38. Calculate the L network components to match a 3-Ω internal resistance in series with an internal inductance of 19 nH to a 72-Ω load impedance in parallel with a stray capacitance of 32 pF at a frequency of 18 MHz.
39. Calculate the components of an *LCC* T network to match a 4-Ω internal resistance to a 52-Ω load at 72 MHz. Assume a *Q* of 12.
40. A transformer has 6 turns on the primary and 18 turns on the secondary. If the generator (source) impedance is 50 Ω, the load impedance should be _____ Ω.
41. A transformer must match a 200-Ω generator to a 50-Ω load. The turns ratio must be _____ .
42. A single-winding transformer is known as a(n) _____ .
43. A doughnut-shaped powdered-iron core widely used to make RF transformers and inductors is called a(n) _____ .
44. An RF transformer with a 1:1 turns ratio called a(n) _____ is connected in a special way to provide 1:4 or 4:1 impedance matching.
45. Untuned RF transformers permit _____ operation over a wide frequency range.

Fig. 6-27   A broadband class A linear power amplifier.

6-4 SPEECH PROCESSING

Speech processing refers to the ways that the voice signal used in communications is dealt with prior to being applied to the modulator. The voice signal from the microphone is usually amplified and applied to circuits that deliberately limit its amplitude and frequency response.

The primary purpose of speech-processing circuits is to ensure that overmodulation does not occur and to restrict the bandwidth of the signal. In AM, an excessively high voice signal will cause overmodulation and severe distortion of the signal. If the amplitude is deliberately limited, overmodulation will not occur.

In an AM transmitter, the distortion caused by overmodulation causes harmonics of the primary frequencies in the voice signal to be generated. As a result, these harmonics will also modulate the carrier and produce sidebands that extend far beyond the assigned bandwidth of the AM signal. These harmonics may interfere with signals on adjacent channels. This interference is generally known as *splatter*. Excessive modulation causes splatter which, in turn, produces interference to other signals. The distortion also causes the signal to be generally less intelligible.

Most transmitters are designed to operate in a specific frequency range. This means that the sidebands may not extend beyond certain assigned limits. In AM transmitters, it is the highest modulating frequency that determines the total bandwidth of the AM signal. For example, in AM broadcasting, the channel width is limited to 10 kHz. For this reason, the upper audio frequency must be limited to one-half of this, or 5 kHz. Speech-processing circuits are designed to limit both the upper amplitude and frequency.

Another purpose of speech processing in AM and SSB transmitters is that it helps keep the average transmitted power higher. Recall that when an AM transmitter is modulated, the power from the modulator which appears in the sidebands increases with the percentage of modulation. The more power transmitted in the sidebands, the greater the transmission distance is and the more reliable communications are. In other words, the strength of the RF signal is determined by the strength of the modulating signal. The same is true of SSB signals. The strength of the modulating signal determines the strength of the transmitted sideband.

Voice signals vary over an extreme amplitude range. It has been determined in the study of speech that the highest peaks of intensity come from vowel sounds. On the other hand, vowel sounds tend to do nothing for intelligibility. As it turns out, it is the consonants, such as B, K, L, S, T, and V, that provide the best detail for good intelligibility. Yet consonant sounds are typically spoken at a much lower amplitude level than vowel sounds. If the vowel sounds could be de-emphasized while consonant sounds were emphasized through increased amplification, the intelligibility of the signal would improve tremendously. At the same time, the total average voice signal would be higher and would, therefore, provide a higher transmitted power. Special-speech processing circuits have been developed to provide for increased amplification of low-level audio content while at the same time limiting the peaks. This process is generally known as *dynamic compression*.

In FM, it is also necessary to deliberately limit the bandwidth and amplitude of the modulating signal. Recall that the frequency deviation of an FM carrier is directly proportional to the amplitude of the modulating signal. Most FM transmitters are allowed to operate up to a certain maximum deviation limit. In mobile communications transmitters, this frequency deviation is ±5 kHz. In order to ensure that the deviation limit is not exceeded, some means must be provided to prevent the modulating signal amplitude from causing overdeviation. This is comparable to overmodulation in AM transmitters.

Additionally, the frequency content of the modulating signal must also be limited. Recall that an FM signal produces multiple sidebands above and below the carrier. These are spaced from the carrier by the frequency of the modulating signal. The higher the modulating signal frequency, the wider the spacing of the sidebands and the greater the spectrum space occupied by the FM signal. As in most applications, the allowed bandwidth of an FM signal is specified. For this reason, the upper frequencies of the modulating signal are usually limited to prevent the bandwidth from becoming too wide.

The basic speech-processing circuit is made up of an amplifier operating in conjunction with a clipper circuit that will limit the amplitude

Overmodulation

Splatter

Compression

Clipper

Fig. 6-28   A basic speech-processing circuit incorporating clipping and filtering.

swings of the modulating signal and a filter that cuts off modulating frequencies above a certain point. Such a basic circuit is illustrated in Fig. 6-28. The low-level signal from the microphone is amplified by two stages of audio amplification to raise the signal to a high enough amplitude to make it compatible with the modulator. The output of the amplifier is applied to a diode clipper. Diodes D_1 and D_2 are low-level silicon diodes that conduct when approximately +0.7 V is applied across them. Notice the diodes are connected in an inverse parallel arrangement; that is, one diode is connected to conduct on positive-going ac signals, while the other will conduct on negative-going ac signals.

As long as the peak-to-peak amplitude of the voice signal is below approximately 1.4 V peak-to-peak, no clipping will occur. However, if voice peaks should cause the signal to be greater, diode D_1 or D_2 will conduct, thereby clipping off the peak and holding the amplitude to ±0.7 V. Regardless of the amplitude of the amplifier output, the signal across the diodes will never be greater than 1.4 V peak-to-peak. Thus with the amplitude deliberately limited, an AM transmitter will not be overmodulated and an FM transmitter will not be overdeviated.

Clipping, however, does cause signal distortion. The deliberate flattening of the signal peaks introduces harmonics. One way to get rid of these harmonics is to pass the clipped signal through a low-pass filter. In Fig. 6-27 this low-pass filter is made up of resistors R_1 and R_2 and capacitors C_1 and C_2. This filter gets rid of those harmonics and thus compensates for the clipping distortion.

The other purpose of the low-pass filter is to deliberately restrict the modulating signal bandwidth. Typically, the cutoff frequency of the filter is chosen so that it is somewhere in the 2.5- to 3-kHz range. This is the upper limit for most two-way communications systems. As you can see then, the filter has a dual purpose:

restricting the bandwidth of the signal and reducing the distortion caused by clipping.

Such circuits are almost always incorporated in FM transmitters and in many AM transmitters as well. However, in AM and SSB transmitters, speech compression is generally preferred over clipping. Speech compression permits the average modulating signal to be higher, thus increasing the average power output. However, this technique is not applicable to FM.

Figure 6-29 shows a typical speech-compression circuit. It consists of two stages of amplification whose gain is controlled automatically by the amplitude of the audio signal. Some of the audio signal is tapped off and applied to a diode rectifier D_1. The audio signal is rectified and filtered by C_2 to produce a dc output. It is amplified by a dc amplifier. The average dc output is proportional to the amplitude of the modulating signal. This dc voltage is fed back to an earlier amplifier stage to control its gain. The gain of a transistor amplifier stage can be controlled by adjusting collector current. This is done by modifying the base bias current. The higher the collector current, the higher the gain. If the modulating signal amplitude is high, the dc output from the rectifier will be high and it, in turn, will reduce the gain of the amplifier thereby limiting the peaks. In this regard, the compression cir-

Fig. 6-29   A simple speech-compression circuit.

cuit effectively limits the peaks, preventing over-modulation and splatter.

If the amplitude of the modulating signal is very low, then the rectifier will produce a very low average dc output. This causes the amplifier gain to be very high, thereby providing extra amplification for low-level signals. This compression circuit is said to have *automatic gain control (AGC)* simply because the gain of the circuit automatically adjusts itself to a level appropriate to the amplitude of the incoming signal. High-level amplitude signals are deliberately limited, while low-level amplitude signals are provided with extra amplification.

The clipping and compression can also be performed in the RF stages of the transmitter rather than in the modulating signal stages. A common technique is to use RF compression for this purpose. This is usually done in SSB transmitters and is commonly referred to as *automatic level control (ALC)*. The technique for RF compression is similar to that used in audio compression. A rectifier circuit is used to tap off a sample of the RF signal at the final power amplifier stage. The RF signal is rectified and converted into a dc voltage that is fed back to control the gain of an earlier stage used to amplify the SSB signal. Again, if the output level should exceed a certain limit, the fed-back dc voltage will reduce the gain of the stage and thus automatically control the amplitude. Low-level signals will receive full amplification.

Self-Test

Answer the following questions.

46. The two techniques used in speech processing are _____ and _____ .
47. Name the two primary purposes of speech processing.
48. A diode _____ circuit is frequently used to restrict voice signal amplitude.
49. Name the two main purposes of a low-pass filter in a speech-processing circuit.
50. Overmodulation in an AM transmitter causes distortion and out-of-band operation which is referred to as _____ .
51. Audio compression is achieved by controlling the _____ of an audio amplifier.
52. Audio compression is used to increase the average _____ in an AM or SSB signal.
53. _____ and _____ circuits are used in a speech compressor to create a dc feedback control signal.
54. The gain of a transistor can be controlled by varying the _____ .
55. The process of adjusting the amplification of a signal based on its amplitude is known as _____ .
56. In an audio compressor, do low-level signals receive more or less amplification?
57. True or false. Speech processing may be done on the RF signal or the audio signal.

SUMMARY

1. A radio transmitter generates the carrier signal, provides power amplification, and applies modulation.
2. The simplest transmitter is an oscillator that is keyed off and on to produce cw Morse code.
3. Most transmitters consist of a crystal oscillator used to generate an accurate and stable carrier frequency, a buffer amplifier that isolates the carrier oscillator from its load, one or more driver amplifiers to increase the RF power level, and a final power amplifier that applies the signal to the antenna.

4. In AM transmitters, class C amplifiers are used to increase the RF power level. The final stage is modulated.
5. Frequency modulation transmitters use class C amplifiers to increase the RF power level.
6. Single-sideband transmitters generate the carrier and modulation at a low frequency and then translate it up with a mixer. Linear amplifiers are used to increase the power level.
7. In a class A amplifier, collector current flows continuously. The output is directly proportional to the input; therefore, it is a linear amplifier.

8. Class B amplifiers are biased at cutoff so that collector current flows for 180° of the input cycle.
9. Class B amplifiers are usually connected in a push-pull circuit where one transistor amplifies each half of the input signal.
10. Class C amplifiers are biased beyond cutoff. Current flows for 90° to 150° of the input cycle.
11. Collector current pulses in a class C amplifier are converted into a continuous sine wave by a resonant circuit.
12. The collector current pulses in a class C amplifier contain many harmonics which are filtered out by the tuned output circuit.
13. A class C amplifier can be used as a frequency multiplier by connecting a resonant circuit tuned to some integer multiple of the input frequency in the output.
14. RF amplifiers may oscillate because of feedback from internal transistor capacitance. This can be eliminated, or prevented, by neutralization, a process that cancels the feedback with out-of-phase feedback.
15. Frequency multipliers can be cascaded to produce higher output frequencies.
16. In an FM transmitter, the frequency multipliers increase the deviation as well as the carrier frequency.
17. Linear amplifiers operating class A or B are used to increase the power level of low-level AM and SSB signals.
18. Impedance-matching networks are used to interconnect RF amplifiers and to couple power to the antenna to ensure the optimum transfer of power.
19. Common impedance-matching circuits and components include LC L networks, LC pi networks, LC T networks, transformers, and baluns. The pi and T networks are preferred since Q can be controlled.

20. Maximum power transfer occurs when the load impedance equals the generator source impedance.
21. Radio-frequency transformers are typically constructed with doughnut-shaped powdered-iron cores called toroids.
22. The impedance-matching ability of a transformer is determined by its turns ratio: $(N_p/N_s)^2 = Z_p/Z_s$.
23. Inductors made with toroids for a given inductance are smaller, use fewer turns of wire, have a higher Q, and do not require shielding.
24. A balun is a transformer connected in a special way to transform circuits from balanced to unbalanced or vice versa and to provide impedance matching.
25. Toroid transformers and baluns are broadband devices that operate over a wide bandwidth.
26. Broadband, linear, untuned RF power amplifiers provide amplification over a broad frequency range.
27. Speech-processing circuits in a transmitter prevent overmodulation, prevent excessive signal bandwidth, and increase the average transmitted power in AM and SSB systems.
28. A voice clipper uses diodes to limit the amplitude of the audio modulating signal. A low-pass filter smoothes out any clipping distortion and prevents excessive sidebands.
29. Voice compressors use automatic-gain-control (AGC) circuits to limit the audio amplitude. The gain of the circuit is inversely proportional to the audio signal amplitude.
30. In AGC circuits, a rectifier and filter convert the audio or RF into a dc voltage that controls the gain of an audio or RF amplifier to prevent overmodulation.
31. The gain of a transistor amplifier can be varied by changing the collector current.

CHAPTER REVIEW QUESTIONS

Choose the letter that best answers each question.

6-1. Which of the following circuits is *not* typically part of every radio transmitter?
a. Carrier oscillator
b. Driver amplifier
c. Mixer
d. Final power amplifier

6-2. Class C amplifiers are *not* used in which type of transmitter?
    a. AM   *c.* CW
    b. SSB   *d.* FM

6-3. A circuit that isolates the carrier oscillator from load changes is called a
    a. Final amplifier.   *c.* Linear amplifier.
    b. Driver amplifier.   *d.* Buffer amplifier.

6-4. A class B amplifier conducts for how many degrees of an input sine wave?
    a. $90°$ to $150°$ *c.* $180°$ to $360°$
    b. $180°$ *d.* $360°$

6-5. Bias for a class C amplifier produced by an input RC network is known as
    a. Signal bias.   *c.* Fixed external bias.
    b. Self bias.   *d.* Threshold bias.

6-6. An FM transmitter has a 9-MHz crystal carrier oscillator and frequency multipliers of 2, 3, and 4. The output frequency is
    a. 54 MHz.   *c.* 216 MHz.
    b. 108 MHz.   *d.* 288 MHz.

6-7. The most efficient RF power amplifier is which class amplifier?
    a. A   *c.* B
    b. AB   *d.* C

6-8. Collector current in a class C amplifier is a
    a. Sine wave.   *c.* Pulse.
    b. Half sine wave.   *d.* Square wave.

6-9. The maximum power of typical transistor RF power amplifiers is in what range?
    a. Milliwatts   *c.* Hundreds of watts
    b. Watts   *d.* Kilowatts

6-10. Self-oscillation in a transistor amplifier is usually caused by
    a. excessive gain.   *c.* internal capacitance.
    b. stray inductance.   *d.* unmatched impedances.

6-11. Neutralization is the process of
    a. canceling the effect of   *c.* reducing gain.
        internal device capacitance.   *d.* eliminating harmonics.
    b. bypassing undesired
        alternating current.

6-12. Maximum power transfer occurs when what relationship exists between the generator impedance Z_i and the load impedance Z_l?
    a. $Z_i = Z_l$ *c.* $Z_i < Z_l$
    b. $Z_i > Z_l$ *d.* $Z_i = 0\ \Omega$

6-13. Which of the following is *not* a benefit of a toroid RF inductor?
    a. No shielding required   *c.* Higher Q
    b. Fewer turns of wire   *d.* Self-supporting

6-14. A toroid is a
    a. Type of inductor.   *c.* Magnetic core.
    b. Transformer.   *d.* Coil holder.

6-15. Which of the following is *not* commonly used for impedance matching in a transmitter?
    a. Resistive attenuator   *c.* L network
    b. Transformer   *d.* T network

6-16. To match a 6-Ω amplifier impedance to a 72-Ω antenna load, a transformer must have a turns ratio N_p/N_s of
    a. 0.083.   *c.* 3.46.
    b. 0.289.   *d.* 12.

6-17. Impedance matching in a broadband linear RF amplifier is handled with a(n)
    a. L network.   *c.* Pi network.
    b. Parallel tuned circuit.   *d.* Balun.

6-18. A class C amplifier has a supply voltage of 24 V and a collector current of 2.5 A. Its efficiency is 80 percent. The RF output power is
  a. 24 W.
  b. 48 W.
  c. 60 W.
  d. 75 W.

6-19. Which of the following is *not* a benefit of speech-processing circuits?
  a. Improved frequency stability
  b. Increased average output power
  c. Limited bandwidth
  d. Prevention of overmodulation

6-20. In an AM transmitter, a clipper circuit eliminates
  a. Harmonics.
  b. Splatter.
  c. Overdeviation.
  d. Excessive gain.

6-21. In a speech-processing circuit, a low-pass filter prevents
  a. Overdeviation.
  b. Overmodulation.
  c. High gain.
  d. Excessive signal bandwidth.

6-22. The gain of a transistor amplifier is
  a. Inversely proportional to collector current.
  b. Directly proportional to frequency.
  c. Directly proportional to collector current.
  d. Inversely proportional to frequency.

6-23. What values of L and C in an L network are required to match a 10-Ω transistor amplifier impedance to a 50-Ω load at 27 MHz?
  a. $L = 47$ nH, $C = 185$ pF
  b. $L = 118$ nH, $C = 236$ pF
  c. $L = 0.13$ μH, $C = 220$ pF
  d. $L = 0.3$ μH, $C = 330$ pF

CRITICAL THINKING QUESTIONS

6-1. Refer to Sec. 7-6, in Chap. 7, which covers phase-locked loops (PLLs) and frequency synthesizers. Explain how you could use a PLL in a transmitter. What other circuits would it replace?

6-2. If transistor RF amplifiers are limited to several hundred watts of power, how is the high power in AM, FM, and TV broadcasting stations produced?

Answers to Self-Tests

1. oscillator
2. continuous-wave, CW
3. crystal oscillator
4. buffer
5. drivers
6. final
7. speech-processing
8. frequency multipliers
9. mixer

10. linear
11. class C
12. AM, SSB
13. class C
14. A, B, AB
15. 27
16. 360
17. base-collector capacitance
18. parasitic oscillation

19. neutralization
20. true
21. bridge
22. true
23. push-pull
24. 90, 150
25. pulses
26. tuned or resonant circuit
27. 60, 85

28. harmonics
29. frequency multiplier
30. 180 [2(3)(4)(5) = 120; 120(1.5) = 180 MHz]
31. 50.4 [28(1.8) = 50.4 W]
32. equals
33. L network
34. pi (π), T
35. low-pass
36. Q
37. true
38. $L = 108$ nH, $C = 558$ pF
39. $L = 106$ nH, $C_1 = 13.5$ pF, $C_2 = 34$ pF

40. 450
41. 2:1
42. autotransformer
43. toroid
44. balun
45. broadband
46. clipping, compression
47. prevent overmodulation and distortion, eliminate out-of-band operation
48. clipper
49. smooth out distortion caused by clipping,

restrict signal bandwidth
50. splatter
51. gain
52. power
53. rectifying, filtering
54. collector current
55. automatic gain control
56. more
57. true

CHAPTER 7

Communications Receivers

∎

CHAPTER OBJECTIVES

This chapter will help you to:

1. *Define* the terms selectivity and sensitivity as they apply to receivers.
2. *Calculate* the *Q*, bandwidth, and shape factor of a tuned circuit.
3. *Explain* the basic concept and benefits of a superheterodyne receiver.
4. *Calculate* the intermediate frequency (IF) and image frequencies of a superheterodyne receiver.
5. *Draw* block diagrams of single- and dual-conversion receivers and *explain* the function and operation of each major section.
6. *Define* noise and *state* the three basic types of noise.
7. *Calculate* the thermal noise, noise figure, and noise temperature in a receiver.
8. *Explain* the need for and operation of AGC circuits.
9. *Define* squelch and *explain* the operation of a basic squelch circuit.
10. *Draw* a block diagram of a phase-locked loop frequency synthesizer, *name* the major circuits and *explain* their operation, and *calculate* the output frequencies given the input frequencies and frequency divider ratio.

In radio communications systems, the transmitted signal is very weak when it reaches the receiver, particularly if it has traveled any distance. Further, the signal has had to share the free-space transmission media with thousands of other radio signals. Various kinds of noise also get added to the signal. The job of the radio receiver then is to have the sensitivity and selectivity to fully reproduce the modulating signal at its output. The kind of radio receiver best suited to this task is known as the superheterodyne. The superheterodyne receiver was conceived in the early 1900s, and today virtually every communications receiver is of that design. The various circuits in the superheterodyne receiver will be discussed here in detail. Many special circuits including frequency synthesizers will also be covered. Finally, we will take a look at transceivers that combine both a transmitter and a receiver in a single package.

∎

7-1 THE SUPERHETERODYNE RECEIVER

The primary requirement for any communications receiver is that it have the ability to select the desired signal from among thousands of others present and to provide sufficient amplification to recover the modulating signal.

These two requirements are generally referred to as selectivity and sensitivity.

Selectivity refers to the ability of a receiver to select a signal of a desired frequency while rejecting those on closely adjacent frequencies. A receiver with good selectivity will isolate the desired signal in the RF spectrum and eliminate all other signals.

Selectivity in a receiver is obtained by using tuned circuits. These are *LC* circuits tuned to resonate at a desired signal frequency. The Q of these tuned circuits determines the selectivity. Recall that Q is the ratio of inductive reactance to resistance ($Q = X_L/R$). The bandwidth of a tuned circuit is a measure of its selectivity. The bandwidth is the difference between the upper f_2 and lower f_1 cutoff frequencies which are located at the 3-dB-down or 0.707 points on the selectivity curve as shown in Fig. 7-1. This bandwidth is determined by the resonant frequency f_r and the Q according to the relationship BW $= f_r/Q$. The higher the Q, the narrower the bandwidth and the better the selectivity. High-Q tuned circuits are used to keep the bandwidth narrow to ensure that only the desired signal is passed.

Some examples will illustrate these concepts. Assume that a 10-μH coil with a resistance of 20 Ω is connected in parallel with a 101.4-pF variable capacitor. The circuit resonates at

$$f_r = \frac{1}{2\pi\sqrt{LC}}$$

$$= \frac{1}{6.28\sqrt{(10 \times 10^{-6})(101.4 \times 10^{-12})}}$$

$$= 5 \text{ MHz}$$

The response curve is as shown in Fig. 7-1. The peak occurs at 5 MHz. The Q of the coil and circuit is

Fig. 7-1 **Selectivity curve of a tuned circuit.**

$$Q = \frac{X_L}{R}$$

The inductive reactance is

$$X_L = 2\pi fL$$

$$= 6.28(5 \times 10^6)(10 \times 10^{-6})$$

$$= 314 \ \Omega$$

The coil resistance was given as 20 Ω. Therefore, the Q is

$$Q = \frac{X_L}{R}$$

$$= \frac{314}{20}$$

$$= 15.7$$

Now we can compute the bandwidth.

$$\text{BW} = \frac{f_r}{Q}$$

$$= \frac{5 \times 10^6}{15.7}$$

$$= 318{,}471.3 \text{ Hz } or \text{ 318 kHz}$$

This bandwidth is centered around 5 MHz, one-half of it appearing on each side of 5 MHz. The upper and lower cutoff frequencies are 5 MHz plus or minus one-half the bandwidth.

$$\frac{318 \text{ kHz}}{2} = 159 \text{ kHz } or \text{ 0.159 MHz}$$

The upper cutoff frequency is

$$f_2 = 5 + 0.159$$

$$= 5.159 \text{ MHz}$$

The lower cutoff frequency is

$$f_1 = 5 - 0.159$$

$$= 4.841 \text{ MHz}$$

The bandwidth can be verified by computing it with the cutoff frequencies.

$$\text{BW} = f_2 - f_1$$

$$= 5.159 - 4.841$$

$$= 0.318 \text{ MHz } or \text{ 318 kHz}$$

To improve the selectivity of the circuit, we need to narrow the bandwidth. Let's say we want a bandwidth of 40 kHz. With this information, we can calculate the needed circuit Q.

$$Q = \frac{f_r}{\text{BW}}$$

$$= \frac{5 \times 10^6}{40 \times 10^3}$$

$$= 125$$

To increase the Q, we must lower the coil resistance. One way to do this is to use much larger gage wire. Alternately, a higher inductance could be used, but the capacitor value would also have to be changed. Assume we simply wind a new 10-μH coil with larger wire. If $Q = X_L/R$, then $R = X_L/Q$, and the resistance then must be

$$R = \frac{X_L}{Q}$$
$$= \frac{314}{125}$$
$$= 2.512 \ \Omega$$

Although very high Q circuits can be obtained, keep in mind that the bandwidth of the receiver must be such that it will pass not only the carrier but also the sideband frequencies that contain the transmitted information. It is possible to make a tuned circuit too selective where the higher-frequency sidebands are eliminated or greatly attenuated. This, of course, distorts the transmitted information. As you can see, the bandwidth is a compromise and must be adjusted to pass all signal components.

The ideal receiver selectivity curve would have perfectly vertical sides as shown in Fig. 7-2(a). Such a curve cannot be obtained with tuned circuits or any other electronic circuit. The response curve of a tuned circuit as Fig. 7-1 shows has gradual attenuation on either side of the center frequency. The sides of the curve are referred to as *skirts*. The objective is to obtain steep skirts and thus better selectivity. Improved selectivity is achieved by cascading tuned circuits or by using crystal or mechanical filters. Both methods are widely used in communications receivers.

A measure of the steepness of the skirts, or the skirt selectivity of a receiver, is the shape factor. The *shape factor* is the ratio of the 60-dB-down bandwidth to the 6-dB-down bandwidth of a tuned circuit or filter. This is illustrated in Fig. 7-2(b). The bandwidth at the 60-dB-down points is $f_4 - f_3$. The bandwidth of the 6-dB-down points is $f_2 - f_1$. The shape factor is their ratio:

$$\text{Shape factor} = \frac{f_4 - f_3}{f_2 - f_1}$$

Assume the 60-dB bandwidth is 12 kHz and the 6-dB bandwidth is 3 kHz. Then,

$$\text{Shape factor} = \frac{12}{3}$$
$$= 4 \ or \ 4{:}1$$

The lower the shape factor, the better the skirt selectivity. The ideal, of course, is 1 which cannot be obtained in practice. Only the ideal curve in Fig. 7-2(a) has a shape factor of 1

The *sensitivity* of a communications receiver refers to the receiver's ability to pick up weak signals. Sensitivity is primarily a function of the overall receiver gain. *Gain*, of course, is the factor by which an input signal is multiplied to produce the output signal. In general, the higher the gain of a receiver, the better its sensitivity. The more gain that a receiver has, the smaller the input signal necessary to produce a desired level of output. High gain in communications receivers is obtained by using multiple stages of amplification.

The sensitivity of a communications receiver

(a) Ideal response curve

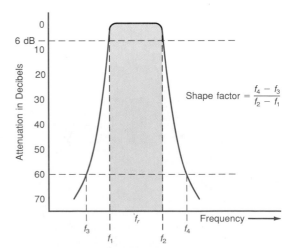

(b) Practical response curve showing shape factor

Fig. 7-2   Receiver selectivity.

is usually expressed as the minimum amount of signal voltage input that will produce an output signal that is 10 dB higher than the receiver background noise. Some specifications state a 20-dB signal-to-noise ratio. A typical sensitivity figure might be 1-μV input. The lower this figure, the better the sensitivity. Good communications receivers typically have a sensitivity of 0.2 to 1 μV. Consumer AM and FM receivers designed for receiving strong local stations have much lower sensitivity. Sensitivities of 5 to 10 μV are typical for FM receivers, while the sensitivity of an AM receiver could be 100 μV or more.

As we discuss the various receiver configurations and circuits, keep in mind that selectivity and sensitivity are the primary requirements of any receiver and all designs attempt to enhance these two capabilities.

The simplest radio receiver is the crystal set consisting of a tuned circuit, a diode (crystal) detector, and earphones. Refer to Fig. 7-3(a). The antenna picks up the signal and causes current to flow in the primary winding of coupling transformer T_1. This induces a voltage into the secondary winding which is the inductance in a series resonant circuit. When signal current flows in the primary, a varying magnetic field cuts the turns of the secondary winding, inducing a voltage in them. Each turn of the secondary, therefore, acts like a tiny voltage generator. The total effect is the same as having a signal generator in series with the secondary. See Fig. 7-3(b). The generator, secondary winding, and C_1 together form a series resonant circuit. Current flows in the secondary winding causing a voltage to be developed across capacitor C_1. When the circuit is tuned to resonance, a high voltage is developed across the capacitor. This is known as the *resonant rise* or *resonant step-up voltage*. This

voltage is significantly higher than the actual voltage induced into the secondary winding.

It is the tuned circuit that provides both selectivity and sensitivity. The tuned circuit, if it has a high Q, can select a desired frequency when it is tuned to resonance. The resonant step-up voltage across the capacitor provides some gain. It is the high Q of the tuned circuit that will permit good selectivity and, in this case, also good sensitivity.

The diode or crystal rectifies the signal, while capacitor C_2 filters out the carrier, leaving only the original information, which is heard in the earphones.

The simple crystal receiver in Fig. 7-3 does not provide the kind of selectivity and sensitivity necessary for modern communications. Only the strongest of signals will produce an output, and selectivity is often insufficient to separate incoming signals. The use of earphones is also inconvenient. However, this simple receiver does make one very important point. The minimum communications receiver is a demodulator. The demodulator strips the RF carrier from the incoming signal, leaving only the original modulating information. All other circuits used in a receiver are designed to improve the sensitivity and selectivity so that the demodulator may perform better.

The earliest communications receivers were, of course, improvements upon the basic crystal receiver. The obvious way to improve sensitivity is to add stages of amplification, both before and after the demodulator. The result is a tuned radio frequency (TRF) receiver shown in Fig. 7-4 on the next page. Here three stages of RF amplification are used between the antenna and the detector. This is followed by two stages of audio amplification. The RF amplifier stages increase the gain of the received signal tremendously before it is applied

Resonant voltage rise

TRF receiver

Fig. 7-3 (a) The simplest receiver—a "crystal" set, and (b) the series resonant circuit formed by the secondary.

Fig. 7-4   **Tuned radio-frequency (TRF) receiver.**

to the detector. The recovered signal is amplified further by the audio amplifiers which provide sufficient gain to operate a loudspeaker.

Another important factor is that the RF amplifiers use tuned circuits. Whenever resonant *LC* circuits tuned to the same frequency are cascaded, the overall selectivity is improved. The greater the number of tuned stages cascaded, the narrower the bandwidth and the steeper the skirts. This is illustrated in Fig. 7-5. A signal above or below the resonant frequency will be attenuated by one tuned circuit. It will be further attenuated by a second tuned circuit and even more by a third. The effect is to steepen the skirts.

The TRF receivers were widely used for many years, but they suffered from a number of annoying problems. The main problem was tracking of the tuned circuits. In a receiver, the tuned circuits must be made variable so that they can be set to the frequency of the desired signal. In most receivers, the capacitors in the tuned circuits are made variable. By rotating a knob, the capacitance is varied, thereby changing the resonant frequency. In early receivers, each tuned circuit had a separate capacitor and, therefore, multiple dials had to be adjusted to tune in a signal. Needless to say, this was a difficult and trying process.

The solution to the problem was to "gang" all the tuning capacitors so that they all would be changed simultaneously when the tuning knob was rotated. See Fig. 7-4. Although this essentially solved the problem, tracking errors still occurred. Differences in capacitors caused the resonant frequency of each tuned circuit to be slightly different, thereby increasing the bandwidth. This problem was offset by connecting a small "trimmer" capacitor in parallel with each of the tuning capacitors. The trimmers could then be adjusted to ensure that these minor differences were compensated for.

Another problem with TRF receivers was that the selectivity varied with frequency. In general, the bandwidth of a tuned circuit will increase with its resonant frequency. Selectivity was good at the low frequencies but less at the higher frequencies. These basic problems ultimately led to the development of a receiver that eliminated them. This receiver configuration is known as the *superheterodyne* or *superhet* (both terms are used interchangeably throughout this book).

The basic process of a superheterodyne receiver is to convert all incoming signals to a lower frequency known as the *intermediate frequency (IF)* where a single set of amplifiers can be used to provide a fixed level of sensitivity and selectivity. Most of the gain and selectivity in a superheterodyne receiver is obtained in the IF amplifiers.

Figure 7-6 shows a general block diagram of a superheterodyne receiver. The key circuit, of course, is the mixer which produces the frequency translation of the incoming signal down

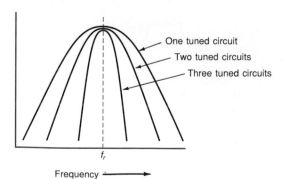

Fig. 7-5   **How the cascading of tuned circuits improves selectivity.**

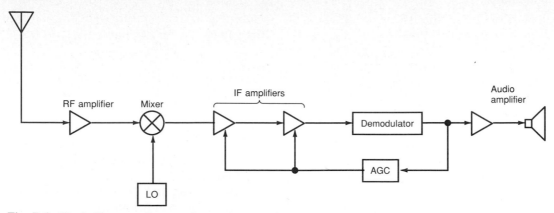

Fig. 7-6 **Block diagram of a superheterodyne receiver.**

RF amplifier

Mixer

Local oscillator

IF amplifier

to the IF. The incoming signal is mixed with an LO signal to produce this conversion. Now let's take a look at each of the circuits in the receiver to understand their basic function.

The antenna picks up the weak radio signal and feeds it to the RF amplifier. The purpose of the RF amplifier is to provide some initial gain and selectivity. Because the RF amplifier provides some initial selectivity, it is sometimes referred to as a *preselector*. Tuned circuits help select the desired signal or at least the frequency range in which the signal resides. In fixed-tuned receivers, the tuned circuits can be given a very high Q so that excellent selectivity is obtained. However, in receivers that must tune over a broad frequency range, selectivity is somewhat more difficult to obtain. The tuned circuits must resonate over a wide frequency range. Therefore, the Q, bandwidth, and selectivity of the amplifier changes with frequency.

Some communications receivers do not use an RF amplifier. Instead, the antenna is connected directly to a tuned circuit at the input to the mixer. The input tuned circuit will provide the desired initial selectivity.

The reason for omitting the RF amplifier is that in many designs, extra gain is simply not needed, particularly at low frequencies. Most of the receiver gain is in the IF amplifier section, and if relatively strong signals are to be received, additional RF gain is not necessary.

In general, it is better to have an RF amplifier. It improves sensitivity because of the extra gain and improves selectivity because of the added tuned circuits. The signal-to-noise ratio is also improved. Further, image and spurious signals are better rejected, thereby minimizing unwanted heterodyning effects in the mixer.

One major reason for including the RF amplifier is to minimize radiation of the LO signal through the receiving antenna. The LO signal is relatively strong, and some of it can leak through and appear at the input to the mixer. If the mixer input is connected directly to the antenna, some of the LO signal will be radiated, and it could cause interference to other nearby receivers. The RF amplifier between the mixer and the antenna isolates the two and thereby effectively eliminates any LO radiation.

Although both bipolar and field-effect transistors can be used as RF amplifiers, in most modern receivers MOSFETs are used. The reason for this is that they contribute less noise than bipolar transistors.

The output of the RF amplifier is applied to the input of the mixer. The mixer also receives an input from an LO. The output of the mixer will be the input signal, the LO signal, as well as their sum and difference frequencies. Usually a tuned circuit at the output of the mixer selects the difference frequency. This frequency is known as the IF. The mixer may be a diode, a balanced modulator, or a transistor. A MOSFET is the preferred mixer because of its low noise characteristics.

The LO is made tunable so that its frequency can be adjusted over a relatively wide range. By tuning the LO frequency, the mixer will translate a wide range of input frequencies to the IF.

In most receivers, the mixer and LO are separate circuits. But in some simple low-frequency receivers, the mixer and LO functions are combined within a single stage called a converter. Here, a bipolar transistor is normally used as the mixer.

The output of the mixer is a signal at the IF containing the same modulation that appeared on the input RF signal. This signal is amplified by one or more IF amplifier stages, and most of the receiver gain is obtained in these IF stages. Selective tuned circuits provide fixed selectivity. Since the IF is usually much lower than the input signal frequency, the IF amplifiers are easier to design and good selectivity is easier to obtain with tuned circuits. Crystal, mechanical, and ceramic filters are used in some IF sections to obtain better selectivity.

The highly amplified IF signal is finally applied to the demodulator or detector. This circuit recovers the original modulating information. The output of the detector is then usually fed to an audio amplifier with sufficient voltage and power gain to operate a speaker.

Another important circuit in the superheterodyne receiver is the AGC circuit. The output of the detector is usually the original modulating signal whose amplitude is directly proportional to the amplitude of the received signal. The recovered signal, which is usually ac, is rectified and filtered into a dc voltage by the AGC circuit. This dc voltage is fed back to the IF amplifiers, and sometimes to the RF amplifier, to control the gain of the receiver. The whole purpose of the AGC circuit is to help maintain a constant output voltage level over a wide range of RF input signal levels.

The amplitude of the RF signal at the antenna of a receiver can vary from a fraction of a microvolt to thousands of microvolts. This represents a very wide signal range, known as the dynamic range. Typically a receiver is designed with very high gain so that weak signals may be reliably received. However, when a very high amplitude signal is applied to a receiver, the circuits will be overloaded, causing distortion and reducing intelligibility.

By using AGC, the overall gain of the receiver can be automatically adjusted depending upon the input signal level. The signal amplitude at the output of the detector will be proportional to the amplitude of the input signal. If it is very high, the AGC circuit will produce a high dc output voltage. This will cause the gain of the IF amplifiers to be reduced. With reduced gain, the distortion normally produced by a large input signal will be eliminated.

When the incoming signal is weak, the detector output will be low. The output of the AGC will then be only a very small dc voltage. This will cause the gain of the IF amplifiers to remain high thereby providing maximum amplification.

Virtually all superheterodyne receivers use some form of AGC. In low-frequency broadcast receivers, it is sometimes referred to as

automatic volume control (AVC). In all cases, the gain of the IF amplifiers is controlled.

Self-Test

Answer the following questions.

1. The ability of a receiver to choose a desired signal frequency while rejecting closely adjacent signal frequencies is known as _____ .
2. Decreasing the Q of a resonant circuit causes its bandwidth to _____ .
3. Good selectivity usually means _____ (wide, narrow) bandwidth.
4. True or false. A tuned circuit can provide voltage gain.
5. Cascading tuned circuits cause the selectivity to _____ (increase, decrease).
6. If the selectivity of a tuned circuit is too sharp, the _____ of the received signal may be attenuated.
7. A tuned circuit has a Q of 100 at its resonant frequency of 500 kHz. Its bandwidth is _____ kHz.
8. A parallel LC tuned circuit has a coil of 3 μH and a capacitance of 75 pF. The coil resistance is 10 Ω. The circuit bandwidth is _____ MHz.
9. A tuned circuit has a resonant frequency of 10 MHz and a bandwidth of 100 kHz. The upper and lower cutoff frequencies are f_1 (upper) = _____ MHz, f_2 (lower) = _____ MHz.
10. To achieve a bandwidth of 3 kHz at 4 MHz, a Q of _____ is required.
11. To narrow the bandwidth of a tuned circuit, the coil resistance must be _____ (increased, decreased).
12. A filter has a 6-dB bandwidth of 500 Hz and a 60-dB bandwidth of 1200 Hz. The shape factor is _____ .
13. The greater the gain of the receiver, the better its _____ .
14. A receiver that uses only amplifiers and a detector is known as a(n) _____ receiver.
15. A receiver that uses a mixer to convert the received signal to a lower frequency is called a(n) _____ .
16. The acronym IF means _____ .
17. Tuning a superheterodyne is done by varying the frequency of its _____ .
18. Most of the gain and selectivity in a superhet is obtained in the _____ .
19. The _____ circuit in a receiver com-

pensates for a wide range of input signal levels.

20. The mixer output is usually the difference between the _____ frequency and the _____ frequency.

21. The AGC voltage controls the gain of the _____ .

7-2 INTERMEDIATE FREQUENCY SELECTION AND IMAGES

The choice of the IF is usually a design compromise. The main objective is to obtain good selectivity. Narrowband selectivity is best obtained at lower frequencies, particularly when conventional *LC* tuned circuits are used. A typical IF is 455 kHz. This has become a standard in AM broadcast receivers and in many other radio receivers. Intermediate frequencies as low as 50 kHz have been used successfully to obtain extremely narrow bandwidths.

The benefits of using a low IF are that the design is not only simpler, but also it is less troublesome. At low frequencies, the circuits will be far more stable with high gain. At higher frequencies, the design is complicated by stray inductances and capacitances as well as the need for shielding. This makes the layouts of the circuits critical if undesired feedback paths are to be avoided. With very high circuit gain, some of the signal can be fed back in phase and cause oscillation. Oscillation is not as much of a problem at lower frequencies.

When low IFs are selected, another problem occurs particularly if the signal to be received is very high in frequency. This is the problem of images. An *image* is an RF signal that is spaced from the desired incoming signal by a frequency that is 2 times the IF above or below the incoming frequency. Expressed mathematically, this is

$$f_i = \begin{cases} f_s + 2f_{IF} \\ f_s - 2f_{IF} \end{cases}$$

where f_i = image frequency
f_s = desired signal frequency
f_{IF} = intermediate frequency

This is illustrated graphically in Fig. 7-7. Only one of these images will occur, depending upon whether the LO frequency f_o is above or below the signal frequency.

Recall that the mixer in a superhet produces the sum and difference frequencies of the incoming and LO signals. Normally, it is the difference frequency that is selected as the IF.

Fig. 7-7 **Relationship of the signal and image frequencies.**

The frequency of the LO is usually chosen to be higher than the incoming signal by an amount equal to the IF. However, the LO frequency could also be made lower than the incoming signal frequency by the same amount. Either choice will work and will produce the desired difference frequency. Let's assume that our LO frequency is higher than the incoming signal frequency.

Now, what happens if an image signal ($f_i = f_s + 2f_{IF}$) appears at the input of the mixer? The mixer will, of course, produce the sum and difference frequencies regardless of the inputs. Therefore, the mixer output will again be the difference frequency at the IF value. Let's look at an example.

Assume a desired signal frequency f_s of 90 MHz and an LO frequency f_o of 100 MHz. The IF is the difference $100 - 90 = 10$ MHz. The image frequency is

$$f_i = f_s + 2f_{IF}$$
$$= 90 + 2(10)$$
$$= 90 + 20$$
$$= 110 \text{ MHz}$$

If the image appears at the mixer input, the output will be the difference $110 - 100 = 10$ MHz. The IF amplifier would pass it. Now look at Fig. 7-8 on the next page. This shows that the mixer will generate the difference between the LO and the received signals, if it is the desired one or an image, since each will produce the same IF.

What this means is that a signal spaced from the desired signal by 2 times the IF will also be picked up by the receiver and converted to the IF. When this occurs, the image signal will interfere with the desired signal.

What are the chances that there will be a signal on the image frequency? In the modern crowded RF spectrum, the chances are quite

Fig. 7-8 Signal, LO, and image frequencies in a superheterodyne.

large. The interference from the image could be so great as to make the desired signal unintelligible.

Images are a real problem in superheterodyne receivers. The design of a superhet, therefore, must take this problem into consideration. And, as you will see, the choice of the IF is critical to minimizing the problem.

The only reason why image interference occurs is that the image signal is allowed to appear at the mixer input. This is the reason for using high-Q tuned circuits ahead of the mixer or a selective RF amplifier. If the selectivity of the RF amplifier and tuned circuits is good enough, then the image will be rejected. In a fixed-tuned receiver designed for a specific frequency, it is possible to optimize the receiver front end for the good selectivity necessary to eliminate images. But many receivers have broadband RF amplifiers that pass many frequencies within a specific band. Other receivers must be made tunable over a wide frequency range. In such cases, the selectivity is a problem. An example will illustrate this.

Assume that the receiver is designed to pick up a signal at 20 MHz. The IF is 500 kHz, or 0.5 MHz. The LO is adjusted to a frequency above the incoming signal by an amount equal to the IF, or 20 + 0.5 = 20.5 MHz. When the LO and signal frequencies are mixed, the difference, of course, is 0.5 MHz as desired.

The image frequency can be computed using the expression given earlier:

$$f_i = f_s + 2f_{IF}$$
$$= 20 + 2(0.5)$$
$$= 21 \text{ MHz}$$

The image frequency of 21 MHz will cause interference to the desired signal at 20 MHz

unless it is rejected. The signal, LO, and image frequencies are shown in Fig. 7-9.

Now, assume that a tuned circuit ahead of the mixer has a Q of 10. Knowing this and the resonant frequency, we can calculate the bandwidth of the resonant circuit from the familiar expression:

$$\text{BW} = \frac{f_R}{Q}$$

where f_R is the resonant frequency of the tuned circuit.

$$\text{BW} = \frac{20}{10}$$
$$= 2 \text{ MHz}$$

The response curve for this tuned circuit is shown in Fig. 7-9.

Fig. 7-9 The tuned-circuit bandwidth is such that it allows the image to get to the mixer.

As you can see, the bandwidth of the resonant circuit is relatively wide. The bandwidth is centered on the signal frequency of 20 MHz and will, therefore, extend from 19 to 21 MHz. Remember that the bandwidth is measured at the 3-dB-down points on the tuned circuit response curve. The upper cutoff frequency is 21 MHz, and the lower cutoff frequency is 19 MHz.

As you can see, the upper cutoff frequency is equal to the image frequency. What this means is that the image frequency would essentially be passed by the tuned circuit and would, therefore, cause interference.

Cascading tuned circuits and making them with higher Q would indeed help solve the problem. However, this usually not only complicates the design, but in some cases, compromises it. To overcome this problem, the usual solution is to choose a higher IF.

Let's assume that we choose an IF of 5 MHz. Now, the image frequency will be

$$f_i = 20 + 2(5)$$
$$= 30 \text{ MHz}$$

The image frequency is now 30 MHz. A signal at this frequency would interfere with our desired signal at 20 MHz if it was allowed to pass into the mixer, but as you can see, 30 MHz is well out of the bandpass of our tuned circuit. The selectivity, even though it is poor, is sufficient to adequately reject the image. Of course, by choosing the higher IF, we make the design more difficult as indicated earlier.

The selection of an IF is usually a compromise. The IF is made as high as possible to effectively eliminate the image problem, while at the same time it is made as low as possible to simplify the design. The IF in most receivers generally varies in proportion to the frequencies that must be covered. At low frequencies, low values of IF are used. A value of 455 kHz is common for AM broadcast band receivers and for others covering that general frequency range. At higher frequencies, up to about 30 MHz, 3385 kHz, and 9 MHz are common IF frequencies. In FM radios that receive frequencies in the range of 88 to 108 MHz, 10.7 MHz is a standard IF. In TV receivers, an IF in the 40- to 50-MHz range is common. Radar receivers typically use an IF in the 60-MHz range, and satellite communications gear uses 70-MHz and 140-MHz IFs.

Another way to get the desired selectivity while at the same time eliminating the image problem is to use what is known as a dual-conversion superheterodyne receiver. A block diagram of a dual-conversion receiver is shown in Fig. 7-10. Notice that it uses two mixers and two LOs. The result is that the receiver has two IFs. The first mixer converts the incoming signal to a relatively high IF for the purpose of eliminating the images. Then a second mixer converts that IF down to a much lower frequency where good selectivity is easier to obtain.

Following the frequencies given in Fig. 7-10, you can see how each is obtained. Each mixer produces the difference frequency. Note that the first LO is variable and provides the tuning for the receiver. The second LO is fixed in frequency. Since it must convert only one fixed IF to a lower IF, the LO need not be tunable. In most cases, its frequency is set by a quartz crystal. In some receivers, the first mixer is driven by the fixed-frequency LO, while tuning is done with the second LO.

In some special situations, triple-conversion receivers can be used to further minimize the image problem. A triple-conversion receiver uses three mixers and three different IF values. Such receivers are not common, but they have been used in unusual or critical applications. Dual-conversion receivers are relatively common. Most shortwave receivers and those at VHF, UHF, and microwave frequencies use dual conversion. For example, a CB receiver typically uses a first IF of 10.7 MHz and a second IF of 455 kHz.

Self-Test

Answer the following questions.

22. For best selectivity and stability, the IF should be
    a. High.
    b. Medium.
    c. Low.
23. An interfering signal that is spaced from the desired signal by twice the IF is called a(n) _____ .
24. A superhet has an input signal of 15 MHz. The LO is tuned to 18.5 MHz. The IF is _____ MHz.
25. Images are caused by the lack of _____ at the mixer input.

Fig. 7-10 **A dual-conversion superheterodyne.**

26. A desired signal at 27 MHz is mixed with an LO frequency of 27.5 MHz. The image frequency is _____ MHz.
27. True or false. The LO may be above or below the signal frequency.
28. The main feature of a dual-conversion superhet is that it has two _____ circuits.
29. The image problem can be solved by proper choice of the _____ .
30. A dual-conversion superhet has an input frequency of 50 MHz and LO frequencies of 59 MHz and 9.6 MHz. The two IFs are _____ MHz and _____ kHz.

7-3 NOISE

One of the most important considerations in any communications system is noise. Noise is random energy that interferes with the information signal. You can hear what noise sounds like by turning on any AM or FM receiver and tuning it to some position between stations. The hiss or static that you hear in the speaker is noise. Noise also shows up in a television picture as snow or colored snow known as confetti. If the noise level is too high or the signal is too weak, the noise may actually dominate and make reception unreliable. The noise may be so high as to completely obliterate the signal. Noise that occurs on transmissions of digital data makes itself known as bit errors. Thus the information being transmitted is lost or incorrect.

Noise is a problem in communications systems simply because the received signals are so low in amplitude. When the transmission is over short distances or when very high power transmitters are used, signal strength is good and noise is not a problem. But in most communications systems, weak signals are normal, so noise becomes a significant factor in the design of communications equipment. Noise is most problematic in the receiver because the receiver has the responsibility for amplifying the weak signal and recovering the information reliably.

In any discussion of communications systems and receiver performance, the term signal-to-noise ratio is used. *Signal-to-noise (S/N) ratio* is simply a number that indicates the relative strengths of the signal and the noise. The stronger the signal and the weaker the noise, the higher the S/N ratio. If the signal is weak and the noise is strong, the S/N ratio will be low and reception will be less reliable. The design of communications equipment has as its objective to produce the highest S/N ratio possible.

Noise comes from two basic sources. First, there is noise that is generated external to the receiver. Then there is internal noise that is produced within the receiver itself. Both types affect the S/N ratio.

External noise usually comes from industrial, atmospheric, or extraterrestrial or space sources. Industrial noise is produced by manufactured equipment, such as automotive ignition systems and electric motors and generators. Any electrical equipment that causes high voltages or currents to be abruptly switched produces "transients" that create noise. Fluorescent and other forms of gas-filled lights are another common noise source.

Regardless of its source, noise shows up as a random ac voltage. You can actually view it on an oscilloscope. The amplitude varies over a wide range as does the frequency. Industrial noise is mostly produced by voltage transients which in general are signals containing a tremendous amount of harmonic energy. You might say that noise in general contains all frequencies, varying randomly.

Another source of noise is electrical disturbances that occur naturally in the earth's atmosphere. You will often hear atmospheric noise referred to as static. Static usually comes from lightning. Lightning is the general term used to refer to electric discharges that occur between clouds or between the earth and clouds. Huge static charges build up, and when the potential difference is great enough, an arc occurs where electricity literally flows through the air. Lightning is very much like the static charges that you experience during a dry spell in the winter, only the voltages involved are enormous. This results in a transient electrical signal that generates harmonic energy that can travel extremely long distances.

Like industrial noise, atmospheric noise shows up primarily as amplitude variations that add to a signal and interfere with it. Atmospheric noise has its greatest impact on signals at frequencies less than 30 MHz.

Extraterrestrial noise comes from sources in space. For example, one of the primary sources of noise is the sun. The sun radiates a wide range of signals in a broad noise spectrum. The noise intensity produced by the sun varies with time. In fact, the sun has a repeating 11-year noise cycle. During the peak of the cycle, the sun produces an awesome amount of noise that causes tremendous radio signal interference, making many frequencies unusable for communications. During other years, the noise is at a minimum level.

Noise is also generated by stars, which are also suns. Noise from the stars is generally known as cosmic noise, and its level is not as great because of the great distances between the stars and earth. Nevertheless, it is an important source of noise that must be considered. It shows up primarily in the 10-MHz to 1.5-GHz range but causes the greatest disruptions in the 15- to 150-MHz range.

External noise is a fact of life and must be dealt with. Atmospheric and space noise simply cannot be eliminated. Industrial noise, because it is manufactured, can sometimes be eliminated at the source, if you have control over the source. In general, because there are so many sources of industrial noise, there is no way to control it. The key to reliable communications then is simply to generate the signal at a high enough power so that it overcomes this noise.

As if external noise wasn't enough, designers of communications systems must also contend with internal noise, that noise generated within the communications receiver. Electronic components such as resistors, diodes, and transistors are major sources of noise. Although this is a low-level noise, it is often great enough to interfere with weak signals. However, since the sources of internal noise are well known, the designer has some control over it.

Most internal noise is caused by a phenomenon known as thermal agitation. Thermal agitation refers to the random motion of the atoms and electrons in an electronic component caused by heat. Increasing the temperature causes this motion to increase. Since the components are conductors, the movement of electrons constitutes a current flow that causes a small voltage to be produced across that component.

You can actually observe this noise by simply connecting a resistor to a very high gain oscilloscope. The motion of the electrons (due to room temperature) in the resistor will cause a voltage to appear across it. The voltage variation is completely random and at a very low level. You will often hear thermal agitation referred to as white noise or Johnson noise.

The noise power developed across a resistor is directly proportional to the temperature.

In a relatively large resistor at room temperature or higher, the noise voltage across it can be as high as several microvolts. This is the same order of magnitude as many weak RF signals. Some signals will have even less amplitude and thus will be totally masked by this noise.

Since noise is a very broad band signal containing a tremendous range of random frequencies, its level can be reduced by limiting the bandwidth. By feeding a noise signal into a selective tuned circuit, many of the noise frequencies will be rejected and the overall noise level will go down. The noise level is directly proportional to the bandwidth of any circuit to which it is applied. Although filtering does reduce the noise level, it does not eliminate it entirely.

The amount of noise voltage appearing across a resistor or the input impedance to a receiver can be calculated with the expression

$$v_n = \sqrt{4kTBR}$$

where v_n = rms noise voltage
k = Boltzmann's constant or 1.38×10^{-23} joule/kelvin (J/K)
T = temperature, K
B = bandwidth, Hz
R = resistance, Ω

Let's look at an example. Assume that a receiver has a 50-Ω input resistance, a bandwidth of 6 MHz, and a temperature of 29°C or 29° + 273° = 302 K. (The relationship between degrees Celsius and degrees kelvin is discussed on page 136.) The input thermal noise voltage then is

$$v_n = \sqrt{4kTBR}$$
$$= \sqrt{4(1.38 \times 10^{-23})(302)(6 \times 10^6)(50)}$$
$$= 2.24 \ \mu V$$

This noise voltage would, of course, mask signals less than this value. The best way to reduce the noise voltage is to reduce the bandwidth to the minimum value acceptable to the application.

Electronic components such as transistors and tubes are also major contributors of noise. The most common type of noise is referred to as *shot noise* which is produced by the random arrival of electrons or holes at the output element, at the plate in a tube, or at the collector or drain in a transistor. Shot noise is also produced by the random movement of electrons or holes across a PN junction. Even though current flow is established by external bias voltages, there will still be some random movement of electrons or holes due to discontinuities in the device. An example of such a discontinuity is the contact between the copper lead and the semiconductor material. The interface between the two creates a discontinuity that causes random movement of the current carriers.

Another kind of noise that occurs in transistors is called transit time noise. *Transit time* is the duration of time that it takes for a current carrier such as a hole or electron to move from the input to the output. The devices themselves are very tiny, so the distances involved are minimal. Yet the time it takes for the current carriers to move even a short distance is finite. At low frequencies this time is negligible. But when the frequency of operation is high and the period of the signal being processed is the same order of magnitude as the transit time, then problems can occur. The transit time shows up as a kind of random noise within the device, and this noise is directly proportional to the frequency of operation.

Noise is usually expressed as a power because the received signal is also expressed in terms of power. By knowing the signal and noise powers, the S/N ratio can be computed. Rather than express the S/N ratio as simply a number, you will usually see it expressed in terms of decibels.

A receiver has an input signal power of 1.2 μW. The noise power is 0.8 μW. The S/N ratio is

$$S/N = 10 \log \frac{1.2}{0.8}$$
$$= 10 \log 1.5$$
$$= 10(0.176)$$
$$= 1.76 \text{ dB}$$

A variety of methods are used to express the noise quality of the receiver. One of these methods is known as the noise figure which is the ratio of the S/N power at the input to the S/N power at the output. The device under consideration can be the entire receiver or a single amplifier stage. The *noise figure F* also called the *noise factor*, can be computed with the expression

$$F = \frac{S/N \text{ input}}{S/N \text{ output}}$$

You can express the noise figure as a number, but more often you will see it expressed in decibels.

An amplifier or receiver will always have more noise at the output than at the input. The reason for this is that the amplifier or receiver generates internal noise which will be added to the signal. And even though the signal may be amplified along the way, that noise will be amplified along with it. The S/N ratio at the output will be less than the S/N ratio of the input; therefore, the noise figure will always be greater than 1. A receiver that contributes zero noise to the signal would have a noise figure of 1, or 0 dB. Such a noise figure is not attainable in practice. A transistor amplifier in a communications receiver will usually have a noise figure of several decibels. The lower the noise figure, the better the amplifier.

Another method of expressing the noise in an amplifier or receiver is to use noise temperature. Most of the noise produced in a device is thermal noise that is directly proportional to temperature. Therefore, temperature is an appropriate unit of measurement.

The noise temperature is expressed in degrees kelvin. Normally we express temperature in terms of degrees Fahrenheit or Celsius. The kelvin temperature scale is related to the Celsius scale by the relationship

$$T_K = T_C + 273°$$

where T_K is the temperature in degrees kelvin and T_C is the temperature in degrees Celsius. The kelvin temperature scale is referred to as the absolute temperature scale. The letter K is used to designate that the temperature is expressed in terms of kelvin. If the temperature is in Celsius, all you have to do is add 273° to it to get the temperature in kelvin.

Remember too that the temperatures in Fahrenheit and Celsius are related by the expression

$$T_C = \frac{5}{9}(T_F - 32°)$$

To compute the noise temperature, the following expression is used.

$$T_N = 290(F - 1)$$

Here F is the noise factor discussed earlier. This expression shows that the noise temperature T_N is related to the noise figure F. In the expression above, the ratio rather than the decibel value is used for F. For example, if the noise factor is 1.5, the equivalent noise temperature is

$$T_N = 290(1.5 - 1)$$
$$= 290(0.5)$$
$$= 145 \text{ K}$$

By examining the expression above, you can see that if the amplifier or receiver contributes no noise, then the noise factor F will be 1 as indicated before. Plugging this value into the expression gives an equivalent noise temperature of 0 K. If the noise figure is greater than 1, then an equivalent noise temperature will be produced.

Noise temperature is used only in circuits or equipment that operate at VHF, UHF, or microwave frequencies. The noise figure or factor is used at lower frequencies. A good low-noise transistor or amplifier stage will typically have a noise temperature of less than 100 K, the lower the better. Often you will see the noise temperature of a transistor given in the data sheet. Noise figures of less than about 3 dB are excellent.

Another method of comparing the quality of communications receivers is called SINAD, which is an acronym for *si*gnal plus *n*oise *a*nd *d*istortion. All receivers will contribute both noise and distortion to an incoming signal. By determining the ratio of the composite signal plus noise plus distortion to the noise plus distortion, a relative figure of merit is obtained.

$$\text{SINAD} = \frac{\text{signal} + \text{noise} + \text{distortion}}{\text{noise} + \text{distortion}}$$

Amplifiers and other circuits in a receiver are not perfectly linear and will, therefore, introduce some distortion. The distortion takes the form of very low levels of signal harmonics. The harmonics add to the signal along with the noise. The SINAD ratio makes no attempt to discriminate between or separate the noise and distortion signals.

To obtain the SINAD figure an RF signal modulated by an audio signal usually of 400 Hz or 1 kHz is applied to the input of an amplifier or receiver. The composite output is then measured. This gives the numerator value.

Next, a highly selective notch (band reject) filter is used to eliminate the modulating audio signal from the output. The result is only the noise and the distortion. This gives the denominator value. The SINAD ratio can then be calculated. The SINAD ratio is also used to express the sensitivity of a receiver.

Although noise is an important consideration at all communications frequencies, it is particularly critical in the microwave region. The rea-son for this is that the noise increases with bandwidth and will impact high-frequency signals more than lower-frequency signals. The limiting factor in most microwave communications systems, such as satellites and radar, is internal noise.

The noise level in a system is proportional to the temperature and the bandwidth. It is also proportional to the amount of current flowing in a component, the gain of the circuit, and the resistance of the circuit. Increasing any of these factors will increase the noise. Therefore, low noise is best obtained by using low-gain circuits, low dc, low resistance values, and narrow bandwidths. Keeping the temperature low can also help. In fact, in some special microwave receivers, the noise level is reduced by cooling the input stages to the receiver.

Noise, of course, has its greatest effect at the input to a receiver simply because that is the point at which the signal level is lowest. The noise performance of a receiver is invariably determined in the very first stage of the receiver, usually an RF amplifier or mixer. Care should be taken in the design of these to ensure the use of very low noise components and to take into consideration the current, resistance, and gain figures in this circuit. Beyond the first stage, noise is basically no longer a problem.

Self-Test

Answer the following questions.

31. List three sources of external noise.
32. List three main types of internal noise.
33. Noise from the sun and stars is called _____ noise.
34. Atmospheric noise comes primarily from _____ .
35. List four sources of industrial noise.
36. The main source of internal noise is _____ .
37. The S/N ratio is usually expressed in _____ .
38. For best reception, the S/N ratio should be _____ (low, high).
39. Increasing the temperature of a component causes its noise power to _____ .
40. Thermal noise is sometimes called _____ or _____ noise.
41. Narrowing the bandwidth of a circuit causes the noise level to _____ .

42. The noise voltage produced across a 75-Ω input resistance at a temperature of 25°C with a bandwidth of 1.5 MHz is _____ μV.

43. Two types of noise caused by tubes or transistors are _____ noise and _____ noise.

44. True or false. The noise at the output of a receiver will be less than the noise at the input.

45. True or false. The receiver amplifies noise as well as the signal.

46. The ratio of the S/N power at the input to the S/N power at the output is called the _____ .

47. Noise temperature in degrees kelvin is used to express the noise in a system at _____ (low, microwave) frequencies.

48. The noise figure of an amplifier is 2.6. The noise temperature is _____ K.

49. The SINAD method considers the signal, noise, and _____ levels in a receiver.

50. Noise is more of a problem at _____ frequencies.

51. The stages of a receiver that contribute the most noise are the _____ and the _____ .

52. True or false. An amplifier with a noise temperature of 170 K is better than one with a rating of 235 K.

7-4 TYPICAL RECEIVER CIRCUITS

Now let's take a more detailed look at some of the circuits used in communications receivers. We have already discussed mixers and demodulators in previous chapters, so we will not discuss them here. Instead, we will focus on RF and IF amplifiers, AGC and AFC circuits, and other special circuits found in receivers.

The most critical part of a communications receiver is the front end. The front end usually consists of the RF amplifier, mixer, and related tuned circuits. This is the part of the receiver that processes the very weak input signal. It is essential that low-noise components be used to ensure a sufficiently high S/N ratio. Further, the selectivity should be such that it effectively eliminates images.

In many communications receivers, an RF amplifier is not used. This is particularly true in receivers designed for frequencies lower than about 30 MHz. The extra gain is not nec-

essary, and its only contribution would be more noise. Therefore, the RF amplifier is usually eliminated, and the antenna is connected directly to the mixer input through one or more tuned circuits. The tuned circuits must provide the input selectivity necessary for image rejection. In a receiver of this kind, the mixer must also be of the low-noise variety. Today, most mixers are MOSFETs, which provide the lowest noise contribution.

Receivers used at frequencies above approximately 100 MHz, however, do typically use RF amplifiers. And RF amplifiers are found in some lower-frequency communications systems as well. The primary purpose of this amplifier is to increase the weak signal amplitude prior to mixing. The RF amplifier also provides some selectivity for image rejection.

In most of these receivers, a single RF stage is used, usually providing a voltage gain in the 10- to 30-dB range. This is easily obtainable with a single transistor. Bipolar transistors are used at the lower frequencies, while FETs are preferred at VHF, UHF, and microwave frequencies. Typically, FETs have a lower noise figure than bipolar transistors and, therefore, give better performance.

The RF amplifier is usually a simple class A circuit. A typical bipolar circuit is shown in Fig. 7-11(a), and a typical FET circuit is shown in Fig. 7-11(b). Note that the bipolar circuit does not have a tuned input. The antenna is connected directly to the base of the transistor. This makes the circuit very broad band as the transistor will amplify virtually any signal picked up by the antenna. However, note that the collector is tuned with a parallel resonant circuit which provides the initial selectivity for the mixer input. An untuned broadband transformer could also be used.

The FET circuits are particularly effective because their high input impedance minimizes loading on tuned circuits, thereby permitting the Q of the circuit to be higher and selectivity to be sharper. Most FETs also have a lower noise figure.

At microwave frequencies (those above 1 GHz), a newer type of FET is preferred. Known as a metal-semiconductor FET or MESFET, the device is a junction FET (JFET) made with gallium arsenide (GaAs). They are also referred to as GASFETs. A cross section of a typical MESFET is shown in Fig. 7-12. The gate junction is a metal-to-semiconductor interface as it is in a Schottky or hot-carrier diode. Like in other JFET circuits, the gate-to-

(a) Broadband
bipolar
RF amplifier

(b) Tuned FET
RF amplifier

Fig. 7-11 The RF amplifiers used in receiver front ends.

Fig. 7-12 The MESFET configuration and symbol.

source is reverse-biased. Also, as in other depletion mode JFETs, the reverse-bias voltage between the source and the gate controls the conduction of current between the source and the drain. The actual transit time of electrons through gallium arsenide is far shorter than in similar silicon transistors, and, therefore, the MESFET provides greater gain at microwave frequencies. In addition, MESFETs have an extremely low noise figure, typically 2 dB or less. Most MESFETs have a noise temperature of less than 200 K.

Another critical part of a superheterodyne receiver is the IF amplifier. This is where most of the gain and selectivity is obtained. The choice of an IF is critical to the design of a receiver. It is a compromise between good selectivity and stability, which is best obtained at lower frequencies, and good image rejection, which is best obtained at the higher frequencies.

One of the most common IF values is 455 kHz. It is low enough to provide good selectivity and to make high gain with minimum instability. With input frequencies of up to about 10 MHz, the image rejection is satisfactory. But beyond that frequency, the input tuned circuits do not provide sufficient selectivity to reduce the images to an acceptable level. When operating above about 10 MHz, a higher IF frequency is selected. A value in the 1500- to 2000-kHz range is satisfactory for frequencies up to about 30 MHz.

When the input frequency is in the VHF range and beyond, a very high value of IF is chosen. For example, most FM receivers operating in the 88- to 108-MHz band use an IF of 10.7 MHz. In many communications receivers operating at higher frequencies, double conversion is used to solve the image and selectivity problems. In some high-frequency receivers operating in the 3- to 30-MHz range, single conversion is used, but a very high IF of 9 MHz is chosen as it effectively solves the image problem.

Like RF amplifiers, IF amplifiers are tuned class A amplifiers capable of providing a gain in the 10- to 30-dB range. Usually two or more IF amplifiers are used to provide adequate overall receiver gain. See Fig. 7-13 on the next page. Ferrite-core transformers are used for coupling between stages. Most IF amplifiers use bipolar transistors, although FETs are used in some designs. In most new designs, IC dif-

Fig. 7-13   A two-stage IF amplifier using double-tuned transformer coupling for selectivity.

ferential amplifiers are used to implement IF amplifiers.

The selectivity in the IF amplifier is provided by the tuned circuits. As indicated earlier, cascading tuned circuits cause the overall circuit bandwidth to be considerably narrowed. High-Q tuned circuits are used, but with multiple tuned circuits, the bandwidth is even narrower.

In designing an IF amplifier, care should be taken so that the selectivity is not too sharp. If the IF bandwidth is too narrow, it will cause sideband cutting. This means that the higher modulating frequencies will be greatly reduced in amplitude, thereby distorting the received signal. The exact nature of the kinds of signals to be received must be well known so that the bandwidth of the IF amplifier can be appropriately set.

It is sometimes necessary when receiving very broad band signals to widen the bandwidth of the IF amplifier. There are several ways of doing this. First, a high value of resistance can be connected across the parallel tuned circuits thereby lowering their Q to a value that will produce the appropriate bandwidth.

Another technique is to use overcoupled tuned circuits. The coupling between IF amplifier stages in some receivers is done with double-tuned ferrite-core transformers as shown in Fig. 7-13. Both the primary and secondary windings are resonated with capacitors. The output voltage versus frequency for such

a double-tuned coupled circuit is strictly dependent upon the amount of coupling or mutual inductance between the primary and secondary windings. That is, the spacing between the windings determines how much of the magnetic field produced by the primary will cut the turns of the secondary. This affects not only the amplitude of the output voltage, but also the bandwidth.

Figure 7-14 shows the effect of different amounts of coupling between the primary and secondary windings. When the windings are

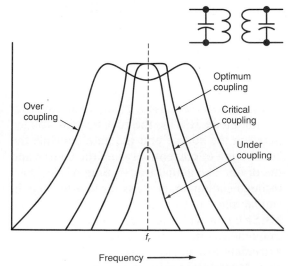

Fig. 7-14   Response curves of a double-tuned air-core transformer for various degrees of coupling.

spaced far apart, the coils are said to be *under-coupled*. The amplitude will be low, and the bandwidth will be relatively narrow.

At some particular degree of coupling, the output will reach a peak value. This is known as *critical coupling*. In most IF designs, critical coupling provides the best gain if the bandwidth provided is adequate.

Moving the coils closer together and increasing the coupling, the actual bandwidth begins to increase. The output signal amplitude is at a maximum and will not increase beyond that obtained at critical coupling, but the bandwidth is considerably widened. This point is usually known as *optimum coupling*.

Increasing the amount of coupling further produces an effect known as *overcoupling*. It produces a doubled peak output response curve with a considerably wider bandwidth. By setting the amount of coupling between the windings in the IF coupling transformers, the desired amount of bandwidth can be obtained.

In many communications receivers where superior selectivity is required, very sharp crystal filters are used to obtain the desired selectivity. These crystal filters are usually of the lattice variety discussed previously in the chapter on SSB. Ceramic and mechanical filters are also used. Such filters are usually packaged as a unit and are connected directly at the output of the mixer but prior to the first IF stage. The desired selectivity is obtained with this filter. The remaining stages of the IF amplifier use tuned circuits whose bandwidths are generally broader than that provided by the filter. In any case, it is the filter alone that sets the bandwidth.

In FM receivers, one or more of the IF amplifier stages are used as limiters. The limiters remove any amplitude variations on the FM signal prior to it being applied to the demodulator.

The limiter is not a special circuit. Typically it is nothing more than a conventional class A IF amplifier. However, any amplifier will act as a limiter if the input signal is high enough. With a very large input signal applied to a single transistor stage, the transistor will alternately be driven between saturation and cutoff. For example, in a bipolar class A amplifier, applying a very large positive input signal to the amplifier will cause the base bias to increase, thereby increasing the collector current. When a sufficient amount of input voltage is supplied, the transistor will reach maximum conduction where both the emitter-base and base-collector

junctions become forward-biased. At this point, the transistor is saturated and the voltage between the emitter and collector drops to some very small value, typically less than 0.1 V. At this time, the amplifier output is approximately equal to the dc voltage drop across any emitter resistor that may be used in the circuit.

When a very large negative-going signal is applied to the base, the transistor can be driven into cutoff. The collector current drops to zero, and the voltage seen at the collector is simply the supply voltage. Figure 7-15 shows the collector current and voltage for both extremes.

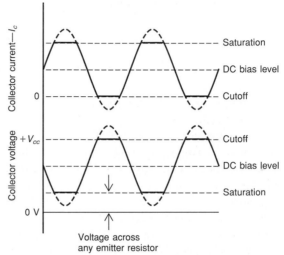

Fig. 7-15 Collector current and voltage in a bipolar limiter IF amplifier circuit.

By driving the transistor between saturation and cutoff, the positive and negative peaks of the input signal are effectively flattened or clipped off. Any amplitude variations are essentially removed. The output signal at the collector, therefore, is a square wave. The most critical part of the limiter design is to set the initial base-bias level to an appropriate point where symmetrical clipping, that is equal amounts of clipping on the positive and negative peaks, will occur. The square wave at the collector, which is made up of many undesirable harmonics, is effectively filtered back into a sine wave by the tuned circuit in the collector.

The selection of the overall gain of a communications receiver is usually based on the weakest signal to be received. In most modern communications receivers the individual gains are as follows: the voltage gain between the antenna and the demodulator is usually in excess of 100 dB; the RF amplifier usually has a gain in the 5- to 15-dB range; the mixer gain is in the 6-to-10-dB range; the IF amplifiers have individual stage gains of 20 to 30 dB; the detector may introduce a typical loss of −2 to −5 dB if it is of the diode type; and the gain of the audio amplifier stage is in the 20- to 40-dB range. As an example, assume the following circuit gains:

RF amplifier	9 dB
Mixer	4 dB
IF amplifiers (3 stages)	27 dB (81 dB total)
Detector	−3 dB
Audio amplifier	30 dB

The total gain then is simply the algebraic sum of the individual stage gains, or

$$\text{Total gain} = 9 + 4 + 27 + 27 + 27 - 3 + 30$$
$$= 121 \text{ dB}$$

In many cases, the total gain is far greater than that required for adequate reception. Excessive gain will usually cause the received signal to be distorted and the transmitted information to be less intelligible. One solution to this problem is to provide a gain control in the receiver. Sometimes a potentiometer is connected at some point in an RF or IF amplifier stage to control the RF gain manually. Further, all receivers include a volume control in the audio circuit. The gain control allows the overall receiver gain to be reduced in order to deal with large signals.

A more effective way of dealing with large signals is to include AGC circuits. Automatic gain control is a feedback system that automatically adjusts the gain of the receiver based on the amplitude of the received signal. Very low level signals cause the gain of the receiver to be high. Large input signals cause the gain of the receiver to be reduced.

The use of AGC results in the receiver having a very wide dynamic range. *Dynamic range* refers to the measure of a receiver's ability to receive both very strong and very weak signals without introducing distortion and is the ratio of the largest signal that can be handled to the lowest, expressed in decibels. The dynamic range of a typical communications receiver with AGC is usually in the 60- to 100-dB range.

An AGC circuit takes the received signal either at the output of an IF amplifier or at the output of the demodulator and rectifies it into a direct current. The amplitude of the dc volt-

age is proportional to the level of the received signal. This dc voltage is applied to one or more IF amplifier stages for the purpose of controlling their gain.

The gain of a bipolar transistor amplifier is proportional to the amount of collector current flowing. Increasing the collector current from some very low level will cause the gain to increase proportionately. At some point, the gain flattens over a narrow collector current range and then begins to decrease as the current increases further. Figure 7-16 shows the typical gain variation versus collector current of a typical bipolar transistor. The gain peaks at 30 dB over the 6- to 12-mA range.

The amount of collector current in the transistor, of course, is a function of the base bias applied. A small amount of base current will produce a small amount of collector current and vice versa. Usually the bias level is fixed by a voltage divider, but in IF amplifiers the bias level is usually controlled by the AGC circuit. In some circuits, a combination of fixed voltage divider bias plus a dc input from the AGC circuit controls the overall gain.

As you can see from Fig. 7-16, the gain can be adjusted in two ways. The gain can be decreased by decreasing the collector current. An AGC circuit that decreases the current flowing in the amplifier in order to decrease the gain is called *reverse AGC*. The gain of the IF amplifier can also be reduced by increasing the collector current. As the signal gets stronger, the AGC voltage increases, which increases the base current. The increase in base current, in turn, increases the collector current, which reduces the gain. This method of gain control is known as *forward AGC*. In general, reverse

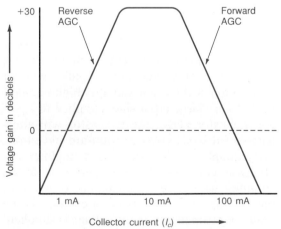

Fig. 7-16 Voltage gain of a bipolar transistor amplifier versus collector current.

AGC is more common in communications receivers. However, forward AGC is widely used in TV sets and typically requires special transistors for optimum operation.

Figure 7-17 shows two typical ways of applying AGC to an IF amplifier. In Fig. 7-17(*a*), the common emitter IF amplifier bias is derived from the voltage divider made up of R_1 and R_2 and the emitter resistor R_3. Resistor R_4 applied to the base accepts a negative dc voltage from the AGC circuit. As the signal amplitude level increases, the negative dc voltage increases which thereby decreases the base current. This in turn decreases the collector current and lowers the circuit gain. This is a reverse AGC circuit.

The circuit in Fig. 7-17(*b*) is similar, but the bias for the stage is derived from the emitter resistor R_1 and the AGC circuit itself. In this case, the AGC dc voltage is positive which sets the bias level. A strong signal will increase the positive voltage and, therefore, will increase the base current and collector current. This will reduce the circuit gain. This is an example of forward AGC.

Integrated-circuit differential amplifiers are widely used as IF amplifiers. The gain of a differential amplifier is directly proportional to the amount of emitter current I_E flowing. Because of this, the AGC voltage can be conveniently applied to the constant-current source transistor in a differential amplifier. A typical circuit is shown in Fig. 7-18 on the next page. The bias on constant-current source Q_3 is adjusted by R_1, R_2, and R_3 to provide a fixed level of emitter current I_E to differential transistors Q_1 and Q_2. Normally in a constant-gain stage, the emitter current value is fixed and the current divides between Q_1 and Q_2. The gain is easy to control by varying the bias on Q_3. In the circuit shown, increasing the positive AGC voltage will increase the emitter current and increase the gain. Decreasing the AGC voltage will decrease the gain so that reverse AGC action is obtained.

Dual-gate MOSFETs are widely used as IF amplifiers because the second control gate on the device provides a convenient means for varying the circuit gain. In the IF amplifier circuit of Fig. 7-19 on the next page, the AGC voltage is applied through R_1 to gate 2. Maximum circuit gain is obtained when the AGC voltage is several volts positive. Reducing the AGC voltage lowers the gain. Minimum gain occurs when the AGC voltage goes negative by several volts.

(a) Reverse AGC

(b) Forward AGC

Fig. 7-17 Methods of applying AGC to an IF amplifier.

The dc voltage used to control the gain is usually derived by rectifying either the IF signal or the recovered information signal after the demodulator. One of the simplest and most widely used methods of AGC voltage genera-

Fig. 7-18 An IF differential amplifier with AGC.

tion in an AM receiver is simply to use the output from the diode detector as shown in Fig. 7-20. The diode detector recovers the original AM information. The voltage developed across R_1 is a negative DC voltage. Capacitor C_1 filters out the IF signal leaving the original modulating signal. The time constant of R_1 and C_1 is adjusted to eliminate the IF ripple but to retain the highest modulating signal. The recovered signal is passed through C_2 to remove the

dc. The resulting ac signal is further amplified and applied to a loudspeaker. The dc voltage across R_1 and C_1 must be further filtered to provide a pure dc voltage. This is done with R_2 and C_3. The time constant of these components is chosen to be very large so that the voltage at the output is pure dc. The dc level, of course, will vary with the amplitude of the received signal. The resulting negative voltage is then applied to one or more IF amplifier stages.

In an FM receiver, the dc voltage can usually be derived directly from the demodulator. Both Foster-Seeley discriminator and ratio detector circuits provide convenient takeoff points for a dc voltage proportional to the signal amplitude. With additional *RC* filtering, a dc level proportional to the signal amplitude is derived for use in controlling the IF amplifier gain.

In many receivers, a special rectifier circuit devoted strictly to deriving the AGC voltage is used. Figure 7-21 shows a typical circuit. The input can either be the recovered modulating signal or the IF signal. It is applied to an AGC amplifier. A voltage doubler rectifier circuit made up of D_1, D_2, and C_1 is used to increase the voltage level high enough for control purposes. The *RC* filter R_1-C_2 removes any signal variations and produces a pure dc voltage. In some circuits, further amplification of the dc control voltage may be necessary. A simple IC op amp like that shown in Fig. 7-21 can be used. The connection of the rectifier and any phase inversion in the op amp will determine the polarity of the AGC voltage. It may be either positive or negative depending upon the

Fig. 7-19 An IF amplifier with a dual-gate MOSFET and AGC.

Fig. 7-20 Deriving the AGC voltage from the diode detector in an AM receiver.

types of transistors used in the IF and their bias connections.

Another feedback-control circuit similar to AGC used in high-frequency receivers is called *automatic frequency control (AFC)*. The purpose of AFC is to keep the LO on frequency. In receivers operating at frequencies above 100 MHz, oscillator stability is a problem. The LOs in superheterodyne receivers must be tunable so that stations of any frequency can be selected. This means creating an *LC* oscillator whose frequency cannot only be varied but also whose frequency stability is quite high. Frequency stability refers to the measure of an oscillator's ability to remain on the frequency to which it is set despite environmental changes. Oscillator frequency variations occur primarily because of temperature variations. Although the oscillator may be set to a particular frequency, it has a tendency to drift off of that frequency because of temperature changes or other circuit conditions. If the oscillator drifts too far off the frequency, the mixer will not convert the incoming signal to the proper IF value. The result is that the desired signal will not be picked up or that the receiver will

be mistuned so that only a portion of the signal passes through. This, of course, causes lower signal amplitude and distortion.

In those receivers where fixed-frequency channel operation is typical, the problem is overcome by using crystal oscillators. But in tunable receivers, normally some form of AFC is used.

In AFC, some of the signal from the output of the demodulator is filtered into a dc voltage and used to control a varactor that will, in turn, control the LO frequency. A typical arrangement is shown in Fig. 7-22 on the next page. The output from the frequency demodulator is a signal whose amplitude varies with frequency deviation. A frequency increase produces a positive output, and a frequency decrease produces a negative output. By using a low-pass filter made up of R_1 and C_1, a dc voltage level is obtained. This voltage is applied to VVC D_1 that is connected across the tuned circuit of the LO. It is reverse-biased with a negative voltage $(-V)$ through R_3.

Now, if the LO frequency should drift, the demodulator will detect this as a frequency variation in the signal. An increase in fre-

Fig. 7-21 An AGC rectifier and amplifier.

Fig. 7-22 Automatic frequency control.

quency will cause the LO frequency and the IF to increase. This produces a positive dc, which offsets the bias on D_1 causing its capacitance to increase and the oscillator frequency to decrease. The original increase is thus corrected for. Regardless of whether the oscillator drifts higher or lower in frequency, the AFC feedback voltage will correct for it.

Many FM radios and TV sets have AFC circuits of some kind. You may have noticed that most FM receivers usually provide a switch that can be used to turn the AFC off and on. For best results, the receiver tuning should be done with AFC off. This allows you to tune the signal right on frequency. Then the AFC can be turned on to keep the receiver on frequency. Tuning with the AFC turned on will cause the receiver to adjust the oscillator so the signal is properly received even though the oscillator has not been manually set to the correct frequency. The AFC circuit corrects for any tuning error. When this occurs, the AFC circuit is not operating at the center of its range. Therefore, it cannot correct for wide frequency variations. By performing the tuning first and getting the signal on channel and then turning on the AFC, full-range frequency control is obtained. Newer FM radios and TV sets do not require AFC because they use frequency synthesis for tuning.

Another circuit found in most communications receivers is a *squelch circuit*. The purpose of this circuit is to keep the receiver audio turned off until an RF signal appears at the receiver input. Most two-way communications are short conversations that do not take place continuously. In most cases, the receiver is left on so that if a call is to be received, it will be heard. With no RF signal at the input to the

receiver, the audio output will simply be background noise. In AM systems such as CB radios, the noise level is relatively high and very annoying. Even the noise level in FM systems is high. Most people do not want to listen to the background noise and would naturally turn the audio volume down. By doing this, when a signal is received, it may be missed.

The squelch circuit provides a means of keeping the audio amplifier turned off during the time that noise is received in the background. When an RF signal appears at the input, the audio amplifier is enabled.

The circuit in Fig. 7-23 shows the basic concept of squelch. The presence of a signal at the input is detected by monitoring the AGC voltage line. The AGC voltage is amplified by a dc amplifier and applied to the base of transistor Q_1 which is simply a switching inverter. This circuit, in turn, drives output switch Q_2. This switch is connected to the collector of one of the input audio amplifier stages.

With no input signal, the AGC voltage will be at a very low level, near zero. The dc amplifier output will, therefore, be low, and Q_1 will not conduct. As a result, Q_2 is turned on by the base current through R_1. Since Q_2 acts like a short circuit, it shunts the audio signal at the collector of Q_3 to ground through diode D_1. As a result, the audio signal from the detector does not get through to the power amplifier stages, and the speaker is quiet.

When an RF signal is received, the output of the dc amplifier is a high positive voltage, and therefore, turns on Q_1. This shunts base current away from Q_2 so that Q_2 cuts off. This allows the audio amplifier Q_3 to operate normally and pass the signal through to the speaker.

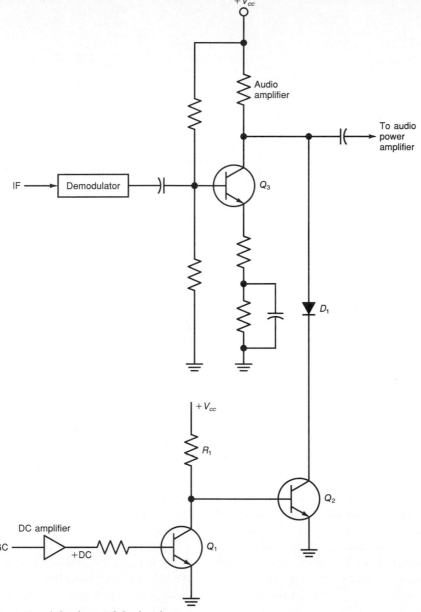

Fig. 7-23 A basic squelch circuit.

Most squelch circuits have a built-in level control that allows the circuit threshold to be adjusted to quiet very weak signals and allow only strong signals to turn the audio on.

Another popular type of squelch circuit amplifies the high-frequency background noise that exists when no signal is present and uses it to keep the audio turned off. When a signal is received, the noise circuit is overridden and the audio amplifier is turned on.

Figure 7-24 on the next page shows a noise-derived squelch circuit common to most communications receivers. The background noise with no signal is taken from one of the IF amplifiers and passed through C_1 and pot R_1 which form a high-pass filter. Only frequen-

cies above 5 kHz are passed. Most noise is of the higher-frequency variety. Resistor R_1 also serves as a squelch level control.

The noise is then further amplified by two transistor stages and is then rectified into a dc control voltage by a rectifier-voltage doubler circuit made up of C_2, C_3, D_1, and D_2. The rectifier output causes squelch gate Q_1 to saturate. The base current of audio amplifier Q_2 is shunted away through Q_1, and therefore, no audio amplification takes place and the receiver is quiet.

If a voice signal occurs, it will blank or mask any high-frequency noise. Thus no noise signal will be applied to the squelch circuit. High-pass filter C_1-R_1 will not pass the audio signal

Fig. 7-24 **A noise-derived squelch circuit.**

since its frequency content is below 3 kHz. As a result, no squelch voltage is developed at the rectifier output. The Q_1 cuts off so that Q_2 is biased normally, and the audio signal is passed through the audio power amplifier to the speaker.

A more sophisticated form of squelch used in some systems is known as *continuous tone control squelch (CTCS)*. This is a squelch system which is activated by a low-frequency tone transmitted along with the audio. The purpose of CTCS is to provide some communications privacy on a particular channel. Although the squelch circuit keeps the speaker quiet when no input signal is received, in some communications systems a particular frequency channel may be active with many calls. Listening to a variety of transmissions may be tiring or undesirable. The objective is to activate the squelch only when the desired signal is received. This is done by having the desired signal transmit a sine wave code tone along with the audio. The code tone is usually a very low frequency sine wave (usually less than 50 Hz) that is added to the audio prior to being applied to the modulator. The low-frequency tone will appear at the output of the demodulator in the receiver but usually will not be heard in

the speaker since the audio response of most communications systems cuts off below 300 Hz. However, the low-frequency tone can be used to activate the squelch circuit.

Figure 7-25 shows a general block diagram of the transmitter and receiver used in a CTCS system. In the transmitter, the low-frequency audio tone has a very specific frequency, and it is linearly mixed with the audio signal prior to being applied to the modulator.

At the receiver, a very selective bandpass filter tuned to the desired tone selects the tone at the output of the demodulator and applies it to a rectifier and filter to generate a dc voltage that will operate the squelch circuit.

Signals which do not transmit the desired tone will, of course, not trigger the squelch. When the desired signal comes along, the low-frequency tone is received, is converted into a dc voltage that operates the squelch, and then turns on the receiver audio. In most systems, multiple tones are available so that the transmitter or receiver may be set up to send to a particular station or to receive a particular station. The overall result is a nearly private communications channel.

Communications receivers designed for re-

(a) Transmitter

(b) Receiver

Fig. 7-25 **Continuous tone control squelch.**

ceiving SSB or CW (Morse code) signals have a built-in oscillator that permits recovery of the transmitted information. This circuit is called the *beat frequency oscillator (BFO)*. Usually it is designed to operate near the IF and is applied to the demodulator along with the IF signal containing the modulation.

Recall from the discussion of SSB that the basic demodulator is a balanced modulator. See Fig. 7-26(*a*) on the next page. The balanced modulator has two inputs: the incoming SSB signal at the IF, and a carrier which will mix with the incoming signal to produce the sum and difference frequencies, the difference being the original audio. The BFO is the oscillator that supplies the carrier signal to the balanced modulator. The term "beat" refers to the difference frequency output.

The BFO will be set to a value above or below the SSB signal frequency by an amount equal to the frequency of the modulating signal. Usually the BFO is made variable so that

its frequency can be adjusted for optimum reception. By varying the BFO over a narrow frequency range, the pitch of the received audio will change from low to high. It is typically adjusted for the sound of a natural voice.

The BFO is also used in receiving CW code. When dots and dashes are transmitted, the carrier is simply turned off and on for short and long periods of time. The amplitude of the carrier does not vary nor does its frequency. However, the on and off nature of the carrier does constitute a form of AM.

Consider for a moment what would happen if a CW signal were applied to a diode detector. The output of the diode detector would simply be pulses of dc voltage representing the dots and dashes. When applied to the audio amplifier, the dots and dashes would quiet the receiver (blank the noise), but they would not be discernable. To make the dots and dashes audible, the IF signal is mixed with the signal from a BFO. The BFO is usually injected di-

(a) SSB demodulation

Diode detector

C_1

(b) CW demodulation

Fig. 7-26   The use of a BFO.

rectly into the diode detector as shown in Fig. 7-26(*b*). In this way, the CW signal at the IF signal output mixes with the BFO in the diode detector. This results in sum and difference frequencies at the output. The sum frequency, of course, is very high, nearly double the IF value. It is filtered out by capacitor C_1 in the detector. The difference frequency will be a low audio frequency. In fact, if the BFO is variable, the difference frequency can be adjusted to any desired audio tone, usually in the 400- to 900-Hz range. Now, when the dots and dashes appear at the input to the diode detector, the output will be an audio tone which is amplified and heard in a speaker or earphones. Of course, the BFO is turned off for standard AM signal reception.

Self-Test

Supply the missing information in each statement.

53. RF amplifiers provide initial _____ and _____ in a receiver but also add _____ .
54. A low-noise transistor preferred at microwave frequencies is the _____ made of _____ .
55. Most of the gain and selectivity in a superhet is obtained in the _____ amplifier.
56. The selectivity in an IF amplifier is usually produced by using _____ between stages.
57. The bandwidth of a double-tuned trans-

former depends upon the degree of _____ between primary and secondary windings.
58. In a double-tuned circuit, minimum bandwidth is obtained with _____ coupling, maximum bandwidth with _____ coupling, and peak output with _____ or _____ coupling.
59. An IF amplifier that clips the positive and negative peaks of a signal is called a(n) _____ .
60. Clipping occurs in an amplifier because the transistor is driven by a high-level signal into _____ .
61. The gain of a bipolar class A amplifier can be varied by changing the _____ .
62. The overall RF-IF gain of a receiver is approximately _____ dB.
63. Using the amplitude of the incoming signal to control the gain of the receiver is known as _____ .
64. AGC circuits vary the gain of the _____ amplifier.
65. The dc AGC control voltage is derived from a(n) _____ circuit connected to the _____ or _____ output.
66. Reverse AGC is where a signal amplitude increase causes a(n) _____ in the IF amplifier collector current.
67. Forward AGC uses a signal amplitude increase to _____ the collector current, which decreases the IF amplifier gain.
68. The AGC of a differential amplifier is produced by controlling the current produced by the _____ transistor.
69. In a dual-gate MOSFET IF amplifier, the dc AGC voltage is applied to the _____ .
70. Another name for AGC in an AM receiver is _____ .
71. In an AM receiver, the AGC voltage is derived from the _____ .
72. Large input signals cause the gain of a receiver to be _____ by the AGC.
73. An AFC circuit corrects for frequency drift in the _____ circuit.
74. The AFC dc control voltage is derived from the _____ circuit in a receiver.
75. A(n) _____ is used in an AFC circuit to vary the LO frequency.
76. A circuit that blocks the audio until a signal is received is called a(n) _____ circuit.
77. Two types of signals used to operate the squelch circuit are _____ .
78. In a CTCS system, a low-frequency _____ is used to trigger the _____ circuit.

79. A BFO is required to receive _____ and _____ signals.

7-5 A TYPICAL COMMUNICATIONS RECEIVER

Now let's combine all you have learned so far and apply it to a real receiver. This will give you practice in analyzing and servicing communications receivers.

A typical receiver circuit is shown in Fig. 7-27 on the next page. This is a receiver designed to receive two-way communication between airplanes and airport controllers. Such communication from air to ground and ground to air takes place in the VHF range of 118 to 135 MHz. Amplitude modulation is used. Like most modern receivers, the circuit is a combination of discrete components and integrated circuits.

The signal is picked up by an antenna and fed through a transmission line to input jack J_1. The signal is coupled through C_1 to a tuned filter consisting of the series- and parallel-tuned circuits made up of L_1 through L_5 and C_2 through C_6. This broad bandpass filter passes the entire 118- to 135-MHz range.

The output of the filter is connected to an RF amplifier through C_7. The RF amplifier is made up of transistor Q_1 and its bias resistor R_4 and collector load R_5. The signal is then applied to the NE602 IC, designated U_1, through C_8. As you will recall, the NE602 contains a balanced mixer and a local oscillator. The local oscillator frequency is set by the circuit made up of inductor L_6 and the related components. C_{14} and D_1 in series form the capacitor that resonates with L_6 to set the frequency of the local oscillator. Tuning of the oscillator is accomplished by varying the dc bias on a varactor D_1. Pot R_1 sets the reverse bias, which in turn varies the capacitance to tune the oscillator.

Remember, to tune a superheterodyne receiver, you should vary the local oscillator frequency which is set to a frequency above the incoming signal by the amount of the intermediate frequency (IF). In this receiver, the IF is 10.7 MHz, a standard value for many VHF receivers. This means that to tune the receiver to the 118 to 135 range, the local oscillator must be varied from 128.7 to 145.7 MHz.

The output of the mixer, which is the difference between the incoming signal frequency and the local oscillator frequency, appears at pin 4 of the NE602 and is fed to a ceramic bandpass filter set to the IF of 10.7 MHz. This filter provides most of the receiver's selectivity. The insertion loss of the filter is offset by an amplifier consisting of Q_2, its bias resistor R_{10}, and collector load R_{11}. The output of this amplifier drives an MC1350 IC through C_{16}. U_2 is an integrated IF amplifier that provides extra gain and selectivity. The selectivity comes from the tuned circuit made up of IF transformer T_1. The MC1350 also contains all the AGC circuitry.

The signal at the secondary of T_1 is then fed to a simple AM diode detector consisting of D_2, R_{12}, and C_{30}. The demodulated audio signal appears across R_{12} and is then fed to an op amp U_{3-b}. This amplifier is a noninverting circuit biased by R_{13} and R_{14}. It provides extra amplification for the demodulated audio and the average direct current at the detector output. This amplifier feeds the volume control, which is pot R_2. The audio signal goes from there through C_{25} and R_{24} to another op amp U_{3-c}. Here the signal is further amplified. It is then fed to the 386 IC power amplifier U_4. This circuit drives the speaker which is connected by means of jack J_2.

The audio signal from the diode detector contains the dc level resulting from detection (rectification). Both the audio and the direct current are amplified by U_{3-b} and are further filtered into almost pure direct current by the low-pass filter made of R_{15} and C_{22}. This DC signal is applied to op amp U_{3-a}, where it is amplified into a dc control voltage. The direct current at the output pin 1 of U_3 is fed back to pin 5 on the MC1350 IC to provide AGC control. This ensures a constant comfortable listening level despite wide variations in the signal strength.

The AGC voltage from U_{3-a} is also fed to an op-amp comparator circuit made from amplifier U_{3-d}. The other input to this comparator is a dc voltage from pot R_3 which is used as a squelch control. Since the AGC voltage from U_{3-a} is directly proportional to the signal strength, it is used as the basis for setting the squelch to a level that will keep the receiver quiet until a signal of a predetermined strength comes along.

If the signal strength is very low or no signal is tuned in, the AGC voltage will be very low or nonexistent. This causes D_3 to conduct and effectively disable amplifier U_{3-c}, which in turn prevents the audio signal from the volume control from passing through to the power amplifier. If a strong signal exists, D_3 will be

Fig. 7-27 The Aviation Receiver—a superheterodyne unit, built around four ICs—is designed to receive AM signals in the 118- to 135-MHz frequency range. (Popular Electronics, January 1991, Gernsback Publications, Inc.)

reverse-biased, and so will not interfere with amplifier U_{3-c}. As a result, the signal from the volume control passes to the power amplifier and is heard in the speaker.

Self-Test

Answer the following questions.

80. What components or circuits in Fig. 7-27 determine the bandwidth of the receiver?

81. As R_1 is varied so that the voltage on the arm of the pot increases toward +9 V, how does the frequency of the local oscillator vary?

82. If there was no audio output from the speaker but you knew that a signal was present, what two controls would you check first?

83. What component provides most of the gain in this receiver?

84. Is the squelch signal or noise derived?

85. Does this receiver contain a BFO? Could it receive CW or SSB signals?

86. If the dc AGC voltage on pin 5 of U_2 was decreased, what would happen to the gain of U_2?

87. Where would you inject an audio signal to test the complete audio section of this receiver?

88. What frequency signal would you use to test the IF section of this receiver, and where would you connect it?

89. What component would be inoperable if C_{31} became shorted?

90. Explain how you would connect a digital counter to this receiver so that it would read the frequency of the signal to which it was tuned.

7-6 TRANSCEIVERS AND FREQUENCY SYNTHESIZERS

At one time, all pieces of communications equipment were individually packaged in units based on their function. In general, transmitters and receivers were always separate units. Today, however, most pieces of two-way radio communications equipment are packaged so that all the functions are contained within a single housing. Typically, this means that both the transmitter and receiver are in the same package. For this reason, this combination of units is referred to as a *transceiver*. Transceivers vary in size and complexity from very large, high-power, desktop units to the very small, pocket-size, hand-held walkie-talkies.

There are many advantages to packaging communications equipment in a single housing. In addition to sharing a common housing and power supply, in many designs the transmitters and receivers can also in many designs share circuits. This results in less circuitry, lower cost, and, in some cases, smaller size. Some of the circuits that can often be shared are antennas, oscillators, power supplies, tuned circuits, filters, and various kinds of amplifiers. These circuits perform a dual function serving both the transmitter and receiver, but not at the same time.

In FM transceivers, both the transmitter and the receiver operate from the same power supply, but usually this is the only shared circuit. The only exception is that in some designs the oscillators or frequency synthesizers used for generating the carrier and LO signals may sometimes be shared.

On the other hand, transceivers designed for AM, CW, and SSB can share many circuits.

Figure 7-28 is a general block diagram of a high-frequency transceiver capable of CW and SSB operation. Because both the receiver and transmitter make use of heterodyning techniques for generating the IF and the final transmission frequencies respectively, proper design enables both transmitter and receiver to share common LOs. Local oscillator 1 is the BFO for the receiver product detector and the carrier for the balanced modulator for producing DSB. Later, crystal LO 2 drives the second mixer in the receiver and the first transmitter mixer used for up conversion. Local oscillator 3 supplies the first receiver mixer and the second transmitter mixer.

Another shared circuit is the crystal filter. In the transmission mode, the crystal filter provides sideband selection after the balanced modulator. In the receiver, this same filter provides selectivity for the IF section of the receiver. In some designs, tuned circuits can also be shared. A particular tuned circuit may serve as the tuned input for the receiver or as the tuned output for the transmitter. Circuit switching may be manual but often is done with relays or diode electronic switches.

In many of the newer transceiver designs, both the transmitter and receiver typically share a circuit known as a frequency synthesizer. A *frequency synthesizer* is a variable-frequency signal generator that replaces one or more LOs or carrier oscillators in a transceiver, providing the frequency stability of a

Fig. 7-28 An SSB transceiver showing circuit sharing.

crystal oscillator but the convenience of continuous tuning over a broad frequency range in small, equal increments. Frequency synthesizers usually provide an output signal that varies in fixed frequency increments over a wide range. They are used in receivers as LOs and perform the receiver tuning function. In a transmitter, a frequency synthesizer provides basic carrier generation for channelized operation.

Over the years, many techniques have been developed for implementing frequency synthesizers with frequency multipliers and mixers. Today most frequency synthesizers are some variation of the phase-locked loop (PLL).

An elementary frequency synthesizer based on a PLL is shown in Fig. 7-29. As with all PLLs, it consists of a phase detector, a low-pass filter, and a VCO. The input to the phase detector is a reference oscillator. The reference oscillator is normally crystal-controlled to provide high-frequency stability. The frequency of the reference oscillator sets the increments in which the frequency may be changed, in this example, 100 kHz.

Note that the VCO output is not connected directly back to the phase detector. The VCO is applied to a frequency divider first. A frequency divider is a circuit whose output frequency is some integer submultiple of the input frequency. A divide-by-ten frequency divider will produce an output frequency that is one-tenth of the input frequency. Frequency dividers can be easily implemented with digital circuits to provide any integer value of frequency division.

Refer to Fig. 7-29. The reference oscillator is set to 100 kHz, or 0.1 MHz. Initially assume that the frequency divider is set for a division by 10. Keep in mind that in order for a PLL to become locked or synchronized, the second input to the phase detector must be equal in frequency to the reference frequency. In other words, for this PLL to be locked, the frequency divider output must be 100 kHz. This means that the VCO output has to be 10 times higher than this, or 1 MHz. One way to look at this circuit is as a frequency multiplier. The 100-kHz input was multiplied by 10 to produce the 1 MHz. In the design of the synthesizer, the VCO frequency is set to 1 MHz so that when it is divided, it will provide the 100-kHz input signal required for the locked condition. The synthesizer output is the output of the VCO. What we have created then is a 1-MHz signal source. Because the PLL is locked to the crystal reference source, the VCO output frequency has the same stability as the crystal oscillator.

To make the frequency synthesizer more useful, some means must be provided to vary its output frequency. This is done by varying the frequency division ratio. Through various switching techniques, the flip-flops in a frequency divider can be arranged to provide any desired frequency division ratio. The frequency division ratio is normally designed to be manually changed in some way. For example, a rotary switch may control logic circuits to provide the correct configuration. In another circuit, a thumbwheel switch may be used. Some designs incorporate a keyboard where the desired frequency division ratio can be keyed in. In most sophisticated circuits, a microprocessor may actually provide the correct frequency division ratio while at the same time controlling a direct frequency readout display.

By varying the frequency division ratio, the output frequency is changed. For example, in our circuit of Fig. 7-29, if the frequency division ratio is changed from 10 to 11, then the VCO output frequency must change to 1.1 MHz. In this

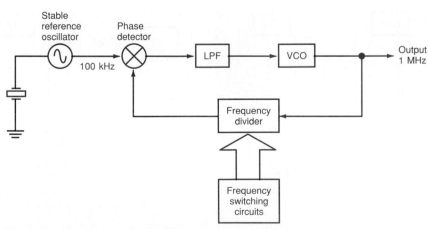

Fig. 7-29 **Basic concept of a PLL frequency synthesizer.**

way, the output of the divider will remain at 100 kHz as necessary to maintain a locked condition. Each increment of frequency division ratio change produces an output frequency change of 0.1 MHz. You can see now how the frequency increment is set by the reference oscillator.

All frequency synthesizers use the basic principles of the circuit in Fig. 7-29. However, more sophisticated versions are typical. Most frequency synthesizers also incorporate mixers for additional frequency conversion capabilities. Also, two or more PLLs may be used as part of a single synthesizer to provide multiple frequencies, say for both LO and carrier frequencies. In most transceivers, different LOs and carrier frequency generators are needed. These can usually be supplied by a synthesizer where channel frequency selection is simultaneous. That is, one manual frequency control will change both LO and carrier frequencies to put them simultaneously on the correct transmit and receive frequencies.

Figure 7-30 shows a PLL synthesizer for a 40-channel CB transceiver. Using two crystal oscillators for reference and a single-loop PLL, it synthesizes the transmitter frequency and the two LO frequencies for a dual-conversion receiver for all 40 CB channels.

The reference crystal oscillator operates at 10.24 MHz. It is divided by a divide-by-2

flip-flop to 5.12 MHz. Then a binary frequency divider that divides by 1024 produces 5.12/1024 MHz = 5 kHz. This is applied to the phase detector. The channel spacing, therefore, is 5 kHz.

The phase detector drives the low-pass filter and a VCO that generates a signal in the 16.27- to 16.71-MHz range. This is the LO frequency for the first receiver mixer. Assume that we want to receive on CB channel 1, or 26.965 MHz. The programmable divider is set to the correct ratio to produce a 5-kHz output when the VCO is 16.27. The first receiver mixer produces the difference frequency of 26.965 − 16.27 = 10.695 MHz. This is the first IF.

The 16.27-MHz VCO signal is also applied to mixer A, whose other input is 15.36 MHz, which is derived from the divided 10.24-MHz reference oscillator and a frequency tripler (×3) (5.12 MHz × 3 = 15.36 MHz). The output of mixer A drives the programmable divider, which feeds the phase detector.

The 10.24-MHz reference output is also used as the LO signal for the second receiver mixer. With a 10.695-MHz first IF, the second IF then is the difference or 10.695 − 10.24 MHz = 0.455 MHz, or 455 kHz.

The VCO output is also applied to mixer B along with a 10.695-MHz signal from a second crystal oscillator. The sum output is selected to produce the transmit frequency 10.695 +

Fig. 7-30 A frequency synthesizer for a CB transceiver.

CHAPTER 7 COMMUNICATIONS RECEIVERS **169**

16.27 = 26.965 MHz. This signal drives the class C drivers and power amplifiers.

Channel selection is done by changing the frequency division ratio on the programmable divider. It is usually done with a rotary switch. Most of the circuitry in the synthesizer is usually contained on one or two IC chips.

Self-Test

Answer the following questions.

91. In an FM transceiver, the only shared circuitry is usually the _____ .
92. In an SSB transceiver, the shared circuits are the _____ and the _____ .
93. The circuit used to generate one or more stable frequencies for the transmitter and receiver is known as a(n) _____ .
94. A basic circuit of a frequency synthesizer is a(n) _____ .
95. The output frequency of a synthesizer is changed by varying the _____ of the divider between the VCO and phase detector.
96. A frequency synthesizer has a phase detector input reference of 10 kHz. The divide ratio is 286. The output frequency is _____ MHz. The frequency change increment is _____ kHz.
97. The PLL is often combined with _____ circuits to produce multiple frequencies.
98. What three frequencies does the synthesizer in a transceiver usually generate?

SUMMARY

1. The simplest form of communications receiver is the tuned radio-frequency (TRF) receiver which provides RF and AF gain, selectivity, and a demodulator.
2. The two primary characteristics of a receiver are selectivity and sensitivity.
3. Selectivity is the ability to separate signals on different frequencies.
4. The bandwidth (BW) of a tuned circuit is the difference between the upper f_2 and lower f_1 3-dB-down frequencies (BW $= f_2 - f_1$).
5. The bandwidth of a tuned circuit is determined by the ratio of the resonant frequency to the Q where Q is X_L/R of the inductor (BW $= f_r/Q$). The Q can be computed from the bandwidth $Q = f_r/BW = f_r/(f_2 - f_1)$.
6. Sharper selectivity with steeper response curve skirts can be obtained by cascading tuned circuits.
7. The shape factor of a response curve is the ratio of the 60-dB-down bandwidth to the 6-dB-down bandwidth. The lower the shape factor, the better the selectivity.
8. Sensitivity is the ability of the receiver to pick up weak signals and is a function of gain.
9. The disadvantages of a TRF receiver are tuning difficulties and selectivity varying with frequency.
10. These problems of a TRF receiver are eliminated by using a superheterodyne (superhet) receiver.
11. A superhet uses a mixer to translate the incoming signal to a lower frequency, known as the intermediate frequency (IF), where fixed gain and selectivity can be obtained.
12. Most of the gain and selectivity in a superhet is obtained in the IF amplifier.
13. Low IFs are preferred because higher selectivity and better stability can be obtained with simple circuits.
14. Low IFs usually cause image interference problems.
15. An image is a signal on a frequency separated from the desired signal frequency by 2 times the IF value that interferes with reception.
16. The mixer will convert both the desired signal and the image to the IF.
17. Image interference is caused by poor receiver input selectivity that does not adequately reject the image.
18. Image interference can be reduced by using a double-conversion superhet that uses two mixers and IFs. The first IF is high to eliminate images, and the second is low to ensure good selectivity.
19. The most critical part of any receiver is the front end which consists of the RF amplifier and mixer as these circuits add the most noise to a weak signal.
20. Noise is any random interference to a weak signal.
21. A measure of a receiver's noise performance is its signal-to-noise (S/N) ratio.

22. External noise comes from industrial, atmospheric, and space sources.

23. Industrial noise sources are ignitions, motors and generators, switching devices, and fluorescent lights.

24. Atmospheric noise comes from lightning and other sources.

25. Space noise comes from the sun as well as other stars and extraterrestrial sources.

26. Internal noise from electronic components also interferes with reception and can totally mask weak signals.

27. Most internal noise comes from thermal agitation—the random movement of electrons and atoms in a material due to heat.

28. The thermal noise voltage across a resistor is proportional to the temperature, the bandwidth, and the resistance value and can be computed with the expression $v_n = \sqrt{4kTBR}$ where k is Boltzmann's constant 1.38×10^{-23}, T is the temperature in degrees kelvin, B is the bandwidth in hertz, and R is the resistance in ohms.

29. Other types of internal noise are shot and transit-time noise in semiconductors.

30. Noise cannot be eliminated, but its effect can be minimized.

31. Noise performance of a receiver or circuit is expressed in terms of the noise figure F, also called the noise factor, which is the ratio of the S/N input to the S/N output. It is usually given in decibels.

32. Noise performance of microwave components is usually expressed in terms of noise temperature in degrees kelvin $K = 290(F - 1)$.

33. Thermal noise is random and is often referred to as white noise or Johnson noise.

34. Since noise is a mixture of all random frequencies, its level can be reduced by narrowing the bandwidth.

35. Most receivers get their selectivity from double-tuned LC circuits.

36. Most receivers have AGC circuits so that a wide dynamic range of input signal amplitudes can be accommodated without distortion.

37. An AGC circuit rectifies the IF or demodulator output into dc to control the IF amplifier gain.

38. The gain of a bipolar transistor can be varied by changing its collector current.

39. In reverse AGC an increase in the AGC voltage decreases the collector current.

40. In forward AGC, an increase in AGC voltage increases collector current.

41. The gain of a dual-gate MOSFET in an IF amplifier is controlled by varying the dc voltage on the second gate.

42. Automatic frequency control (AFC) is a feedback system similar to AGC that is used to correct for frequency drift and instability in the LO of VHF, UHF, and microwave frequency receivers.

43. A squelch circuit is used to cut off the audio output to prevent annoying noise until a signal is received. Either the audio signal or background noise can be used to operate the squelch circuit.

44. Continuous tone control squelch (CTCS) circuits permit selective signaling by allowing only low-frequency tones to trigger the squelch.

45. A beat frequency oscillator (BFO) is used in SSB and CW receivers to provide a carrier that will mix with the input signal in the demodulator to generate the audio output.

46. A transceiver is a piece of communications equipment that combines a receiver and a transmitter in a common package where they share a common housing and power supply. Single-sideband transceivers allow sharing of the filter, LO stages, and other circuits.

47. Many new transceivers contain a frequency synthesizer that eliminates multiple crystal oscillators and LC tuned oscillators in the transmitter and receiver.

48. A frequency synthesizer is a signal generator usually implemented with a PLL that produces LO and transmitter carrier signals in a transceiver.

49. A frequency synthesizer has the stability of a crystal oscillator, but the frequency can be varied in small, equal increments over a wide range.

50. The frequency increments in a synthesizer are set by the frequency of the reference input to the phase detector.

51. The frequency of a synthesizer is changed by varying the divide ratio of the frequency divider between the VCO output and phase detector input.

52. Phase-locked loop synthesizers often incorporate mixers and multipliers to permit more than one frequency to be generated.

Choose the letter which best answers each question.

7-1. The simplest receiver is a(n)
    a. RF amplifier.
    b. Demodulator.
    c. AF amplifier.
    d. Tuned circuit.

7-2. The key conceptual circuit in a superhet receiver is the
    a. Mixer.
    b. RF amplifier.
    c. Demodulator.
    d. AF amplifier.

7-3. Most of the gain and selectivity in a superhet is obtained in the
    a. RF amplifier.
    b. Mixer.
    c. IF amplifier.
    d. AF amplifier.

7-4. The sensitivity of a receiver depends upon the receiver's overall
    a. Bandwidth.
    b. Selectivity.
    c. Noise response.
    d. Gain.

7-5. The ability of a receiver to separate one signal from others on closely adjacent frequencies is called the
    a. Sensitivity.
    b. S/N ratio.
    c. Selectivity.
    d. Gain.

7-6. A mixer has a signal input of 50 MHz and an LO frequency of 59 MHz. The IF is
    a. 9 MHz.
    b. 50 MHz.
    c. 59 MHz.
    d. 109 MHz.

7-7. A signal 2 times the IF away from the desired signal that causes interference is referred to as a(n)
    a. Ghost.
    b. Image.
    c. Phantom.
    d. Inverted signal.

7-8. A receiver has a desired input signal of 18 MHz and an LO frequency of 19.6 MHz. The image frequency is
    a. 1.6 MHz.
    b. 18 MHz.
    c. 19.6 MHz.
    d. 21.2 MHz.

7-9. The main cause of image interference is
    a. Poor front-end selectivity.
    b. Low gain.
    c. A high IF.
    d. A low S/N ratio.

7-10. For best image rejection, the IF for a 30-MHz signal would be
    a. 455 kHz.
    b. 3.3 MHz.
    c. 9 MHz.
    d. 55 MHz.

7-11. A tuned circuit is resonant at 4 MHz. Its Q is 100. The bandwidth is
    a. 400 Hz.
    b. 4 kHz.
    c. 40 kHz.
    d. 400 kHz.

7-12. A crystal filter has a 6-dB bandwidth of 2.6 kHz and a 60-dB bandwidth of 14 kHz. The shape factor is
    a. 0.186.
    b. 5.38.
    c. 8.3.
    d. 36.4.

7-13. Most internal noise comes from
    a. Shot noise.
    b. Transit-time noise.
    c. Thermal agitation.
    d. Skin effect.

7-14. Which of the following is *not* a source of external noise?
    a. Thermal agitation
    b. Auto ignitions
    c. The sun
    d. Fluorescent lights

7-15. Noise can be reduced by
    a. Widening the bandwidth.
    b. Narrowing the bandwidth.
    c. Increasing temperature.
    d. Increasing transistor current levels.

7-16. Noise at the input to a receiver can be as high as several
a. Microvolts.   c. Volts.
b. Millivolts.   d. Kilovolts.

7-17. Which circuit contributes most to the noise in a receiver?
a. IF amplifier   c. AF amplifier
b. Demodulator   d. Mixer

7-18. Which noise figure represents the lowest noise?
a. 1.6 dB   c. 2.7 dB
b. 2.1 dB   d. 3.4 dB

7-19. Which filter shape factor represents the best skirt selectivity?
a. 1.6   c. 5.3
b. 2.1   d. 8

7-20. Which input signal below represents the best receiver sensitivity?
a. 0.5 µV   c. 1.8 µV
b. 1 µV   d. 2 µV

7-21. The transistor with the lowest noise figure in the microwave region is a(n)
a. MOSFET.   c. JFET.
b. Dual-gate MOSFET.   d. MESFET.

7-22. The AGC circuits usually control the gain of the
a. Mixer.   c. IF amplifiers.
b. Detector.   d. Audio amplifiers.

7-23. Selectivity is obtained in most receivers from
a. Crystal filters.   c. Double-tuned circuits.
b. Mechanical filters. d. Audio filters.

7-24. Widest bandwidth in a double-tuned circuit is obtained with
a. Undercoupling.   c. Optimum coupling.
b. Critical coupling.   d. Overcoupling.

7-25. Automatic gain control permits a wide range of signal amplitudes to be accommodated by controlling the gain of the
a. RF amplifier.   c. Mixer.
b. IF amplifier.   d. AF amplifier.

7-26. In an IF amplifier with reverse AGC, a strong signal will cause the collector current to
a. Increase.   c. Remain the same.
b. Decrease.   d. Drop to zero.

7-27. Usually AGC voltage is derived by the
a. RF amplifier.   c. Demodulator.
b. IF amplifier.   d. AF amplifier.

7-28. An AFC circuit is used to correct for
a. Audio distortion.   c. Instability in the IF amplifier.
b. Strong input signals. d. Frequency drift in the LO.

7-29. A circuit that keeps the audio cut off until a signal is received is known as
a. A squelch.   c. AGC.
b. AFC.   d. A noise blanker.

7-30. A BFO is used in the demodulation of which types of signals?
a. AM   c. SSB or CW
b. FM   d. QPSK

7-31. Which of the following circuits are *not* typically shared in an SSB transceiver?
a. Crystal filter   c. Power supply
b. Mixers   d. LO

7-32. The basic frequency synthesizer circuit is a(n)
a. Mixer.   c. Frequency divider.
b. Frequency multiplier. d. PLL.

7-33. The output frequency increment of a frequency synthesizer is determined by the
a. Frequency division ratio.
b. Reference input to the phase detector.
c. Percentage of output frequency.
d. Frequency multiplication factor.

7-34. The output frequency of a synthesizer is changed by varying the
a. Reference frequency input to the phase detector.
b. Frequency division ratio.
c. Frequency multiplication factor.
d. Mixer LO frequency.

7-35. In Fig. 7-28, if the input reference is 25 kHz and the divide ratio is 144, the VCO output frequency is
a. 173.61 Hz.   c. 3.6 MHz.
b. 144 kHz.   d. 5.76 MHz.

7-36. The bandwidth of a parallel LC circuit can be increased by
a. Increasing X_C .
b. Decreasing X_L .
c. Decreasing coil resistance.
d. A resistor connected in parallel.

7-37. The upper and lower cutoff frequencies of a tuned circuit are 1.7 and 1.5 MHz respectively. The circuit Q is
a. 8.   c. 16.
b. 10.   d. 24.

7-38. The noise voltage across a 300-Ω input resistance to a TV set with a 6-MHz bandwidth and a temperature of 30°C is
a. 2.3 μV.   c. 5.5 μV.
b. 3.8 μV.   d. 6.4 μV.

7-39. The stage gains in a superheterodyne are as follows: RF amplifier, 10 dB; mixer, 6 dB; two IF amplifiers, each 33 dB; detector, −4 dB; AF amplifier, 28 dB. The total gain is
a. 73 dB.   c. 106 dB.
b. 82 dB.   d. 139 dB.

7-40. A tuned circuit resonates at 12 MHz with an inductance of 5 μH whose resistance is 6Ω. The circuit bandwidth is
a. 98 kHz.   c. 754 kHz.
b. 191 kHz.   d. 1.91 MHz.

7-41. In a receiver with noise-derived squelch, the presence of an audio signal causes the audio amplifier to be
a. Enabled.   b. Disabled.

CRITICAL THINKING QUESTIONS

7-1. Explain the operation of a superheterodyne receiver designed with a local oscillator that is set to the exact frequency of the received signal.

7-2. How could you design a receiver that can receive transmissions of AM, FM, and SSB signals?

7-3. Explain how you would make a receiver that could cover the frequency range from 500 kHz to 180 MHz. What special circuits and controls would you need?

7-4. How do scanners work? Could you design a receiver that would scan four frequencies? How would you proceed?

Answers to Self-Tests

1. selectivity
2. increase
3. narrow
4. true (resonant step-up)
5. increase (narrower bandwidth)
6. sidebands
7. 5
8. 0.53 MHz or 530 kHz (f_r = 10.6 MHz, X_L = 200Ω, Q = 20)
9. 9.95, 10.05
10. 1333.3
11. decreased
12. 2.4
13. sensitivity
14. tuned radio frequency
15. superheterodyne
16. intermediate frequency
17. local oscillator
18. IF amplifier
19. automatic gain control
20. signal or input, local oscillator
21. IF amplifier
22. c
23. image
24. 3.5
25. selectivity
26. 28
27. true
28. mixer
29. intermediate frequency
30. 9, 600
31. industrial, atmospheric, and space
32. thermal agitation, shot noise, transit-time noise
33. extraterrestrial or space
34. lightning
35. automobiles, motors, generators, fluorescent lights
36. thermal agitation
37. decibels
38. high
39. increase
40. white, Johnson
41. decrease
42. 1.36
43. shot, transit-time
44. false
45. true
46. noise figure
47. microwave
48. 464
49. distortion
50. high
51. RF amplifier, mixer
52. true
53. gain, selectivity, noise
54. MESFET or GASFET, gallium arsenide
55. IF
56. tuned circuits
57. mutual inductance
58. under, over, optimum, critical
59. limiter
60. cutoff, saturation
61. collector current
62. 100
63. automatic gain control
64. IF
65. rectifier, IF amplifier or detector
66. decrease
67. increase
68. constant-current source
69. control gate
70. automatic volume control
71. diode detector
72. reduced
73. local oscillator
74. demodulator
75. voltage-variable capacitor
76. squelch
77. audio, noise
78. tone, squelch
79. SSB, CW
80. the ceramic filter FIL_1
81. As the pot arm voltage goes more positive, the reverse bias across D_1 increases, thus decreasing its capacitance and the overall capacitance of D_1 in series with C_{14}; this in turn increases the local oscillator frequency.
82. volume R_2 and squelch R_3
83. U_2, the MC1350 IF amplifier IC
84. The squelch is signal-derived.
85. With no BFO, the receiver cannot receive CW or SSB.
86. The gain increases.
87. pin 5 of the U_{3-b} op amp
88. 10.7 MHz injected to the ceramic filter FIL_1
89. U_2, the MC1350 IC, would lose supply voltage at pin 2.
90. A digital counter could be connected to monitor the local oscillator output signal at pin 6 or 7 of the NE602 mixer IC. This would read the local oscillator frequency, which is 10.7 MHz higher than the actual input signal frequency. Therefore, the counter would have to be preset to a value that is 10.7 MHz less than zero count. In this way, the IF value would effectively be subtracted from the input count to read the correct frequency. Assume that the counter is set up to count four digits, 000.0. The counter is preset to 100.0 − 10.7 = 89.3, and the input signal frequency is 118 MHz. The local oscillator is tuned to 118 + 10.7 = 128.7 MHz. The counter counts 10.7 to reach 000.0 and then counts 118 more to display a final value of 118.0.
91. power supply
92. crystal filter, local oscillators
93. frequency synthesizer
94. phase-locked loop
95. frequency division ratio
96. 2.86, 10
97. mixer
98. LO for first mixer, LO for second mixer, transmitter

CHAPTER 8

Multiplexing

∎

CHAPTER OBJECTIVES

This chapter will help you to:

1. *Define* the terms multiplexing, demultiplexing, codec, and telemetry.
2. *Explain* the processes of frequency division multiplexing and demultiplexing and *state* three modern applications.
3. *Explain* the processes of time division multiplexing and demultiplexing and *name* two common applications.
4. *Explain* the operation of a pulse-code modulation system and *name* a common application.
5. *Calculate* the lowest possible sampling rate for a pulse-amplitude modulation system.
6. *Define* the terms compression, expansion, and companding.
7. *Explain* the operation and application of a sample and hold amplifier.

A communications link is established whenever a cable is strung between two points or when radio transmitters and receivers at two or more locations make contact. With such arrangements, voice conversations can take place, data can be exchanged, and control functions can be performed. Generally speaking, only one of these functions can be performed at a time. With one communications link, only one transaction can take place. However, there are many occasions when it would be desirable for multiple functions to occur simultaneously using the established link. In fact, many applications require that multiple conversations take place, dozens of data signals be transmitted, or several control operations take place at the same time. Logically, multiple cables or radio links would have to be installed to carry out such applications. Unfortunately, that would usually be impractical because of the cost and complexity. As it turns out, a single cable or radio link can handle multiple signals. Thanks to a process called multiplexing, hundreds or even thousands of signals of any type can be combined and transmitted over a single medium. As a result, multiplexing has made communications more practical and economically feasible. It helps conserve spectrum space and has allowed new sophisticated applications to be implemented. In this chapter, the techniques of multiplexing are introduced and some real world uses explained.

∎

8-1 INTRODUCTION

Multiplexing is the process of simultaneously transmitting two or more individual signals over a single communications channel. Multiplexing has the effect of increasing the number of communications channels so that more information can be transmitted.

There are many instances in communications where it is necessary or desirable to transmit more than one voice or data signal. The application itself may require multiple signals, and money can be saved by using a single communications channel to send multiple information signals. Telemetry and telephone applications are good examples. In satellite communications, multiplexing is essential to making the system practical and for justifying the expense.

Telemetry is a good illustrative example. *Telemetry* is the process of measurement at a distance. Telemetry systems are used to monitor physical characteristics of some application for the purpose of determining its status and operational conditions. This information may also be used as feedback in a closed-loop control system. Most spacecraft and many chemical plants, for example, use telemetry systems for monitoring their operation. Physical characteristics such as temperature, pressure, speed, light level, flow rate, and liquid level are monitored. Sensitive transducers convert these physical characteristics into electrical signals. These electrical signals are then processed in various ways and sent to a central monitoring location.

The most obvious way to send multiple signals from one place to another is to provide a single communications channel for each. For example, each signal could be sent over a single pair of wires. If long distances are involved, the signals will be degraded, and, therefore, special techniques must be used to prevent this. Using multiple wires is also an expensive process. When a very large number of signals must be monitored, many pairs of cables will be needed. This leads to extra cost and complexity. Ideally, it would be more economical if all the telemetry signals could somehow be combined and sent over a single cable.

In a spacecraft with multiple transducers, multiple transmitters would be required to send the signals back to earth. Again this leads to incredible cost and complexity. In the case of a spacecraft, multiple transmitters would weigh much and consume an enormous amount of extra power, making them impractical. Again the ideal situation would be to use a single transmitter and in some way combine all the various information signals and transmit them simultaneously over the single radio channel.

The telephone system is another example of the need for some means to increase the information-carrying capability of a single channel. There are literally hundreds of millions of telephones in this country and each must be capable of being connected to any other telephone. This is what the telephone

system is all about. If each telephone requires a two-wire path, imagine the enormous number of wires required to make all the various interconnections. The problem is further compounded if you want to add in the ability to connect each telephone to any other telephone in the world. On such a large scale, cost is a major factor. Everything must be done to minimize the number of interconnecting wires. This leads to the use of some method of combining multiple telephone conversations in such a way that they can be transmitted over a single pair of wires or a single radio communications channel.

The concept of a simple multiplexer is illustrated in Fig. 8-1. Multiple input signals are combined by the multiplexer into a single composite signal that is transmitted over the communications medium. Alternately, the multiplexed signals may modulate a carrier before transmission. At the other end of the communications link, a demultiplexer is used to sort out the signals into their original form.

Multiplexer (MUX *or* MPX) combines all inputs into a single signal

Demultiplexer (DEMUX) processes input signal by sorting it out into the original individual signals

Fig. 8-1   Concept of multiplexing.

There are two basic types of multiplexing: *frequency division multiplexing (FDM)* and *time division multiplexing (TDM)*. Generally speaking, FDM systems are used to deal with analog information, while TDM systems are used for digital information. Of course, TDM techniques are found in many analog applications as well because the process of analog-to-digital (A/D) and digital-to-analog (D/A) is so common. The primary difference between these techniques is that in FDM, individual signals to be transmitted are assigned a different frequency within a common bandwidth. In TDM, the multiple signals are transmitted in different time slots. In the following sections, we will discuss FDM and TDM in more detail.

From page 176:

Multiplexing

On this page:

Telemetry

Frequency division multiplexing (FDM)

Time division multiplexing (TDM)

Self-Test

Answer the following questions.

1. Which statement is most correct?
   a. Multiplexing uses multiple channels to transmit multiple signals.
   b. Multiplexing uses multiple channels to transmit individual signals.
   c. Multiplexing uses a single channel to transmit multiple signals.
   d. Multiplexing uses a single channel to transmit a single signal.
2. Two common applications of multiplexing are _____ .
3. The primary benefit of multiplexing is _____ .
4. When multiplexing is used to transmit information, then a _____ must be used at the receiving end of the communications link.
5. The two types of multiplexing are _____ .

8-2 FREQUENCY DIVISION MULTIPLEXING

Frequency division multiplexing is based on the idea that a number of signals can share the bandwidth of a common communications channel. The multiple signals to be transmitted over this channel are each used to modulate a separate carrier. Each carrier is on a different frequency. The modulated carriers are then added together to form a single complex signal which is transmitted over the single channel.

Figure 8-2 shows a general block diagram of an FDM system. Each signal to be transmitted feeds a modulator circuit. The carrier for each modulator f_c is on a different frequency. The carrier frequencies are usually equally spaced from one another over a specific frequency range. Each input signal is given a portion of the bandwidth. The result is illustrated in Fig. 8-3. As for the type of modulation, any of the standard kinds can be used including AM, SSB, FM, or PM.

Fig. 8-2   The transmitting end of an FDM system.

Fig. 8-3 Spectrum of an FDM signal.

The modulator outputs containing the sideband information are added together in a linear mixer. In a linear mixer, modulation and the generation of sidebands do not take place. Instead, all the signals are simply added together algebraically. The resulting output signal is a composite of all carriers containing their modulation. This signal is then used to modulate a radio transmitter. Alternately, the composite signal itself may be transmitted over the single communications channel. Another option is that the composite signal may become one input to another multiplexer system.

The receiving portion of the system is shown in Fig. 8-4. A receiver picks up the signal and demodulates it into the composite signal. This is sent to a group of bandpass filters (BPF), each centered on one of the carrier frequencies. Each filter passes only its channel and rejects all others. A channel demodulator then recovers each original input signal.

To be more specific about FDM systems, let's consider two practical examples, telemetry and FM stereo. As indicated earlier, telemetry is one of the most common uses for multiplexing techniques. In telemetry systems, many different physical characteristics are monitored by sensors. These sensors generate electrical signals which change in some way to indicate the amplitude or measurement of the physical characteristics. An example of a sensor is a thermistor used to measure temperature. A thermistor's resistance varies inversely with temperature: As the temperature increases, the resistance decreases. The thermistor is usually connected into some kind of a resistive network, such as a voltage divider or bridge, and is connected to a dc voltage source. The result is a dc output from this network whose voltage varies in accordance with temperature. That varying dc level must then be transmitted to a remote receiver for mea-

Fig. 8-4 The receiving end of an FDM system.

surement, readout, and recording. In this case, the thermistor becomes one channel of an FDM system.

Other sensors have different kinds of outputs. Many simply have varying dc outputs, while others may be ac in nature. Each of these signals is typically amplified, filtered, and otherwise conditioned before being used to modulate a carrier. All the carriers are then added together to form a single multiplexed channel. In such systems, FM is normally used.

The conditioned transducer outputs are used to modulate a subcarrier. The varying dc or ac changes the frequency of an oscillator operating at the carrier frequency. Such a circuit is generally referred to as a *voltage-controlled oscillator (VCO)* or a *subcarrier oscillator (SCO)*. The outputs of the SCOs are added together. A diagram of such a system is shown in Fig. 8-5.

The output of the signal conditioning circuits are fed to the VCOs. Of course, to produce FDM, each VCO operates at a different frequency.

Figure 8-6(a) shows a block diagram of a typical VCO circuit. The VCOs are available as single IC chips. One version made by Signetics is called the 566. It consists of a dual-polarity current source that linearly charges and discharges

an external capacitor C. The current value is set by an external resistor R_1. Together R_1 and C set the operating or center carrier frequency which can be any value up to about 1 MHz.

The current source may be varied by an external signal, either dc or ac. This is the modulating signal from a transducer or other source. The input signal varies the charging and current and, therefore, varies the carrier frequency, producing FM.

The current source output is a linear triangular wave which is buffered by an amplifier for external use. This triangular waveform is also fed to an internal Schmitt trigger which generates a rectangular pulse at the operating frequency. This is fed to a buffer amplifier for external use.

The Schmitt trigger output is also fed back to the current source where it controls whether the capacitor is charged or discharged. For example, the VCO may begin by charging the capacitor. When the Schmitt trigger senses a specific level on the triangular wave, it switches the current source. Discharging then occurs. The waveforms in Fig. 8-6(b) show this action. It is this feedback that creates a free-running, astable oscillator.

Most VCOs are astable multivibrators whose frequency is controlled by the input from the

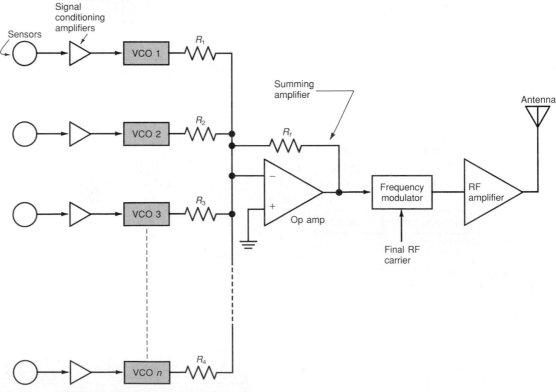

Fig. 8-5   An FDM telemetry transmitting system.

Fig. 8-6 (a) Typical IC VCO circuit and (b) waveforms.

signal conditioning circuits. The frequency of the VCO changes linearly in proportion to the input voltage. Increasing the input voltage causes the VCO frequency to increase.

The rectangular or triangular output of the VCO is usually filtered into a sine wave by a bandpass filter centered on the unmodulated VCO center frequency. This may be either a conventional *LC* filter or an active filter made with an op amp and *RC* input and feedback networks. The resulting sinusoidal output is applied to the linear mixer.

The linear mixing process in the FDM system can be accomplished with a simple resistor network as shown in Fig. 8-7. However, such networks greatly attenuate the signal. Typically, some voltage amplification is required in practical systems. A way to achieve the mixing and amplification at the same time is to use an op-amp summer like that shown in Fig. 8-5. Recall that the gain of each input is a function of the ratio of the feedback resistor R_f to the input resistor value (R_1, R_2, etc.). The output is given by the expression

$$V_o = V_1 \frac{R_f}{R_1} + V_2 \frac{R_f}{R_2}$$

$$+ V_3 \frac{R_f}{R_3} + \cdots + V_n \frac{R_f}{R_n}$$

In most cases, the VCO FM output levels are the same, and, therefore, all input resistors on the summer amplifier are equal. However, if variations do exist, amplitude corrections can be made by making the summer input resistors

Fig. 8-7 Resistive summing network.

adjustable. The output of the summer amplifier does invert the signal, but this has no effect upon the content.

The composite output signal is then typically used to modulate a radio transmitter. Again, most telemetry systems use FM. A system that uses FM of the VCO subcarriers as well as FM of the final carrier is usually called an FM/FM system. However, keep in mind that other kinds of modulation schemes may be used.

Most FM/FM telemetry systems conform to standards established many years ago by an organization known as the Inter-Range Instrumentation Group (IRIG). These standards define specific channels as indicated in Fig. 8-8. The center frequency of each channel is given along with the related channel number. A frequency deviation of ±7.5 percent is used on most of the channels, and this is increased to ±15 percent on the upper frequency channels. Also given in Fig. 8-8 is the upper frequency range that the modulating signal can have on each channel. On channel 1, for example, with a 400-Hz center frequency, the maximum signal frequency that can be used is 6 Hz. Most of the lower frequency channels are used for dc or very low frequency ac signals.

This set of standards is known as the proportional bandwidth FM/FM system. Since a fixed percentage of frequency deviation is specified, this means that the bandwidth is proportional to the carrier frequency. The higher the carrier frequency, the wider the bandwidth over which the modulating signal can occur.

Constant-bandwidth FM telemetry channels are also used. Carrier frequencies in the same approximate range as those given in Fig. 8-8 are used. However, a fixed deviation of ±2 kHz is typically specified, thereby creating multiple channels with a 4-kHz bandwidth. These are spaced throughout the frequency spectrum with some guard space in between channels to minimize interference.

The receiving end of a telemetry system appears as shown in Fig. 8-9. A standard superheterodyne receiver tuned to the RF carrier frequency is used to pick up the signal. An FM demodulator then reproduces the original composite multiplexed signal. This multiplexed signal is then fed to a demultiplexer which divides the signals and reproduces the original inputs.

The output of the first frequency demodulator is fed simultaneously to multiple BPFs, each of which is tuned to the center frequency of one of the specified channels. Each filter passes only its subcarrier and related sidebands and rejects all the others. As you can see, the demultiplexing process is essentially that of using filters to sort the composite multiplex signal back into its original components. The output of each filter is the SCO frequency with its modulation.

These signals then, in turn, are applied to frequency demodulators. Also known as discriminators, these circuits take the FM signal and recreate the original dc or ac signal produced by the transducer. These original signals are then measured and otherwise interpreted to

Band Number	Center Frequency (Hz)	Lower Limit (Hz)	Upper Limit (Hz)	Maximum Deviation (%)	Frequency Response (cps)
1	400	370	430	±7.5	6.0
2	560	518	602	"	8.4
3	730	675	785	"	11
4	960	888	1,032	"	14
5	1,300	1,202	1,399	"	20
6	1,700	1,572	1,828	"	25
7	2,300	2,127	2,473	"	35
8	3,000	2,775	3,225	"	45
9	3,900	3,607	4,193	"	59
10	5,400	4,995	5,805	"	81
11	7,350	6,799	7,901		110
12	10,500	9,712	11,288	"	160
13	14,500	13,412	15,588	"	220
14	22,000	20,350	23,650	"	330
15	30,000	27,750	32,250	"	450
16	40,000	37,000	43,000	"	600
17	52,500	48,562	56,438	"	790
18	70,000	64,750	75,250	"	1,050
A.	22,000	18,700	25,300	±15	660
B.	30,000	25,500	34,500	"	900
C.	40,000	34,000	46,000	"	1,200
D.	52,500	44,625	60,375	"	1,600
E.	70,000	59,500	80,500	"	2,100

Fig. 8-8 The IRIG FM subcarrier bands and specifications.

Fig. 8-9   An FM/FM telemetry receiver.

provide the desired information from the remote transmitting source. In most systems, the multiplexed signal is sent to a data recorder where it is stored for possible future use. The original telemetry output signals may be graphically displayed on a strip chart recorder or otherwise converted into usable outputs.

The demodulator circuits used in typical FM demultiplexers are either of the PLL or pulse-averaging type. Of these two types, the PLL circuit is generally preferred because of its superior noise performance. However, it is typically more complex and expensive than the simpler pulse-averaging type. A PLL discriminator is also used to demodulate the receiver output.

Another example of a commonly used FDM system is the telephone system. Telephone companies have for years been using FDM to send multiple telephone conversations over a

minimum number of cables. The concepts are the same as those previously discussed. Here the original signal is voice in the 300- to 3000-Hz range. The voice is used to modulate a subcarrier. Each subcarrier is on a different frequency. These subcarriers are then added together to form a single channel. This multiplexing process is repeated at several levels so that an enormous number of telephone conversations can be carried over a single communications channel assuming its bandwidth is sufficient.

The frequency plan for a typical telephone multiplex system is shown in Fig. 8-10. The symbols are explained in Fig. 8-11 on the next page. Here the voice signal amplitude modulates one of twelve channels in the 60- to 108-kHz range. The carrier frequencies begin at 60 kHz with a spacing of 4 kHz, slightly higher than the highest frequency used in a typical voice communication.

Fig. 8-10   Basic group frequency plan for FDM telephone system.

(a)

(b)

Fig. 8-11   Meaning of symbols in Fig. 8-10.
(*a*) **Upper and lower sidebands, and** (*b*) **sideband detail showing guard bands.**

Single-sideband, suppressed-carrier modulation is used in telephone multiplex systems. Refer to Fig. 8-12. The voice signal is applied to a balanced modulator along with a carrier. The output of the balanced modulator consists of the upper and lower sideband frequencies. The carrier is suppressed by the balanced mod-

ulator, and a highly selective filter is used to pass either the upper or lower sideband. The upper sidebands are selected here. The output of the filter is the sideband containing the original voice signal. All 12 SSB signals are then summed in a linear mixer to produce a single-frequency multiplexed signal. This set of 12 modulated carriers is generally referred to as a *basic group*.

If more than 12 voice channels are needed, multiple basic groups are used. The outputs of these basic groups can then be further multiplexed onto higher-frequency subcarriers. In the telephone system, as many as five 12-channel basic groups can be combined. Carrier frequencies in the 360- to 552-kHz range are used. These carriers are spaced 48 kHz apart. The frequency plan is shown in Fig. 8-13. The multiplexing process is similar, but here the output of each basic group modulates the higher-frequency carriers which are again summed to create an even more complex single-channel signal. Also SSB is used, and the lower sidebands are selected. Each of the 5 channels in this group carries 12 channels for a total of 60 voice signals. This composite signal is referred to as a *super group*.

Fig. 8-12   An FDM telephone multiplexer using SSB.

Fig. 8-13   Five basic groups are used to form a super group.

The process may again be taken another step farther. Up to 10 super groups can be used to modulate subcarriers in the 60- to 2540-kHz range. This allows a total of 5(12)(10) = 600 voice channels to be carried. The output of these 10 multiplexers is referred to as a *master group*. The process can continue with six master groups being further combined into one jumbo group for a total of 3600 channels. These three jumbo groups can then be multiplexed again into one final output to achieve a total of 10,800 voice channels.

Just keep in mind that as more and more levels of multiplexing are used, the bandwidth required to carry all the signals increases. A bandwidth of many megahertz is required to deal with the 10,800-channel composite signal mentioned above.

The receiving end of the system is shown in Fig. 8-14. Bandpass filters select out the various channels, while balanced modulators are used to reinject the carrier frequency and produce the original voice input.

Another well-known example of FDM is broadcast stereo FM. All FM broadcast stations use frequency multiplexing to transmit two channels of audio to the FM receiver in your car or to your hi-fi system at home. Let's take a look at how it works.

In stereo, two microphones are used to generate two separate audio signals. The two microphones pick up sound from a common source, such as a voice or band, but from different directions. The separation of the two microphones provides sufficient difference in the two audio signals to provide more realistic

Fig. 8-14   Demultiplexing the telephone signals.

reproduction of the original sound. These two independent signals must somehow be transmitted by a single transmitter. This is done by frequency multiplexing techniques.

Figure 8-15 is a general block diagram of a stereo FM multiplex modulator. The two audio signals generally called the left L and right R signals designate the positions of the microphones picking up the original sound. These two signals are fed to a circuit where they are combined to form sum $L + R$ and difference $L - R$ signals. The $L + R$ signal is a linear algebraic combination of the left and right channels. The composite signal it produces is the same as if a single microphone were used to pick up the sound. It is this signal that a monaural receiver will hear. See Fig. 8-16.

The combining circuit inverts the right channel signal thereby subtracting it from the left channel signal to produce the $L - R$ signal. These two signals, $L + R$ and $L - R$, will be transmitted independently and recombined later in the receiver to produce the individual right- and left-hand channels.

The $L - R$ signal is used to amplitude modulate a 38-kHz carrier. This carrier is fed to a

balanced modulator along with the $L - R$ signal. The balanced modulator suppresses the carrier but generates upper and lower sidebands as shown in Fig. 8-16. Since the audio response of an FM signal is in the 0.05- to 15-kHz range, the sidebands are in the frequency range of 38 kHz ± 15 kHz or in the range of 23 to 53 kHz. This DSB signal will be transmitted along with the standard $L + R$ audio signal.

Also transmitted with the $L + R$ and $L - R$ signals is a 19-kHz pilot carrier. This is generated by an oscillator whose output will also modulate the main transmitter. Note that the 19-kHz oscillator drives a frequency doubler to generate the 38-kHz carrier for the balanced modulator.

Some FM stations also broadcast another signal referred to as the *Subsidiary Communications Authorization (SCA)* signal. This is a separate subcarrier of 67 kHz which is frequency-modulated by audio signals, usually music. Special SCA FM receivers can pick up this music. The SCA portion of the system is generally used for broadcasting background music for stores, offices, and restaurants. If the

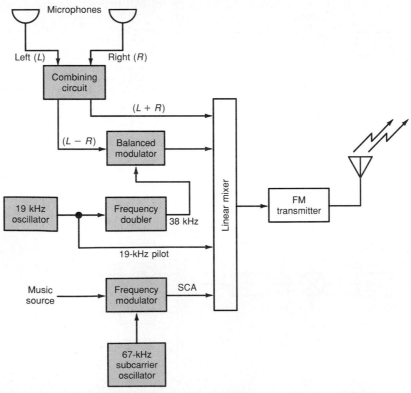

Fig. 8-15 General block diagram of an FM stereo multiplex modulator, multiplexer, and transmitter.

Fig. 8-16 Spectrum of FM stereo multiplex broadcast signal. This signal frequency modulates the RF carrier.

SCA system is used, the 67-kHz subcarrier with its music modulation will also modulate the FM transmitter.

As in other FDM systems, all the subcarriers are added with a linear mixer to form a single signal. The spectrum of that composite signal is shown in Fig. 8-16. This signal is used to frequency modulate the carrier of the broadcast transmitter. Again note that FDM simply provides a portion of the frequency spectrum for such independent signals to be transmitted. In this case there is sufficient spacing between adjacent FM stations so that the additional information can be accommodated. Some FM stations now transmit computer data over other subcarriers.

At the receiving end, the demodulation is accomplished with a circuit similar to that illustrated in Fig. 8-17. The FM superheterodyne receiver picks up the signal, amplifies it, and translates it to an IF, usually 10.7 MHz. It is then demodulated. The output of the demodulator is the original multiplexed signal. The various additional circuits now sort out the various signals and reproduce them in their original form.

The original audio $L + R$ signal is extracted by passing the multiplex signal through a low-pass filter. Only the 50- to 15000-Hz original audio is passed. This signal is fully compatible with monaural FM receivers without stereo capability. In a stereo system, the $L + R$ audio signal is fed to a linear matrix where it is mixed with the $L - R$ signal to create the two separate L and R channels.

The multiplexed signal is also applied to a bandpass filter that passes the 38-kHz suppressed subcarrier with its sidebands. This is the $L - R$ signal that modulates the 38-kHz carrier. This signal is fed to a balanced modulator for demodulation.

The 19-kHz pilot carrier on the multiplexed

Fig. 8-17 Demultiplexing and recovering the FM stereo and SCA signals.

signal is extracted by passing the multiplexed signal through a narrow-bandpass filter. This 19-kHz subcarrier is then fed to an amplifier and frequency doubler circuit which produces a 38-kHz carrier signal. This is fed to the balanced modulator. The output of the balanced modulator, of course, is the $L - R$ audio signal. This is fed to the linear resistive matrix along with the $L + R$ signal.

If the SCA signal is used, a separate bandpass filter centered on the 67-kHz subcarrier would extract the signal and feed it to a frequency demodulator. The demodulator output would then be sent to an audio amplifier and speaker.

At this point, both the $L + R$ and $L - R$ audio signals have been recovered and fed to the linear matrix. The matrix simply performs an algebraic operation on the two signals. The matrix both adds and subtracts these two signals. Adding the signals produces the left-hand channel.

$$(L + R) + (L - R) = 2L$$

Subtracting the two signals produces the right-hand channel.

$$(L + R) - (L - R) = 2R$$

The left- and right-hand audio signals are then sent to separate audio amplifiers and ultimately to the speakers. Similar multiplex systems are used in AM stereo radio and TV stereo.

Self-Test

Supply the missing word(s) in each statement.

6. In FDM, multiple signals share a common channel _____ .
7. The multiple signals in an FDM system are used to modulate separate _____ .
8. To combine the multiple signals in FDM, a _____ is used.
9. The composite multiplex signal is normally used to _____ an RF carrier before being transmitted.

10. Most telemetry multiplex systems used _____ modulation.
11. Telemetry signals are first conditioned by _____ and/or _____ before modulating a subcarrier.
12. Telemetry signals typically frequency modulate a _____ .
13. The key circuit in a demultiplexer is a(n) _____ .
14. The best demodulator for telemetry work is the _____ discriminator.
15. In FDM telephone systems, _____ modulation is used.
16. The monaural signal in FM stereo multiplex is designated _____ .
17. The four signals transmitted in FM stereo multiplex are _____ .
18. The type of modulation used on the $L - R$ channel is _____ .
19. The SCA signals use _____ modulation.

8-3 TIME DIVISION MULTIPLEXING

In FDM, multiple signals are transmitted over a single channel by sharing the channel bandwidth. This is done by allocating each signal a portion of the spectrum within that bandwidth. In TDM, each signal can occupy the entire bandwidth of the channel. However, each signal is transmitted for only a brief period of time. In other words, the multiple signals take turns transmitting over the single channel. This concept is illustrated graphically in Fig. 8-18. Here, four signals are transmitted over a single channel. Each signal is allowed to use the channel for a fixed period of time, one after another. Once all the signals have been transmitted, the cycle repeats again and again.

Time division multiplexing may be used with both digital and analog signals. To transmit multiple digital signals, the data to be transmitted is formatted into serial data words. For example, the data may consist of sequential bytes. One byte of data may be transmitted during the time interval assigned to a particu-

Fig. 8-18 The basic TDM concept.

lar channel. For example, in Fig. 8-18, each time slot might contain 1 byte from each channel. One channel transmits 8 bits and then halts while the next channel transmits 8 bits. The third channel then transmits its data word and so on. One transmission of each channel complete one cycle of operation called a *frame*. The cycle repeats itself at a high rate of speed. In this way, the data bytes of the individual channels are simply interleaved. The resulting single-channel signal is a digital bit stream that must somehow be deciphered and reassembled at the receiving end.

The transmission of digital data by TDM is straightforward in that digital data is incremental and can be broken up into words which can be easily assigned to different time slots. What is not obvious is how TDM can be used to transmit continuous analog signals. Yet, virtually any analog signal, be it voice, video, or telemetry measurements, can readily be transmitted by TDM techniques. This is accomplished by sampling the analog signal repeatedly at a high rate. Sampling is the process of "looking at" an analog signal for a brief instant of time. During this very short sampling interval, the amplitude of the analog signal is allowed to be passed or stored. By taking multiple samples of the analog signal at a periodic rate, most of the information contained in the analog signal will be passed. The resulting signal will be a series of samples or pulses that vary in amplitude according to the variation of the analog signal.

Figure 8-19 shows an analog signal which is being sampled. The resulting output is a series of pulses whose amplitudes are the same as those of the analog signal during the sample period. This process is known as *pulse-amplitude modulation (PAM)*.

Fig. 8-19 Sampling and analog signal to produce pulse-amplitude modulation.

The basic circuit for generating PAM is illustrated in Fig. 8-20. An astable clock oscillator drives a one-shot multivibrator which generates a narrow fixed-width pulse. This pulse is applied to a gate circuit which is essentially a switch that will open and close in accordance with the one-shot signal. When the one shot is off, the gate is closed and the analog signal applied to it will not pass. When the clock triggers the one shot once per cycle, the gate opens for a short period of time allowing the analog signal to pass through. The gate circuit may be constructed with diodes or can be an arrangement of bipolar or field-effect transistors.

Fig. 8-20 A pulse-amplitude modulator.

To recover the original information, the transmitted pulses are simply passed through a low-pass filter. The upper cutoff frequency of the low-pass filter is selected to pass the highest-frequency components contained within the analog signal. All higher frequencies are eliminated. Since the pulses themselves represent a composite of many high-frequency harmonics, these are effectively filtered out. The pulses, therefore, are smoothed into a continuous analog signal that is virtually identical in information content to the original transmitted signal. The process is similar to that used in a simple AM diode detector.

In order for the recovered signal to be an accurate representation of the original, the sampling rate must be high enough to ensure that rapid fluctuations are sampled a sufficient number of times. It has been determined that the sampling rate must be at least 2 times the highest-frequency component of the original signal in order for the signal to be adequately represented. This relationship between the original analog signal and the sampling frequency is known as the sampling theorem. If the upper bandwidth value of the analog signal is known, the minimum sampling rate can be found by simply multiplying it by 2.

If the sampled signal is a simple sine wave, then the minimum sampling frequency can be twice the sine wave frequency. A 2-kHz sine wave would have to be sampled a minimum of 2×2 kHz, or 4 kHz. A more complex signal containing harmonics up to 650 kHz would have to be sampled at a 2×650 kHz = 1300 kHz, or 1.3-MHz, rate or higher. The higher the sampling rate, the better the representation. Most systems sample at a rate higher than the minimum 2 times to ensure good fidelity.

In telephone communications, the upper frequency value of the voice content is assumed to be 3 kHz. This dictates a 2×3 kHz, or 6-kHz, sampling rate. In practice, the sampling rate for audio in telephone systems is 8 kHz. The higher sampling rate provides more faithful reproduction of the audio signal.

Although in many applications a sampling rate of twice the highest-frequency content is satisfactory, usually the sampling rate is made much higher. The actual value depends upon the application, but typically the sampling rate is 4 to 5 times the highest-frequency component in the analog signal. A sampling rate of 10 times the maximum analog bandwidth is ideal. This provides excellent representation of the signal.

Now, by combining the concepts of TDM and PAM, you can see how multiple analog signals can be transmitted over a single channel. This is accomplished by a circuit called a multiplexer (usually abbreviated MUX or MPX). The *multiplexer* is simply a single-pole, multiple-position mechanical or electronic switch that sequentially samples the multiple analog inputs at a high rate of speed. The simple rotary switch shown in Fig. 8-21 is an example. The switch arm dwells momentarily on each contact allowing the input signal to be passed through to the output. It then switches quickly to the next channel and allows that channel to pass for a fixed duration. The remaining channels are sampled in the same way. After each signal has been sampled, the cycle repeats. The result is that four analog signals

will be sampled creating PAM signals that are interleaved with one another. Figure 8-22 illustrates how four different analog signals are sampled by this technique. Be sure that you study the figure so that you can recognize each of the four signals in the composite waveform.

Multiplexers used in early TDM/PAM systems used a form of rotary switch known as a commutator. Multiple switch segments were attached to the various incoming signals, while a high-speed brush rotated by a dc motor rapidly sampled the signals as it passed over the contacts. Such commutators were used in early telemetry systems but have now been totally replaced by electronic circuits.

In practice, the duration of the sample pulses is shorter than the time which is allocated to each channel. For example, assume it takes the commutator or multiplexer switch 1 ms to move from one contact to another. The contacts could be set up so that each sample is 1 ms long. Typically, the duration of the sample is usually made about half that period, or in this example, 0.5 ms.

One complete revolution of the commutator switch is referred to as a frame. In other words, during one frame, each input channel is sampled one time. The number of contacts

Four analog inputs

Fig. 8-21 Simple rotary-switch multiplexer.

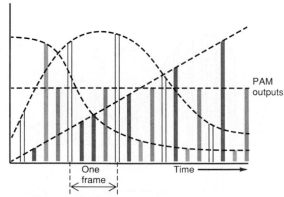

PAM outputs

One frame Time →

Fig. 8-22 Four-channel PAM time division multiplexer.

on the multiplexer switch or commutator sets the number of samples per frame. The number of frames completed in 1 s is called the frame rate. If you multiply the number of samples per frame by the frame rate, you will get the commutation rate or multiplex rate. This is the frequency of the pulses in the final multiplexed signal.

In our example in Fig. 8-22, the number of samples per frame is 4. Assume that the frame rate is 100 frames per second. The period for one frame, therefore, is 1/100 = 0.01 = 10 ms. During that 10-ms frame period, each of the four channels will be sampled once. Assuming equal sample durations, each channel would be allowed 10/4 = 2.5 ms. As indicated earlier, the full 2.5-ms period would not be used. Instead, the sample duration during that interval might only be 1-ms long.

Since there are four samples taken per frame, the commutation rate would be 4(100) or 400 pulses per second. This would be the basic frequency of the composite signal to be transmitted over the communications channel.

In practical TDM/PAM systems, electronic circuits are used instead of mechanical switches or commutators. The multiplexer itself is usually implemented with FETs which are nearly ideal off/on switches but can turn off and on at very high speeds. A complete TDM/PAM circuit is illustrated in Fig. 8-23. Only four channels are used so as to simplify the discussion.

The multiplexer is an op-amp summer circuit with FETs on each input resistor. When the FET is conducting, it has a very low resistance and, therefore, acts as a closed switch. When the transistor is cut off, no current flows through it and, therefore, it acts as an open switch. A digital pulse applied to the gate of the FET turns the transistor on. The absence of a pulse means that the transistor is cut off. The control pulses to the FET switches are such that only one FET is turned on at a time. These FETs are turned on in sequence by the digital circuitry illustrated.

All the FET switches are connected in series with resistors R_1 to R_4 that in combination with the feedback resistor R_f on the op-amp circuit determine the gain. For our discussion here, we will assume that the input and feedback resistor are all equal in value, meaning that the op-amp circuit has a gain of 1. Since this op-amp summing circuit inverts the polarity of the analog signals, it is followed by another op-amp inverter which again inverts and restores the proper polarity.

Fig. 8-23 A PAM mutliplexer.

The digital control pulses are developed by the counter and decoder circuit shown in Fig. 8-23. Since there are four channels, four counter states are needed. Such a counter can be implemented with two flip-flops which can represent four discrete states. These are 00, 01, 10, and 11. These are the binary equivalents of the decimal numbers 0, 1, 2, and 3. We can, therefore, label our four channels as channels 0, 1, 2, and 3.

A clock oscillator circuit triggers the two flip-flop counters. The clock and flip-flop waveforms are illustrated in Fig. 8-24. The flip-flop outputs are applied to the decoder gates. These are AND gates that are connected to recognize the four binary combinations 00, 01, 10, and 11. The output of each decoder gate is applied to one of the multiplexer FET gates.

The one-shot multivibrator shown in Fig. 8-24 is used to trigger all the decoder AND gates at the clock frequency. This one-shot multivibrator produces an output pulse whose duration has been set to the desired sampling interval. Recall in our earlier discussion the sampling interval was 2.5 ms while the actual sample length was set at 1 ms. Here the one shot would have a 1-ms pulse duration.

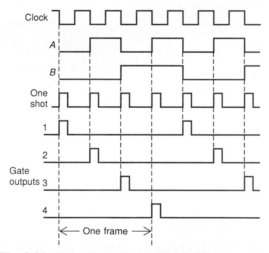

Fig. 8-24 Waveforms for a PAM multiplexer.

Each time the clock pulse occurs, the one shot generates its pulse which is applied simultaneously to all four AND decoder gates. At any given time, only one of the gates is enabled. The output of the enabled gate will be a pulse whose duration is the same as that of the one shot.

When the pulse occurs, it turns on the associated MOSFET and allows the analog sig-

nal to be sampled and passed through the op amps to the output. The output of the final op amp is the multiplexed PAM signal like that in Fig. 8-22.

The varying-amplitude PAM signal is not transmitted as it is over the single channel. Instead, these varying-amplitude pulses are used to modulate a carrier which is then transmitted over the communications medium. In most systems, the PAM signal is used to frequency modulate either an RF carrier for radio transmission or a subcarrier which, in turn, modulates a final RF carrier. Two such arrangements are shown in Fig. 8-25. In the first arrangement, the PAM signals phase modulate a carrier. This system is, therefore, referred to as a PAM/PM system.

In the second arrangement, the PAM signals phase modulate a subcarrier. In fact, two sets of PAM signals are shown, each phase modulating a subcarrier. These subcarriers are then linearly mixed and used to phase modulate the RF carrier which is the final transmitted signal. This system is referred to as PAM/PM/PM. The second system uses a combination of TDM and FDM schemes to create the final composite signal.

Once the composite signal is received, it must be demodulated and demultiplexed. In a PAM/PM/PM system, the signal is picked up by the receiver which ultimately sends the signal to a phase demodulator which recovers the original PAM data. In a PAM/PM/PM system, two levels of phase demodulation are required before the PAM signal is available. Once the composite PAM signal is obtained, it is applied to a demultiplexer (usually abbreviated DEMUX). The DEMUX is, of course, the reverse of a multiplexer. It has a single input and multiple outputs, one for each original input signal. Following with our four-channel example, a DEMUX for this system would have a single input and four outputs. Again, most DEMUXes use FETs driven by a pulse counter arrangement like that shown earlier in Fig. 8-23.

The main problem encountered in demultiplexing is synchronization. That is, in order for the PAM signal to be accurately demultiplexed into the original sampled signals, some method must be used to ensure that the clock frequency used on the DEMUX is identical to that used at the transmitting multiplexer. Further, even though the clock frequencies may be identical, the sequence of the DEMUX must be identical to that of the multiplexer so that when

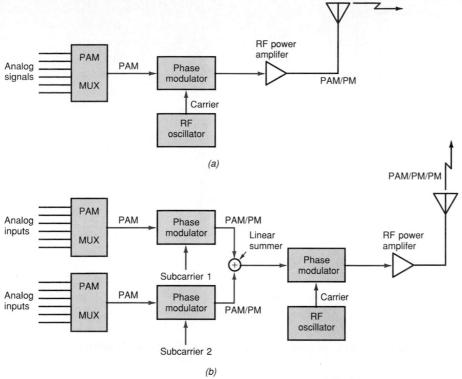

Fig. 8-25 **Combining analog (frequency) and digital (PAM) multiplexing.** (*a*) **PAM/PM,** **and** (*b*) **PAM/PM/PM.**

channel 1 is being sampled at the transmitter, channel 1 will be turned on in the receiver DEMUX at the same time. Such synchronization is usually carried out by a special synchronizing pulse included as a part of each frame. Let's take a look at some of the circuits used for clock frequency and frame synchronization.

Instead of using a free-running clock oscillator set to the identical frequency of the transmitter system clock, the clock for the DEMUX is derived from the received PAM signal itself. The circuits shown in Fig. 8-26 on the next page are typical of those used to generate the DEMUX clock pulses. They are called clock recovery circuits.

In Fig. 8-26(*a*) on the next page, the PAM signal is first applied to an amplifier-limiter circuit. This amplifies all the received pulses to a high level and then clips them off at a fixed level. The result is that the output of the limiter is a constant-amplitude rectangular wave whose output frequency is equal to the commutation rate. This is the frequency at which the PAM pulses occur. This, of course, is determined by the transmitting multiplexer clock.

The rectangular pulses at the output of the limiter are applied to a bandpass filter. This bandpass filter eliminates all the upper harmon-

ics, creating a sine wave signal at the transmitting clock frequency. This signal is applied to the phase detector circuit in a PLL along with the input from a VCO. The VCO is set to operate at the frequency of the PAM pulses. However, the VCO frequency is controlled by a dc error voltage applied to its input. This input is derived from the phase detector output which is filtered by a low-pass filter into a dc voltage.

The phase detector compares the phase of the incoming PAM sine wave to the VCO sine wave. If a phase error exists, the phase detector will produce an output voltage which is translated into dc to vary the VCO frequency. The system is stabilized or locked when the VCO output frequency is identical to that of the sine wave frequency derived from the PAM input. When the PLL is locked, the two sine waves are shifted in phase by 90°.

If the PAM signal frequency changes for some reason, the phase detector picks up the variation and generates an error signal which is used to change the frequency of the VCO to match. Because of the closed-loop feature of the system, the VCO will automatically track frequency changes in the PAM signal. This means that the clock frequency used in the

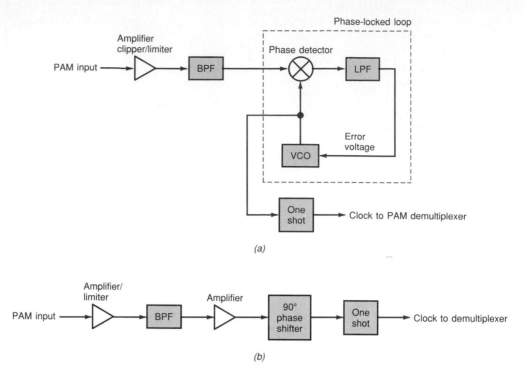

Fig. 8-26   Two PAM clock recovery circuits. (*a*) **Closed loop and** (*b*) **open loop.**

DEMUX will always perfectly match that of the original PAM signal regardless of any frequency changes that occur.

The output signal of the VCO is applied to a one-shot pulse generator which creates rectangular pulses at the proper frequency. These are used to step the counter in the DEMUX from which are derived the gating pulses for the FET DEMUX switches.

A simpler open-loop clock pulse circuit is shown in Fig. 8-26(*b*). Again the PAM signal is applied to an amplifier-limiter and then a bandpass filter, just as it was in the previous circuit. The sine wave output of the bandpass filter is then amplified and applied to a phase-shift circuit which produces a 90° phase shift at the frequency of operation. This phase-shifted sine wave is then applied to a pulse generator which, in turn, creates the clock pulses for the DEMUX. Although this circuit works satisfactorily, the phase-shift circuit is fixed to create a 90° shift at only one frequency, and, therefore, minor shifts in input frequency will produce clock pulses whose timing is not perfectly accurate. In most systems where frequency variations are not great, the circuit operates reliably.

With clock pulses of the proper frequency, some means is now needed to synchronize the multiplexer channels. This is usually done with a special synchronizing pulse that is applied to one of the input channels at the transmitter.

In our example of a four-channel system, only three actual signals would be transmitted. The fourth channel would be used to transmit a special pulse whose characteristics would be unique in some way so that it could be easily recognized. The amplitude of the pulse may be higher than the highest-amplitude data pulse, or the width of the pulse may be wider than those pulses derived by sampling the input signals. Special circuits can then be used to detect the synchronizing (sync) pulse.

Figure 8-27 shows an example of a sync pulse that is higher in amplitude than the maximum pulse value of any data signal. The sync pulse is also the last to occur in the frame. At the receiver, a comparator circuit is used to detect the sync pulse. One input to the comparator is set to a dc reference voltage equal to slightly higher than the maximum amplitude possible for the data pulses. When a pulse occurs that is greater than this amplitude, the comparator will generate an output pulse. The only time this occurs is when the sync pulse occurs. The output of the comparator will be a pulse that occurs at the same time as the sync pulse. This pulse can then be used for synchronization purposes.

Another method of providing sync is not to transmit a pulse during one channel interval. This leaves a blank space in each frame. This blank space can then be detected and used for

Fig. 8-27 Frame sync pulse and comparator detector.

synchronizing purposes. Such a circuit for doing this is illustrated in Fig. 8-28. Here the PAM multiplex signal is applied to an amplifier-limiter as it was in the clock circuits. The output is a series of pulses occurring at the pulse repetition rate of the PAM signal. However, since no pulse is transmitted during the sync interval, a blank space occurs. The resulting signal is inverted, so the blank space actually appears as a wider pulse as shown in Fig. 8-28. This signal is then added to a series of narrow clock pulses. These pulses are delayed for one-half the clock period before they are linearly added to the clipped PAM signal. This delay is accomplished with a one-shot multivibrator whose pulse duration is set to one-half the clock period. The first one shot generates a delay, while another one shot generates clock pulses at the same frequency but at a predetermined width. These are then added in a linear resistive circuit to form the composite

shown. The output of the linear mixer is applied to a comparator. The comparator threshold is set so that only the clock pulse added to the blank portion of the original signal is passed. Again, this synchronizing signal is used by the DEMUX.

As indicated, the sync pulse is usually the last one transmitted within a given frame. This sync pulse when detected at the receiver is used as a reset pulse for the counter in the DEMUX circuit. At the end of each frame, the counter is reset to zero, meaning that channel 0 is selected. Now when the next PAM pulse occurs, the DEMUX will be set to the proper channel. Clock pulses then step the counter in the proper sequence for demultiplexing.

Finally, at the output of the DEMUX, separate low-pass filters are applied to each channel to recover the original analog signals. Figure 8-29 on the next page shows the complete PAM DEMUX.

Fig. 8-28 A PAM sync detector circuit.

Pulse-code
modulation (PCM)

Analog-to-digital
converter (ADC)

Digital-to-analog
converter (DAC)

Fig. 8-29 Complete PAM demultiplexer.

Self-Test

Supply the missing information in each statement.

20. In TDM, multiple signals share a channel by transmitting in different _____ .

21. The process of sampling an analog signal at a high rate of speed is called _____ .

22. The minimum sampling rate for a signal with a 7-kHz bandwidth is _____ kHz.

23. The basic switching element in a PAM multiplexer is a _____ .

24. The name of the circuit used to recreate the clock pulses at the receiver in a PAM system is called a _____ .

25. The multiplexer and demultiplexer are kept in step with one another by a(n) _____ .

26. The time period during which all channels are sampled once is called a(n) _____ .

27. A PAM signal is demodulated with a(n) _____ .

28. A multiplexer is an electronic multiposition _____ .

29. The circuit that samples an analog signal in a PAM system is called a(n) _____ .

30. In a PAM/FM system, the RF carrier modulation is _____ .

31. A clock recovery circuit that tracks PAM frequency variations uses a(n) _____ .

8-4 PULSE-CODE MODULATION

The most popular form of pulse modulation used in TDM systems is *pulse-code modulation (PCM)*. Pulse-code modulation is a form of digital modulation where the code refers to a binary word that represents digital data. Multiple channels of serial digital data are trans-

mitted with TDM by allowing each channel a time slot in which to transmit one binary word of data. The various channel data are interleaved and transmitted sequentially. Instead of transmitting a single pulse whose amplitude is the same as the analog signal being sampled, in PCM a binary number representing the amplitude of the analog waveform at the sampling point is transmitted. As this statement implies, analog signals may be transmitted by PCM. The analog signal is sampled as in PAM and is then converted into digital format by an *analog-to-digital converter (ADC)*. The ADC converts the analog signal into a series of binary numbers where each number is proportional to the amplitude of the analog signal at the various sampling points. These binary words are converted from parallel to serial format and are then transmitted.

At the receiving end, the various channels are demultiplexed and the original sequential binary numbers are recovered. These are usually stored in a digital memory and then transferred to a *digital-to-analog converter (DAC)* which reconstructs the analog signal. Of course, the original data may be strictly digital in format, in which case no D/A conversion is required.

Pulse-code modulation systems allow the transmission of any form of digital data regardless of what it represents. Pulse-code modulation is used in telephone systems to transmit analog voice conversations, binary data for use in digital computers, and even video data. Most long-distance space probes such as the Mariner and Voyager have on-board video cameras whose output signals are digitized and transmitted back to earth in binary format. Such PCM video systems make possible the transmission of pictures over incredible distances.

Figure 8-30 shows a general block diagram of the major components in a PCM system. We will assume analog voice signals are the initial inputs. These are applied to ADCs as shown. The output of each ADC is an 8-bit parallel binary word. Since the digital data must be transmitted serially, the ADC output is fed to a shift register which produces a serial data output from the parallel input. The clock oscillator circuit driving the shift register operates at the desired frequency.

The multiplexing is done with a simple digital multiplexer. Since all the signals to be transmitted are binary in nature, a multiplexer constructed of standard AND or NAND gates can be used. A binary counter drives a decoder that selects the desired input channel.

Fig. 8-30   A PCM system.

The multiplexer output is a serial data waveform of the interleaved binary words. This binary signal is used to modulate a carrier. Either AM or FM may be used in typical systems. This creates either PCM/AM or PCM/FM. The output of the modulator is then fed to a transmitter for radio communications or can otherwise be transmitted by wire or fiber-optic cable. Additional levels of modulation may also be used.

At the receiving end of the communications link, the RF signal is picked up by a receiver and then demodulated. Refer to Fig. 8-31 on the next page. The original serial PCM binary waveform is recovered. This is fed to a shaping circuit, such as a Schmitt trigger, to clean up and rejuvenate the binary pulses. The original signal is then demultiplexed. This is done with a digital DEMUX using AND or NAND gates. The binary counter and decoder driving the DEMUX are kept in step with the receiver through a combination of clock recovery and sync-pulse detector circuits similar to those used in PAM systems. The demultiplexed serial output signals are fed to a shift register for conversion to parallel data and are then sent to a DAC followed by a low-pass filter. The result is a very accurate reproduction of the original voice signal.

Quantizing is the name given to the process of translating amplitude samples of an analog waveform into a binary code word. Quantizing is really the same thing as A/D conversion. You will also hear the term digitizing used to designate the same process.

The first step in the quantizing process is virtually identical to that in PAM. The analog signal is sampled at periodic intervals. This is done as described previously, with a gate. Another method of sampling is to use a sample and hold (S/H) amplifier. An S/H amplifier, also called a track/store circuit, accepts the analog input signal and passes it through unchanged during its sampling mode. In the hold mode, the amplifier remembers a particular voltage level at the instant of sampling. The output of the S/H amplifier is a fixed dc level whose amplitude is the value at the sampling time.

Figure 8-32 on the next page shows a simplified drawing of an S/H amplifier. A high-gain dc differential op amp is the basic element. The amplifier is connected as a follower with 100 percent feedback. Any signal applied to the noninverting (+) input will be passed through unaffected. The amplifier has unity gain and no inversion.

A storage capacitor is connected across the very high input impedance of the amplifier. The input signal is applied to the storage capacitor and the amplifier input through a MOSFET gate. A depletion mode MOSFET is normally used. This MOSFET acts as an on/off switch. When the gate is at 0 V, the transistor turns on, acting as a very low resistance and connecting the input signal to the amplifier. The charge on the capacitor follows the input signal. This is the sample or track mode for the amplifier. The output is simply equal to the input. When the gate voltage is made positive with a pulse, the MOSFET cuts off. In this mode, it acts as an open switch.

During the sample mode, the charge on the capacitor and the op-amp output simply follows the input signal. When the S/H control

Fig. 8-31 A PCM receiver-demultiplexer.

signal goes high, the transistor is cut off. The charge on the capacitor remains. The very high input impedance of the amplifier allows the capacitor to retain the charge for a relatively long period of time. The output of the S/H amplifier then is the dc voltage value of the input signal at the instant the S/H control pulse switches from low (sample) to high (hold). It is this voltage that is applied to the ADC for conversion into a proportional binary number.

The primary benefit of an S/H amplifier is that it stores the analog voltage during the sam-

pling interval. In some high-frequency signals, the analog voltage may actually change during the sampling interval, either increasing or decreasing. This is an undesirable condition since it will confuse the A/D converter and introduce some error. The S/H amplifier, however, stores the voltage on the capacitor which remains constant during the sampling interval and ensures more accurate quantizing.

In the quantizing process, we are effectively dividing the total analog signal amplitude range into a number of equal amplitude increments.

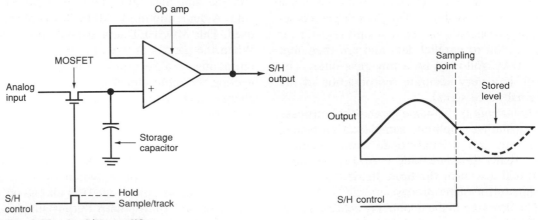

Fig. 8-32 An S/H amplifier.

Each one of these increments will be represented by a specific binary code. For example, assume that the total analog amplitude voltage range is 0 to 15 V. We could represent each voltage increment by a 4-bit binary number where 0 V was represented by 0000 and 15 V was represented by 1111. During the sampling of the analog wave, the amplitudes of the samples can assume any one of a number of infinite values between 0 and 15 V. In the quantizing process, each of those values will be converted into an even or integer value. For example, one of the analog samples may be 9.2 V. This is closest to the integer 9 and, therefore, the 9.2-V value will be represented by the value 9 in binary form, or 1001. An analog value of 12.7 V might be represented as the integer 13, or 1101. As you can see, the quantizing process introduces some error. This is called *quantizing error*. The result is that the analog signal is somewhat distorted by the process. The quantized analog signal is only an approximation of the real thing.

Although fifteen levels provide only crude quantization, improved representations of the analog signal can be used by providing more quantizing increments. The greater the number of individual voltage increments or levels provided in the quantizer, the more closely the analog signal can be approximated.

It has been determined that the range of voice amplitude levels in the telephone system is approximately 1000 to 1. In other words, the largest-amplitude voice peak is approximately 1000 times the smallest voice signal. This voltage ratio of 1000:1 represents a 60-dB range. If a quantizer with 1000 increments were used, very high quality analog signal representation would be achieved. For example, an ADC with a 10-bit word can represent 1024 individual levels. A 10-bit ADC would provide excellent signal representation. If the maximum peak audio voltage were 1 V, then the smallest voltage increment would be one-thousandth of this, or 1 mV.

In practice, it has been found that it is not necessary to use this many quantizing levels for voice. In most practical PCM systems, a 7- or 8-bit ADC is used for quantizing. One popular format is to use an 8-bit code where 7 bits represent 128 amplitude levels and the eighth bit designates polarity (0 = + , 1 = −). Overall, this provides 256 levels, one-half positive, the other half negative.

As indicated earlier, the analog voltage range of a typical voice signal is approximately 1000 to 1. As it turns out, however, it is the lower-level signals that predominate. Most of the conversations take place at a normal low level. For this reason, the upper end of the quantizing scale is not often used. It may be reached during momentary peaks of loud talking, shouting, or emotional outbursts, but for most general conversations, the lower-level signals will be more typical.

Since most of the signals are low-level, the quantizing error will be larger. In other words, the smallest increment of quantization becomes a larger percentage of the lower-level signal. It is a smaller percentage of the peak amplitude value, of course, but that is irrelevant when the signals are much lower in amplitude. The increased quantizing error can produce garbled or distorted sound.

In addition to increased quantizing error, low-level signals are also more susceptible to noise. Noise represents random spikes or voltage impulses added to the signal. The result is static that interferes with the low-level signals and makes intelligibility difficult.

The most common means of overcoming the problems of quantizing error and noise is to use a process of signal compression and expansion known as *companding*. At the transmitting end, the voice signal to be transmitted is compressed. That is, its dynamic range is decreased. The lower-level signals are emphasized, while the higher level signals are de-emphasized. This compression can take place prior to quantizing. But in some systems, companding is accomplished digitally in the ADC by having unequal quantizing steps, small ones at low levels and larger ones at higher levels.

At the receiving end, the recovered signal is fed to an expander circuit that does the opposite, de-emphasizing the lower-level signals and emphasizing the higher-level signals, thereby returning the transmitted signal to its original condition. Companding greatly improves the quality of the signal being transmitted.

One type of compression circuit is a nonlinear amplifier that amplifies lower-level signals more than it does upper-level signals. Figure 8-33 on the next page shows a graph illustrating the companding process. The curve shows the relationship between the input and output of the compander. Note that at the lower input voltages, the gain of the amplifier is high and produces high output voltages. As the input voltage increases, the curve begins to flatten producing proportionately lower gain. The nonlinear curve compresses the upper-level signals while bringing the lower-level signals up to a higher amplitude. Such

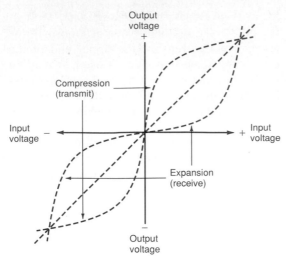

Fig. 8-33 Compression and expansion curves.

compression greatly reduces the dynamic range of the audio signal. Instead of an amplitude ratio of approximately 1000:1, compression reduces this to approximately 60:1. The actual degree of compression, of course, can be controlled by carefully designing the gain characteristics of the compression amplifier. Thus the 60-dB voice range is reduced to around a 36-dB range.

In addition to minimizing quantizing error and the effects of noise, compression also lowers the dynamic range so that fewer binary bits are required to digitize the audio signal. A 64:1 voltage ratio could be easily implemented with a 6-bit ADC. In practice, a 7-bit ADC is used.

Figure 8-34 shows a simplified diagram of a compression amplifier. Two biased diodes and their resistors are connected as the feedback elements in an op amp. Although only two diodes are shown, typically multiple diodes each biased to a different voltage level are used. At the lower-

level signals, the amplifier output is basically linear. But as the input signals get larger, the amplifier output will grow larger and at some point the diodes will begin to turn on. When the diodes turn on, they begin to shunt the feedback resistor RF and thereby reduce the amplifier gain. Once the audio signal has been compressed, it is then applied to the ADC for quantizing.

At the receiving end, the digital signals are translated into analog signals. An analog signal is then passed through an expander amplifier that performs the opposite function of the compressor. A typical circuit is shown in Fig. 8-35. Again, biased diodes are used to shape the amplifier gain curve. Here the biased diodes are used on the input resistance to an op amp. Figure 8-33 shows the variation of the output voltage with respect to the input voltage during expansion. The lower-level signals are amplified less than the higher-level signals. If the compression and expansion curves are equal and opposite, the result will be a highly accurate reproduction of the original transmitted signal. In Fig. 8-35, as the input increases, the output increases up to a certain level. Beyond that point, the diodes cut in, shunting the input resistor R_i, thereby increasing the gain of the circuit in steps at the higher-input levels.

Fig. 8-35 Expansion amplifier.

There are two basic types of companding used in telephone systems. One is called the μ-*law compander* and the other is called the *A-law compander*. The μ-law compander has a slightly different compression and expansion curve than the A-law compander. The μ-law compander is used in telephone systems in the United States and Japan, while the A-law compander is used in European telephone networks. The two are incompatible, but conversion circuits have been developed to make the two systems compatible. These circuits convert μ-law to A-law and vice versa. According to international telecommunications regulations, those using μ-law are responsible for the conversions.

Although much companding is analog in nature as described, digital companding can also

Fig. 8-34 Compression amplifier.

be accomplished. One method is simply to use a nonlinear ADC. These converters provide a greater number of quantizing steps at the lower levels than they do at the higher levels, providing compression. On the receiving end, a matching nonlinear DAC is used to provide the opposite, compensating, expansion effect.

Now let's take a look at a typical PCM system. The most commonly used PCM system is the T-1 system developed by Bell Telephone for transmitting telephone conversations by high-speed digital links. The T-1 system multiplexes 24 voice channels onto a single line using TDM techniques. Each analog voice channel is sampled at an 8-kHz rate. In other words, the analog waveform is sampled every 125 μs. Those samples are then converted to serial digital words by the ADCs. These serial digital words from each of the 24 channels are then transmitted one after another. Each sample is an 8-bit word, 7 bits of magnitude and 1 bit representing the sign of polarity.

During the 125-μs interval between analog sampling on each channel, twenty-four 8-bit words occur, each representing a sample from each of the channels. The channel sampling rate is (125 μs)/24 = 5.2 μs per cycle or 192 kHz. This represents a total of 24(8) = 192 bits. One additional bit is then added to this stream for synchronization purposes. That bit is a sync pulse used to keep the transmitting and receiving signals in step with one another. This represents a total of 193 bits. The twenty-four 8-bit words and the synchronizing bit form one frame. The sequence is then repeated again and again. The 193 bits multiplied by the 8-kHz sampling rate produce a total bit rate for the multiplexed signal of 193 × 8 kHz = 1544 kHz = 1.544 MHz. Figure 8-36 shows one frame of a T-1 signal.

Since telephone conversations are two-way in nature, both ends of the communication link have transmitting (multiplexing) and receiving (demultiplexing) capability. All the A/D and

D/A conversion and related functions, such as serial-to-parallel and parallel-to-serial conversion as well as companding, are usually taken care of by a single large-scale IC chip known as a *codec*. One codec is used per channel at each end of the communication channel. The codecs are combined with the digital multiplexers and DEMUXes and the clock and synchronizing circuits to complete the system.

Figure 8-37 on the next page shows a simplified block diagram of a codec. The analog input is sampled by the S/H amplifier at an 8-kHz rate. The samples are quantized by the successive-approximations-type of ADC. Compression is done digitally in the ADC. The parallel ADC output is sent to a shift register to create the serial data output which goes to one input of a digital multiplexer.

The serial digital input is derived from a digital DEMUX. The clock shifts it into a shift register for serial-to-parallel conversion. The 8-bit parallel word is sent to the DAC. This DAC has digital expansion built in. The analog output is then buffered. It may be filtered externally. Most codecs are made with CMOS and come in a 16-pin dual-in-line IC package.

Today PAM is rarely used in telephone, telemetry, or other systems requiring multiplexing. Neither are pulse-width modulation (PWM) or pulse-position modulation (PPM), both of which were once popular in TDM systems. Today, virtually all TDM systems use PCM. Because it is all digital, PCM is highly immune to noise, very reliable, and inexpensive.

In PCM, the transmitted binary pulses all have the same amplitude; therefore, like an FM signal, they may be clipped to reduce noise. Further, even if the signals are highly degraded due to noise, attenuation, or distortion, all the receiver has to do is determine if a pulse was transmitted or not. It does not have to worry about amplitude, width, frequency, phase shape, or otherwise. It just determines if the pulse is there or not. Such signals are easily

Fig. 8-36   The T-1 frame format, serial data.

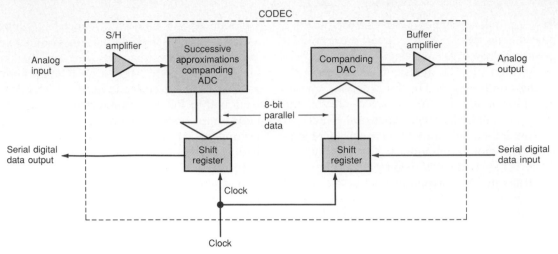

Fig. 8-37 **Simplified block diagram of an IC codec.**

recoverable and rejuvenated, even under the most severe attenuation, noise, or distortion. Pulse-code modulation is so superior to most other forms of modulation that it has replaced them all. Even FM, the long-standing favorite is being replaced by PCM in analog systems. This is especially true in satellite communications systems. Thanks to low-cost digital ICs, PCM is now cheap and easy to implement.

Self-Test

Supply the missing information in each statement.

32. In PCM systems, analog signals are transmitted as _____ .
33. Quantizer is another name for a(n) _____ .
34. The standard audio sampling rate in a PCM telephone system is _____ kHz.
35. The standard word size in a PCM audio system is _____ bits.
36. The name of a widely used PCM telephone system is _____ .
37. The process of emphasizing lower-level signals and de-emphasizing higher-level signals is called _____ .
38. An LSI circuit containing an ADC, DAC, compander, expander, and serial/parallel converters is called a(n) _____ .
39. The main benefit of PCM over PAM, FM, and other modulation techniques is its immunity to _____ .
40. A T-1 multiplexer has _____ channels and a total bit rate of _____ MHz.

SUMMARY

1. Multiplexing is the process of transmitting multiple signals over a single communications channel.
2. The primary benefit of multiplexing is economic since multiple communications can take place for the cost of a single link plus the multiplexing equipment.
3. The two major types of multiplexing are frequency division and time division multiplexing.
4. Some examples of multiplexing include telemetry, telephone systems, satellite communications, and radio/TV stereo broadcasting.
5. In frequency division multiplexing (FDM), multiple signals share the common bandwidth of a single communications channel, each occupying a separate portion of the bandwidth.
6. In FDM, each signal modulates a subcarrier on a different frequency. The subcarriers are then linearly mixed to form a composite signal that is usually used to modulate a final carrier for transmission.
7. In FDM telemetry systems, transducers frequency modulate subcarrier oscillators whose outputs are combined and used to frequency modulate a final carrier (FM/FM).
8. A subcarrier oscillator is a VCO whose frequency varies linearly in proportion to the amplitude of a modulating dc or ac signal.
9. Recovering the individual signals at the re-

ceiver is done with a demultiplexer whose main components are bandpass filters tuned to the individual subcarrier frequencies.

10. In FM/FM telemetry systems, PLL and pulse-averaging discriminators are used to demodulate the subcarriers; PLL discriminators are preferred because of their superior noise performance.

11. The telephone system routinely uses FDM systems to allow many conversations to be carried on a single cable.

12. Single-sideband suppressed-carrier modulation is used in telephone FDM systems to minimize bandwidth requirements.

13. Stereo broadcasts from FM stations use FDM techniques. Two channels of audio L and R are combined to form $L + R$ and $L - R$ signals. The $L - R$ signal modulates a 38-kHz subcarrier by DSB. The $L + R$ signal, DSB signal, and a 19-kHz pilot carrier are combined and used to frequency modulate the final transmitter carrier.

14. In some FM broadcast systems, music is broadcast on a separate 67-kHz subcarrier which is frequency modulated. This is referred to as Subsidiary Communication Authorization (SCA) signal or channel.

15. In time division multiplexing (TDM), each channel is assigned a time slot and may transmit for a brief period using the entire bandwidth of the medium. Signal sources take turns transmitting.

16. Both digital and analog signals may be transmitted by TDM.

17. Analog signals are transmitted by converting them into a series of pulses whose amplitudes approximate the shape of the analog signal. This process is called pulse-amplitude modulation (PAM).

18. Pulse-amplitude modulation is produced by sampling the analog signal. This is done by periodically opening a gate for a brief period, allowing a narrow portion of the analog signal to pass through.

19. Samples must be taken fast enough in order for high-frequency components to be recognized and adequately represented.

20. A minimum sampling rate is 2 times the highest frequency component or upper bandwidth limit of the analog signal. This is called the sampling theorem.

21. Pulse-amplitude modulation signals may be multiplexed by allowing samples of several signals to be interleaved into adjacent time slots.

22. Field-effect transistor switches are commonly used in sampling gates and PAM multiplexers. They are controlled by digital circuits that set the sampling intervals and pulse rates.

23. The period of time during which each channel in a PAM system is sampled once is called a frame.

24. Pulse-amplitude modulation signals are normally used to frequency modulate another carrier creating PAM/FM.

25. Demultiplexing PAM signals requires some means of synchronization to ensure matching clock frequencies and channel timing at the receiver.

26. Special clock recovery circuits use the PAM signal itself to derive the clock signal at the receiver rather than generating it independently. This ensures perfect frequency and phase relationships.

27. A special sync pulse with a unique shape to distinguish it from the PAM pulses is used to keep the demultiplexer in synchronization with the multiplexer. The sync pulse usually occurs as the last pulse in a frame.

28. Most TDM systems in use today use pulse-code modulation (PCM) to transmit analog signals.

29. Pulse-code modulation uses A/D conversion techniques to translate analog signals into binary form for serial transmission.

30. In a multichannel PCM system, each signal is furnished with an analog-to-digital converter (ADC) and parallel-to-serial converter. The resulting binary outputs are digitally multiplexed using TDM techniques.

31. Voice, telemetry data, video, and other analog signals may be transmitted via PCM.

32. Pulse-code modulation signals are recovered at the receiver by demultiplexing and D/A conversion plus appropriate filtering.

33. Pulse-code modulation is generated by periodically sampling the analog signal as in PAM systems. Then the varying amplitude pulses are converted into proportional binary words by an ADC. This process is called quantizing.

34. Quantizing means dividing a given signal voltage range into a number of discrete increments each represented by a binary code. Each analog sample is matched to the nearest binary level.

35. Sampling in a PCM system is done by a sample/hold (S/H) amplifier that stores the analog value on a capacitor at the instant of measurement.

36. Most ADCs have an 8-bit word providing a quantizing resolution of 1 in 256.
37. Most PCM systems, especially for voice transmission, use companding to compress the voice signal dynamic range by emphasizing lower-level signals and de-emphasizing higher-level signals. Companding minimizes quantization error and improves noise immunity.
38. Companding may be done by analog or digital techniques. Analog companders use diode circuits in amplifiers for compression and expansion. Digital companders use special nonlinear ADCs and DACs.
39. A common PCM system using TDM techniques is the Bell T-1 used in telephone work.
40. A T-1 system multiplexes 24 voice channels, each represented by an 8-bit word. The sampling interval is 125 μs. A frame consists of twenty-four 8-bit words and a single sync pulse. The frame data rate is 1.544 MHz.
41. Most PCM systems like T-1 use a special LSI IC called a codec to handle conversion and companding. A codec contains the ADC, DAC, serial-to-parallel conversion circuits, companders, and related timing circuits.
42. The sampling rate of a codec is usually 8 kHz.
43. Pulse-width modulation (PWM) and pulse-position modulation (PPM) systems are no longer used.
44. Most TDM systems use PCM though some use PAM.
45. Pulse-code modulation is preferred over PAM because of its superior noise immunity.

CHAPTER REVIEW QUESTIONS

Choose the letter which best answers each question.

8-1. Multiplexing is the process of
   a. Several signal sources transmitting simultaneously to a receiver on a common frequency.
   b. Sending the same signal over multiple channels to multiple destinations.
   c. Transmitting multiple signals over multiple channels.
   d. Sending multiple signals simultaneously over a single channel.

8-2. In FDM, multiple signals
   a. Transmit at different times.
   b. Share a common bandwidth.
   c. Use multiple channels.
   d. Modulate one another.

8-3. Each signal in an FDM system
   a. Modulates a subcarrier.
   b. Modulates the final carrier.
   c. Is mixed with all the others before modulation.
   d. Serves as a subcarrier.

8-4. Frequency modulation in FDM systems is usually accomplished with a
   a. Reactance modulator.   *c.* VCO.
   b. Varactor.   *d.* PLL.

8-5. Which of the following is *not* a typical FDM application?
   a. Telemetry   *c.* Telephone
   b. Stereo broadcasting   *d.* Secure communications

8-6. The circuit that performs demultiplexing in an FDM system is a(n)
   a. Op amp.   *c.* Discriminator.
   b. Bandpass filter.   *d.* Subcarrier oscillator.

8-7. Most FDM telemetry systems use
   a. AM.   *c.* SSB.
   b. FM.   *d.* PSK.

8-8. The best frequency demodulator is the
  a. PLL discriminator.
  b. Pulse-averaging discriminator.
  c. Foster-Seeley discriminator.
  d. Ratio detector.

8-9. The modulation used in FDM telephone systems is
  a. AM.   *c*. SSB.
  b. FM.   *d*. PSK.

8-10. The FDM telephone systems accommodate many channels by
  a. Increasing the multiplexer size.
  b. Using many final carriers.
  c. Narrowing the bandwidth of each.
  d. Using multiple levels of multiplexing.

8-11. In FM stereo broadcasting, the $L + R$ signal
  a. Double-sideband modulates a subcarrier.
  b. Modulates the FM carrier.
  c. Frequency modulates a subcarrier.
  d. Is not transmitted.

8-12. In FM stereo broadcasting, the $L - R$ signal
  a. Double-sideband modulates a subcarrier.
  b. Modulates the FM carrier.
  c. Frequency modulates a subcarrier.
  d. Is not transmitted.

8-13. The SCA signal if used in FM broadcasting is transmitted via
  a. A 19-kHz subcarrier.   *c*. A 67-kHz subcarrier.
  b. A 38-kHz subcarrier.   *d*. The main FM carrier.

8-14. In TDM, multiple signals
  a. Share a common bandwidth.
  b. Modulate subcarriers.
  c. Are sampled at high speeds.
  d. Take turns transmitting.

8-15. In TDM, each signal may use the full bandwidth of the channel.
  a. True
  b. False

8-16. Sampling an analog signal produces
  a. PAM.   *c*. FM.
  b. AM.   *d*. PCM.

8-17. The maximum bandwidth that an analog signal can use with a sampling frequency of 108 kHz is
  a. 27 kHz.   *c*. 108 kHz.
  b. 54 kHz.   *d*. 216 kHz.

8-18. Pulse-amplitude modulation signals are multiplexed by using
  a. Subcarriers.   *c*. A/D converters.
  b. Bandpass filters.   *d*. FET switches.

8-19. In PAM demultiplexing, the receiver clock is derived from
  a. Standard radio station WWV.
  b. A highly accurate internal oscillator.
  c. The PAM signal itself.
  d. The 60-Hz power line.

8-20. In a PAM/TDM system, keeping the multiplexer and DEMUX channels step with one another is done by a
  a. Clock recovery circuit.   *c*. Sampling.
  b. Sync pulse.   *d*. Sequencer.

8-21. Transmitting data as serial binary words is called
  a. Digital communications.   *c*. PAM.
  b. Quantizing.   *d*. PCM.

8-22. Converting analog signals to digital is done by sampling and
  a. Quantizing.   c. Pre-emphasis.
  b. Companding.   d. Mixing.

8-23. A quantizer is a(n)
  a. Multiplexer.   c. A/D converter.
  b. Demultiplexer.   d. D/A converter

8-24. Emphasizing low-level signals and compressing higher-level signals is called
  a. Quantizing.   c. Pre-emphasis.
  b. Companding.   d. Sampling.

8-25. Which of the following is *not* a benefit of companding?
  a. Minimizes noise   c. Minimizes quantizing error
  b. Minimizes number of bits   d. Minimizes signal bandwidth

8-26. A telephone system using TDM and PCM is called
  a. PBX.   c. T-1.
  b. RS-232.   d. Bell 212.

8-27. An IC that contains A/D and D/A converters, companders, and parallel-to-serial converters is called a
  a. Codec.   c. Multiplexer.
  b. Data converter.   d. Modem.

8-28. Pulse-code modulation is preferred to PAM because of its
  a. Resistance to quantizing error.
  b. Simplicity.
  c. Lower cost.
  d. Superior noise immunity.

CRITICAL THINKING QUESTIONS

8-1 How does increasing the number of signals to be multiplexed affect the bandwidth of the transmitted composite signal? Explain your answer for both FDM and TDM.

8-2. Assume that AM and FM broadcast stations could transmit signals other than their main programming on higher-frequency subcarriers. Name some applications you can think of for this capability.

Answers to Self-Tests

1. *c*
2. telephone, telemetry
3. economy
4. demultiplexer
5. time division, frequency division
6. bandwidth
7. subcarriers
8. linear mixer
9. modulate
10. frequency
11. amplifying, filtering
12. subcarrier oscillator or VCO
13. bandpass filter
14. PLL
15. SSB
16. $L + R$
17. $L + R, L - R$, 19-kHz pilot, SCA
18. DSB
19. frequency
20. time slots or intervals
21. pulse-amplitude modulation
22. 14
23. FET or MOSFET
24. clock recovery circuit
25. sync pulse
26. frame
27. low-pass filter
28. switch
29. gate
30. FM
31. phase-locked loop
32. binary codes
33. A/D converter
34. 8
35. 8
36. T-1
37. companding
38. codec
39. noise
40. 24, 1.544

CHAPTER 9

Antennas and Transmission Lines

∎

CHAPTER OBJECTIVES

This chapter will help you to:

1. *Define* and *state* the importance of standing waves and compute standing-wave ratio.
2. *Explain* how quarter- and half-wave transmission lines can act as tuned circuits at high frequencies.
3. *State* the characteristics of a radio wave and *explain* how antenna polarization is determined.
4. *Compute* the length of quarter-, half-, and full-wavelength antennas and transmission lines given the operating frequency.
5. *Name* three widely used types of antenna arrays that exhibit gain.
6. *Name* the three main types of radio wave propagation and *state* how each behaves.

All electronic communications systems consist of a transmitter, a receiver, and a communications medium. In some systems, the transmission medium is a direct link such as a wire, cable, or fiber-optic cable. In radio communications systems, there are no direct connections. The system is "wireless." The RF signal generated by the transmitter is sent into free space and is eventually picked up by the receiver. The processes of launching the signal into space and receiving it are the functions of the antenna. The antenna is a device that acts as the interface between the transmitter and free space and between free space and the receiver. It converts the transmitter RF power into electromagnetic signals that can propagate over long distances, and it is also the device that picks up the electromagnetic signals and converts them into signals for the receiver. Since the antennas are typically located remotely from the transmitters and receivers, some means must be used to get the power to and from the antenna. This is the job of the transmission line, a special kind of cable. In this chapter we introduce you to antenna and transmission line principles. We will also briefly examine the propagation characteristics of electromagnetic signals in free space to see how they propagate over long distances.

9-1 TRANSMISSION LINES

A *transmission line* is a two-wire cable that connects the transmitter to the antenna or the antenna to the receiver. The purpose of the transmission line is to carry the RF energy for the desired distance. Ideally, the antenna should be connected directly to the transmitter or the receiver, but this is not practical. Antennas must be located out of doors as high as possible for best radiation or reception. The equipment, of course, is usually located inside a building or vehicle. The transmission line, therefore, is a very important link in the communications system.

There are two basic types of wire transmission line. These are balanced line and coaxial cable. *Balanced line* is made up of two parallel conductors spaced from one another by a distance of ½ inch (in.) up to several inches. Insulating spacers may be used to keep the wires separated, or the spacing may be maintained by a continuous plastic insulator that is part of the conductor insulation. Figure 9-1(*a*) shows a two-wire balanced line.

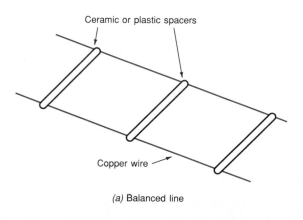

Ceramic or plastic spacers

Copper wire

(a) Balanced line

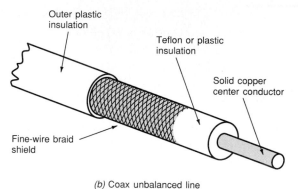

Outer plastic insulation

Teflon or plastic insulation

Solid copper center conductor

Fine-wire braid shield

(b) Coax unbalanced line

Fig. 9-1 **The two most common types of transmission lines.**

The term "balanced line" means that the same current flows in each wire with respect to ground, although the direction of current in one wire is 180° out of phase with the current in the other wire. Neither wire is connected to ground. In an unbalanced line, one conductor is connected to ground.

The other type of transmission line is *coaxial cable,* usually just called *coax.* Refer to Fig. 9-1(*b*). It consists of a solid-center conductor surrounded by a plastic insulator such as Teflon. Over the insulator is a second conductor, a tubular braid or shield made of fine

wires. An outer plastic sheath protects and insulates the braid. Coax is an unbalanced line since the current in the center conductor is referenced to the braid which is connected to ground. Coax comes in a variety of sizes from approximately ¼ in. to several inches in diameter. Of the two types of transmission lines, coax is the most widely used.

Two-wire cables are very common in electricity and electronics. The cables that carry the 60-Hz power line signal into your house are a form of transmission line. The wires connecting the audio output of your stereo receiver to your speaker are also a form of transmission line. At these low frequencies, the transmission line acts simply as a carrier of the ac signal. In these applications, the only characteristic of the cable we are interested in is its resistive loss. Such cables are basically noncritical in their size and characteristics except for conductor size which determines the current-carrying capability and the voltage drop over long distances.

However, when such cables are used to carry RF energy, strange things happen. The cables are no longer simply resistive conductors but instead appear to be complex interconnections of inductors and capacitors as well as resistors. Further, whenever the length of the cable becomes the same order of magnitude or greater than the wavelength of the transmitted signal, then the transmission line takes on special characteristics which require a more complex analysis.

Wavelength (λ) is a distance. It is the length or distance of one cycle of an ac wave. See Fig. 9-2. Wavelength is also the distance that an ac wave travels in the time required for one cycle of that signal.

Mathematically, wavelength (λ) is expressed

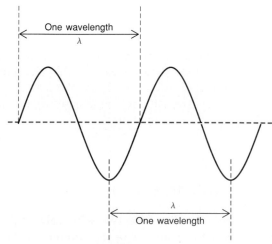

One wavelength
λ

λ
One wavelength

Fig. 9-2 **One wavelength.**

as the ratio of the speed of light to the frequency of the signal.

$$\lambda = \frac{300{,}000{,}000}{f}$$

In this formula, the large number is the speed of light in meters per second. This is also the speed of a radio signal. If the distance is converted to miles, you get the familiar number 186,000 miles per second (mi/s). The frequency f in the above expression is given in hertz. As you can see then, the wavelength of a 60-Hz power line signal is

$$\lambda = \frac{300{,}000{,}000}{60}$$
$$= 5{,}000{,}000 \text{ m}$$

That's an incredibly long distance and, in fact, represents a distance of several thousand miles. Power lines are typically much shorter than a wavelength at 60 Hz.

But at RF, say those frequencies of 3 MHz or more, the wavelength gets considerably shorter. The wavelength at 3 MHz is

$$\lambda = \frac{300{,}000{,}000}{3{,}000{,}000}$$
$$= 100 \text{ m}$$

That's a distance of a little more than 300 ft. This is a very practical distance. As the frequency gets higher, the wavelength gets shorter. The wavelength formula can be simplified at these higher frequencies.

$$\lambda = \frac{300}{f}$$

Here, the frequency is expressed in megahertz. A 50-MHz signal will have a wavelength of 6 m. Converting meters to feet, the wavelength formula becomes

$$\lambda = \frac{984}{f} \qquad (1 \text{ m} = 3.281 \text{ ft})$$

where f is again in megahertz

If you know the wavelength and wish to compute frequency, the above formulas can be rearranged:

$$f = \frac{300}{\lambda} \qquad or \qquad f = \frac{984}{\lambda}$$

When the length of a transmission line is longer than several wavelengths at the signal frequency, the two parallel conductors of the transmission line appear as a complex impedance. The wires exhibit considerable series inductance at high frequencies. Further, there is a capacitance between the parallel conductors. The result is that to a high-frequency signal, the transmission line appears as a distributed low-pass filter as shown in Fig. 9-3 consisting of series inductors, resistors, and shunt capacitors. An RF generator connected to the transmission line sees an impedance that is a function of the inductance and capacitance in the circuit. This impedance is known as the *characteristic impedance Z_o*. It is also referred to as the surge impedance. This impedance can be calculated with the simple expression

$$Z_o = \sqrt{\frac{L}{C}}$$

where Z_o = characteristic impedance of transmission line, Ω
$\quad L$ = inductance of transmission line for a given length
$\quad C$ = capacitance for that same length

Of course, knowing or measuring the inductance and capacitance for a given length of line is impractical. However, we do know that the inductance and capacitance depend upon the physical characteristics of the line just as the inductance of a coil and capacitance of a capacitor depend upon the physical characteristics of the coil and the capacitor, respectively. Therefore, we can easily express the characteristic impedance in terms of the physical sizes of the cable. The basic expression for the characteristic impedance of a parallel two-wire transmission line is

$$Z_o = 276 \log \frac{2S}{D}$$

In this expression, S is the spacing between the centers of the parallel conductors and D is the diameter of the conductors. See Fig. 9-4 on the next page.

For example, assume two wires are spaced 0.5 in. apart. Each wire has a diameter of

Fig. 9-3 Equivalent electric circuit of a two-wire transmission line.

CHAPTER 9 ANTENNAS AND TRANSMISSION LINES **209**

$$Z_o = 276 \log \frac{2S}{D}$$

Fig. 9-4 The characteristic impedance Z_o of a two-wire transmission line depends upon its physical dimensions.

0.082 in. The impedance is found by substituting 0.5 in. and 0.082 in. in the formula above.

$$Z_o = 276 \log \frac{2(0.5)}{0.082}$$

$$= 276 \log 12.22$$

$$= 276(1.087)$$

$$= 300 \ \Omega$$

The characteristic impedance of a coax is given by the expression

$$Z_o = 138 \log \frac{D}{d}$$

Here, D is the inside diameter of the outer braid shield, and d is the diameter of the inner conductor. See Fig. 9-5.

$$Z_o = 138 \log \frac{D}{d}$$

Fig. 9-5 The characteristic impedance of a coax.

Assume that the shield inner diameter is 0.2 in. and that the center conductor has a diameter of 0.087 in. The impedance is

$$Z_o = 138 \log \frac{0.2}{0.087}$$

$$= 138 \log 2.3$$

$$= 138(0.3617)$$

$$= 50 \ \Omega$$

Most transmission lines come with standard fixed values of characteristic impedance. For example, a widely used balanced transmission line is the popular TV twin lead in which two stranded wires are separated by a flat continuous plastic insulator as shown in Fig. 9-6. The typical characteristic impedance is 300 Ω. Coaxial cables are commonly available with characteristic impedances of 50 and 75 Ω.

Fig. 9-6 A 300-Ω twin-lead balanced line widely used on TV antennas.

Figure 9-7 gives a summary of the characteristics of several popular types of transmission lines. Note that coaxes are designated by an alphanumeric code beginning with the letters RG. The primary specifications are characteristic impedance and attenuation. The attenuation is the amount of power lost per 100 ft of cable expressed in decibels. This attenuation is directly proportional to the cable length and increases with frequency. The loss is significant at very high frequencies. However, the larger the cable, the lower the loss. The characteristics of 300-Ω twin lead are also listed in Fig. 9-7 for contrast. Note the low loss compared to coax.

Assume that a cellular telephone cell site antenna is mounted on a tower 65 ft high. The operating frequency is 900 MHz. Type RG-214/U coax cable is used. What is the attenuation of the cable?

Figure 9-7 gives the attenuation of 7.8 dB/100 ft. The attenuation per foot is:

$$7.8/100 = 0.078 \text{ dB/ft}$$

A cable of 65 ft has an attenuation of:

$$0.078 \times 65 = 5.07 \text{ dB}$$

To use a transmission line properly, it must be terminated in a resistive impedance equal in value to its characteristic impedance. For example, a 50-Ω coax must be terminated with a 50-Ω impedance as shown in Fig. 9-8. If the load is an antenna, then that antenna must look

Type	Characteristic Impedance (Ω)	Outside Diameter (in.)	Frequency (MHz)	Attenuation (dB/100 ft)	Applications
RG-59/U	73	0.242	100 400	3.4 7.1	TV antennas, cable TV, HF antennas
RG-11/U	75	0.405	100 400	2.5 5.5	VHF antennas
RG-214/U	50	0.405	100 400 900	2.0 4.7 7.8	Satellite TV antennas, UHF and microwave
RG-58/U	53.5	0.195	100	5.3	CB antennas
Twin lead	300	—	100	0.55	TV antennas

Fig. 9-7 **Common types and characteristics of transmission line.**

Fig. 9-8 **A transmission line must be terminated in its characteristic impedance for proper operation.**

Forward power

Incident power

Reflected power

Standing wave

like a resistance of 50 Ω. When the load impedance and characteristic impedance match, the transmission operates properly and maximum power transfer takes place minus any resistive losses in the line. The line can be any length. If you could measure the voltage (or current) at any point on the line, you would see a constant value, disregarding losses. A correctly terminated transmission line, therefore, is said to be flat.

If the load impedance is different from the line characteristic impedance, then not all the power transmitted will be absorbed by the load. The power not dissipated in the load will actually be reflected back toward the source. The power sent down the line toward the load is called the *forward* or *incident power,* while any power that is not absorbed by the load is known as *reflected power.* The signal actually on the line is simply the algebraic sum of the forward and reflected signals.

The reflected power represents a significant loss. Not all the power generated at the transmitter is taken advantage of. The reflected power is dissipated in the transmission line, and, in some cases when the mismatch is great, the reflected power can be high enough to actually cause damage to the transmitter or the line itself. For proper application then, the load impedance should equal the characteristic impedance

as closely as possible. One of the key objectives in designing antenna and transmission line systems is to ensure this impedance match.

The forward and reflected signals on an incorrectly terminated transmission line produce what is referred to as a *standing wave.* A standing wave is the unique distribution of voltage and current along a transmission line which is not terminated in its characteristic impedance.

The concept of standing waves is best illustrated by considering the two worst cases of impedance mismatch at the load or antenna end of the transmission line: an open circuit and a short circuit. The condition for a short circuit is illustrated in Fig. 9-9. The graph below the

Fig. 9-9 **Standing waves on a shorted transmission line.**

CHAPTER 9 ANTENNAS AND TRANSMISSION LINES **211**

transmission line shows a plot of the voltage and current at each point on the line. If you had a voltmeter and ammeter that you could move along the line and measure the voltage and current values and if you then plotted these values, you would obtain the curves shown.

As you would expect with a short at the end of the line, the voltage is zero when the current is maximum. All the power is reflected back toward the generator. Working back toward the generator, you see the voltage and current variations distribute themselves according to the wavelength of the signal. The pattern repeats every one-half wavelength. The voltage and current levels at the generator will be dependent upon the signal wavelength and the actual line length. Just remember that this standing wave pattern is fixed and is the result of a composite of the forward and reflected signals.

Figure 9-10 shows the standing waves on an open-circuited line. With an infinite impedance, the voltage at the end of the line is maximum and the current is zero. All the energy is reflected thereby setting up this stationary pattern of voltage and current standing waves.

Now in practice you won't actually have a short or open. Instead, the load impedance will not equal the transmission line impedance. Further, the antenna will probably have a reactive component, either inductive or capacitive, as well as its resistance. The mismatch will produce standing waves, but their amplitude will be lower and their distribution will look something like that shown in Fig. 9-11. Note that the voltage never goes to zero.

The magnitude of the standing waves on a transmission line is determined by the ratio of the maximum current to the minimum current along the line, or the ratio of the maximum

voltage to the minimum voltage. These ratios are referred as the *standing-wave ratio (SWR)*. See Fig. 9-11.

$$SWR = \frac{I_{max}}{I_{min}} = \frac{V_{max}}{V_{min}}$$

With the shorted and open conditions described earlier, the current or voltage minimums are zero. This produces an SWR of infinity. It means that no power is dissipated in the load since it is all reflected.

On a properly terminated line, there are no standing waves. The voltage and current are constant along the line, so there are no maximums or minimums, or else you could say the maximums and minimums are the same. Therefore, the SWR is 1. Of course, this is the ideal condition.

The SWR can also be computed by knowing the impedance of the transmission line and the actual impedance of the load. The SWR is the ratio of the load impedance Z_l to the characteristic impedance Z_o or vice versa.

$$SWR = \frac{Z_l}{Z_o} \text{ if } Z_l > Z_o$$
$$\frac{Z_o}{Z_l} \text{ if } Z_l < Z_o$$

Since the standing wave is really the composite of the original incident wave added to the reflected wave, we can also define SWR in terms of these waves. If V_i is the incident voltage wave and V_r is the reflected voltage wave, then their ratio will tell us what is happening along the line. This ratio R is called the *reflection coefficient*.

$$R = \frac{V_r}{V_i}$$

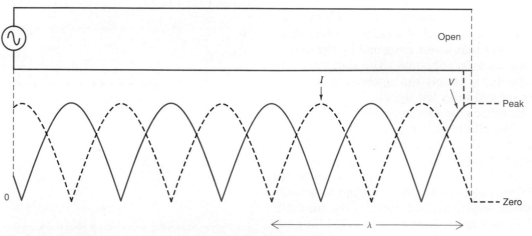

Fig. 9-10 **Standing waves on an open-circuit transmission line.**

Fig. 9-11 Transmission line with mismatched load and the resulting standing waves.

If the line is terminated in its characteristic impedance, then there is no reflected voltage, so $R = 0$. If the line is open or shorted, then total reflection occurs. This means that V_r and V_i are the same. In this case, $R = 1$. The reflection coefficient really expresses the percentage of reflected voltage to incident voltage. If $R = 0.5$, it means that 50 percent of the incident power is reflected.

If the load is not matched but is also not an open or short, the line will have voltage minimums and maximums as described previously. These can also be used to obtain the reflection coefficient.

$$R = \frac{V_{max} - V_{min}}{V_{max} + V_{min}}$$

To obtain the SWR from the reflection coefficient, we use the formula

$$SWR = \frac{1 + R}{1 - R}$$

If the load matches the line impedance, then $R = 0$. The formula above gives an SWR of 1 as you would expect. With an open or shorted load, $R = 1$. This produces an SWR of infinity.

The reflection coefficient can also be derived from the line and load impedances.

$$R = \frac{Z_l - Z_o}{Z_l + Z_o}$$

As before, Z_l is the load impedance and Z_o is the characteristic impedance of the line.

The SWR is always expressed as a number greater than 1. If the characteristic and load impedances are equal, the SWR is 1, as it should be.

The importance of the SWR is that it gives a relative indication of just how much power is lost in the transmission line. This assumes that none of the reflected power is re-reflected by the generator. In a typical transmitter, some power is reflected and sent to the load again. The curve in Fig. 9-12 shows the relationship between the percentage of reflected power and

Fig. 9-12 Percent reflected power on a transmission line for different values of the SWR.

the SWR. Naturally when the SWR is 1, the percent reflected power is zero. But as a line and load mismatch grows, reflected power increases. When the SWR is 1.5, for example, the percent reflected power is 4 percent. For SWR values of less than about 2, the reflected power is less than 10 percent. For values greater than this, the percent increases dramatically.

The best way to prevent a mismatch between the antenna and transmission line is to have a correct design. But, in practice, mismatches will occur. If they do, some corrections are possible. One solution is to tune the antenna, usually by adjusting its length. You can also insert an impedance-matching circuit or antenna tuner between the transmitter and the transmission line. This can be a balun or LC, L, T, or pi network like those described in a previous chapter. These circuits will make the impedances equal so that the mismatch is minimized or does not occur at all. The goal is to achieve SWR = 1. Any SWR value below 2 is usually acceptable.

The standing-wave conditions described earlier with open- and short-circuited loads must usually be avoided when working with transmission lines. On the other hand, the unique conditions that occur can be used for other purposes. For example, short lengths of transmission line, up to one-half wavelength long, with an open or short circuit can be used as resonant or reactive circuits. Consider the shorted quarter-wavelength ($\lambda/4$) line shown in Fig. 9-13. The voltage is zero and the current is maximum at the end, as you saw earlier. But at the generator, one-quarter wavelength back, the voltage is maximum while the current is zero. To the generator, the line appears as an open circuit or at least a very high impedance.

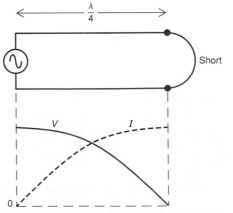

Fig. 9-13   A shorted quarter-wavelength line acts like a parallel resonant circuit.

The key point here is that this condition exists at only one frequency, the frequency at which the line length is exactly one-quarter of the wavelength. Because of this frequency sensitivity, the line acts like an LC tuned or resonant circuit, in this case, a parallel resonant circuit because of its very high impedance.

If you make the line one-half wavelength long, you get the standing-wave pattern shown in Fig. 9-14. The generator sees the same conditions as at the end of the line: zero voltage and maximum current. This represents a short or very low impedance. That condition occurs only if the line is exactly one-half wavelength long at the generator frequency. In this case, the line looks like a series resonant circuit to the generator.

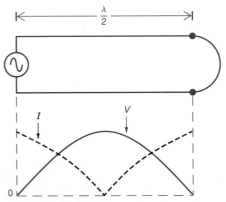

Fig. 9-14   A shorted half-wavelength line acts like a series resonant circuit.

If the line length is less than one-quarter wavelength at the operating frequency, the shorted line looks like an inductor to the generator. If the shorted line length is between one-quarter and one-half wavelength, the line looks like a capacitor to the generator. All these conditions repeat with multiple quarter or half wavelengths of shorted line.

A similar set of conditions is obtained with an open line as shown in Fig. 9-15. To the generator, a quarter-wavelength line looks like a series resonant circuit, whereas a half-wavelength line looks like a parallel resonant circuit, just the opposite of a shorted line. If the line is less than one-quarter wavelength, the generator sees a capacitance. If the line is between one-quarter and one-half wavelength, the generator sees an inductance. Again, all these characteristics repeat for multiple quarter or half wavelengths. Figure 9-16 is a complete summary of the conditions represented by open and shorted lines.

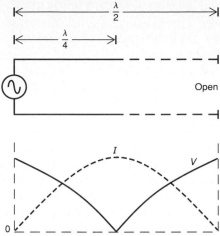

Fig. 9-15 One-quarter and one-half wavelength open lines also look like resonant circuits to a generator.

it is difficult to realize them physically with standard coils and capacitors. At UHF and microwave frequencies, transmission lines become useful components and circuits. They are used as tuned circuits, filters, phase shifters, reactors, and impedance-matching circuits.

An important factor in transmission line applications is that the speed of the signal in the transmission line is slower than the speed of a radio signal in free space. Light and radio signals propagate through free space at a speed of 300,000,000 m/s or 186,000 mi/s as indicated earlier. This speed is considerably less in a transmission line. This leads to the development of a velocity factor F which is the ratio of the transmission speed in the transmission line V_L and the transmission speed in free space V_S.

$$F = \frac{V_L}{V_S}$$

At low frequencies, these characteristics of open and shorted lines have minimum significance. However, at UHF (more than 300 MHz) and microwaves (more than 1 GHz), the length of one-half wavelength becomes less than 1 ft. Because of this, we can actually use transmission lines as tuned circuits and reactances. The values of inductance and capacitance become so small at these frequencies that

Velocity factors in transmission lines vary from approximately 0.8 to 0.6. The velocity factor of coax is typically in the 0.6 to 0.7 range. Open-wire line and twin lead have a velocity factor near 0.8. This velocity factor must be taken into consideration when computing the length of a transmission line in wavelengths.

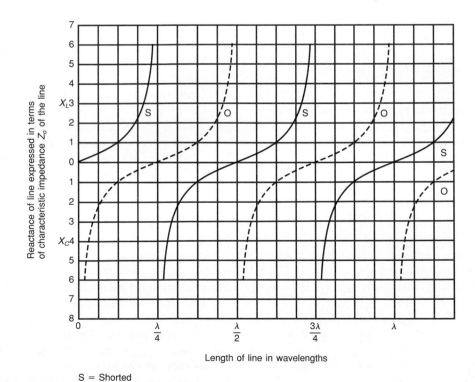

S = Shorted
O = Open

Fig. 9-16 Summary of impedance and reactance variations of shorted and open lines for lengths up to one wavelength.

Recall earlier that one wavelength of a transmission line is given by the expression

$$\lambda = \frac{984}{f}$$

However, this is not a true wavelength of a transmission line. The above formula must be modified by the velocity factor F in order to get the correct length. The new formula is

$$\lambda = 984\,\frac{F}{f}$$

For example, what is the actual length in feet of one-quarter wavelength of a coax with a velocity factor of 0.65 at 30 MHz?

$$\lambda = 984\,\frac{F}{f}$$
$$= 984\,\frac{0.65}{30}$$
$$= 21.32 \text{ ft}$$

One wavelength is 21.32 ft. One-quarter wavelength then is one-fourth of this, or 21.32/4 = 5.33 ft.

It is often necessary to know the length of the transmission line in wavelengths; the above formula can give it to you. Just be sure to use the correct velocity factor for the desired transmission line. This can be obtained from the manufacturer's literature and various handbooks.

Self-Test

Supply the missing information in each statement.

1. The two basic types of transmission line are _____ .
2. If one wire of a transmission line is connected to ground, the line is said to be _____ .
3. The distance that a signal travels during one cycle is called the _____ .
4. One wavelength at a frequency of 450 MHz is _____ m.
5. A line 4 in. long represents one-half wavelength at a frequency of _____ GHz.
6. The physical dimensions of a transmission line determine its _____ .
7. To a generator, a transmission line looks like a(n) _____ made up of distributed _____ and _____ .
8. A coax line has a shield braid with an inside diameter of 0.2 in. and a center conductor with a diameter of 0.057 in. The characteristic impedance is _____ Ω.
9. The attenuation of 250 ft of RG-11U coax at 100 MHz is _____ dB. (See Fig. 9-7.)
10. For optimum transfer of power from a generator to a load of 52 Ω, the transmission line impedance should be _____ Ω.
11. The current and voltage along a properly matched line are _____ .
12. If a transmission line is not terminated in its characteristic impedance, _____ _____ will develop along the line.
13. If a load and line have mismatched impedances, power not absorbed by the load will be _____ .
14. Patterns of voltage and current variations along a transmission line with a mismatched load are known as _____ .
15. A 52-Ω coax has a 36-Ω antenna load. The SWR is _____ .
16. If the load and line impedances are matched, the SWR will be _____ .
17. All incident power on a line will be reflected if the line is _____ or _____ at its end.
18. The ratio of the reflected voltage to the incident voltage on a transmission line is called the _____ .
19. The maximum voltage along a transmission line is 150 V, and the minimum voltage is 90 V. The SWR is _____ . The reflection coefficient is _____ .
20. The reflection coefficient of a transmission line is 0.75. The SWR is _____ .
21. An open or shorted transmission line will have a reflection coefficient of _____ and an SWR of _____ .
22. Transmission lines, one-quarter or one-half wavelength, can be used as _____ .
23. A transmission line has an SWR of 1.75. The power applied to the line is 90 W. The amount of reflected power is _____ W.
24. A shorted quarter-wave line looks like a(n) _____ impedance to the generator.
25. The following lines look like a series resonant circuit: a _____ $\lambda/4$ line; a _____ $\lambda/2$ line.
26. Transmission lines less than $\lambda/4$ or between $\lambda/4$ and $\lambda/2$ act like _____ or _____ .
27. An open transmission line 6 in. long acts as a _____ resonant circuit at a frequency of 492 MHz.
28. A coax has a velocity factor of 0.68. One-half wavelength of this coax at 120 MHz is _____ ft long.

9-2 ANTENNA FUNDAMENTALS

An antenna, or aerial as it is sometimes called, is one or more electrical conductors of a specific length that radiate radio waves generated by a transmitter or that collect radio waves at the receiver. There are literally hundreds of different types of antennas in use today. In this section we will introduce you to the most common and popular types.

A radio wave is generally called an electromagnetic wave because it is made up of a combination of both electric and magnetic fields. Whenever voltage is applied to the antenna, an electric field will be set up. At the same time, this voltage will cause current to flow in the antenna. This current flow will produce a magnetic field. The electric and magnetic fields are at right angles to one another. These electric and magnetic fields are emitted from the antenna and propagate through space over very long distances. Figure 9-17 shows the general concept of electric and magnetic fields making up a radio signal.

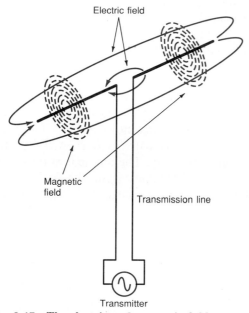

Fig. 9-17   The electric and magnetic fields produced by an antenna.

A closer approximation to what an electromagnetic field may look like is illustrated in Fig. 9-18. Note that the amplitude and direction of the magnetic and electric fields vary in a sinusoidal manner depending upon the frequency of the signal being radiated. The electric and magnetic fields are perpendicular to

(a) Horizontal polarization

(b) Vertical polarization

Fig. 9-18   The orientation of the electric and magnetic fields determines the signal polarization.

one another, and they are also both perpendicular to the direction of propagation of the wave.

An important consideration in the transmission and reception of radio waves is the orientation of the magnetic and electric fields with respect to the earth. The direction of the electric field specifies the polarization of the antenna. If the electric field is parallel to the earth as in Fig. 9-18(*a*), the electromagnetic wave is said to be horizontally polarized. However, if the electric field is perpendicular to the earth as in Fig. 9-18(*b*), then the wave is vertically polarized. As you would expect, antennas that are horizontal to the earth produce horizontal polarization, and antennas that are vertical to the earth produce vertical polarization.

The most important factor is that both the transmitting and receiving antennas must be of the same polarization for optimum transmission and reception. Theoretically, a vertically polarized wave will produce 0 V in a horizontal antenna and vice versa. But during transmission over long distances, the waves have their polarization changed slightly due to the various propagation effects in free space. For this reason, even though the polarization of the transmitting and receiving antennas may be mismatched, a signal will typically be received, although optimum results will take place when the polarization of the two antennas are matched.

As indicated earlier, an antenna is some form of electrical conductor. It may be a length of wire, a metal rod, or a piece of tubing. Many different sizes and shapes are used. The length of the conductor is dependent upon the frequency of transmission. Antennas radiate most effectively when their length is directly related to the wavelength of the transmitted signal. Most antennas have a length that is some fraction of a wavelength. The most common lengths are one-half and one-quarter wavelengths.

One of the most widely used antenna types is the half-wave dipole shown in Fig. 9-19. Also called a doublet, it is simply a piece of wire, rod, or tubing that is one-half wavelength long at the operating frequency. The antenna is actually cut into two quarter-wavelength sections. The transmission line is connected at the center. The dipole has an impedance of 73 Ω at its center.

Fig. 9-19   A half-wave dipole antenna.

As you saw earlier, it is important for maximum power transfer that the impedance of the transmission line match the load. For the dipole, a 73-Ω coax like RG-59/U is a perfect transmission line; RG-11/U coax with an impedance of 75 Ω is also an excellent match.

The length of a half-wavelength dipole antenna is computed with the formula

$$L = \frac{468}{f}$$

As an example, an antenna for a frequency of 18 MHz would have a length of

$$L = \frac{468}{18}$$
$$= 26 \text{ ft}$$

To create a half-wave dipole for 18 MHz, you would cut two 13-ft lengths of wire. Copper wire size no. 12 or 14 gage is commonly used. Physically the antenna would be suspended between two points as high as possible off the ground. See Fig. 9-20. The wire conductors themselves would be connected to glass or ceramic insulators in each end and at the middle of the antenna to provide good insulation between the antenna itself and its supports. The transmission line would be attached to the two conductors at the center insulator. The transmission line should leave the antenna at a right angle so that it will not interfere with the antenna's radiation.

Most half-wave dipole antennas are mounted horizontally to the earth and, therefore, will be horizontally polarized. Although the antenna in general will radiate energy in many different directions, typically that radiation is concentrated into a very specific geometric radiation pattern. The radiation pattern of a half-wave dipole has the shape of a doughnut. Figure 9-21 shows this pattern with half the doughnut cut away. The dipole is at the center hole of the doughnut, while the doughnut itself represents the radiated energy. If you look

Fig. 9-20   A half-wave dipole for 18 MHz.

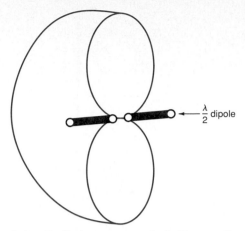

Fig. 9-21   Radiation pattern of a half-wave dipole.

down on the top of the dipole, the radiation pattern appears to be a figure eight as shown in Fig. 9-22. This is the horizontal radiation pattern. It is plotted on a polar coordinate graph. The center of the antenna is assumed to be at the center of the graph. The dipole is assumed to be aligned with the 90°–270° axis. As you can see, the maximum amount of energy is radiated at right angles to the dipole, at 0° and 180°. For that reason, a dipole is what is known as a directional antenna. For optimum transmission and reception, the antenna should be aligned broadside to the signal destination. The transmitting and receiving antennas must be parallel to one another for maximum signal transmission.

Whenever you point a dipole receiving antenna toward a transmitter or vice versa, it must be broadside to the direction of the trans-

mitter. If the antenna is at some angle, the maximum signal will not be received. As you can see from the radiation pattern in Fig. 9-22, if the end of the receiving antenna is pointed directly at the transmitting antenna, then zero signal will be received. As indicated earlier, this is a theoretical condition as the radiated wave does undergo some shifts during propagation and, therefore, some signal will be received off the ends of the antenna, but it is a very minimum amount. In practice, if you observed the signal strength as you rotate the receiving antenna, the maximum signal will be obtained when the antenna is broadside, while the minimum signal will be obtained when the end of the antenna is pointed toward the transmitter.

Although most half-wave dipoles are mounted horizontally to the earth, they can also be mounted vertically. Now the antenna is vertically polarized. Its radiation pattern is still a doughnut, but now looking down upon the antenna you see a radiation pattern that is a perfect circle. This means that the antenna will transmit an equal amount of energy in all directions. Such an antenna is said to be omnidirectional.

A popular variation of the half-wave dipole is the folded dipole shown in Fig. 9-23(a). Like

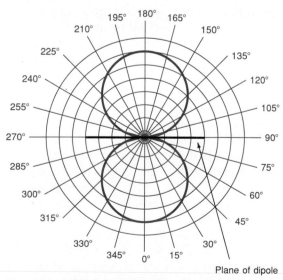

Fig. 9-22   Horizontal radiation pattern of a half-wave dipole.

(a) Basic configuration

(b) Construction with twin lead

Fig. 9-23   Folded dipole.

the standard dipole, it is one-half wavelength long. However, note that it consists of two parallel conductors connected at the ends with one side open at the center for connection to the transmission line. The reason for the popularity of this particular antenna is the fact that its impedance is 300-Ω, and, therefore, it makes an excellent match to the widely available 300-Ω twin lead. The actual spacing between the two parallel conductors is not particularly critical, although it is typically inversely proportional to the frequency. For very high frequency antennas, spacing is less than 1 in. For low-frequency antennas, the spacing may be 2 or 3 in.

An easy way to make a folded dipole antenna is to construct it entirely of 300-Ω twin lead. A piece of twin lead is cut to a length of one-half wavelength, and the two ends are soldered together as shown in Fig. 9-23(b). When figuring the half wavelength, be sure to include the velocity factor of twin lead (0.8) in the calculation. The center of one conductor is then cut open and a 300-Ω twin-lead transmission line is soldered to the two wires. The result is an effective, low-cost antenna that can be used for both transmitting and receiving purposes. Such antennas are commonly used for both TV and FM radio reception.

If vertical polarization and omnidirectional characteristics are required, it is not necessary to use a half-wave length dipole. The same effect can be achieved by using a quarter-wavelength vertical radiator as illustrated in Fig. 9-24(a). This antenna is known as a quarter-wave vertical or a ground plane antenna. It is usually fed with coax, the center conductor connected to the vertical radiator and the shield connected to earth ground. With this arrangement, the earth acts as a type of electrical "mirror," effectively providing the other quarter wavelength of the antenna and making it the equivalent of a vertical dipole. The result is a vertically polarized omnidirectional antenna.

The effectiveness of this antenna depends upon a good electrical ground. Making good electrical contact with the earth is a tricky process. Usually a reasonable ground can be obtained by driving a copper rod 5 to 10 ft long into the earth. In some cases, even this is insufficient. When a good electrical connection is made to the earth, the earth becomes what is known as a ground plane. If a good electrical connection cannot be made to the earth, then an artificial ground plane can be constructed of several quarter-length wires laid

(a) Quarter wave vertical

(b) Using radials as a ground plane

Fig. 9-24   Ground plane antenna.

horizontally on the ground or buried in the earth. Usually four wires are sufficient, but in some antenna systems, more are used. See Fig. 9-24(b). These horizontal wires at the base of the antenna are referred to as radials. The entire collection of radials is often referred to as a counterpoise or ground plane.

At very high frequencies where antennas are short, any large, flat metallic surface can serve as an effective ground plane. For example, vertical antennas are widely used on cars, trucks, boats, and other vehicles. The metallic roof of a car makes a superior ground plane for VHF and UHF antennas. In any case, the ground plane must be large enough so that it has a radius of greater than one quarter wavelength.

The impedance of a vertical ground plane antenna is exactly one-half the impedance of the dipole or approximately 36.5 Ω. Since there is no such thing as 36.5-Ω coax, 50-Ω coax is commonly used to feed power to the ground plane antenna. This represents a mismatch with an SWR of 1.39. This is a relatively low SWR and will not be the cause of any significant loss in power.

One way to adjust the antenna's impedance is to use "drooping" radials as shown in Fig. 9-25. At some angle depending upon the height of the antenna above the ground, the antenna's impedance will be near 50 Ω making it a near perfect match for most 50-Ω coax.

Fig. 9-25 Drooping radials adjust the antennas impedance to near 50 Ω.

In addition to its vertical polarization and omnidirectional characteristics, another major benefit of the quarter-wave vertical antenna is its length. It is one-half the length of a standard dipole which represents a significant savings at lower RFs. For example, a half-wave antenna for a frequency of 5 MHz is 93.6 ft long. Creating a vertical antenna of that length represents a major structural problem as it requires a support at least that long. By using a quarter-wave vertical antenna, however, the necessary length is cut in half, or in this case the length 46.8 is ft. Creating a structure 50 ft high instead of 100 ft high is a lot easier. Most low-frequency transmitting antennas use verticals for this reason. Amplitude modulation broadcast stations use vertical antennas because they are shorter, less expensive, and not as visually obtrusive. Additionally, they provide an equal amount of radiation in all directions.

One of the primary advantages of a vertical antenna or any antenna with omnidirectional characteristics is that it can send messages in any direction or receive them from any direction. In many communications systems this is a highly desirable advantage. On the other hand, there are communications systems in which it is more advantageous to restrict the direction in which signals were sent or received. This requires an antenna with directivity. Directivity refers to the ability of an antenna to send or receive signals over a narrow horizontal directional range. In other words, the antenna has a response or directivity curve that makes it highly directional based on its physical orientation. The advantage of a directional antenna is that it eliminates interference from other signals being received from all directions except the direction of the desired signal. A highly directional antenna acts as a type of filter to provide selectivity based on the direction of the signal. The receiving antenna is pointed directly at the station to be received, thereby effectively rejecting signals from transmitters in all other directions.

When using a directional antenna, there is greater efficiency of power transmission. When a transmitter uses an omnidirectional antenna, the transmitted power goes off into all directions. Only a small portion of it is received by the desired station. In effect, all the other power is wasted. By making the antenna directional, the transmitter power can be focused into a narrow beam that can be directed toward the station of interest.

The conventional half-wave dipole has some directivity in that it sends or receives signals in those directions perpendicular to the line of the antenna. You saw this earlier in the figure-eight response curve of the dipole in Fig. 9-22. The antenna is directional in that no signal is radiated from or picked up off the ends of the antenna. Such an antenna is referred to as bidirectional since it receives signals best in two directions. An antenna can be made unidirectional with proper design. Unidirectional means that the antenna sends or receives signals in only one direction.

The measure of an antenna's directivity is beam width. Beam width refers to the angle of the radiation pattern over which a transmitter's energy is directed or received. The beam width is measured on the antenna's radiation pattern. Figure 9-26 on the next page shows the horizontal radiation pattern of a typical directional antenna plotted on a polar coordinate graph. The antenna is assumed to be at the center of the graph. The concentric circles extending outward from the pattern indicate the relative strength of the signal as it moves away from the antenna. The beam width is measured between the points on the radiation curve that are down 3 dB from the maximum amplitude of the curve. The maximum amplitude of the pattern

Fig. 9-26 **Measuring the beam width of a directional antenna.**

occurs at 0°. The 3-dB-down points are 70.7 percent of the maximum. Then an angle is formed with two lines extending from the center of the curve to the 3-dB-down points. The angle formed by these two lines is the beam width. In this example, the beam width is 30°.

Note in Fig. 9-26 the three smaller patterns or loops in different directions from the main larger pattern we have been discussing. They are called minor lobes. Few antennas are perfectly directive. Because of various imperfections, some power is radiated in other directions. The goal is to eliminate or at least minimize these minor lobes through various antenna adjustments to put more power into the main lobe.

The beam width on a standard half-wave dipole is approximately 70°. This is not a highly directional antenna. The narrower the beam width, of course, the better the directivity and the more highly focused the signal becomes. At microwave frequencies, antennas with beam widths of less than a degree have been built for pinpoint communications accuracy.

When a highly directive antenna is used, all the transmitted power is focused in one direction. Because the power is concentrated into a small beam, the effect is as if the antenna amplified the transmitted signal. Directivity, because it focuses the power, causes the antenna to exhibit gain. Gain as you know is a form of amplification. And although the antenna does not have the capability of actually amplifying the signal, since it can focus the energy in a

single direction, the effect is as if the amount of radiated power were substantially higher than the power output of the transmitter. The antenna has power gain.

The gain of an antenna can be expressed as the ratio of the power transmitted P_t to the input power of the antenna P_i. This is the power gain. But usually, the gain is expressed in decibels of power gain.

$$\text{dB} = 10 \log \frac{P_t}{P_i}$$

The total amount of power radiated by the antenna is called the *effective radiated power (ERP)* and is simply the power applied to the antenna multiplied by the antenna gain. Power gains of 10 or more are easily achieved, especially at the higher RFs. This means that a 100-W transmitter can be made to perform like a 1000-W transmitter when applied to an antenna with gain.

To create an antenna with directivity and gain, two or more antenna elements are combined to form what is known as an array. There are two basic types of antenna arrays used to achieve gain and directivity. These are parasitic arrays and driven arrays.

A *parasitic array* consists of a basic antenna connected to a transmission line plus one or more additional conductors that are not connected to the transmission line. These extra conductors are referred to as parasitic elements. The antenna itself is referred to as the driven element. Typically the driven element is a half-wave dipole or some variation. The parasitic elements are slightly longer than and slightly less than one-half wavelength long. These parasitic elements are placed in parallel with and near the driven element. A common arrangement is illustrated in Fig. 9-27.

One type of parasitic element is known as a reflector. It is typically about 5 percent longer than the half-wave dipole driven element. The reflector is spaced from the driven element by a distance of 0.15 to 0.25 wavelength. The physical arrangement is shown in Fig. 9-27. When the signal radiated from the dipole reaches the reflector, it will induce a voltage into the reflector. The reflector will produce some radiation of its own. Because of the spacing, the reflector's radiation will be mostly in phase with the driven-element radiation. As a result, the reflected signal will add to the dipole signal and create a stronger, more highly focused beam. The reflector minimizes the radiation to the rear of the driven element and

Fig. 9-27 **A parasitic array known as a Yagi antenna.**

Labels in figure:
Direction of radiation
Director
Boom
Insulator
Driven element $\left(\dfrac{\lambda}{2}\right)$
Reflector
Coax feed line

reinforces the radiation to the front of the driven element.

Another kind of parasitic element is called a director. It is approximately 4 percent shorter than the half-wave driven element and is mounted in front of the driven element. See Fig. 9-27.

The directors are placed in front of the driven element and are spaced by some distance between approximately one-tenth to two-tenths of a wavelength from the driven element. The signal from the driven element will be induced into the director. The signal radiated by the director will then add in phase to that from the driven element. The result is increased focusing of the signal, a narrow beam width, and a higher antenna gain. Note the direction of radiation in Fig. 9-27.

An antenna made up of a driven element, a reflector, and one or more directors is generally referred to as a Yagi antenna, named after one of its Japanese inventors, Yagi and Uda. The antenna elements are usually made of aluminum tubing and are mounted on an aluminum cross member or boom. Since the center of the parasitic elements are neutral electrically, they may be connected directly to the boom. The boom itself may then be connected to a metal mast and electrical ground for the best lightning protection. The resulting antenna is often referred to as a beam antenna because it is very highly directional and has very high gain. Gains of 6 to 15 dB are possible with beam angles of 40° to 20°. The three-element Yagi shown in Fig. 9-27 has a gain of about 8 dB when compared to a half-wave dipole. Most Yagis have a driven element, a reflector, and from 1 to 20 directors. The greater the

number of parasitic elements, the higher the gain and the narrower the beam angle.

Yagis are widely used communications antennas because of their directivity and gain. At one time they were widely used for TV reception, but since they are tuned to only one frequency, they are not good for reception or transmission over a wide frequency range. Amateur radio operators are major users of beam antennas. Many other communications services use them because of their excellent performance and low cost.

Because of its physical size, a beam antenna is used mainly at VHF and UHF. For example, at a frequency of 450 MHz, the elements of a Yagi are only about 1 ft long. This makes the antenna relatively small and easy to handle. The lower the frequency, the larger the elements and the longer the boom. In general, such antennas are only practical above frequencies of about 15 MHz. At this frequency, the elements are in excess of 35 ft long. Although such antennas are large and difficult to work with, they are still widely used in some communications services.

The other major type of directional antenna is the *driven array*. This is an antenna that has two or more driven elements. Each element receives RF energy from the transmission line. The elements may be arranged in various ways to produce different degrees of directivity and gain.

There are three basic types of driven arrays, the collinear, the broadside, and the end-fire. The *collinear antenna* usually uses two or more half-wave dipoles mounted end to end as shown in Fig. 9-28(*a*) on the next page. The lengths of the transmission lines connecting the various driven elements are carefully selected so that the energy reaching each antenna is in phase with all the other antennas. In this way, the individual antenna signals will combine and produce a more focused beam. Collinears still have a bidirectional radiation pattern like that of a dipole, but the two beam widths are much narrower, providing greater directivity and gain.

The collinears in Fig. 9-28(*b*) and (*c*) use half-wave sections separated by shorted quarter-wave matching stubs which ensure that the signals radiated by each half-wave section are in phase. The greater the number of half-wave sections used, the higher the gain and the narrower the beam width.

Collinears are usually used only on VHF and UHF bands because their length becomes prohibitive at the lower frequencies. Collinears can

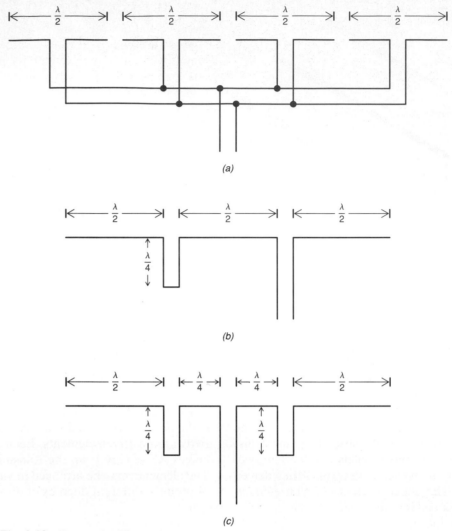

Fig. 9-28 Types of collinear antennas.

be mounted vertically at the high frequencies to provide an omnidirectional antenna with gain.

A *broadside array* is essentially a stacked collinear antenna. It consists of half-wave dipoles spaced from one another by one-half wavelength as shown in Fig. 9-29. Two or more elements can be combined. Each is connected to the other and to the transmission line. The crossover transmission line ensures the correct signal phasing. The resulting antenna produces a highly directional radiation pattern, not in the line of the elements as in a Yagi, but broadside or perpendicular to the plane of the array. Like the collinear, the broadside is bidirectional in radiation, but the radiation pattern has a very narrow beam width and high gain.

The *end-fire array* shown in Fig. 9-30(*a*) uses two half-wave dipoles spaced one-half wavelength apart. Both elements are driven by the transmission line. The antenna has a bidirectional radiation pattern but with narrower beam

Fig. 9-29 A broadside array.

(a) Bidirectional

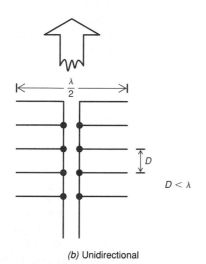

$D < \lambda$

(b) Unidirectional

Fig. 9-30   End-fire antennas.

The lengths of the elements set the operating frequency. When more than one frequency is to be used, of course, multiple antennas can be used. Most TV antennas in use today are of the log-periodic variety so that they can provide high gain and directivity on both VHF and UHF TV channels. Log-periodic antennas are also used in other two-way communications systems where multiple frequencies must be covered.

All the antennas described in this section will operate effectively at their design frequency. They can be used in the 500-kHz to 1-GHz range. The higher the frequency, the shorter the antenna. At frequencies above 1 GHz (microwaves), different types of antennas are used. These include the horn, parabolic, and helix. These will be discussed in the next chapter on microwaves.

One of the most critical aspects of any antenna system is ensuring that the maximum amount of power is transferred from the transmitter to the antenna. An important part of this, of course, is the transmission line itself. The characteristic impedance of the transmis-

widths and gain. The radiation is in the plane of the driven elements as in a Yagi. The end-fire array in Fig. 9-30(*b*) uses five driven elements spaced some fraction of a wavelength *D*. By selecting the desired number of elements and a related spacing, a highly directional unidirectional antenna is created. The spacing causes the lobe in one direction to be canceled so that it adds to the other lobe, creating high gain and directivity.

A special version of the driven array is the wide bandwidth log-periodic antenna shown in Fig. 9-31. The lengths of the driven elements vary from long to short and are related logarithmically. The longest element has a length of one-half wavelength at the lowest frequency to be covered, while the shortest element is one-half wavelength at the highest frequency. Each element is fed with a special short transmission line segment to properly phase the signal sent to each element. The result is a highly directional antenna with excellent gain. Its great advantage over a Yagi or other array is its very wide bandwidth. Most Yagis and other driven arrays are designed for a very specific frequency or a narrow band of frequencies.

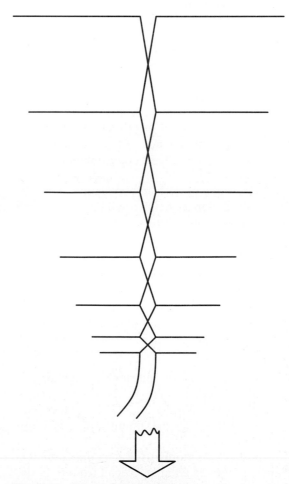

Fig. 9-31   Log-periodic antenna.

CHAPTER 9   ANTENNAS AND TRANSMISSION LINES   **225**

sion line must match the output impedance of the transmitter and the impedance of the antenna itself. If this is the case, the SWR will be 1 and maximum power transfer will take place.

Inevitably, there will be situations where a perfect match between an antenna, transmission line, and transmitter is not possible. In these applications, special techniques are used. Collectively, they are generally referred to as antenna tuning or antenna matching and involve numerous techniques for ensuring maximum power output.

Most of the techniques are aimed at impedance matching. That is, making one impedance look like another through the use of tuned circuits or other devices. For example, one technique is to use what is known as a matching stub or Q section. This is a quarter wavelength of coax or balanced transmission line of a specific impedance that is connected between a load and a source for the purpose of matching impedances. See Fig. 9-32. It has been found that a quarter-wave transmission line can be used to make one impedance look like another. This is expressed in the following relationship:

$$Z_Q = \sqrt{Z_L Z_O}$$

where Z_Q = characteristic impedance of quarter-wave matching stub or Q section

Z_O = characteristic impedance of transmission line or transmitter at input of Q section

Z_L = impedance of load, usually antenna feed point impedance

An example of such an application is the use of the Q section to match a standard 73-Ω coax transmission line to the 36.5-Ω impedance of a quarter-wave vertical antenna. To compute the impedance of the transmission line used as the matching stub, the expression above is used.

$$Z_Q = \sqrt{36.5(73)}$$
$$= \sqrt{2664.5}$$
$$= 51.6$$

This tells us that a quarter-wavelength section of 50-Ω coax will make the 73-Ω transmission line look like the 36.5-Ω antenna or vice versa. In this way maximum power is transferred.

Another commonly used impedance-matching technique is a balun. Recall from previous

Fig. 9-32 A quarter-wave matching stub or Q section.

discussions that a balun is simply a type of transformer used to match impedances. Most baluns are made of a ferrite core, either a toroid or rod, and windings of copper wire. Baluns have a very wide bandwidth and, therefore, are essentially independent of frequency. Baluns can be created for producing impedance-matching ratios of 4:1, 9:1, and 16:1.

A balun can also be constructed from coax as shown in Fig. 9-33. A half wavelength of coax is connected as shown between the antenna and the feed line. When figuring the length of this half-wave section, be sure to take into account the velocity factor of the coax. Such a balun provides a 4:1 impedance transformation. For example, it can easily convert a 300-Ω antenna to a 75-Ω load and vice versa.

When baluns and matching sections cannot do the job, antenna tuners are used. An antenna tuner is a variable inductor, one or more variable capacitors, or a combination of these components connected in various configurations. Here L, T, and pi networks are all widely used. The inductor and capacitor values are adjusted until the impedances match as indicated by minimum SWR.

Fig. 9-33 A coax balun with a 4:1 impedance ratio.

Self-Test

Answer the following questions.

29. Radio waves are made up of _____ and _____ fields.
30. The polarization of a radio wave depends upon the position of its _____ with respect to the earth's surface.
31. The antenna is connected to the transmitter or receiver by a(n) _____ .
32. A radio wave has its magnetic field horizontal to the earth. It is, therefore, _____ polarized.
33. One of the most widely used and simplest antennas is the half-wave _____ .
34. The length of a doublet antenna at 150 MHz is _____ ft.
35. The feed impedance of a dipole antenna is approximately _____ Ω.
36. The horizontal radiation pattern of a dipole looks like a(n) _____ .
37. The measure of an antenna's directivity is _____ .
38. An antenna that radiates equally well in all horizontal directions is said to be _____ .
39. A quarter-wave vertical antenna is commonly known as a(n) _____ antenna.
40. The length of a quarter-wave vertical antenna at 890 MHz is _____ in.
41. For proper operation of a vertical antenna, the shield of the feed coax must make a good connection to _____ or a set of quarter-wave wires called _____ .
42. The horizontal radiation pattern of a quarter-wave vertical is a(n) _____ .
43. The feed impedance of a quarter-wave vertical is approximately _____ Ω.
44. An antenna that transmits or receives equally well in two opposite directions is said to be _____ .
45. A unidirectional antenna transmits best in _____ direction(s).
46. A directional antenna that focuses the energy into a narrow beam has _____ since it effectively amplifies the signal.
47. To have gain and directivity, an antenna must have two or more _____ .
48. List two basic types of antenna arrays.
49. The three basic elements in a Yagi antenna are _____ .
50. The two parasitic elements in a beam antenna are the _____ .
51. A Yagi may have one or more _____ elements.
52. The beam width of a Yagi is usually in

the range of _____ to _____ degrees.
53. The length of the driven element in a Yagi at 222 MHz is _____ ft.
54. List three kinds of driven arrays.
55. True or false. Yagis and driven arrays may be operated either horizontally or vertically.
56. A popular wideband driven array is the _____ array.
57. An impedance-matching circuit used to make the antenna, transmission line, and transmitter impedances match is the _____ .
58. A transformer used for impedance matching is the _____ .
59. A quarter-wavelength section of transmission line used for impedance matching is called a(n) _____ .
60. A coax balun has an impedance-matching ratio of _____ .
61. A quarter-wavelength of coax with a velocity factor of 0.7 at 220 MHz is _____ in.

9-3 RADIO-FREQUENCY WAVE PROPAGATION

Once a radio signal has been radiated by the antenna, it will travel or propagate through space and will ultimately reach the receiving antenna. As you would expect, the energy level of the signal decreases rapidly as the distance from the transmitting antenna is increased. Further, the electromagnetic signal can take one or more of several different paths to the receiving antenna. The path that a radio signal takes depends upon many factors including the frequency of the signal, atmospheric conditions, and the time of day.

The three basic paths that a radio signal can take through space are the ground wave, the sky wave, and the space wave.

The *ground* or *surface wave* leaves the antenna and remains close to the earth. See Fig. 9-34 on the next page. The ground wave will actually follow the curvature of the earth and can, therefore, travel at distances beyond the horizon.

Ground-wave propagation is strongest at the low- and medium-frequency ranges. That is, ground waves are the main signal path for radio signals in the 30-kHz to 3-MHz range. The signals can propagate for hundreds and sometimes thousands of miles at these low frequencies. Amplitude modulation broadcast signals are propagated primarily by ground waves.

Fig. 9-34   Ground- or surface-wave radiation from an antenna.

At the higher frequencies beyond 3 MHz, the earth begins to attenuate the radio signals. Objects on the earth and terrain features become the same order of magnitude in size as the wavelength of the signal and will, therefore, absorb and otherwise affect the signal. For this reason, the ground-wave propagation of signals above 3 MHz is insignificant except within several miles of the antenna.

A *sky-wave signal* is one that is radiated by the antenna into the upper atmosphere where it is bent or reflected back to earth. This bending of the signal is caused by a region in the upper atmosphere known as the ionosphere. Ultraviolet radiation from the sun causes the upper atmosphere to ionize, that is, to become electrically charged. The atoms take on extra electrons or lose electrons to become positive and negative ions, respectively. Free electrons are also present. This results in a relatively thick but invisible layer that exists above the earth. The ionosphere is illustrated in Fig. 9-35. At its lowest point, the ionosphere is approximately 30 mi above the earth and extends to as much as 250 mi above the earth. The ionosphere is generally considered to be divided into three basic layers, designated the D layer, the E layer, and the F layer.

The D and E layers are only weakly ionized areas primarily because they are the farthest from the sun. Primarily they exist only during daylight hours. During daylight hours they tend to absorb radio signals in the medium-frequency range from 300 kHz to 3 MHz.

The F layer is generally considered to be divided into two sublayers designated F1 and F2. These are the most highly ionized because they are closest to the sun and, as a result, will have the most effect on radio signals. The F layer exists both during daylight and nighttime hours.

The primary effect of the F layer is to cause refraction of the radio signal. Refraction is the deflection or bending of electromagnetic waves such as radio waves, light, or even sound when the waves cross a boundary line between two mediums with different characteristics. In the case of the ionosphere, the boundary is the interface between different levels of ionization. When a radio signal goes into the ionosphere, the different levels of ionization will cause the radio waves to be gradually bent. The direction of bending depends upon the angle at which the radio wave enters the ionosphere and the different degrees of ionization of the layers. Figure 9-35 shows the effects of refraction with different angles of radio signals entering the ionosphere. When the angle is large with respect to the earth, the radio signals are bent slightly and pass on through the ionosphere and are lost in space. If the angle of entry is smaller, the radio wave will actually be bent and sent back to earth. Because of this effect, it actually appears as though the radio wave has been reflected by the ionosphere. In reality, the ionosphere does not act like a mirror and cause reflection, but it does cause refraction, which produces an equivalent effect.

The radio waves are sent back to earth with minimum signal loss. The result is that the signal is propagated over an extremely long distance. This effect is most pronounced in the 3- to 30-MHz range which permits extremely long distance communications.

In some cases, the signal refracted back from the ionosphere strikes the earth and is reflected back up to the ionosphere again to be bent and sent back to earth. When this occurs, this is known as multiple-skip or multiple-hop transmission. For strong signals and ideal ionospheric conditions, two, three, or more hops may be possible. This can extend the communications range by many thousands of miles. The maximum distance of a single hop is usually no more than about 2000 mi, but with multiple hops, transmissions halfway around the world are possible.

Radio signals can be reflected by large objects. The earth reflects signals upward as indicated above, but smaller metallic reflectors at least one-half wavelength long at the signal frequency can also reflect signals. Tall buildings,

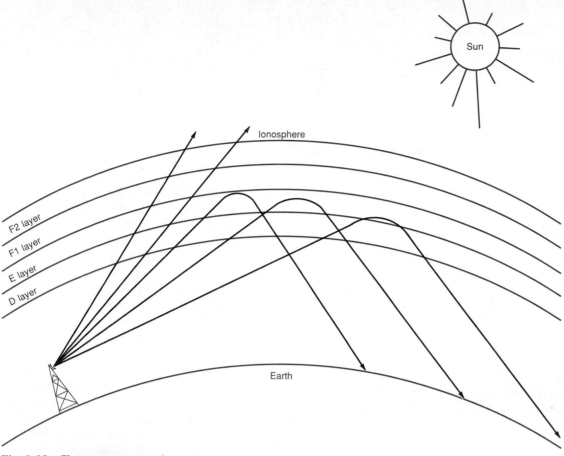

Fig. 9-35 Sky-wave propagation.

mountains, water towers, and even air-planes can reflect radio signals and thus change their direction. Such reflectors act just like mirrors.

The third method of radio signal propagation is by *direct* or *space waves*. A direct wave travels in a straight line directly from the transmitting antenna to the receiving antenna. You will often hear direct-wave radio signaling referred to as line-of-sight communications. Direct or space waves are not refracted nor do they follow the curvature of the earth.

Because of their straight-line nature, direct waves will at some point be blocked because of the curvature of the earth. The signals will travel horizontally from the antenna until they reach the horizon at which point they are blocked. This is illustrated in Fig. 9-36 on the next page. If the signal is to be received beyond the horizon, then the antenna must be high enough to intercept the straight-line radio waves.

Obviously, the transmitting distance with direct waves is limited to relatively short distances and is strictly a function of the height of the transmitting and receiving antennas. For example, the formula for computing the distance between a

transmitting antenna and the horizon is

$$d = \sqrt{2h_t}$$

where h_t = the height of the transmitting antenna in feet
d = the distance from the transmitter to the horizon in miles

For a complete communications system, the height of both the transmitting and receiving antennas must be taken into consideration. The distance between transmitting and receiving antennas then is given by the formula

$$D = \sqrt{2h_t} + \sqrt{2h_r}$$

In this formula, h_r is the height of the receiving antenna in feet.

For example, if a transmitting antenna is 150 ft high and the receiving antenna is 40 ft high, the greatest transmission distance is

$$D = \sqrt{2(150)} + \sqrt{2(40)}$$
$$= \sqrt{300} + \sqrt{80}$$
$$= 17.32 + 8.9$$
$$= 26.22 \text{ mi}$$

$$d = \sqrt{2h_t}$$
$$D = \sqrt{2h_t} + \sqrt{2h_r}$$

Fig. 9-36   Line-of-sight communications by direct or space waves.

Line-of-sight communications by direct wave is characteristic of most radio signals with a frequency above approximately 30 MHz. This is particularly true of VHF, UHF, and microwave signals. Such signals pass through the ionosphere and are not bent. Transmission distances at these frequencies are extremely limited. Antenna height is of utmost importance. You can see why FM and TV broadcasts cover only limited distances. This is the reason why very high transmitting antennas must be used. Transmitters and receivers operating at the very high frequencies typically locate their antennas on top of tall buildings or on mountains to greatly increase the range of transmission and reception.

To extend the communications distance at VHF, UHF, and microwave frequencies, special techniques have been adopted. The most important of these is the use of repeater stations. The concept of a repeater is illustrated in Fig. 9-37. A repeater is a combination of a receiver and a transmitter. The function of the repeater is to pick up the signal from a transmitter, amplify it, and retransmit it on another frequency to the receiver. Usually the repeater will be located between the transmitting and receiving stations and will, therefore, extend the communications distance. Repeaters have extremely sensitive receivers, high-power transmitters, and antennas located at a high point.

Repeaters are extremely effective in increasing the communications range for mobile and hand-held radio units. The antennas on cars and trucks are naturally not very high off the ground, and, therefore, their transmitting and receiving range is extremely limited. However, by operating through a repeater located at some high point, the communications range can be increased considerably.

This concept can be taken further by using a whole series of repeater stations as shown in Fig. 9-38. Each repeater contains a receiver and a transmitter. The original signal is picked up, amplified, and retransmitted on a different frequency to a second repeater which does the same thing. A whole string of repeater stations can relay signals for great distances. Typically such relay stations are located 20 to 60 mi apart, mostly at high elevations to ensure reliable communications over long distances. A

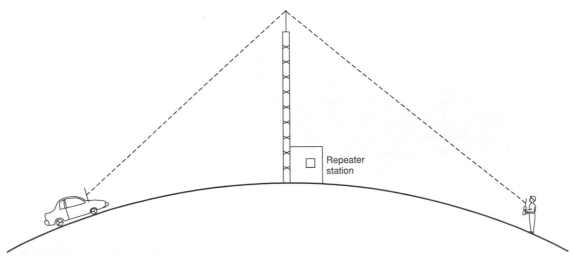

Fig. 9-37   How a repeater extends the communications distance of mobile radio units at VHF and UHF.

Fig. 9-38 Using repeater stations to increase communications distances at microwave frequencies.

good example of such a series of repeaters are the microwave relay stations used by many telephone companies for long-distance communications.

The ultimate repeater or relay station, of course, is the communications satellite. Most communications satellites are located in a geostationary orbit 22,300 mi out in space. At this distance, when rotating about the equator, they rotate in synchronism with the earth and, therefore, appear stationary above the earth. They act as fixed repeater stations. Signals sent to the satellite are amplified and retransmitted back to earth from long distances away. The receiver-transmitter combination within the satellite is known as a transponder. Most satellites have many transponders so that multiple signals can be relayed. With such an arrangement, worldwide communication is possible at microwave frequencies. You will learn more about satellite communications in a later chapter.

Self-Test

Answer the following questions.

62. List the three paths that a radio signal may take through space.
63. A radio wave that propagates near the surface of the earth is called a(n) _____ or _____ wave.
64. The radio wave that is refracted by the ionosphere is known as a(n) _____ wave.
65. A radio wave that propagates only over-line-of-sight distances is called a(n) _____ or _____ wave.
66. The surface wave is effective only at frequencies below about _____ MHz.
67. The upper part of the earth's atmosphere ionized by the sun that affects radio waves is called the _____ .
68. The _____ layer has the greatest effect on a radio signal.
69. The ionized atmosphere causes radio waves at some frequencies to be _____ .
70. True or false. Radio waves are easily reflected by large objects.
71. True or false. The ionosphere reflects radio waves.
72. Only signals in the _____ to _____ MHz range are significantly affected by the ionosphere.
73. Worldwide radio communications is possible thanks to _____ transmission.
74. The VHF, UHF, and microwave signals travel in a(n) _____ .
75. To increase transmission distances at VHF and above, special stations called _____ are used.
76. A microwave relay station contains a(n) _____ and a(n) _____ operating on different frequencies.

SUMMARY

1. A transmission line is a two-wire cable used to carry RF energy between two different pieces of communications equipment or between an antenna and a receiver or transmitter.
2. The two most common types of transmission lines are balanced and coaxial.
3. The primary feature of a transmission line is its characteristic or surge impedance Z_o which is a function of the distributed in-

ductance L and the capacitance C per unit length ($Z_o = L/C$).

4. The characteristic impedance of a balanced line is determined by its physical dimensions. [$Z_o = 276 \log (2S/D)$ where S is the center-to-center spacing of the conductors and D is the diameter of the conductors.]

5. The characteristic impedance of coax also depends on its physical dimensions. [$Z_o = 138 \log (D/d)$ where D is the inside diameter of the shield and d is the diameter of the inner conductor.]

6. The proper use of a transmission line is to terminate it in a load impedance equal to its surge impedance. All the power applied to the line will be absorbed by the load.

7. Wavelength is the distance between adjacent peaks of an RF wave. It is also the distance traveled by a signal in one cycle. Wavelength (λ) is computed with the expression $\lambda = 300/f$ where f is the frequency in megahertz and λ is in meters, Or $\lambda = 984/f$ where λ is in feet.

8. If a transmission line is not terminated in its characteristic impedance, the load will not absorb all the power. Some of it will be reflected back toward the generator.

9. If the load on a transmission line is an open or short, all the power applied to the line will be reflected back to the generator.

10. The forward or incident power applied to the line combines with the reflected power to produce a pattern of voltage and current variations along the line known as standing waves.

11. If the load impedance matches the line impedance, there are no standing waves.

12. A measure of the mismatch between line and load impedances or the maximum and minimum voltage and current variations along the line is the standing-wave ratio (SWR) which is a number always greater than 1.

13. The SWR indicates how much power is delivered to the load and lost in the line. With SWR = 1, all power is delivered to the load.

14. The ratio of the reflected voltage V_r to the incident voltage V_i on a transmission line is called the reflection coefficient R ($R = V_r/V_i$). A properly terminated line will have $R = 0$. A shorted or open line will have $R = 1$.

15. The SWR in terms of the reflection coefficient is

$$\text{SWR} = \frac{1 + R}{1 - R}$$

16. The SWR $= Z_o/Z_l$ or Z_l/Z_o where Z_o is the characteristic impedance and Z_l is the load impedance.

17. Every effort is made to reduce the SWR by using impedance-matching circuits to ensure that maximum power is delivered to the load.

18. Transmission lines, one-quarter or one-half wavelength long and either shorted or open, act like resonant or reactive circuits.

19. At UHF and microwave frequencies where one-half wavelength is less than 1 ft, transmission lines are commonly used to replace conventional LC tuned circuits.

20. A shorted quarter wave and an open half wave act like a parallel resonant circuit.

21. Both an open quarter-wave circuit and a shorted half-wave circuit act like a series resonant circuit.

22. The velocity of propagation of a radio signal is slower in a transmission line than in free space. This difference is expressed as the velocity factor F for different types of lines. Coax has a velocity factor of 0.6 to 0.7. The velocity factor of open wire line or twin lead is in the 0.7 to 0.8 range. In computing the length of a transmission line at a specific frequency, the velocity factor must be considered ($\lambda = 984 \, F/f$).

23. An antenna or aerial is one or more conductors used to transmit or receive radio signals.

24. A radio signal is electromagnetic energy made up of electric and magnetic fields at right angles to one another and to the direction of signal propagation.

25. The polarization of a radio signal is defined as the orientation of the electric field with respect to the earth and is either vertical or horizontal.

26. The most common antenna is the half-wave dipole or doublet that has a characteristic impedance of approximately 73 Ω at the center. Its length in feet is $468/f$ where f is the frequency in megahertz.

27. The dipole has a bidirectional figure-eight radiation pattern and is usually mounted horizontally but may also be used vertically.

28. A popular variation of the dipole is the folded dipole which is one-half wavelength long and has an impedance of 300 Ω.

29. Another popular antenna is the quarter-wave vertical. The earth acts as the other quarter wave to simulate a half-wave vertical dipole.

30. The quarter-wave vertical is referred to as a ground plane antenna. It is fed with coax with the center conductor connected to the antenna and the shield connected to earth

ground, to an array of quarter-wave wires called radials, or to a large, flat, metal surface. Its length in feet is 234/f where f is in megahertz.

31. The characteristic impedance of a ground plane is about 36.5 Ω. It has an omni-directional radiation pattern that sends or receives equally well in all directions.

32. A directional antenna is one that transmits or receives over a narrow range in only one direction.

33. Directional antennas made up of two or more elements focus the radiation into a narrow beam, thus giving the antenna gain.

34. The gain of the antenna is the power amplification resulting from the concentration of power in one direction. The gain may be expressed as a power ratio or in decibels.

35. The effective radiated power (ERP) of an antenna is the power input multiplied by the antenna power gain.

36. Directional antennas with two or more elements are called arrays. There are two types of arrays: parasitic and driven.

37. Parasitic elements called reflectors and directors when spaced parallel to a half-wave dipole driven element help focus the signal into a narrow beam.

38. The measure of the directivity of an antenna is the beam width or beam angle measured in degrees.

39. A parasitic array made up of a driven element, reflector, and one or more directors is known as a Yagi or beam antenna and has a gain of 10 to 20 dB with a beam width of 40° to 20°.

40. Driven arrays consist of two or more half-wavelength elements, each receiving power from the transmission line.

41. The three most popular driven arrays are the collinear, end-fire, and broadside.

42. A widely used driven array is the log-periodic antenna which exhibits gain, directivity, and a wide operating frequency range.

43. A radio wave propagates through space in one of three ways: ground wave, sky waves, or direct waves.

44. The ground or surface wave leaves the antenna and follows the curvature of the earth. The ground wave is only effective on frequencies below 3 MHz.

45. The sky wave propagates from the antenna upward where it is bent back to earth by the ionosphere.

46. The ionosphere is a portion of the earth's atmosphere 30 to 250 mi above the earth that has been ionized by the sun.

47. The ionosphere is made up of three layers of different ionization density: the D, E, and F layers. The F layer is the most highly ionized and causes refraction or bending of radio waves back to earth.

48. The refraction of the ionosphere causes a radio signal to be bent back to earth with little or no attenuation long distances from the transmitter. This is known as a skip or hop.

49. Multiple skips or hops between the ionosphere and earth permit very long distances, even worldwide, communications. This effect is useful over the 3- to 30-MHz range.

50. At frequencies above 30 MHz, propagation is primarily by the direct or space wave which travels in a straight line between transmitting and receiving antennas. This is known as line-of-sight communications.

51. Radio waves are easily blocked or reflected by large objects. This is particularly true of VHF, UHF, and microwave signals.

52. The communications distance at VHF, UHF, and microwave frequencies is limited to the line-of-sight distance between transmitting and receiving antennas.

53. The line-of-sight distance D is limited by the curvature of the earth and is dependent upon the heights h_t and h_r of the transmitting and receiving antennas, respectively.

$$D = \sqrt{2h_t} + \sqrt{2h_r}$$

54. To extend transmission distances at VHF, UHF, and microwave frequencies, relay stations known as repeater stations receive and retransmit signals.

CHAPTER REVIEW QUESTIONS

Choose the letter which best answers each question.

9-1. The most commonly used transmission line is a
  a. Two-wire balance line.
  b. Single wire.
  c. Three-wire line.
  d. Coax.

9-2. The characteristic impedance of a transmission line does *not* depend upon its
  a. Length.
  b. Conductor diameter.
  c. Conductor spacing.
  d. None of the above.

9-3. Which of the following is *not* a common transmission line impedance?
  a. 50 Ω
  b. 75 Ω
  c. 120 Ω
  d. 300 Ω

9-4. For maximum absorption of power at the antenna, the relationship between the characteristic impedance of the line Z_o and the load impedance Z_l should be
  a. $Z_o = Z_l$
  b. $Z_o > Z_l$
  c. $Z_o < Z_l$
  d. $Z_o = 0$

9-5. The mismatch between antenna and transmission line impedances *cannot* be corrected for by
  a. Using an *LC* matching network.
  b. Adjusting antenna length.
  c. Using a balun.
  d. Adjusting the length of transmission line.

9-6. A pattern of voltage and current variations along a transmission line not terminated in its characteristic impedance is called
  a. An electric field.
  b. Radio waves.
  c. Standing waves.
  d. A magnetic field.

9-7. The desirable SWR on a transmission line is
  a. 0.
  b. 1.
  c. 2.
  d. Infinity.

9-8. A 50-Ω coax is connected to a 73-Ω antenna. The SWR is
  a. 0.685.
  b. 1.
  c. 1.46.
  d. 2.92.

9-9. The most desirable reflection coefficient is
  a. 0.
  b. 0.5.
  c. 1.
  d. Infinity.

9-10. A ratio expressing the percentage of incident voltage reflected on a transmission line is known as the
  a. Velocity factor.
  b. Standing-wave ratio.
  c. Reflection coefficient.
  d. Line efficiency.

9-11. The minimum voltage along a transmission line is 260 V, while the maximum is 390 V. The SWR is
  a. 0.67.
  b. 1.0.
  c. 1.2.
  d. 1.5.

9-12. Three feet is one wavelength at a frequency of
  a. 100 MHz.
  b. 164 MHz.
  c. 300 MHz.
  d. 328 MHz.

9-13. At very high frequencies, transmission lines are used as
  a. Tuned circuits.
  b. Antennas.
  c. Insulators.
  d. Resistors.

9-14. A shorted quarter-wave line at the operating frequency acts like a(n)
  a. Series resonant circuit.
  b. Parallel resonant circuit.
  c. Capacitor.
  d. Inductor.

9-15. A shorted half-wave line at the operating frequency acts like a(n)
  a. Capacitor.
  b. Inductor.
  c. Series resonant circuit.
  d. Parallel resonant circuit.

9-16. A popular half-wavelength antenna is the
  a. Ground plane.
  b. End-fire.
  c. Collinear.
  d. Dipole.

9-17. The length of a doublet at 27 MHz is
    a. 8.67 ft.   *c.* 18.2 ft.
    b. 17.3 ft.   *d.* 34.67 ft.

9-18. A popular vertical antenna is the
    a. Collinear.   *c.* Ground plane.
    b. Dipole.   *d.* Broadside.

9-19. The magnetic field of an antenna is perpendicular to the earth. The antenna's polarization
    a. Is vertical.
    b. Is horizontal.
    c. Is circular.
    d. Cannot be determined from the information given.

9-20. An antenna that transmits or receives equally well in all directions is said to be
    a. Omnidirectional.   *c.* Unidirectional.
    b. Bidirectional.   *d.* Quasidirectional.

9-21. The horizontal radiation pattern of a dipole is a
    a. Circle.   *c.* Clover leaf.
    b. Figure eight.   *d.* Narrow beam.

9-22. The length of a ground plane vertical at 146 MHz is
    a. 1.6 ft.   *c.* 2.05 ft.
    b. 1.68 ft.   *d.* 3.37 ft.

9-23. The impedance of a dipole is about
    a. 50 Ω.   *c.* 93 Ω.
    b. 73 Ω.   *d.* 300 Ω.

9-24. A directional antenna with two or more elements is known as a(n)
    a. Folded dipole.   *c.* Loop.
    b. Ground plane.   *d.* Array.

9-25. The horizontal radiation pattern of a vertical dipole is a
    a. Figure eight.   *c.* Narrow beam.
    b. Circle.   *d.* Clover leaf.

9-26. In a Yagi antenna, maximum direction of radiation is toward the
    a. Director.   *c.* Reflector.
    b. Driven element.   *d.* Sky.

9-27. Conductors in multielement antennas that do not receive energy directly from the transmission line are known as
    a. Parasitic elements.   *c.* The boom.
    b. Driven elements.   *d.* Receptors.

9-28. A coax has an attenuation of 2.4 dB per 100 ft. The attenuation for 275 ft is
    a. 2.4 dB.   *c.* 4.8 dB.
    b. 3.3 dB.   *d.* 6.6 dB.

9-29. An antenna has a power gain of 15. The power applied to the antenna is 32 W. The effective radiated power is
    a. 15 W.   *c.* 120 W.
    b. 32 W.   *d.* 480 W.

9-30. Which beam width represents the best antenna directivity?
    a. 7°   *c.* 19°
    b. 12°   *d.* 28°

9-31. The radiation pattern of collinear and broadside antennas is
    a. Omnidirectional.
    b. Bidirectional.
    c. Unidirectional.
    d. Clover-leaf shaped.

9-32. Which antenna has a unidirectional radiation pattern and gain?
    a. Dipole   *c.* Yagi
    b. Ground plane   *d.* Collinear

9-33. A wide-bandwidth multielement driven array is the
 a. End-fire.
 b. Log-periodic.
 c. Yagi.
 d. Collinear.

9-34. Ground-wave communications is most effective in what frequency range?
 a. 300 kHz to 3 MHz
 b. 3 to 30 MHz
 c. 30 to 300 MHz
 d. Above 300 MHz

9-35. The ionosphere causes radio signals to be
 a. Diffused.
 b. Absorbed.
 c. Refracted.
 d. Reflected.

9-36. The ionosphere has its greatest effect on signals in what frequency range?
 a. 300 kHz to 3 MHz
 b. 3 to 30 MHz
 c. 30 to 300 MHz
 d. Above 300 MHz

9-37. The type of radio wave responsible for long-distance communications by multiple skips is the
 a. Ground wave.
 b. Direct wave.
 c. Surface wave.
 d. Sky wave.

9-38. Microwave signals propagate by way of the
 a. Direct wave.
 b. Sky wave.
 c. Surface wave.
 d. Standing wave.

9-39. Line-of-sight communications is *not* a factor in which frequency range?
 a. VHF
 b. UHF
 c. HF
 d. Microwave

9-40. A microwave-transmitting antenna is 550 ft high. The receiving antenna is 200 ft high. The maximum transmission distance is
 a. 20 mi.
 b. 33.2 mi.
 c. 38.7 mi.
 d. 53.2 mi.

9-41. To increase the transmission distance of a UHF signal, which of the following should be done?
 a. Increase antenna gain.
 b. Increase antenna height.
 c. Increase transmitter power.
 d. Increase receiver sensitivity.

9-42. A coax has a velocity factor of 0.68. What is the length of a half wave at 30 MHz?
 a. 11.2 ft
 b. 12.9 ft
 c. 15.6 ft
 d. 16.4 ft

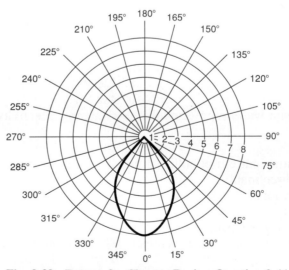

Fig. 9-39   Pattern for Chapter Review Question 9-44.

9-43. Which transmission line has the lowest attenuation?

    a. Twin lead   *c.* RG-59/U

    b. RG-11/U   *d.* RG-214/U

9-44. Refer to Fig. 9-39. The beam width of this antenna pattern is approximately

    a. 30°.   *c.* 45°.

    b. 38°.   *d.* 60°.

9-45. A receiver-transmitter station used to increase the communications range of VHF, UHF, and microwave signals is called a(n)

    a. Transceiver.   *c.* Repeater.

    b. Remitter.   *d.* Amplifier.

CRITICAL THINKING QUESTIONS

9-1. Assume that you wanted to transmit audio directly by antenna without modulation. If the signal to be transmitted has a frequency of 2 kHz, calculate the length of a quarter-wave vertical antenna. Explain why this is an impractical antenna.

9-2. What do you think happens to the gain of a directional antenna as its beam width is decreased? Explain.

9-3. Does the length of a properly terminated and matched transmission line affect the SWR?

9-4. How can a satellite work if radio signals are affected by the ionosphere, which extends only 250 miles out?

9-5. Name two sure things that you can do to improve transmission distance and communications reliability of VHF and UHF signals.

Answers to Self-Tests

1. balanced, coax
2. unbalanced
3. wavelength
4. 0.667
5. 1.475
6. characteristic or surge impedance
7. low-pass filter, capacitors, inductors
8. 75.23
9. 6.25
10. 52
11. constant or flat
12. standing waves
13. reflected
14. standing waves
15. 1.44
16. 1
17. open, shorted
18. reflection coefficient
19. 1.67, 0.25
20. 7
21. 1, infinity
22. tuned or resonant circuits
23. 6.3 [0.07(90) = 6.3] (See Fig. 9-12.)
24. high or infinite
25. open, shorted

26. inductors, capacitors
27. series
28. 2.788
29. electric, magnetic
30. electric field
31. transmission or feed line
32. vertically
33. dipole or doublet
34. 3.12
35. 73
36. figure eight
37. beam width
38. omnidirectional
39. ground plane
40. 3.16
41. earth ground, radials
42. circle
43. 36.5
44. bidirectional
45. one
46. gain
47. elements
48. parasitic, driven
49. director, reflector, driven element
50. director, reflector
51. director
52. 20, 40

53. 2.1
54. collinear, end-fire, broadside
55. true
56. log-periodic
57. antenna tuner
58. balun
59. *Q* section, matching stub
60. 4:1
61. 9.39
62. ground or surface wave, sky wave, direct or space wave
63. ground, surface
64. sky
65. direct or space
66. 3
67. ionosphere
68. F
69. refracted or bent
70. true
71. false
72. 3, 30
73. multiple hop
74. straight line
75. repeaters
76. transmitter, receiver

CHAPTER 10

Microwave Techniques

■

CHAPTER OBJECTIVES

This chapter will help you to:

1. *State* the importance of microwaves to the future of communications.
2. *Define* the term waveguide, *explain* how a waveguide works, and *calculate* the cutoff frequency of a waveguide.
3. *Define* microstrip and stripline and *tell* how they are used.
4. *Explain* the physical structure of a cavity resonator and *tell* what cavity resonators are used for.
5. *Name* seven microwave diode types and *explain* how they operate and where they are used.
6. *Name* the three commonly used microwave vacuum tubes, *explain* their operation, and *state* where they are used.
7. *Name* four common types of microwave antennas and *calculate* the gain and beam width of horn and parabolic dish antennas.

The electromagnetic frequency spectrum is a finite natural resource. As the use of electronic communications has increased over the years, the frequency spectrum normally used for radio signals has become extremely crowded. Many new communications services have been added, and the increased number of users has crowded the airwaves. Thanks to technological developments, more efficient usage of the spectrum is being made. But the future holds only more growth and further overcrowding. One of the primary solutions to this problem has been to extend radio communications into higher frequencies in the spectrum. Initially this expansion was in the VHF and UHF ranges, but now, the primary expansion of radio communications services is in those frequencies beyond 1 GHz. Known as microwaves, these frequencies offer tremendous bandwidth for communications and at least temporarily resolve the problem of spectrum crowding. In this chapter, we introduce you to microwaves and show you some of the special techniques and components unique to this field.

■

10-1 MICROWAVES IN PERSPECTIVE

Microwaves are signals with a frequency greater than 1 GHz. The microwave region is generally considered to extend to 30 GHz, although some definitions include frequencies up to 300 and 600 GHz. In the 1- to 30-GHz range, these signals have wavelengths of 30 centimeters (cm) (about 1 ft) to 1 cm (or about 0.4 in.).

Operating in the microwave region solves many of the problems of the overcrowding in the radio spectrum. At the same time, it introduces additional benefits but also causes some unique problems. Working with microwave equipment requires special knowledge and skills considerably different from those needed for conventional electronic equipment.

Every electronic signal used in communications has a finite bandwidth. When a carrier is modulated by the information signal, sidebands are produced. The resulting signal occupies a certain amount of space in the RF spectrum. We call that small segment of spectrum a *channel*. As the number of communications signals and channels increase, more and more of the spectrum space is used up. Over the years as the need for electronic communication has increased, the number of radio communications stations has increased dramatically, and the radio spectrum has become extremely crowded.

Use of the RF spectrum is typically regulated by the government. In the United States, this job is assigned to the Federal Communications Commission (FCC). The FCC establishes various classes of radio communications and regulates the assignment of spectrum space. For example, certain areas of the spectrum are set aside for radio and TV broadcasting, and frequency assignments are given to the stations. For two-way radio communications, other portions of the spectrum are used. All the various classes of radio communications are given specific areas in the spectrum in which they can operate. Over the years, the available spectrum space has essentially been used up. In many cases, communications services share frequency assignments. In some areas, new licenses are no longer being granted since the current spectrum space for that service is full. Yet the demand for new electronic communications channels continues.

One of the solutions to this problem has been to let technology help solve the problem. For example, great improvements in communications receivers have been made to improve selectivity so that adjacent channel interference will not be as great. This permits stations to operate on more closely spaced frequencies.

On the transmitting side, new techniques have helped squeeze more signals into the same frequency spectrum. A classic example is the use of SSB signals where, only one sideband is used rather than two, thereby cutting the spectrum usage in half. Limiting the deviation of FM signals also helps to control the bandwidth. In data communications new modulation techniques such as phase-shift keying have been developed to help narrow the bandwidth of transmitted information. Multiplexing techniques help put more signals into a single channel.

Although excellent progress has been made,

the available spectrum is still overcrowded and future expectations are that the problem will worsen. The spectrum space between 0 and approximately 300 MHz is absolutely full.

To solve this problem, higher and higher frequencies are being used. Initially, the VHF and UHF bands were tapped. Today, most new communications services are assigned to the microwave region simply because more channels and wider bandwidths are available.

To give you some idea why more bandwidth is available at the higher frequencies, let's take an example. A standard AM broadcast station operating on 1000 kHz has a maximum bandwidth of 10 kHz. The bandwidth represents only $10/1000 = 0.01 = 1$ percent of the spectrum space at that frequency. Now consider a microwave carrier frequency of 3 GHz. One percent of 3 GHz is $0.01(3,000,000,000) = 30,000,000$ Hz, or 30 MHz. A bandwidth of 30 MHz is incredibly wide. In fact, it represents the low-frequency, medium-frequency, and high-frequency portions of the spectrum. This is the space that might be occupied by a 3-GHz carrier modulated by a 15-MHz information signal. Obviously, most information signals do not require that kind of bandwidth. A voice signal would take up only a tiny fraction of that. Our 10-kHz bandwidth AM signal represents only $10,000/3,000,000,000 = 0.000333$ percent of 3 GHz. Up to 3000 AM broadcast stations with 10-kHz bandwidths could be accommodated.

Obviously then, the higher the frequency, the greater the bandwidth available for the transmission of information. This gives not only more space for individual stations, but also allows wide-bandwidth information signals such as video to be accommodated. The average TV signal has a bandwidth of approximately 6 MHz. It is impractical to transmit video signals on low frequencies because they use up entirely too much spectrum space. That's why most TV transmission is in the VHF and UHF ranges. There is even more space for video in the microwave region.

The wide bandwidth also makes it possible to use various multiplexing techniques to transmit more information. When signals are multiplexed, the resulting bandwidth is generally wide, but these multiplexed signals can be easily handled in the microwave region. Finally, data communications of binary information often requires relatively wide bandwidths. These are easily transmitted on microwave frequencies.

The biggest benefit of microwave radio transmission then is the enormous amount of spectrum space available for new communications services. Today most new communications equipment is being developed for the microwave region. It is the only solution to the growing need for more spectrum space.

Although the use of microwaves offers some advantages, it also presents numerous problems. For example, the higher the frequency, the more difficult it becomes to analyze electronic circuits. The analysis of electronic circuits at lower frequencies, say those below 30 MHz, is based upon current-voltage relationships. Such relationships are simply not usable at microwave frequencies. Instead, most components and circuits are analyzed in terms of electric and magnetic fields. As a result, those techniques commonly used for analyzing antennas and transmission lines also become effective for understanding microwave electronic circuits.

Measuring techniques are also different. In low-frequency electronics, we regularly measure currents and voltages. In microwave circuits, the measurements are those of electric and magnetic fields. Power measurements are more common than voltage and current measurements.

Another problem is that at microwave frequencies, conventional components become more difficult, if not impossible, to implement. For example, a common resistor may look like a pure resistance at lower frequencies, but at microwave frequencies, it exhibits other characteristics. The short leads of a resistor, although they may be less than an inch, represent a significant amount of inductive reactance at very high frequencies. A small capacitance also exists between the leads. As a result, at microwave frequencies a simple resistor looks like a complex *LCR* circuit. The same goes for any inductor or capacitor as well.

In addition, to create resonant circuits at microwave frequencies, the values of inductance and capacitance must be smaller and smaller. Physical limits become a problem. Even a half-inch piece of wire represents a significant amount of inductance at some microwave frequencies.

To overcome these problems, lumped components are not used at microwave frequencies. Instead, distributed circuit elements are used. Transmission lines are examples of distributed circuit elements where they constitute inductance and capacitance. When cut to the appropriate length, transmission lines can act

as inductors, capacitors, and resonant circuits. Special versions of transmission lines known as strip lines, microstrips, waveguides, and cavity resonators are widely used to implement tuned circuits and reactances.

Conventional semiconductor devices also do not work properly at microwave frequencies. Because of inherent capacitances and inductances, ordinary diodes and transistors simply will not function. Sufficient gain is lacking.

Another serious problem is transit time in a transistor. Transit time refers to the amount of time it takes for the current carrier (holes or electrons) to move through a device. At low frequencies, the transit times are totally negligible, but at microwave frequencies, transit times become a high percentage of the actual signal period.

This problem has been solved by designing special smaller microwave diodes and transistors and using special materials such as gallium arsenide, in which transit time is significantly less than in silicon. In addition, new and different components are used. This is particularly true for power amplification, where special vacuum tubes known as klystrons, magnetrons, and traveling-wave tubes are the primary components used for power amplification.

Another problem is that microwave signals like light waves travel in perfectly straight lines. This means that the communications distance is limited to line-of-sight range. Antennas must be very high for long-distance transmission. Further, because of the very short wavelength, microwave signals are easily reflected and otherwise diverted by both large and small objects. Even rain, snow, fog, and other atmospheric conditions can greatly absorb and attenuate microwave signals, especially those above about 20 GHz.

Presently, the microwave frequency spectrum is used primarily for telephone communications and radar. Many long-distance telephone systems use microwave relay links for carrying telephone calls. By using multiplexing techniques, thousands of two-way communications are modulated on a single carrier and relayed from one station to another over long distances. Satellite repeater stations are also widely used for long-distance telephone communications.

Another communications technique that is used primarily in the microwave region is radar. Radar is an acronym for *ra*dio *d*etection *a*nd *r*anging. It is a method of detecting the presence of a distant object and determining its distance and direction. Radar systems transmit a high-frequency signal which is then reflected from the distant object: an airplane, a missile, a ship, an automobile, or any other object. The reflected signal is picked up by the radar unit and compared to the transmitted signal. The time difference between the two gives the distance to the object. Since highly directional antennas are used, naturally the direction or azimuth of the object can also be determined.

The greatest use of radar, of course, is in military applications, but it is also used for safety purposes in aviation and marine fields. Another major use of radar is in weather forecasting. It has also found application in other areas such as in highway law enforcement.

Television stations and networks use microwave relay links to transmit TV signals over long distances rather than rely on long, lossy coax cables. Cable TV networks use satellite communications to transmit programs from one location to another. There are even consumer microwave satellite TV receivers capable of picking up TV transmissions directly from the satellite.

A growing application for microwave communications is space communications. Communications with satellites, deep-space probes, and other spacecraft is usually done by microwave transmission. The reason for this is that microwave signals are not reflected or absorbed by the ionosphere as are many lower-frequency signals. A microwave signal penetrates all the ionospheric layers directly into outer space with minimum attenuation.

Electromagnetic radiations from the stars are also primarily in the microwave region. Through the use of radio telescopes, astronomers can study the stars and other bodies in this galaxy and beyond. Sensitive radio receivers and large antennas operating in the microwave region are capable of plotting outer space even more effectively than can be done with visual (optical) telescopes.

Finally, microwaves are also used for heating. Microwave ovens are a common appliance in most kitchens today, and special microwave equipment known as diathermy machines are used in medicine for heating body muscles and tissues without hurting the skin. Microwave heating is also widely used in industry for a variety of heating and melting operations.

In the remainder of this chapter, we will introduce you to the major types of microwave components and circuits, including waveguides and cavity resonators, microwave semiconductors, transmitting tubes, and antennas.

Self-Test

Answer the following questions.

1. Microwaves are frequencies above _____ GHz.
2. The main advantage of microwaves is that more of the _____ is available for signals.
3. List seven reasons why microwaves are more difficult to work with than lower-frequency signals.
4. The applications most commonly found in the microwave region are _____ and _____ .
5. List four popular uses for microwaves.
6. The government agency that regulates radio communications in the United States is the _____ .
7. Name four techniques that have helped squeeze more signals into the given spectrum.
8. Two reasons why conventional transistors won't work at microwave frequencies are _____ and _____ .
9. At microwave frequencies, conventional components like resistors, capacitors, and inductors act like _____ circuits.
10. The TV channels 2 to 13 occupy a total bandwidth of about 72 MHz. At a frequency of 200 MHz, this represents a bandwidth of _____ percent of the spectrum space. At 2 GHz, it represents _____ percent.

10-2 TRANSMISSION LINES, WAVEGUIDES, AND CAVITY RESONATORS

Transmission lines are the means for carrying electromagnetic energy from one place to another. At frequencies below microwaves, parallel wire transmission lines and coax are the primary means of carrying radio signals. But at microwave frequencies, these kinds of transmission lines are less effective. Long parallel transmission lines, such as 300-Ω twin lead, will actually radiate some of the electromagnetic energy while transporting it from one place to another. As the frequency of operation gets higher, the amount of radiation from the line increases. At microwave frequencies, virtually all the energy is radiated; therefore, little energy ever reaches the end of the transmission line.

Coaxial cable eliminates the radiation problem, but it still has significant loss at the higher frequencies. The attenuation at microwave frequencies is so large as to make coax virtually unusable at all but the lower microwave frequencies (less than 10 GHz). Special types of coax using larger inner conductors and shields have been developed to minimize loss, although they are more difficult to use because in most cases they are rigid rather than flexible like most other cable. Special coax has been developed for the lower microwave frequencies and is used in some applications, such as consumer satellite TV. These special cables can be used up to approximately 6 GHz if the length is kept short. Above this frequency, coax loss is too great. Short lengths of coax, several feet or less, can be used to interconnect pieces of equipment that are close together, but for longer runs other methods of transmission must be used.

As indicated in Chap. 9, transmission lines can be used as tuned circuits, filters, and even reactive components. And this is exactly how these circuits and components are implemented at microwave frequencies. Lengths of one-quarter and one-half wave are only inches or less, making them a practical size. But instead of coax or balanced line, special transmission lines are constructed by using printed circuit boards (PCBs) or IC manufacturing techniques.

Recall that a PCB is a flat insulating base made of phenolic or fiberglass to which is bonded copper on one or both sides and sometimes in several layers. Teflon is used as the base for some PCBs in microwave applications. The copper is etched away in patterns to form the interconnections for transistors, ICs, resistors, and other components. Thus point-to-point connections with wire are eliminated.

The PCB techniques can be used to create special microwave transmission lines called microstrip and stripline. *Microstrip* is a flat conductor separated from a large conducting ground plane by an insulating dielectric as shown in Fig. 10-1. The length of the microstrip is usually one-quarter or one-half wavelength. The ground plane is the circuit common.

Stripline is a flat conductor sandwiched between two ground planes as shown in Fig. 10-2. It is more difficult to make, but it will not radiate as microstrip does. The length is one-quarter or one-half wave.

Fig. 10-1   Microstrip.

Both stripline and microstrip are widely used to form the tuned circuits used in microwave receiver front ends and in the low-power amplifier sections of transmitters. Diodes, transistors, and other components are mounted right on the PCB and are connected directly to the formed microstrip or stripline.

Even tinier microstrips and striplines can be made using monolithic, thin-film, and hybrid IC techniques. When combined with diodes, transistors, and other components, microwave integrated circuits (MIC) are formed.

Most microwave energy transmission is handled by waveguides. A *waveguide* is a hollow metal tube designed to carry microwave energy from one place to another. Waveguides may be used to carry energy between pieces of equipment or over longer distances to carry transmitter power to an antenna or microwave signals from an antenna to a receiver.

Waveguides are made from copper, aluminum, or brass. These metals are extruded into long rectangular or circular pipes. Often the insides of these waveguides are plated with silver to reduce their resistance to a very low level.

A microwave signal to be carried by a waveguide is injected into one end of the waveguide. This is done with an antennalike device which creates an electromagnetic wave that propagates through the waveguide. The electric and magnetic fields associated with the signal bounce off the inside walls back and forth as it progresses down the waveguide. The

Fig. 10-2   Stripline.

waveguide completely contains the signal so that none escapes by radiation.

A common method of applying a microwave signal to a waveguide is done by using a probe as shown in Fig. 10-3(*a*). The probe is a quarter-wavelength vertical antenna at the signal frequency which is inserted in the waveguide one-quarter wavelength from the end which is closed. The signal is usually coupled to the probe through a coaxial connector. The probe sets up a vertically polarized electromagnetic wave in the waveguide, and the wave is then propagated down the line. The location of the probe one-quarter wavelength from the closed end of the waveguide causes the signal from the probe to be reflected from the closed end of the line back toward the open end. Over a quarter-wave distance, the reflected signal will appear back at the probe in phase to aid the signal going in the opposite direction. Remember that a radio signal consists of both electric and magnetic fields at right angles to one another. The electric and magnetic fields established by the probe propagate down the waveguide at a right angle to those two fields. An alternative probe position is shown in Fig. 10-3(*b*). The position of the probe determines whether the signal is horizontally or vertically polarized.

Figure 10-3(*c*) shows how a loop can be used to introduce electromagnetic radiation into the waveguide. Here the loop is mounted in the closed end of the waveguide. Microwave energy applied through a short piece of coax causes a magnetic field to be set up in the loop. The magnetic field also establishes an electric field which is then propagated down the waveguide.

Extracting a signal from a waveguide is also done with probes and loops. When the signal strikes a probe or loop, a signal will appear across the coaxial connector which can then be fed to other circuitry.

Most waveguides are of the rectangular variety. It is the size of the waveguide that determines its operating frequency range. Figure 10-4 on the next page shows the most important dimensions of a waveguide. The width of the waveguide is designated *a*, and the height is designated *b*. Note that these are the inside dimensions of the waveguide. The frequency of operation is determined by the *a* dimension. This dimension is usually made equal to one-half the wavelength at the lowest frequency of operation. This frequency is known as the waveguide cutoff frequency. At the cutoff frequency and below, the waveguide will not transmit energy. At frequencies above the cutoff frequency, the waveguide will propagate electromagnetic energy. As you can see then, the waveguide is essentially a high-pass filter with a cutoff frequency equal to

$$f_{co} = \frac{300}{2a}$$

Here f_{co} is in megahertz and *a* is in meters.

As an example, a waveguide has an *a* width of 0.8 in. What is the cutoff? To convert 0.8 in to meters, multiply by 2.54 to get centimeters and divide by 100 to get meters. So, 0.8 in. = 0.02032 m.

$$f_{co} = \frac{300}{2(0.02032)}$$

$$= 7382 \text{ MHz}$$

$$= 7.382 \text{ GHz}$$

Normally, the height of the waveguide is made equal to approximately one-half the *a* dimension, in this case about 0.4 in.

When a probe or loop launches energy into the waveguide, the electromagnetic fields will bounce off the side walls of the waveguide as illustrated in Fig. 10-5 on the next page. The angles of incidence and reflection depend upon the operating frequency. At high frequencies, the angles are large, and, therefore,

(*a*)　　　　(*b*)　　　　(*c*)

Fig. 10-3 **Coupling RF energy to a waveguide.**

$$f_{co} = \frac{300}{2a}$$

Fig. 10-4 **Dimensions of a waveguide determine its operating frequency range.**

the path between the opposite walls is relatively long as shown in Fig. 10-5(a). As the operating frequency gets lower, the angles decrease and the path between the sides shortens. When the operating frequency reaches the cutoff frequency of the waveguide, the signal will simply bounce back and forth directly between the side walls of the waveguide and will have no forward motion. At the cutoff frequency and below, no energy is propagated.

Rectangular waveguides usually come in a variety of standard sizes. The exact size is selected based on the desired operating frequency. The size of the waveguide is chosen so that its rectangular width is greater than one-half the wavelength but less than one wavelength at the operating frequency. This gives a cutoff frequency that is below the operating frequency, thereby ensuring that the signal will be propagated down the line. For example, a typical waveguide to transmit a 7-GHz signal has dimensions of $a = 1.37$ in. and $b = 0.62$ in. A waveguide of this size has a cutoff frequency of 4.3 GHz.

Whenever a microwave signal is launched into a waveguide by a probe or loop, the electric and magnetic fields are set up in various patterns depending upon the method of energy coupling, the frequency of operation, and the size of the waveguide.

Figure 10-6 shows typical fields in a waveguide. Looking at the end view, you see the lines which represent the electric (E) field. The X's and dots represent the magnetic (H) field. The X's represent H field lines going into the page, while the dots represent lines coming out of the page.

In the top view, the dashed lines represent the H field, and the X's and dots represent the E field lines.

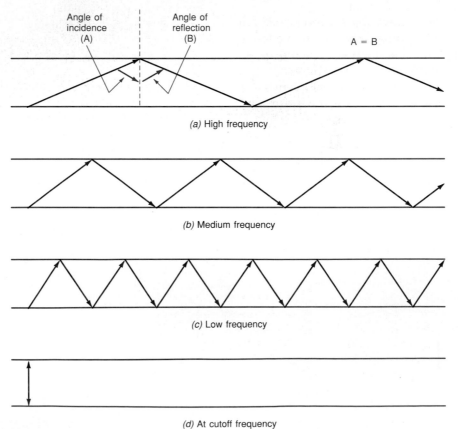

(a) High frequency

(b) Medium frequency

(c) Low frequency

(d) At cutoff frequency

Fig. 10-5 **Wave paths in a waveguide at various frequencies.**

Top view   End view

Magnetic lines

Electric
field lines

Magnetic
lines

Electric
field lines

✕ means into page
● means out of page

Fig. 10-6 Electric (E) and magnetic (H) fields in a rectangular waveguide.

The pattern of the electromagnetic fields within a waveguide takes many forms. Each form is called an operating mode. Either the H or E field must be perpendicular to the direction of propagation of the wave. Modes are classified as either *transverse electric (TE)* or *transverse magnetic (TM)*. In the TE mode, the E field exists across the guide and no E lines extend lengthwise along the guide. In the TM mode, the H lines form loops in planes perpendicular to the walls of the guide, and no part of an H line is lengthwise along the guide.

Subscript numbers are used along with the TE and TM designations to further describe the E and H field patterns. A typical designation is $TE_{0,1}$. The first number indicates the number of half-wave patterns of transverse lines that exist along the short dimension of the guide through the center of the cross section. Transverse lines are those lines perpendicular to the walls of the guide. The second number indicates the number of transverse half-wave patterns that exist along the long dimension of the guide through the center of the cross section. If there is no change in the field intensity of one dimension, a zero is used.

The waveguide in Fig. 10-6 is TE because the E lines are perpendicular (transverse) to the sides of the guide. Looking at the end view in Fig. 10-6 along the short length, there is no field intensity change, so the first subscript is 0. Along the long dimension of the end view, the E lines spread out at the top and bottom but are close together in the center. The actual field intensity is sinusoidal, zero at the ends, maximum in the center. This is one-half

of a sine wave variation. The second subscript, therefore, is 1. So the mode of the line in Fig. 10-6 is $TE_{0,1}$. This, by the way, is the main or dominant mode of most rectangular waveguides. Of course, there are many other patterns. Two others are shown in Fig. 10-7.

Waveguides are available in a variety of different lengths. As many sections as necessary can be interconnected to form a straight-line path between the microwave generator and its ultimate destination. When a waveguide is used as a transmission line between a transmitter or receiver and the antenna, relatively long straight sections are common. However, in

Fig. 10-7 Other waveguide operating modes.

most installations it is necessary for the waveguide to make turns. Special curved waveguide sections are available for making 90° bends. Curved sections introduce reflections and power loss, but these are kept small by proper design. When the radius of the curved section is greater than two wavelengths at the signal frequency, losses are minimized. Figure 10-8 shows several different configurations.

Radius greater than (>) 2λ for minimum loss

(a) Right angle bends

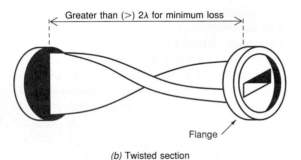

Greater than (>) 2λ for minimum loss

Flange

(b) Twisted section

Fig. 10-8   Curved waveguide sections.

Most of these special sections are accompanied by flanges which are used to interconnect them with straight runs. Most waveguides and their fittings are precision-made so that the dimensions match perfectly. Any mismatch in dimensions or misalignment of pieces that fit together will introduce significant losses and reflections.

A waveguide is a basic microwave component. Not only is it used as a transmission line, but special shorter segments are used for a variety of other purposes. These include simulating reactive components and forming resonant circuits. Other applications include the coupling of multiple units and impedance matching. Some of the special waveguide-

derived components include cavity resonators, directional couplers, circulators, hybrids, and duplexers. Let's take a brief look at each of these.

One of the most important microwave devices is the cavity resonator. A cavity resonator is a short segment of waveguide that acts as a high-Q parallel resonant circuit. A simple cavity resonator can be formed with a short piece of waveguide one-half wavelength long as illustrated in Fig. 10-9. A small probe at the center injects microwave energy. The short piece of waveguide is closed at each end with a metallic short. The result is that when microwave energy is injected into the cavity, the signal will bounce off the shorted ends of the waveguide and reflect back toward the probe. Because the probe is located a quarter wavelength from each shorted end, the reflected signal will reinforce the signal at the probe. The result is that the signal will bounce back and forth off the shorted ends. If the signal is removed, the wave will continue to bounce back and forth for a continuous length of time until losses cause it to die out.

Fig. 10-9   Side view of a half-wavelength waveguide used as a cavity resonator.

This effect is pronounced at a frequency where the length of the waveguide is exactly one-half wavelength. The cavity is said to resonate and acts like a parallel resonant circuit. A brief burst of energy applied to the probe will cause the cavity to oscillate until losses cause it to die out. Cavities such as this have an extremely high Q, as high as 30,000. For this reason, they are commonly used to create resonant circuits and filters at microwave frequencies.

A cavity can also be formed by using a short section of circular waveguide as shown in Fig. 10-10. The diameter should be one-half wavelength at the operating frequency. Other cavity shapes are also possible. Further, a coupling loop rather than a probe is more commonly used in cavity resonators. In some cavity resonators two probes or two coupling loops are used, one for the input and one for

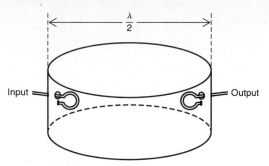

Fig. 10-10 Circular resonant cavity with input-output loops.

(a) Cylindrical cavity with adjustable disk

(b) Cavity with adjustable plugs

Fig. 10-11 Adjustable or tunable cavities.

the output. In this way, the cavity can serve as a filter.

Cavities can be constructed in a variety of ways. Although short segments of waveguide can be used, typically cavities are specially designed components. They are often hollowed out sections in a block of metal. By machining the cavity, very precise dimensions can be obtained for a specific frequency. The internal walls of the cavity are often plated with silver or some other low-loss material to ensure minimum losses and a very high Q.

Some cavities are also tunable. One wall of the cavity is made movable as shown in Fig. 10-11(*a*). An adjustment screw moves the end wall in and out to adjust the resonant frequency. The smaller the cavity, the higher the operating frequency. Cavities can also be tuned with adjustable plugs in the side of the cavity as Fig. 10-11(*b*) illustrates. As the plug is screwed in, more of it protrudes into the cavity and the operating frequency goes up.

Self-Test

Answer the following questions.

11. True or false. Twin lead is not used at microwave frequencies.
12. The main disadvantage of using coax for microwave signals is its high _____ .
13. Coax is not used beyond frequencies of about _____ GHz.
14. A hollow metal pipe used to carry microwaves is called a(n) _____ .
15. Two types of transmission line made of PCB material used to produce tuned circuits and filters are _____ .
16. The two ways to couple or extract energy from a waveguide are by using a _____ or a _____ .
17. A waveguide acts as a(n) _____ filter.

18. A rectangular waveguide has a width of 1.2 in. and a height of 0.7 in. The waveguide will pass all signals above _____ GHz.
19. A waveguide has a cutoff frequency of 5 GHz. (**a**) True or false. It will pass a signal of 8 GHz. (**b**) True or false. It will pass a signal of 3 GHz.
20. The magnetic and electric fields in a waveguide are designated by the letters _____ and _____ , respectively.
21. The basic operating mode of most waveguides is designated _____ .
22. If the H field in the waveguide is perpendicular to the direction of signal travel, the mode is said to be _____ .
23. A one-half wavelength section of waveguide shorted at both ends creates a(n) _____ .
24. A cavity resonator acts as a(n) _____ circuit.
25. Resonant cavities are used as _____ .
26. Mechanically varying the cavity's dimensions allows its _____ to be adjusted.

10-3 MICROWAVE SEMICONDUCTORS

Most conventional diodes and transistors do not operate at microwave frequencies. The lengths of the diode and transistor leads present a significant amount of inductance at microwave frequencies and, therefore, introduce losses. The internal capacitances of diodes and transistors are also very high and will actually prevent operation at microwave frequencies. Additionally, the transit time of the carriers in a diode or transistor is sufficiently high to prevent normal operation. Transit time refers to the amount of time it takes for an electron or hole in a diode or transistor to pass from cathode to anode, emitter to base, emitter to collector, or source to drain.

These problems have been overcome by designing special semiconductor devices for use at microwave frequencies. In addition to new semiconductor materials other than silicon, the geometry is different. In this section we want to review briefly the most common semiconductor devices used in microwave equipment.

A wide variety of semiconductor diodes have been developed for microwave use. Those diodes used for signal detection and mixing are the most common. Two types are widely used: the point-contact diode and Schottky barrier (or hot-carrier) diode.

The typical semiconductor diode is a junction formed of P- and N-type semiconductor materials. Because of the relatively large surface area of the junction, these diodes exhibit a high capacitance which prevents normal operation at microwave frequencies. For this reason, standard PN junction diodes are not used in the microwave region. Instead, special microwave diodes are used.

Perhaps the oldest microwave semiconductor device is the *point-contact* diode. This is nothing more than a piece of semiconductor material and a fine wire which makes contact with the semiconductor material. Because the wire makes contact with the semiconductor over a very small surface area, the capacitance is extremely low. Current flows easily from the fine wire (the cathode) to the semiconductor material (the anode). However, current will not flow easily in the opposite direction.

Most early point-contact diodes used germanium as the semiconductor material, but today point-contact diodes are made of P-type silicon with a fine tungsten wire as the cathode.

Fig. 10-12   A point-contact diode.

See Fig. 10-12. The forward threshold voltage is extremely low.

Point-contact diodes are ideal for low-signal applications and are widely used in microwave mixers and detectors. They are extremely delicate and cannot withstand high power. Also, they are easily damaged and, therefore, must be used in such a way as to avoid shock and vibration.

Although point-contact diodes will be found in some older-design microwave equipment, for the most part they have been replaced by *Schottky* or *hot-carrier* diodes. You will see the terms Schottky and hot carrier both used, but they are the same thing. However, the term Schottky diode is the most widely used. A Schottky diode has a metal semiconductor junction.

Most Schottky diodes are made with N-type silicon on which has been deposited a thin metal layer. The semiconductor forms the cathode, and the metal forms the anode. Typical anode materials include nickel chromium and aluminum, although other metals such as gold are also used.

The Schottky diode, like the point-contact diode, is extremely small and, therefore, has a tiny junction capacitance. It also has a low bias threshold voltage. It too is ideal for mixing, signal detection, and other low-level signal operations. They are widely used in balanced modulators and mixers. Because of their very high frequency response, Schottky diodes are also used as fast switches at microwave frequencies.

The most important use of microwave diodes is as mixers. Most microwave diodes are installed as part of a waveguide or cavity resonator tuned to the incoming signal frequency. An LO signal is injected with a probe or loop.

The diode mixes the two signals and produces the sum and difference output frequencies. The difference frequency is usually selected with another cavity resonator or with a low-frequency LC tuned circuit. The mixer is usually the input circuit in a microwave receiver since it is desirable to convert the microwave signal down to a lower frequency level as early as possible where amplification and demodulation can take place with simpler, more conventional electronic circuits.

Another widely used microwave semiconductor device is the *varactor* diode. As you saw earlier, a varactor diode is basically a VVC. When a reverse bias is applied to the diode, it acts like a capacitor. Its capacitance depends upon the value of the reverse bias. Varactor diodes are VVCs made with gallium arsenide and optimized for use at microwave frequencies. Their main application is as frequency multipliers.

Figure 10-13 shows a varactor frequency multiplier circuit. When an input signal is applied across the diode, it alternately conducts and cuts off. The result is a nonlinear or distorted output containing many harmonics. By using a tuned circuit in the output, the desired harmonic is selected and others are rejected. Since the lower harmonics produce the greatest amount of energy, varactor multipliers are usually used only for doubling and tripling operations. In Fig. 10-13, the input tuned circuit L_1-C_2 resonates at the input frequency f_i, while the output tuned circuit L_2-C_3 resonates at 2 or 3 times the input frequency as desired. In practice, the tuned circuits are not actually made up of individual inductors or capacitors. Instead, they are microstrip, stripline, or cavity resonators. Both C_1 and C_4 are used for impedance matching.

A varactor frequency multiplier does not have gain, in fact, it produces a signal power loss. But it is a relatively efficient circuit, and the output can be as high as 80 percent of the input. Typical efficiencies are in the 50 to 80 percent range. Note also in Fig. 10-13 that no external source of dc power is required for this circuit. Only the RF input power is required for proper operation. Outputs up to 20 W are obtainable with special high-power varactors.

Varactors are used in those applications where it is difficult to generate microwave signals. Usually it is a lot easier to generate a VHF or UHF signal and then use a series of frequency multipliers to put it into the desired microwave region. Varactor diodes are available for producing relatively high power outputs at frequencies up to 100 GHz.

Another diode used in microwave frequency multiplier circuits like that in Fig. 10-13 is the *step-recovery diode* or *snap-off varactor*. It is a PN junction diode made with gallium arsenide or silicon. When it is forward-biased, it conducts as any diode, but a charge is stored in the depletion layer. When reverse bias is applied, the charge keeps the diode on momentarily. Then suddenly the diode turns off abruptly. This snap-off produces an extremely high intensity reverse current pulse with a duration of about 100 picoseconds (ps). (1 ps $= 10^{-12}$ s.) It is extremely rich in harmonics. Even the higher harmonics are of relatively high amplitude.

Step-recovery diodes are used in the circuit of Fig. 10-13 to produce multipliers with factors up to 5 and 10. Power ratings of 50 W can be obtained. Operating frequencies up to 10 GHz are possible with an efficiency of 80 percent or more.

Another very popular microwave diode is the *Gunn* diode. It is also called a transferred-electron device (TED). The Gunn diode is not a diode in the usual sense as it does not have a junction. Instead, it is a thin piece of N-type gallium arsenide (GaAs) or indium phosphide (InP) semiconductor which forms a special resistor when voltage is applied to it. This device exhibits a negative resistance character-

Fig. 10-13   A varactor frequency multiplier.

istic. That is, over some voltage range, an increase in voltage results in a decrease in current and vice versa, just the opposite of what Ohm's law predicts for a linear resistance. When so biased, the time it takes for electrons to flow across the material is such that the current will be 180° out of phase with the applied voltage. If the Gunn diode is so biased and connected to a cavity which is resonant near the frequency determined by the electron transit time, the resulting combination will oscillate. The Gunn diode, therefore, is used primarily as a microwave oscillator.

Gunn diodes are available to oscillate at frequencies up to about 50 GHz. In the lower microwave range, power outputs up to several watts are possible. It is the thickness of the semiconductor that determines the frequency of oscillation. However, if the cavity is made variable, the Gunn oscillator frequency can be adjusted over a narrow range.

Gunn diodes are widely used as LOs in receivers and as primary frequency sources in transmitters. Police radar guns use a Gunn diode oscillator.

Two other microwave diodes are also widely used as oscillators. These are the *IMPATT* and *TRAPATT* diodes. Both are PN junction diodes made of silicon, GaAs, or InP. They are designed to operate with a high reverse bias that causes them to avalanche or break down. A high current flows. Over a narrow range, a negative resistance characteristic is produced. This will produce oscillation when the diode is mounted in a cavity. The IMPATT diodes are available with power ratings up to about 25 W in the low gigahertz range.

The *tunnel* diode is another negative resistance diode. It produces a narrow range of negative resistance when forward-biased. Tunnel diodes are used to produce low-power microwave oscillators.

Self-Test

Answer the following questions.

27. List three reasons why conventional diodes and transistors do not work in the microwave region.
28. A point-contact diode has a(n) _____ anode and a(n) _____ cathode.
29. A microwave diode with an N-type silicon cathode and a metal anode forming a junction is called a(n) _____ diode.
30. The most common application of microwave signal diodes is in _____ circuits.
31. A diode that acts like a VVC is a(n) _____ diode.
32. Two types of diodes widely used as frequency multipliers are the _____ and _____ diodes.
33. A diode with no junction that is widely used with a cavity resonator to form an oscillator is the _____ diode.
34. The _____ , _____ , and _____ diodes are also used as oscillators.
35. Microwave diodes that cause a current decrease for a voltage increase have what is known as a(n) _____ characteristic.
36. An IMPATT diode operates with _____ (forward, reverse) bias.
37. A tunnel diode operates with _____ (forward, reverse) bias.

10-4 MICROWAVE TUBES

Before transistors were invented, all electronic circuits were implemented with vacuum tubes. Like transistors, tubes are devices used for controlling a large signal with a smaller signal to produce amplification, oscillation, switching, and other operations. Today, vacuum tubes are no longer used in most circuits. Yet they have survived in special applications. The cathode-ray tube (CRT) used in TV sets, oscilloscopes, and other display devices is a special form of vacuum tube which has no semiconductor equivalent. Vacuum tubes are also still found in microwave equipment. This is particularly true in microwave transmitters used for producing high output power. Although both bipolar and field-effect transistors are available to produce power in the microwave region up to approximately 50 W, many applications require more power. In satellite earth stations, TV stations, and in some military equipment such as radar, very high output powers are needed. Special microwave tubes developed during World War II are still widely used for microwave power amplification. These tubes are the klystron, the magnetron, and the traveling-wave tube. In this section we take a brief look at each.

A *klystron* is a microwave vacuum tube using cavity resonators to produce velocity modulation of the electron beam and produce amplification. Figure 10-14 shows the concept of a two-cavity klystron. The vacuum tube

Fig. 10-14   **A two-cavity klystron amplifier.**

contains a cathode which is heated by a filament. At a very high temperature, the cathode emits electrons. These negative electrons are attracted by a positive plate or collector. Thus, current flow is established between the cathode and the collector inside the evacuated tube.

The electrons emitted by the cathode are focused into a very narrow stream by electrostatic and electromagnetic focusing techniques. In electrostatic focusing, special elements called focusing plates to which have been applied high voltages force the electrons into a narrow beam. Electromagnetic focusing makes use of coils around the tube through which current is passed to produce a magnetic field. This magnetic field helps focus the electrons into a narrow beam.

The sharply focused beam of electrons is then forced to pass through the centers of two cavity resonators. The cavity is open in the center area. The microwave signal to be amplified is applied to the lower cavity through a coupling loop. This sets up electric and mag-

netic fields in the cavity which affect the electron beam. Specifically they cause the electrons to speed up and slow down as they pass through the cavity. On one half cycle, the electrons are speeded up, while on the next half cycle of the input, they are slowed down. The effect of this is to create bunches of electrons that are one-half wavelength apart in the drift space between the cavities. This speeding up and slowing down of the electron beam is known as velocity modulation. Since the input cavity produces bunches of electrons, it is commonly referred to as the buncher cavity.

The bunched electrons move on through the tube since they are attracted by the positive collector. They then pass through the center of another cavity known as the catcher cavity. These bunched electrons move toward the collector in clouds of alternately dense and sparse areas. We can say that the electron beam is density-modulated.

As the bunches of electrons pass through the catcher cavity, it excites the cavity into oscillation at the resonant frequency. Therefore, the dc energy in the electron beam is converted into RF energy at the cavity frequency. Thus, amplification takes place. The output is extracted from the catcher cavity with a loop.

Klystrons are also constructed with additional cavities between the buncher and catcher cavities. These intermediate cavities produce further bunching which causes increased amplification of the signal. If the buncher cavities are tuned off the center frequency from the input and output cavities, they have the effect of broadening the bandwidth of the tube. The frequency of operation of a klystron is set by the sizes of the input and output cavities. Since cavities typically have high Q's, their bandwidth is limited. By lowering the Q's of the cavities and by introducing intermediate cavities, wider bandwidth operation can be achieved.

Klystrons are available in a wide range of sizes. Small hand-held units produce only milliwatts of power amplification, while larger, human-size klystrons produce many thousands of watts of power. Klystrons are used at frequencies as low as UHF and as high as 100 GHz in the microwave region.

A special variation of the basic klystron tube known as a reflex klystron is commonly used as a microwave oscillator. A reflex klystron uses only a single cavity as shown in Fig. 10-15. The cathode emits electrons which are accelerated forward by an accelerating grid with a positive voltage on it and focused into a narrow beam as previously described. The electrons pass through the cavity and undergo velocity modulation which produces electron bunching. The electron bunches move toward the repeller plate which is made negative with respect to the cathode. This causes the electrons to be repelled rather than attracted as they are by a positive collector in a standard klystron. The result is that the bunched electrons are forced back through the cavity. The distance between the repeller and the cavity is chosen such that the repelled electron bunches will reach the cavity at the proper time to be in synchronization. Because of this, they deliver energy to the cavity. The result is os-

Fig. 10-15 A reflex klystron used as an oscillator.

cillation at the cavity frequency. The cavity itself is made positive so that the electrons are ultimately attracted by the cavity and cause direct current flow in the external circuit. A coupling loop in the cavity removes the RF energy.

Reflex klystrons are small, low-power devices. They typically generate output power from 100 mW to several watts of power. They are commonly used for microwave signal generators and for LOs in microwave receivers. Reflex klystrons can also be easily frequency-modulated by simply adding the ac modulating signal in series with the repeller supply voltage. This causes the repelled electron bunches to reach the cavity at different times, thereby producing a variable output frequency (FM).

Klystrons are gradually being replaced in most microwave equipment. Gunn diodes are replacing the smaller reflex klystrons in signal-generating applications because they are smaller, lower in cost, and do not require high dc supply voltages. The larger multicavity klystrons are being replaced by traveling-wave tubes in higher-power applications.

Another widely used microwave tube is the magnetron. The *magnetron* is the combination of a simple diode vacuum tube with built-in cavity resonators and an extremely powerful permanent magnet. The typical tube assembly is shown in Fig. 10-16. It consists of a circular anode into which has been machined an even number of resonant cavities. The diameter of

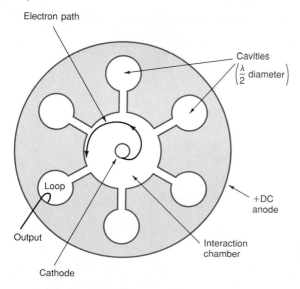

Electron path

Cavities
$\left(\frac{\lambda}{2}\ \text{diameter}\right)$

Loop

Output

+DC anode

Interaction chamber

Cathode

Note: Magnetic field is perpendicular to the page.

Fig. 10-16 A magnetron tube used as an oscillator.

each cavity is equal to one-half wavelength of the desired operating frequency. The anode is usually made of copper and is connected to a high-voltage positive dc source.

In the center of the anode, called the interaction chamber, is a circular cathode that emits electrons when heated. As in a normal diode vacuum tube, the electrons would flow directly from the cathode straight to the anode causing a high current to flow. However, the direction of the electrons is modified by surrounding the tube with a strong magnetic field. This is usually supplied by a C-shaped permanent magnet. In Fig. 10-16, the lines of force could be coming out of the page or going into the page depending upon the construction of the tube.

When an electron moves, it has a magnetic field associated with it. The magnetic fields of the electrons will, therefore, interact with the strong field supplied by the magnet. The result is that the path for electron flow from the cathode will not be directly to the anode. Instead, the path will be curved. By properly adjusting the anode voltage and the strength of the magnetic field, the electrons can be made to bend such that they rarely reach the anode and cause current flow. The path becomes circular loops as illustrated in Fig. 10-16. At some point, the electrons will eventually reach the anode and cause current flow. By adjusting the dc anode voltage and the strength of the magnetic field, the electron path is made circular. In making their circular passes in the interaction chamber, the electrons excite the resonant cavities into oscillation. The magnetron, therefore, is an oscillator, not an amplifier. A takeoff loop in one cavity provides the output.

Magnetrons are capable of developing extremely high levels of microwave power. Thousands and even millions of watts of power can be produced by a magnetron. When operated in a pulsed mode, magnetrons can generate several megawatts of power in the microwave region. Pulsed magnetrons are commonly used in radar systems. Continuous-wave magnetrons are also used and can generate hundreds and even thousands of watts of power. A typical application for a CW magnetron is for heating purposes in microwave ovens.

One of the most versatile microwave RF power amplifiers is the *traveling-wave tube (TWT)*. The main virtue of the TWT is its extremely wide bandwidth of operation. In addition to its ability to generate hundreds and even thousands of watts of microwave power, it is not resonant at a single frequency.

Figure 10-17 shows the basic structure of a TWT. It consists of a cathode and filament heater plus an anode that is biased positively to accelerate the electron beam forward and to focus it into a narrow beam. The electrons are attracted by a positive plate called the collector to which is applied a very high dc voltage. The length of the tube can be anywhere from approximately 1 ft to several feet. In any case, the length of the tube is usually many wavelengths at the operating frequency. Surrounding the tube is either permanent magnets or electromagnets that keep the electrons tightly focused into a narrow beam.

The unique feature of the TWT is a helix or coil that surrounds the length of the tube. The electron beam passes through the center or axis of the helix. The microwave signal to be amplified is applied to the end of the helix near the cathode, and the output is taken from the end of the helix near the collector. The purpose of the helix is to provide a path for the RF signal that will slow down its propagation. The propagation of the RF signal along the helix is made approximately equal to the velocity of the electron beam from cathode to collector. The structure of the helix is such that the wave traveling along it is slightly slower than that of the electron beam.

The passage of the microwave signal down the helix produces electric and magnetic fields that will interact with the electron beam. The effect on the electron beam is similar to that in a klystron. The electromagnetic field produced by the helix causes the electrons to be speeded up and slowed down. This produces velocity modulation of the beam which pro-duces density modulation. Density modulation, of course, causes bunches of electrons to group together one wavelength apart. These bunches of electrons travel down the length of the tube toward the collector. Since the density-modulated electron beam is essentially in step with the electromagnetic wave traveling down the helix, the electron bunches induce voltages into the helix which reinforce the voltage already present there. The result is that the strength of the electromagnetic field on the helix increases as the wave travels down the tube toward the collector. At the end of the helix, the signal is considerably amplified. Coaxial cable or waveguide structures are used to extract energy from the helix.

As indicated earlier, the primary benefit of the TWT is its extremely wide bandwidth. Tubes can be made to amplify signals from UHF to hundreds of gigahertz. Most TWTs have a frequency range of approximately 2-to-1 in the desired segment of the microwave region to be amplified. The TWTs can be used in both continuous and pulsed modes of operation with power levels up to several thousand watts. One of the most common applications of TWTs is as power amplifiers in satellite transponders.

Self-Test

Determine whether each statement is true or false.

38. Klystrons can be used as amplifiers or oscillators.
39. A magnetron operates as an amplifier.

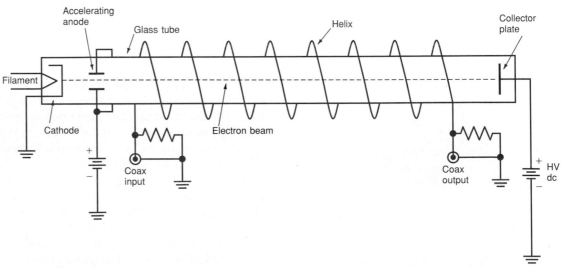

Fig. 10-17   A traveling-wave tube (TWT).

Supply the missing word(s) in each statement.

40. A TWT is an _____ (amplifier, oscillator).

41. The operating frequency of klystrons and magnetrons is determined by _____ .

42. The cavities in a klystron produce _____ modulation of the electron beam.

43. In a klystron amplifier, the input is applied to the _____ cavity and the output taken from the _____ cavity.

44. A one-cavity klystron that oscillates is known as a(n) _____ klystron.

45. Low-power klystrons are being replaced by _____ , and high-power klystrons are being replaced by _____ .

46. A(n) _____ causes the electrons in a magnetron to travel in circular paths.

47. Two major applications of magnetrons are in _____ .

48. The TWT is a microwave _____ .

49. Density modulation of the electron beam in a TWT is produced by a(n) _____ .

50. The major benefit of a TWT is its wide _____ .

10-5 MICROWAVE ANTENNAS

All the antennas we discussed in Chap. 9 can also be used at microwave frequencies. However, these antennas will be extremely small. A half-wave dipole at 5 GHz is slightly less than 1 in. long. A quarter-wave vertical at this frequency would be slightly less than ½ in. long. Naturally, these antennas will radiate microwave signals but inefficiently. Because of the line-of-sight transmission of microwave signals, highly directive antennas are preferred because they do not waste the radiated energy and provide an increase in gain which helps offset the path loss at these frequencies. For these important reasons, special high-gain, highly directive antennas are normally used in microwave applications.

The most widely used microwave antenna is the horn. A *horn* antenna is nothing more than a flared waveguide. The horn exhibits gain and directivity. However, its performance is improved even more by using it in combination with a parabolic reflector. This is the familiar "dish" normally associated with microwave and satellite transmissions. In this section we introduce you to the horn antenna and para-

bolic reflectors. The popular helical antenna is also discussed.

Waveguides are the most predominant type of transmission line used with microwave signals. Below approximately 6 GHz, special coax can be used effectively if the distances are kept short. In most microwave systems, waveguides are preferred because of their low loss. Microwave antennas, therefore, must be some extension of or compatible with a waveguide.

Actually, a waveguide will act as an inefficient radiator if it is simply left open at the end as shown in Fig. 10-18. The problem with using a waveguide as a radiator is that it makes a poor impedance match with free space. This represents a mismatch that causes standing waves and reflected power. The result is tremendous power loss of the radiated signal.

Fig. 10-18 A waveguide will act as an inefficient radiator.

It has been discovered that this mismatch can be overcome by simply flaring the end of the waveguide. Flaring of the waveguide ends creates a horn antenna as shown in Fig. 10-19. The longer and more gradual the flare, the better the impedance match and the lower the loss. Horn antennas have excellent gain and directivity. The longer the horn, the greater its gain and directivity.

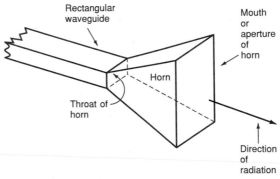

Fig. 10-19 Basic horn antenna.

Different kinds of horn antennas can be created by flaring the end of the waveguide in different ways. For example, flaring the waveguide in only one dimension creates a *sectoral horn* as shown in Fig. 10-20(*a*) and (*b*). Two sides of the horn remain parallel with the sides of the waveguide, while the other dimension is flared.

Flaring both dimensions of the horn produces a *pyramidal horn* as shown in Fig. 10-20(*c*). If a circular waveguide is used, the flare produces a *conical horn* as shown in Fig. 10-20(*d*).

The gain and directivity of a horn is a direct function of its various dimensions. The important dimensions are illustrated in Fig. 10-21. They include horn length, aperture area, and flare angle.

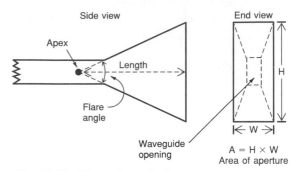

Fig. 10-21   **Dimensions of a horn.**

The length of a typical horn is usually 2 to 15 wavelengths at the operating frequency. Assume an operating frequency of 10 GHz. The length of one wavelength at this frequency is

$$\lambda = \frac{300}{f}$$

$$= \frac{300}{10,000}$$

$$= 0.03 \text{ m}$$

The frequency f is in megahertz. A length of 0.03 m is the same as 3 cm. One inch equals 2.54 cm. Therefore, a wavelength at 10 GHz is

$$\lambda = \frac{3}{2.54}$$

$$= 1.18 \text{ in.}$$

or just a shade short of 1¼ in. This means that the length of a typical horn at this frequency could be anywhere from about 2½ to 18 in. The longer horns are, of course, more difficult to mount and work with but otherwise provide higher gain and better directivity.

The aperture area is the area of the rectangle formed by the opening of the horn and is simply the product of the height and width of the horn as shown in Fig. 10-21. The greater this area, the higher the gain and directivity.

The flare angle also affects gain and directivity. Typical flare angles vary from about 20° to 60°. You can see how all these dimensions are interrelated. Increasing the flare angle, of course, increases the aperture area. For a given size of aperture area, decreasing the length increases the flare angle. Each of these dimensions is adjusted to achieve the desired design objective.

(a) Sectoral horn

(b) Sectoral horn

(c) Pyramidal horn

(d) Conical horn

Fig. 10-20   **Types of horn antennas.**

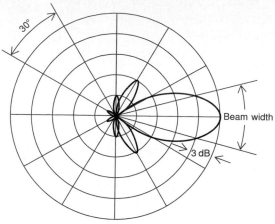

Fig. 10-22   Directivity of an antenna as measured by beam width.

Recall that we indicated that the directivity of an antenna is measured in terms of its beam width. The beam width is the angle formed by extending lines from the center of the antenna response curve to the 3-dB-down points as shown in Fig. 10-22. In this example, the beam width is approximately 30°. Horn antennas typically have a beam angle somewhere in the 10° to 60° range.

Figure 10-23 shows another way of plotting the beam angle of an antenna. Instead of using polar coordinates, standard x and y coordinates are used. The horizontal, or x axis, represents the beam width in degrees, and the vertical, or y axis, represents the gain in decibels.

Remember that the signal radiated from an antenna is three-dimensional in nature. The directivity patterns indicate the horizontal radiation pattern of the antenna. The antenna also has a vertical radiation pattern. A typical plot of the vertical radiation pattern of a horn antenna is illustrated in Fig. 10-24 on the next page. Usually the vertical beam width is approximately the same angle as the horizontal beam width on a horn.

The horizontal beam width of a pyramidal horn may be computed with the simple expression

$$B = \frac{80}{w/\lambda}$$

where B = beam width, degrees
w = width of horn, meters
λ = wavelength of operating frequency, meters

Assume an operating frequency of 10 GHz = 10,000 MHz. This gives us a wavelength as

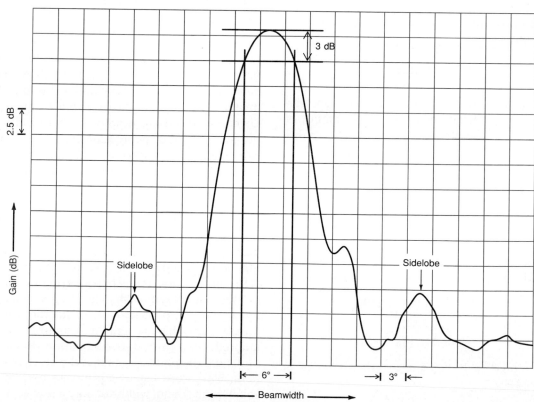

Fig. 10-23   Plot of beam width versus gain of a horn antenna.

Power gain

Bandwidth

Parabolic reflector

Fig. 10-24   Vertical directivity of an antenna.

computed earlier of 0.03 m. Assume that the dimensions of the pyramidal horn are 10 cm high and 12 cm wide. The beam width then is

$$B = \frac{80}{0.12/0.03}$$

$$= \frac{80}{4}$$

$$= 20°$$

The gain of a pyramidal horn can also be computed from its dimensions. The expression below gives the approximate power gain of the antenna.

$$G = \frac{4\pi KA}{\lambda^2}$$

where K = constant derived from how uniformly the phase and amplitude of the electromagnetic fields are distributed across the aperture (Typical values are in the 0.5 to 0.6 range. We will assume 0.5 here.)
A = aperture of horn, m²
λ = wavelength, meters

Using the previously described horn with a height dimension of 10 cm and a width of 12 cm, the aperture area is

$$A = \text{height} \times \text{width}$$

$$= 10(12)$$

$$= 120 \text{ cm}^2$$

Since there is 10,000 cm²/m², then 120 cm² is

$$\frac{120}{10,000} = 0.012 \text{ m}^2$$

Assuming an operating frequency of 10 GHz, one wavelength is equal to 0.03 m or 3 cm. The gain, therefore, is

$$G = \frac{4(3.14)(0.5)(0.012)}{(0.03)^2}$$

$$= \frac{0.07536}{0.0009}$$

$$= 83.7$$

To find the gain in decibels, use the standard power formula

$$\text{dB} = 10 \log P$$

where P is the power ratio or gain.

$$\text{dB} = 10 \log 83.7$$

$$= 10(1.923)$$

$$= 19.23$$

This is the power gain of the horn over a standard half-wave dipole or quarter-wave vertical.

An important aspect of a microwave antenna is its bandwidth. Most antennas, as you know, have a narrow bandwidth because they are resonant at only a single frequency. Their dimensions determine the frequency of operation. Bandwidth is an important consideration at microwave frequencies because the spectrum transmitted on the microwave carrier is usually very wide so that a considerable amount of information can be carried.

Luckily, horn antennas do have a relatively large bandwidth. Horns are essentially nonresonant or aperiodic, which means they will operate over a wide frequency range. The bandwidth of a typical horn antenna is approximately 10 percent of the operating frequency. Using our 10-GHz example, the bandwidth of our horn would be approximately 1 GHz. This is an enormous bandwidth and certainly wide enough to accommodate almost any kind of complex modulating signal.

Horn antennas are used by themselves in many microwave applications, but many times, higher gain and directivity is desirable. This can easily be obtained by using a horn in conjunction with a parabolic reflector. A parabolic reflector is a large dish-shaped structure made of metal or screen mesh. The energy radiated by the horn is pointed at the reflector, which focuses the radiated energy into a narrow beam and reflects it toward its destination. Because of the unique parabolic shape, the electromagnetic waves are narrowed into an extremely small beam. Beam widths of only a few degrees are typical with parabolic reflectors. Of course,

258 CHAPTER 10 MICROWAVE TECHNIQUES

course, such narrow beam widths also represent extremely high gains.

A parabola, illustrated in Fig. 10-25, is a common geometric figure. A key dimension of the parabola is a line drawn from its center at point Z to a point on the axis labeled F, which is the focal point. The ends of the parabola could extend outwardly for an infinite distance, but usually they are limited. The limits are shown by the dashed vertical line with the end points labeled X and Y.

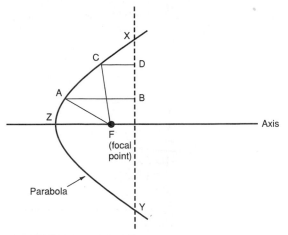

Fig. 10-25   A parabola.

The parabola has a unique characteristic. The distance between the focal point and the parabola and then to the vertical dashed line is a constant value. For example, in Fig. 10-25, the distance represented by the sum of the lines FA to AB and FC to CD are equal. This effect causes a parabolic-shaped surface to collimate electromagnetic waves into a narrow beam of energy. Placing an antenna at the focal point F will cause it to radiate waves from the parabola in parallel lines. If used as a receiver, the parabola will pick up the electromagnetic waves and reflect them to the antenna located at the focal point.

The key thing to remember about a parabola is that it is not a two-dimensional figure as shown in Fig. 10-25. If you rotate the parabola about its axis, a three-dimensional dish-shaped structure results.

Figure 10-26 shows how a parabolic reflector is used in conjunction with a horn antenna for both transmission and reception. The horn antenna is placed at the focal point. In transmitting, the horn radiates the signal toward the reflector which bounces the waves off and collimates them into a narrow parallel beam. When used for receiving, the reflector picks up the electromagnetic signal and bounces the

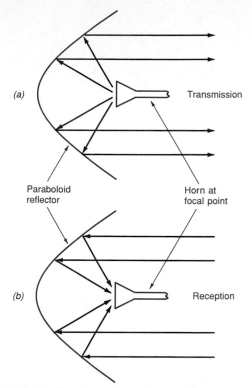

Fig. 10-26   Sending and receiving with a parabolic reflector antenna.

waves toward the antenna at the focal point. The result is an extremely high-gain, narrow beam width antenna.

The gain of a parabolic antenna is directly proportional to the aperture of the parabola. The aperture, of course, is the area of the outer circle of the parabola. This area is

$$A = 3.14R^2$$

The gain of a parabolic antenna is given by the simple expression

$$G = 6\left(\frac{D}{\lambda}\right)^2$$

where G = gain expressed as a power ratio
D = diameter of dish, m
λ = wavelength, m

Most parabolic reflectors are designed so that the diameter is no less than 10 wavelengths at the lowest operating frequency.

Assume that we are using a 5-m-diameter dish at 10 GHz. The 10 GHz represent a wavelength of 0.03 as computed earlier. The gain, therefore, is

$$G = 6\left(\frac{5}{0.03}\right)^2$$
$$= 6(166.67)^2$$
$$= 6(27778.9)$$
$$= 166,673$$

This is the power gain which can be expressed in decibels.

$$dB = 10 \log 166{,}673$$
$$= 10(5.22)$$
$$= 52.2$$

The beam width of a parabolic reflector can also be computed by knowing the parabolic reflector dimensions. The beam width is given by the expression

$$B = \frac{58}{D/\lambda}$$

Here the beam width is inversely proportional to the diameter.

Let's compute the beam width of our 5-m, 10-GHz antenna. Its beam width is

$$B = 58\left(\frac{0.03}{5}\right)$$
$$= \frac{1.74}{5}$$
$$= 0.348°$$

With a beam width of less than ½° the signal radiated from an antenna such as this is a pencil-thin beam. Naturally it must be pointed with great accuracy in order for the signal to be picked up. Usually both the transmitting and receiving antennas are of a parabolic design and have extremely narrow beam widths. For that reason, they must be accurately pointed if contact is to be made. Naturally, the beam spreads out and grows in size with distance. Nevertheless, the directivity is good and the gain is high. The directivity helps to prevent interference from signals coming in at angles outside of the beam width.

Keep in mind that the parabolic dish is not the antenna, only a part of it. The antenna is the horn at the focal point. There are many physical arrangements used in positioning the horn. One of the most common is that shown in Fig. 10-27. The waveguide feeds through the center of the parabolic dish and is curved around so that the horn is positioned exactly at the focal point. Such a structure is inconvenient, but it is a widely used arrangement.

One of the more popular methods of feeding a parabolic antenna is to use a second reflector positioned at the focal point as shown in Fig. 10-28. Here the horn antenna is positioned at the center of the parabolic reflector. At the focal point is another small reflector with either a parabolic or hyperbolic shape. The electromagnetic radiation from the horn strikes the small reflector, which then reflects

Fig. 10-27   Standard waveguide and horn feed.

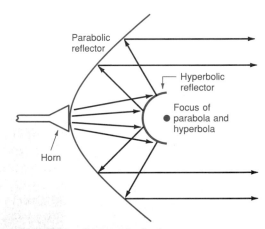

Fig. 10-28   Cassegrain feed.

the energy toward the large dish. The large dish then radiates the signal in parallel beams. This arrangement is known as a Cassegrain feed.

The Cassegrain feed has several advantages over the more obvious feed arrangement shown earlier. The first is that the waveguide transmission line is shorter. In addition to being shorter, the radical bends in the waveguide are eliminated. Both add up to less signal attenuation. The noise figure is also improved somewhat. Most large earth station antennas use a feed arrangement of this type.

Many other feed arrangements have been developed for parabolic reflectors. In those antennas used for satellite TV reception, a waveguide is not used. Instead, a horn antenna mounted at the focal point is usually fed with large microwave coax. In other large antenna systems, various mechanical arrangements are used to permit the antenna to be rotated or its position otherwise physically changed. Many earth station antennas must be set up so that their azimuth and elevation can be changed to ensure proper orientation for the receiving antenna. This is particularly true of antennas used in satellite communications systems.

Some sophisticated systems use multiple horns on a single reflector. Such multiple feeds permit several signals on different frequencies to be either radiated or received with a single large reflecting structure.

Although the horn and the parabolic reflector are the most commonly used microwave antennas, two other antennas find wide use in the microwave region. These are the helical antenna and the bicone antenna.

A *helical antenna,* as its name implies, is a wire helix as shown in Fig. 10-29. A center insulating support is used to hold a heavy wire or tubing formed into a circular coil or helix as shown. The diameter of the helix is typically one-third wavelength, and the spacing between turns is approximately one-quarter wavelength. Most helical antennas use from six to eight turns. A circular or square ground plane reflector is also used behind the helix. Figure 10-29 shows a coaxial feed line. This antenna is also widely used in the VHF and UHF range.

The gain of a helical antenna is typically in the 12- to 20-dB range. Beam widths vary from approximately 12° to 45°. Although the gain and beam width do not compare with horns and parabolic reflectors, the helical antenna is favored in many applications because of its simplicity and low cost.

A peculiar characteristic of a helical antenna is that the signal that it radiates is circularly polarized. Most antennas transmit either a vertical or horizontally polarized electromagnetic field. With the helix, the electromagnetic field is caused to rotate. This is known as circular polarization. Either right-hand (clockwise) or left-hand (counterclockwise) circular polarization may be produced depending upon the di-

rection of winding of the helix. Because of the rotating nature of the magnetic field, a circularly polarized signal will be received by either a horizontally or vertically polarized receiving antenna, but with a 3-dB loss. A helical receiving antenna will also receive horizontally or vertically polarized signals, again with a 3-db loss. Although this is true, a right-hand circularly polarized (RHCP) signal cannot be picked up by a left-hand circularly polarized (LHCP) antenna and vice versa. Therefore, when helical antennas are used at both transmitting and receiving ends of the communication link, both antennas must have the same polarization.

Most microwave antennas are highly directional. But in some applications an omnidirectional antenna may be required. One of the most widely used omnidirectional microwave antennas is the *bicone,* shown in Fig. 10-30 on the next page. The signal is fed to the antenna through a circular waveguide which ends in a flaired cone. The upper cone acts as a reflector and causes the signal to be radiated equally in all directions with a very narrow vertical beam width.

Self-Test

Supply the missing information in each statement.

51. The most commonly used microwave antenna is the _____ antenna.
52. Increasing the length of a horn antenna causes its gain to _____ and its beam width to _____ .
53. A horn antenna is six wavelengths long at 8 GHz. That length is _____ in.

Fig. 10-29 **The helical antenna.**

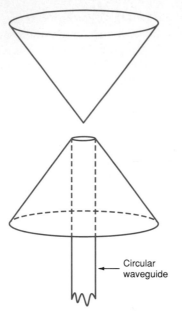

← Circular
waveguide

Fig. 10-30　**The omnidirectional bicone antenna.**

54. The beam width of a horn antenna is usually in the _____ to _____ degree range.
55. The width of a horn antenna is 8 cm. The height is 6 cm. The operating frequency is 8 GHz. The beam angle is _____ degrees. The gain is _____ dB. Assume $k = 0.5$.
56. Horn antennas have _____ (narrow, wide) bandwidth.
57. The geometric shape of a microwave reflector is a(n) _____ .

58. In order for a dish reflector to work, the antenna must be located at the _____ of the dish.
59. The antenna used with a dish reflector is usually a(n) _____ .
60. A parabolic reflector antenna has a diameter of 9 m. The frequency of operation is 8 GHz. The gain is _____ dB. The beam width is _____ °.
61. The effect of a parabolic reflector on an antenna is to _____ gain and _____ beam width.
62. A dish with a horn at its center and a small reflector at the focal point is said to use _____ feed.
63. The benefits of a helical antenna are its _____ and _____ .
64. The gain of a helical antenna is typically in the _____ to _____ -dB range.
65. The beam width of a helical antenna is in the _____ to _____ degree range.
66. The helical antenna radiates a(n) _____ polarized wave.
67. The acronym RHCP means _____ .
68. A popular omnidirectional microwave antenna is the _____ .

Determine whether each statement is true or false.

69. A helical antenna will receive either vertical or horizontally polarized signals.
70. An RHCP antenna can receive a signal from an LHCP antenna.

SUMMARY

1. Microwaves are radio signals in the frequency range from 1 to 300 GHz.
2. The RF spectrum below UHF is mostly already fully occupied leaving little or no room for the growth of new radio services.
3. At microwave frequencies, tremendous bandwidth is available for new radio services as well as for wide-bandwidth signals such as TV, multiplexed signals, or computer data.
4. The microwave frequencies are used primarily for telephone communications, radar, and satellite communications.
5. Other microwave applications include cable TV, space communications, radio astronomy, and heating.

6. The primary benefit of microwaves is wide-bandwidth availability.
7. The main disadvantages of microwaves are that they are limited to line-of-sight transmission distances, conventional components are not usable, and circuits are more difficult to analyze and design.
8. The allocation of the RF spectrum is handled by the Federal Communications Commission (FCC) in the United States.
9. Balanced transmission line is not used for microwaves because of radiation losses. Coaxial cable is not used because of its high attenuation.
10. The preferred transmission line for microwaves is waveguides.

11. Because of the short physical length of transmission lines at microwave frequencies, quarter- and half-wave lines are commonly used for tuned circuits and filters.

12. Two printed circuit board implementations of transmission lines, called stripline and microstrip, are widely used to create resonant circuits and filters.

13. A waveguide is a hollow metal pipe with a circular or rectangular cross section used for carrying microwave signals from one place to another.

14. A waveguide acts like a high-pass filter, passing all frequencies above its cutoff frequency and rejecting those below it.

15. The cutoff frequency f_{co} of a waveguide depends upon its physical size. For a rectangular waveguide, it is $300/2a$, where a is the wide dimension of the waveguide in meters.

16. The microwave signal carried by a waveguide is made up of electric (E) and magnetic (H) fields that bounce off the walls of the waveguide as they propagate along its length.

17. The modes of a waveguide describe the various patterns of electric and magnetic fields that are possible.

18. A transverse electric (TE) mode is one where the electric field is transverse or perpendicular to the direction of propagation.

19. A transverse magnetic (TM) mode is one where the magnetic field is perpendicular to the direction of propagation.

20. Waveguides are available in standard lengths and sizes, and special pieces are used for right-angle bends and 90° twists.

21. Half-wavelength sections of waveguides with shorted or closed ends are known as resonant cavities since they "ring" or oscillate at the frequency determined by their dimensions.

22. Cavity resonators are metallic chambers of various shapes and sizes that are used as parallel-tuned circuits and filters. They have a Q of up to 30,000.

23. Point-contact and Schottky or hot-carrier diodes are widely uses as mixers in microwave equipment as they have low capacitance and inductance.

24. Varactor diodes are widely used as microwave frequency multipliers. Multiplication factors of 2 and 3 are common with power levels up to 20 W and efficiencies up to 80 percent.

25. Step-recovery or snap-off diodes are also widely used as frequency multipliers with multiplication factors up to 10, power ratings up to 50 W, and efficiencies approaching 80 percent.

26. A Gunn diode is a microwave semiconductor device used to generate microwave energy. When combined with a microstrip, stripline, or resonant cavity, simple low-power oscillators with frequencies up to 50 GHz are easily implemented.

27. Both IMPATT and TRAPATT diodes are GaAs devices operated with high reverse bias to produce avalanche breakdown. Both are used in microwave oscillators.

28. A klystron is a vacuum tube used for microwave amplification and oscillation.

29. Klystrons use a cavity resonator to velocity modulate an electron beam which imparts energy to another cavity, producing power amplification. Klystrons are available which produce from a few to many thousands of watts.

30. A single-cavity reflex klystron is used as a microwave oscillator.

31. Klystrons are being gradually replaced by Gunn diodes and traveling-wave tubes.

32. A magnetron is a diode vacuum tube used as a microwave oscillator in radar and microwave ovens to produce powers up to the megawatt range.

33. In a magnetron, a strong magnetic field creates circular paths of electron flow to excite cavities into oscillation.

34. A traveling-wave tube (TWT) is a microwave power amplifier with very wide bandwidth.

35. A microwave signal applied to a helix around the TWT produces velocity and density modulation of the electron beam over a long distance which induces a higher-power signal in the helix.

36. The most commonly used microwave antenna is the horn, which is essentially a rectangular waveguide with a flared end.

37. A pyramidal horn flares in both waveguide dimensions. A sectoral horn flares in only one dimension.

38. Horn antennas are directional and produce a beam width in the 10° to 60° range with a gain in the 10- to 20-dB range, depending upon dimensions.

39. A parabolic or dish-shaped reflector is used with most microwave antennas to focus the RF energy into a narrow beam and increase gain.

40. The parabolic reflector usually has a diameter that is no less than 10 wavelengths at the operating frequency.
41. The gain and directivity of a parabolic reflector antenna is directly proportional to its diameter.
42. Parabolic reflector antennas are fed by placing a horn antenna at the focal point or by placing the horn at the center of the reflector and placing a small reflector at the focal point. The latter is known as Cassegrain feed.
43. A helical antenna is made up of six to eight turns of heavy wire or tubing to form a coil or helix. It is fed with coax and is backed up with a reflector.

44. Helical antennas are used at UHF and microwave frequencies and have a gain in the 12- to 20-dB range and a beam width in the 12° to 45° range.
45. Helical antennas produce circular polarization where the electric and magnetic fields rotate. The polarization may be right-hand or left-hand depending upon the direction in which the helix is wound.
46. Helical antennas can receive either vertically or horizontally polarized signals but can only receive a circularly polarized signal of the same direction.
47. A popular omnidirectional microwave antenna is the bicone.

CHAPTER REVIEW QUESTIONS

Choose the letter which best answers each question.

10-1. The main benefit of using microwaves is
 a. Lower-cost equipment.
 b. Simpler equipment.
 c. Greater transmission distances.
 d. More spectrum space for signals.

10-2. Radio communications are regulated in the United States by the
 a. Federal Trade Commission.
 b. Congress.
 c. Federal Communications Commission.
 d. Military.

10-3. Which of the following is *not* a disadvantage of microwaves?
 a. Higher-cost equipment
 b. Line-of-sight transmission
 c. Conventional components are not usable.
 d. Circuits are more difficult to analyze.

10-4. Which of the following is a microwave frequency?
 a. 1.7 MHz   c. 0.98 GHz
 b. 750 MHz   d. 22 GHz

10-5. Which of the following is *not* a common microwave application?
 a. Radar
 b. Mobile radio
 c. Telephone
 d. Spacecraft communications

10-6. Coaxial cable is not widely used for long microwave transmission lines because of its
 a. High loss.   c. Large size.
 b. High cost.   d. Excessive radiation.

10-7. Stripline and microstrip transmission lines are usually made with
 a. Coax.   c. Twisted pair.
 b. Parallel wires.   d. PCBs.

10-8. The most common cross section of a waveguide is a
 a. Square.   c. Triangle.
 b. Circle.   d. Rectangle.

10-9. A rectangular waveguide has a width of 1 in. and a height of 0.6 in. Its cutoff frequency is
    a. 2.54 GHz.   *c.* 5.9 GHz.
    b. 3.0 GHz.   *d.* 11.8 GHz.

10-10. A waveguide has a cutoff frequency of 17 GHz. Which of the signals will *not* be passed by the waveguide?
    a. 15 GHz   *c.* 22 GHz
    b. 18 GHz   *d.* 25 GHz

10-11. Signal propagation in a waveguide is by
    a. Electrons.   *c.* Holes.
    b. Electric and magnetic fields.   *d.* Air pressure.

10-12. When the electric field in a waveguide is perpendicular to the direction of wave propagation, the mode is said to be
    a. Vertical polarization.
    b. Horizontal polarization.
    c. Transverse electric.
    d. Transverse magnetic.

10-13. The dominant mode in most waveguides is
    a. $TE_{0,1}$.   *c.* $TM_{0,1}$.
    b. $TE_{1,2}$.   *d.* $TM_{1,1}$.

10-14. A magnetic field is introduced into a waveguide by a
    a. Probe.   *c.* Stripline.
    b. Dipole.   *d.* Capacitor.

10-15. A half-wavelength, closed section of a waveguide that acts as a parallel resonant circuit is known as a(n)
    a. Half-wave section.   *c.* *LCR* circuit.
    b. Cavity resonator.   *d.* Directional coupler.

10-16. Decreasing the volume of a cavity causes its resonant frequency to
    a. Increase.   *c.* Remain the same.
    b. Decrease.   *d.* Drop to zero.

10-17. A popular microwave mixer diode is the
    a. Gunn.   *c.* Hot carrier.
    b. Varactor.   *d.* IMPATT.

10-18. Varactor and step-recovery diodes are widely used in what type of circuit?
    a. Amplifier   *c.* Frequency multiplier
    b. Oscillator   *d.* Mixer

10-19. Which diode is a popular microwave oscillator?
    a. IMPATT   *c.* Varactor
    b. Gunn   *d.* Schottky

10-20. Which type of diode does *not* ordinarily operate with reverse bias?
    a. Varactor   *c.* Snap-off
    b. IMPATT   *d.* Tunnel

10-21. Low-power Gunn diodes are replacing
    a. Reflex klystrons.   *c.* Magnetrons.
    b. TWTs.   *d.* Varactor diodes.

10-22. Which of the following is *not* a microwave tube?
    a. Traveling-wave tube   *c.* Klystron
    b. Cathode-ray tube   *d.* Magnetron

10-23. In a klystron amplifier, velocity modulation of the electron beam is produced by the
    a. Collector.   *c.* Cathode.
    b. Catcher cavity.   *d.* Buncher cavity.

10-24. A reflex klystron is used as a(n)
    a. Amplifier.   *c.* Mixer.
    b. Oscillator.   *d.* Frequency multiplier.

10-25. For proper operation, a magnetron must be accompanied by a
- *a.* Cavity resonator.
- *b.* Strong electric field.
- *c.* Permanent magnet.
- *d.* High dc voltage.

10-26. The operating frequency of klystrons and magnetrons is set by the
- *a.* Cavity resonators.
- *b.* DC supply voltage.
- *c.* Input signal frequency.
- *d.* Number of cavities.

10-27. A magnetron is used only as a(n)
- *a.* Amplifier.
- *b.* Oscillator.
- *c.* Mixer.
- *d.* Frequency multiplier.

10-28. A common application for magnetrons is in
- *a.* Radar.
- *b.* Satellites.
- *c.* Two-way radio.
- *d.* TV sets.

10-29. In a TWT, the electron beam is density-modulated by a
- *a.* Permanent magnet.
- *b.* Modulation transformer.
- *c.* Helix.
- *d.* Cavity resonator.

10-30. The main advantage of a TWT over a klystron for microwave amplification is
- *a.* Lower cost.
- *b.* Smaller size.
- *c.* Higher power.
- *d.* Wider bandwidth.

10-31. High-power TWTs are replacing what in microwave amplifiers?
- *a.* MESFETs
- *b.* Magnetrons
- *c.* Klystrons
- *d.* IMPATT diodes

10-32. The most widely used microwave antenna is a
- *a.* Half-wave dipole.
- *b.* Quarter-wave probe.
- *c.* Single loop.
- *d.* Horn.

10-33. What happens when a horn antenna is made longer?
- *a.* Gain increases
- *b.* Beam width decreases
- *c.* Both *a* and *b*
- *d.* Neither *a* nor *b*

10-34. A pyramidal horn used at 5 GHz has an aperture that is 7 by 9 cm. The gain is about
- *a.* 10.5 dB.
- *b.* 11.1 dB.
- *c.* 22.6 dB.
- *d.* 35.8 dB.

10-35. Given the frequency and dimensions in Question 10-34 above, the beam width is about
- *a.* 27°.
- *b.* 53°.
- *c.* 60°.
- *d.* 80°.

10-36. The diameter of a parabolic reflector should be at least how many wavelengths at the operating frequency?
- *a.* 1
- *b.* 2
- *c.* 5
- *d.* 10

10-37. The point where the antenna is mounted with respect to the parabolic reflector is called the
- *a.* Focal point.
- *b.* Center.
- *c.* Locus.
- *d.* Tangent.

10-38. Using a small reflector to beam waves to the larger parabolic reflector is known as
- *a.* Focal feed.
- *b.* Horn feed.
- *c.* Cassegrain feed.
- *d.* Coax feed.

10-39. Increasing the diameter of a parabolic reflector causes which of the following:
- *a.* Decreased beam width
- *b.* Increased gain
- *c.* Increased beam width
- *d.* *a* and *b*
- *e.* *b* and *c*
- *f.* None of the above

10-40. A helical antenna is made up of a coil and a
- *a.* Director.
- *b.* Reflector.
- *c.* Dipole.
- *d.* Horn.

10-41. The output of a helical antenna is
    a. Vertically polarized.   *c.* Circularly polarized.
    b. Horizontally polarized.   *d.* Both *a* and *b*.

10-42. A common omnidirectional microwave antenna is the
    a. Horn.   *c.* Helical.
    b. Parabolic reflector.   *d.* Bicone.

CRITICAL THINKING QUESTIONS

10-1. At microwave frequencies, the values of inductors and capacitors required for resonance, bypassing, and coupling become very small. Components like this are not available commercially. Explain how you could obtain or realize such components physically. (Hint: See Chapter 9.)

10-2. What is the main reason that vacuum tubes are still used in some microwave applications?

10-3. Assume that all of the radio and other communications signals in the spectrum from 100 kHz to 30 MHz are frequency-multiplexed on a single coax cable. Could this composite signal be used to frequency-modulate the carrier of a 30-GHz transmitter? Draw a block diagram of the transmitter and pay attention to the modulator. Explain what the resulting output signal might look like in the frequency domain.

Answers to Self-Tests

1. 1
2. bandwidth or spectrum
3. more difficult to analyze; different measurement techniques; resistors, capacitors, and inductors act like *LCR* circuits; conventional semiconductors do not work owing to internal capacitances and long transit time; special, expensive vacuum tubes are used for power amplification; line-of-sight transmission differences; excessive signal reflection and absorption
4. telephone, radar
5. television signal relay, space communications, radio telescopes, microwave heating
6. Federal Communications Commission
7. improved receiver selectivity, SSB, multiplexing, reduced FM deviation
8. internal capacitance, long transit time
9. *LCR*
10. 36, 3.6
11. true
12. attenuation
13. 6
14. waveguide
15. microstrip, stripline
16. probe, loop
17. high-pass
18. 4.921
19. (a) true, (b) false
20. H, E
21. $TE_{0,1}$
22. transverse magnetic
23. cavity resonator
24. parallel resonant
25. tuned circuits, filters
26. resonant frequency
27. high capacitance, high inductance, long transit time
28. P-type silicon, tungsten whisker
29. Schottky barrier, or hot-carrier
30. mixer
31. varactor
32. varactor, snap-off or step-recovery diode
33. Gunn
34. IMPATT, TRAPATT, tunnel
35. negative resistance
36. reverse
37. forward
38. true
39. false
40. amplifier
41. cavity resonators
42. velocity
43. buncher, catcher
44. reflex
45. Gunn diodes, TWTs
46. permanent magnet
47. radar, microwave ovens
48. amplifier
49. helix
50. bandwidth
51. horn
52. increase, decrease
53. 11.8
54. 10, 60
55. 37.5, 13.3
56. wide
57. parabola
58. focal point
59. horn
60. 55.39, 0.242
61. increase, decrease
62. Cassegrain
63. simplicity, low cost
64. 12, 20
65. 12, 45
66. circularly
67. right-hand circular polarization
68. bicone
69. true
70. false

CHAPTER 11

Introduction to Satellite Communications

∎

CHAPTER OBJECTIVES

This chapter will help you to:

1. *Name* the most common communication satellite orbit and *tell* how it is used and how it is accomplished.
2. *State* the type of coordinates used in locating a communications satellite in orbit.
3. *Draw* a block diagram of the basic electronic unit in a satellite, *give* its name, and *tell* how it operates.
4. *Name* the six main subsystems of a satellite and *tell* what the function of each is.
5. *Draw* a basic block diagram of a satellite earth station, *name* the five basic subsystems, and *explain* the function of each.
6. *Name* four common applications for satellites.

———

A satellite is a physical object that orbits or revolves around some celestial body. Satellites occur in nature; our own solar system is a perfect example. The earth and other planets are satellites which revolve around the sun, and the moon is a satellite of the earth. A balance between the inertia of the revolving satellite and the gravitational pull of the orbited body keeps the satellite in orbit. Artificial satellites can be launched into orbit for a variety of purposes. One of these major applications is communications. In this chapter, we will discuss why satellites are such an important development in electronics communications and will briefly discuss how communications satellites work.

∎

11-1 SATELLITE ORBITS

In this section we will talk about the orbital dynamics of a satellite, that is, the forces that keep the satellite in orbit and the physical and mathematical laws that it follows. With this information you can better appreciate the technological achievement involved in putting a satellite into orbit and using it for communications and other purposes.

If a satellite were simply launched vertically from the earth and then released, it would fall back to earth because of gravity. In order for the satellite to go into orbit around the earth, it must have some forward motion. For that reason, the satellite is launched with both vertical and forward motion. The forward motion produces inertia which tends to keep the satellite moving in a straight line. However, gravity tends to pull the satellite toward the earth. The combined effect is called *centripetal acceleration*. The centripetal force is caused by the inertia of the satellite balanced by the earth's gravitational pull. If the satellite's velocity was high enough, it would be carried away from the earth into space. It takes a speed of approximately 25,000 mi/h, the escape velocity of the earth, to cause a spacecraft to break the gravitational pull of the earth. At lower speeds gravity constantly pulls the satellite toward the earth. The acceleration of the satellite caused by the gravity pull exactly balances the effect of the satellite's own velocity.

The closer the satellite is to earth, the stronger the effect of the earth's gravitational pull. So in low orbits, the satellite must travel faster to avoid falling back to earth. The farther the satellite is from the earth, the lower its orbital speed. The lowest practical earth orbit is approximately 100 mi. At this height, the satellite's speed must be approximately 17,500 mi/h in order to stay in orbit. With this speed, the satellite orbits the earth in approximately 1½ h. Communications satellites are usually much farther from earth. A typical distance is 22,300 mi. At this distance, a satellite need travel only about 6800 mi/h in order to stay in orbit. With this speed, the satellite revolves about the earth in approximately 24 h, the earth's own rotational time.

As it turns out, other factors beside just the velocity and gravitational pull determine a satellite's orbit. It is also affected by the satellite's own weight and the gravitational pulls of the moon and sun. Although these factors have less effect than the velocity and gravitational pull, over the long term these factors are important in determining the location and performance of a satellite.

A satellite revolves around the earth in either a circular or elliptical path as shown in Fig. 11-1. Circles and ellipses are, of course, special geometric figures that can be accurately described mathematically. Figure 11-2 on the next page summarizes the mathematical characteristics of both circles and ellipses. Because the orbit is either circular or elliptical, it is possible to calculate the position of a satellite at any given time.

A satellite revolves in an orbit that forms a plane which passes through the center of gravity of the earth or the *geocenter*. This is illustrated in Fig. 11-3 on the next page. In addition, the direction of the satellite's revolution may be either in the same direction as the earth's rotation or against the direction of the earth's rotation. In the former case, the orbit is said to be *posigrade* and in the latter case, *retrograde*. Most orbits are posigrade. In a circular orbit the speed of revolution is constant. However, in an elliptical orbit the speed changes depending upon the height of the satellite above the earth. Naturally the speed of the satellite is higher when it is close to the earth than when it is far away.

Now let's talk about some of the specific characteristics of a satellite orbit. Some of these characteristics include height, speed or period, angle of inclination, and angle of elevation.

In a circular orbit, the height is simply the distance of the satellite from the earth. However, in geometrical calculations, the height is really the distance between the center of the earth and the satellite. In other words, the distance includes the radius of the earth which

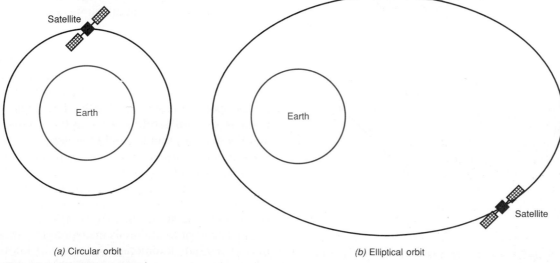

(a) Circular orbit

(b) Elliptical orbit

Fig. 11-1 Satellite orbits.

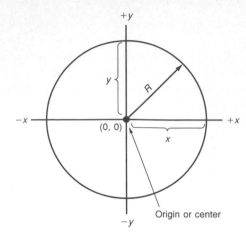

Circle

Radius (R): Distance from origin to circle

Diameter (D): Distance across the circle through the origin
$$D = 2R$$

Circumference (C): Distance around outside edge of circle
$$C = 2\pi R = \pi D$$

Equation: $x^2 + y^2 = R^2$

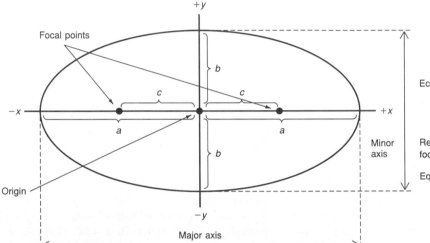

Ellipse

Eccentricity (e): The measure of elongation

$$e = \sqrt{1 - \left(\frac{b}{a}\right)^2}$$

Relationship between axes and focal points: $c = \sqrt{a^2 - b^2}$

Equation: $\dfrac{x^2}{a^2} + \dfrac{y^2}{b^2} = 1$

Fig. 11-2   Circles and ellipses.

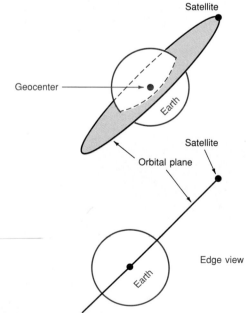

Fig. 11-3   The orbital plane passes through the geocenter.

is generally considered to be about 3960 mi or 6370 km. A satellite that is 5000 mi above the earth in a circular orbit is 3960 + 5000 = 8960 mi from the center of the earth. See Fig. 11-4.

When the satellite is in an elliptical orbit, the center of the earth is one of the focal points of the ellipse. Refer to Fig. 11-5. In this case, the distance of the satellite from the earth varies depending upon its position. Typically the two points of greatest interest are the highest and lowest points above the earth. The highest point is generally referred to as the *apogee,* and the lowest point is called the *perigee*. The apogee and perigee distances typically are measured from the geocenter of the earth and, therefore, include the earth's radius. To determine the height above the earth's surface, the radius must be subtracted from these figures.

The speed of a satellite is measured in either miles per hour, kilometers per hour, or knots (nautical miles per hour). As indicated earlier,

UNIVERSITY OF
WOLVERHAMPTON
Harrison Learning Centre

Items ISSUED to: 7604287614

Title: Communication electronics
ID: 7608688109
Due: 12/03/2006 23:59

Total items: 1
15/03/2006 15:56

Thank You for using Self Service.
Please keep your receipt.

Overdue books are fined at 40p per day for
1 week loans, 10p per day for long loans.

Fig. 11-4   Satellite height.

Sidereal period

Synodic period

Angle of inclination

volving around the earth, the earth itself is rotating.

Another method of expressing the time for one orbit is the *revolution* or *synodic period*. One revolution is the period of time that elapses between the successive passes of the satellite over a given meridian of earth longitude. Naturally, the synodic and sidereal periods differ from one another because of the earth's rotation. The time difference is determined by the height of the orbit, the angle of the plane of the orbit, and whether the satellite is in a posigrade or retrograde orbit. The period is generally expressed in hours. Typical revolutional periods range from approximately 1½ h for a height of 100 mi to 24 h for a height of 22,300 mi.

The *angle of inclination* of a satellite orbit is the angle formed between the line that passes through the center of the earth and the north pole and a line that passes through the center of the earth but which is also perpendicular to the orbital plane. This is illustrated in Fig. 11-6 on the next page. Satellite orbits can have inclination angles of 0° through 180°.

Another definition of inclination is the angle between the equatorial plane and the satellite orbital plane as the satellite enters the northern hemisphere. This definition may be a little bit easier to understand, but it means the same thing as the preceding definition.

When the angle of inclination is 0° or 180°, the satellite will be directly above the equator. When the angle of inclination is 90°, the satellite will pass over both the north and south poles once during each orbit. Orbits with a 0°

the speed varies depending upon the distance of the satellite from the earth. For a circular orbit the speed is constant, but for an elliptical orbit the speed varies depending upon the height. Low earth satellites with a height of about 100 mi have a speed in the neighborhood of 17,500 mi/h. Very high satellites such as communications satellites which are approximately 22,300 mi out revolve much slower, a typical speed being in the neighborhood of 6800 mi/h.

The time that it takes for a satellite to complete one orbit is called the *sidereal period*. Some fixed or apparently motionless external object, such as the sun or a star, is used for reference in determining a sidereal period. The reason for this is that while the satellite is re-

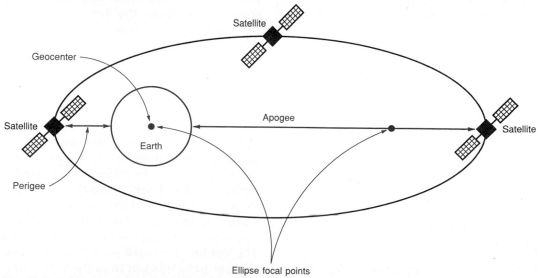
Fig. 11-5   Elliptical orbit showing apogee and perigee.

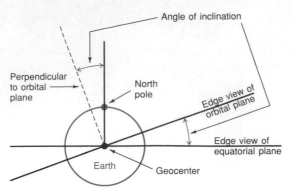

Fig. 11-6 **Angle of inclination.**

inclination are generally called *equatorial,* while orbits with inclinations of 90° are referred to as *polar.*

The *angle of elevation* of a satellite is that angle that appears between the line from the earth station's antenna to the satellite and the line between the earth station's antenna and the earth's horizon. This is shown in Fig. 11-7. If the angle of elevation is too small, the signals between the earth station and the satellite have to pass through much more of the earth's atmosphere. Because of the very low powers used and the high absorption of the earth's atmosphere, it is desirable to minimize the amount of time that the signals spend in the atmosphere. Noise in the atmosphere also contributes to poor performance. The lower the angle of radiation, the more time that this signal spends in the atmosphere. Five degrees is generally regarded as the minimum practical angle of elevation for good satellite performance. The higher the angle of elevation, the better.

To use a satellite for communications relay or repeater purposes, the ground station antenna must be able to follow or track the satellite as it passes overhead. Depending upon

the height and speed of the satellite, the earth station will only be able to use it for communications purposes for that short period of time when it is visible. The earth station antenna will track the satellite from horizon to horizon, but at some point, the satellite will disappear around the other side of the earth. At this time, it can no longer support communications.

One solution to this problem is to launch a satellite with a very long elliptical orbit where the earth station can "see" the apogee. In this way the satellite stays in view of the earth station for most of its orbit, and the satellite is useful for communications for a longer period of time. It is only that short duration when the satellite disappears on the other side of the earth (perigee) that it can no longer be used.

Because of these orbital characteristics, intermittent communications is highly undesirable in many communications applications. Some means of providing continuous communication must be found. This can be done by using more than one satellite. Typically three satellites, if properly spaced in the correct orbits, can provide continuous communications at all times. However, multiple tracking stations and complex signal switching or "handoff" systems between stations are required. This is expensive and inconvenient.

The solution to this problem is easy. Simply launch a synchronous or geostationary satellite. This is a satellite which orbits the earth around the equator at a distance of 22,300 mi or 35,860 km. A satellite at that distance revolves around the earth in exactly 24 h. In other words, the satellite revolves in exact synchronism with the earth's rotation. For this reason, it appears to be fixed or stationary, from which come the terms *synchronous, geosynchronous,* or *geostationary* orbit. Since the satellite remains apparently fixed, no special earth station tracking antennas are required. The antenna can simply be pointed at the satellite and remain in a fixed position. With this arrangement, continuous communications are possible. Most communications satellites in use today are of the geosynchronous variety. Approximately 40 percent of the earth's surface can be "seen" or accessed from such a satellite. Users inside that area can use the satellite for communications.

An eclipse is a condition where the earth or moon gets between the sun and the satellite. The shadow thus cast on the satellite causes sunlight to be blocked from the solar cell panels. Since these panels supply the main power

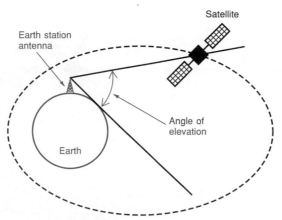

Fig. 11-7 **Angle of elevation.**

to the satellite, an eclipse virtually shuts off all power to the satellite.

Eclipses of geostationary satellites occur on the autumnal and vernal equinoxes, the forty-fourth day of fall and spring, respectively, and last from several minutes to over an hour. Occasionally, the moon's shadow will block sunlight to the satellite. During eclipse conditions, backup batteries take over so that operation can continue to take place.

Once a satellite is in orbit, the forces acting on it tend to keep it in place. If the satellite's height and speed during launch are accurately controlled, the satellite will enter the proper orbit and remain there. However, even with a very good launch, the satellite will drift somewhat in its orbit. This drift is particularly undesirable in a geosynchronous satellite whose position is supposed to remain fixed for reliable continuous communications. Orbital drift is caused by a variety of forces. The gravitational pull of the sun and the moon affect the satellite's position. Further, the earth's gravitational field is not perfectly consistent at all points on the earth. This is due to the fact that the earth itself is not a perfect sphere. Instead, the earth is a little bit fatter around the equator and is flattened somewhat at the poles. Even though the earth appears to be a nearly perfect sphere, in reality, it is more accurately described as oblate.

Because of this drift, the orbit of the satellite must be periodically adjusted. Most satellites contain small rockets or thruster jets for that purpose. These rockets, placed at various positions on the satellite, can be used to speed up or slow down the satellite for the purpose of compensating for orbital drift. Depending upon how accurate the orbit must be, these rockets may be fired as often as every several weeks or as little as once per year.

The rockets can be conventional rocket motors of the type used to launch the satellite and put it into orbit. Smaller corrections are made with thrusters which simply cause motion by releasing a gas under pressure. The most common jet thruster uses a gas called hydrazine which when released with a catalyst causes an explosive jet force that will move the satellite. Most satellites have several of these thruster jets to make various satellite position adjustments possible. The process of firing the rockets under ground control to maintain or adjust the orbit is referred to as *station keeping*.

In addition to maintaining the position of the satellite in orbit, some means must be provided to position the satellite for optimum performance. This is called *attitude control*. The attitude of the satellite must be controlled so that the antennas can be pointed toward the correct locations on earth. Attitude control is also necessary in some satellites to keep the satellite's solar panels pointed toward the sun so that maximum power is produced at all times. It may also be necessary in some applications where selected sensors or a TV camera must be pointed in the right direction. Attitude control is maintained by a combination of satellite stabilization techniques and jet thrusters for correctional purposes.

The attitude of the satellite is first determined right after the satellite is put into stable orbit. At some point, the various jet thrusters on the satellite are actuated to move the satellite in such a way that the correct attitude is assumed. For example, a thruster may be fired to rotate the satellite so that its antenna points toward earth. Once the initial attitude of the satellite is set, it must be maintained in this position. This is done by either of two stabilization methods. The most common stabilization method is spin stabilization.

Many satellites are cylindrical in shape as shown in Fig. 11-8 on the next page. Such satellites may be made to spin on their axes. Once the satellite is in the proper orbit, a jet thruster is fired to begin spinning the satellite. A typical spin-stabilized satellite rotates at approximately 100 r/min. Once the satellite is spinning, it remains very stable. The spinning causes a gyroscopic or flywheel effect that keeps the satellite pointed in one direction. A typical position may be with the axis of the satellite pointed toward earth. An antenna mounted at the axis will remain fixed in that position.

Sometimes the antenna system must be despun. This means that the portion of the satellite around the axis on the ends of the cylinder are independent of the outer spinning cylinder of the satellite. If the antennas themselves were attached to the spinning outer body, they would revolve and not provide a fixed position with respect to the earth. Despinning allows the antennas to remain oriented to a fixed position on earth as the satellite spins. The gyroscopic effect holds the satellite in position.

On noncylindrical-shaped satellites, spin stabilization is also used. This is done by including a large flywheel at some point on the satellite body. Once the satellite is in position and its antennas, solar panels, and sensors are

Axis

Helical antenna
(does not spin)

De-spin
mechanism

Outer surface
of satellite
covered with
solar cells

Jet thruster
fired to
start rotation
or adjust it

Spin direction
of satellite

Axis of
satellite

Fig. 11-8   A spin-stabilized satellite.

oriented, the flywheel is put into motion. Again, the gyroscopic effect of the flywheel keeps the satellite oriented with the proper attitude.

Another form of attitude control or stabilization is called *three-axis stabilization*. The three axes are referred to as pitch, roll, and yaw. Any aircraft or spacecraft exhibits the properties of pitch, roll, and yaw. These axes are illustrated in Fig. 11-9. In a three-axis stabilized system, three heavy flywheels or reaction wheels, one for each axis, are spun by motors to provide a gyroscopic effect to stabilize the satellite. Any pitch, roll, or yaw is corrected for by firing thruster jets in the proper direction and by controlling flywheel motor speed. The three-axis stabilization system is far more accurate in attitude control and positioning than a spin-stabilized satellite. In those applications where pinpoint accuracy of antenna pointing must be maintained, a three-axis stabilization system is used.

A three-axis stabilization system uses sensors to observe external reference points such as the sun, the earth, or some remote star. These optical or infrared sensors operate an electronic control system that determines when the attitude of the satellite needs adjustment. If adjustment is required, one or more thrusters are fired to make corrections.

In a three-axis stabilized satellite, attitude corrections are made automatically by the attitude-control subsystem of the satellite. However, most satellites can also be controlled from the ground station. By the use of command signals from the earth station, thrusters may be fired to make minor adjustments when required.

In order to use a satellite, you must be able to locate its position in space. That position is usually predetermined by the design of the satellite and is achieved during the initial launch and subsequent position adjustments. Once the position is known, the earth station antenna can be pointed at the satellite for optimum transmission and reception. For geosynchronous satellites, the earth station antenna will be adjusted once and will remain in that position except for occasional minor adjustments. For other satellites, their positions above the earth will vary depending upon their orbital characteristics. To use these satellites, special tracking systems must be used. A tracking system is essentially an antenna whose position can be changed in order to follow the satellite across the sky. In order to maintain optimum transmission and reception, the antenna must be continually pointed at the satellite as it revolves. We discuss methods of locating and tracking satellites in the following paragraphs.

Fig. 11-9 Pitch, yaw, and roll.

The location of a satellite is generally specified in terms of latitude and longitude, as any point on earth can be. The satellite location is specified by a point on the surface of the earth directly below the satellite. This point is known as the *subsatellite point (SSP)*. The SSP is located using conventional latitude and longitude designations.

Latitude and longitude form a system for locating any given point on the surface of the earth. This system is widely used for navigational purposes. If you have ever studied a globe of the earth, you have seen the lines of latitude and longitude drawn on it. The lines of *longitude,* or *meridians,* are those lines drawn on the surface of the earth from the north pole to the south pole. Lines of *latitude* are those lines drawn on the surface of the earth from east to west parallel to the equator. The centerline of latitude is the *equator* which separates the earth into north and south hemispheres.

Latitude is defined as the angle between the line drawn from a given point on the surface of the earth to the geocenter and the line between the geocenter and the equator. This is illustrated in Fig. 11-10. Zero-degree latitude

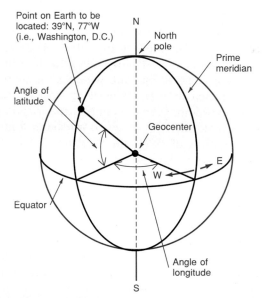

Fig. 11-10 Tracking and navigation by latitude and longitude.

would be at the equator, whereas 90° latitude would be at either the north or south pole. Usually an *N* or an *S* is added to the latitude angle in order to designate whether the point is in the northern or southern hemisphere.

Fig. 11-11 The location of some of the many communications satellites in geosynchronous orbit.

A line drawn on the surface of the earth between the north and south pole is generally referred to as a meridian. One special meridian called the prime meridian is used as a reference point for measuring longitude. The *prime meridian* is the line on the surface of the earth drawn between the north and south poles that passes through Greenwich, England. The longitude of a given point is the angle between the line connecting the geocenter to the point where the prime meridian and equator intersect and the line connecting the geocenter and the point where the meridian containing the given point of interest and the equator intersect. This is illustrated in Fig. 11-10. The designation east or west is usually added to the longitude angle to indicate whether the angle is being measured to the east or west of the prime meridian. As an example, the location of Washington, D.C. is given by the latitude and longitude of 39°N and 77°W.

To show how latitude and longitude are used to locate a satellite, refer to Fig. 11-11. This illustrates some of the many geosynchronous communications satellites serving the United States and other parts of North America. Since geosynchronous satellites orbit over the equator, their SSP is on the equator. For that reason, all geosynchronous satellites have a latitude of 0°.

Only geosynchronous satellites have a fixed SSP on the surface of the earth. The SSP of other satellites will move with respect to a given reference point on the earth. Their SSP traces a line on the earth known as the *subsatellite path* or *ground track*. The ground track for most satellites crosses the equator twice per orbit. The point where the SSP

crosses the equator headed in the northerly direction is called the *ascending node*. The point where the SSP crosses the equator headed in the southerly direction is called the *descending node*. The ascending and descending nodes are designated by longitudinal angles. With these two points known, the satellite path can be traced across the surface of the earth between them. The ascending node is sometimes designated by the abbreviation EQX and is used as a reference point for locating and tracking a satellite.

The location of the satellite at any given time is specified by the SSP in terms of latitude and longitude. Except for geostationary satellites, this point changes continuously as the satellite orbits. The exact position of a satellite is usually designated by an orbit calendar. This is a listing that usually consists of the orbit number and the time of the occurrence of the ascending node (EQX). Usually the number of orbits that a satellite makes is tracked from the very instant it is put into orbit. By using various formulas involving the height, speed, and elliptical characteristics of the orbit, the time of occurrence of the ascending node can be computed for each orbit. With the orbital calendar and various maps and plotting devices, the ground track can be traced for each orbit. This allows the satellite user to determine whether or not the satellite is within a usable range.

Just knowing the location of the satellite is insufficient information for most earth stations that must communicate with the satellite. What the earth station really needs to know is the azimuth and elevation settings of its antenna to intercept the satellite. Most earth station satellite antennas are highly directional and must

Fig. 11-12 **Azimuth and elevation: azimuth = 90°, elevation = 40°.**

be accurately positioned to "hit" the satellite. The azimuth and elevation designations in degrees tell where to point the antenna.

Take a look at Fig. 11-12. *Azimuth* refers to the direction, where north is equal to 0°. The azimuth angle is measured clockwise with respect to north. The angle of elevation is the angle between the horizontal plane and the pointing direction of the antenna.

Once the azimuth and elevation are known, the earth station antenna can be pointed in that direction. For a geosynchronous satellite, the antenna will simply remain in that position. For any other satellite, the antenna must then be moved as the satellite passes overhead.

For geosynchronous satellites, the angles of

azimuth and elevation are relatively easy to determine. Because geosynchronous satellites are fixed in position over the equator, special formulas and techniques have been developed to permit easy determination of azimuth and elevation for any geosynchronous satellite for any point on earth. A chart used for this purpose is shown in Fig. 11-13. An example will illustrate its use.

Assume that a ground station is located in Washington, D.C. with coordinates of 39°N and 77°W and that the location of the satellite is 100°W longitude. Now take the difference between the longitude of the satellite and the longitude of Washington, D.C., in this case, $100 - 77 = 23°$. This longitude and the latitude of Washington, D.C., 39°N, define a point on the chart in Fig. 11-13 (see Point A).

The lines drawn on the chart in Fig. 11-13 show various levels of both azimuth and elevation. The azimuth or radial lines originate at the lower left-hand 0° point and emerge upward and outward to the right. The other lines designate elevation. If the point defined falls within these sets of lines, then the satellite is within range of the ground station.

Next, note the elevation and radial values at the point defined on the chart. The angle of el-

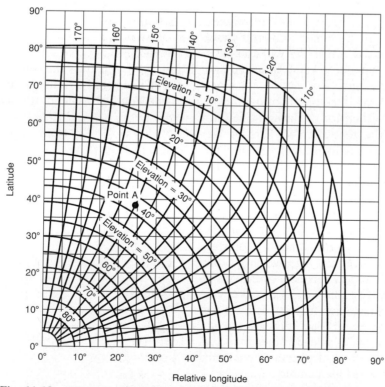

Fig. 11-13 **Chart for obtaining azimuth and elevation directions from a ground station to a geostationary satellite.**

evation is approximately 37°, while the radial value is approximately 146°. For a northern-hemisphere station, the azimuth reading is determined by subtracting the radial value from 360° or, in this case, 360° − 146° = 214°. If the earth station is in the southern hemisphere, the azimuth value is equal to 180° plus the radial value or 180° minus the radial value depending upon whether the satellite is east or west of the ground station. In this case, the earth station antenna will be pointed with an azimuth reading of 214° or approximately south-southwest. The angle of elevation is very low, being only about 37° above the horizon. Once the antenna has been set, minor adjustments can be made in both azimuth and elevation to optimize signal reception.

Tracking a nonsynchronous satellite is a more difficult job. Normally a computer is set up to calculate the orbit calendar, and various graphical devices are used to trace the ground track. From this information, azimuth and elevation angles for the tracking antenna can be determined.

Satellites are placed into their orbits by mounting them on top of rockets which literally shoot them into space. Most communications satellites today are large and heavy, weighing many thousands of pounds. A huge multistage rocket is required to put it into orbit. We will describe the launch procedure in detail so that you can understand the complex steps required to put a satellite into orbit about the earth. For this example, we will assume a communications satellite is being launched into geosynchronous orbit.

The satellite is mounted on top of the rocket and is covered by a fairing or shroud. The fairing is usually the top, pointed part of the rocket. Its purpose is to protect the satellite during the initial part of the flight and aerodynamically streamline the rocket.

Occasionally, the rocket will contain more than one satellite. The main satellite to be launched is called the *initial payload,* whereas other, smaller satellites are generally referred to as *secondary payloads*. The weight of the main payload and of any secondary payloads is accurately known. In this way, the rocket can be properly fueled and fine-tuned to achieve the desired thrust and speed. Since it is easier to adjust the weight of the payload rather than the thrust, ballast in the form of secondary payloads is added or subtracted as required.

The main booster rocket is ignited and lifts off. The rocket overcomes the earth's gravita-

tional pull and rises slowly. Then a gyroscopic guidance system points the rocket in a specific direction. Typically, the rocket is not launched directly vertically upward. Instead, it is launched at a slight angle that will help the satellite achieve the desired orbit. During this time the rocket and its payload pass through the lower atmosphere of the earth.

If during the launch process the speed of the launch vehicle exceeds 25,000 mi/h, the satellite will not go into orbit. At this speed, the spacecraft will break away from the gravitational pull of the earth and go out into deep space. A spacecraft such as this, of course, is no longer called a satellite. Typically it is called a space probe. Examples of space probes are the special satellitelike craft that were constructed to survey the moon, Venus, Mars, and Saturn.

At approximately 50 to 100 mi above the earth, the large booster rocket flames out and is jettisoned. The second-stage rocket is ignited to carry the satellite farther upward into the higher parts of the atmosphere. At some point this rocket also burns out and is jettisoned. A third-stage rocket may also be required. This third and final stage, although not always used, is sometimes needed to achieve the final speed required to put the satellite into orbit. Also during this period the fairing is jettisoned. This exposes the satellite which no longer requires protection as the upper atmosphere is not harmful.

At some point the final rocket burns out and the satellite begins to coast, ultimately going into orbit. It enters the orbit near the perigee. Refer to Fig. 11-14. This is not the final orbit of the satellite. Instead, the satellite is first put

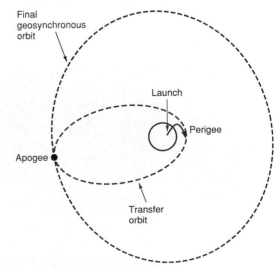

Fig. 11-14 Launching a satellite.

into what is called a transfer orbit, a highly elliptical orbit that permits adjustments to the satellite to be made prior to its being placed into final position.

The satellite speed is adjusted to put it into a transfer orbit whose apogee is the final desired height for the geosynchronous satellite, usually 22,300 mi. The satellite is allowed to stay in this transfer orbit for several passes in order to permit various alignment maneuvers. For example, a geosynchronous satellite must be in position over the equator. The initial transfer orbit will not be over the equator because the launch site is typically not located on the equator. As a result, during the transfer orbit, an on-board rocket is fired at the appropriate time to put the orbit over the equator. Typically when a satellite is launched from Cape Canaveral, it goes into an elliptical orbit with an inclination of 28.5° with respect to the equatorial plane.

After several revolutions in the transfer orbit, the satellite is placed into a circular orbit around the equator by firing the apogee kick motor. This is a small rocket motor built into the satellite and specifically designed to be fired for positioning purposes. The apogee kick motor is fired at the apogee of the elliptical transfer orbit. The satellite is oriented by ground station control with its jet thrusters so that the apogee kick motor fires in the proper direction to move the satellite in the direction that will achieve the desired equatorial orbit. The firing of the apogee kick motor removes the angle of inclination and circularizes the orbit. The satellite is allowed to drive in this orbit for several revolutions.

Finally, attitude stabilization is initiated. If spin stabilization is used, thrusters are fired to start the spinning. If three-axis stabilization is used, thrusters are fired until the sensors pick up the external reference points at which time the satellite's internal attitude-control system goes into operation to position the satellite properly. During this time, antennas are properly aligned and solar cell panels are adjusted to produce optimum power. The satellite is allowed to operate for several days and is tracked from the ground. At some point additional firing of the jet thrusters may be necessary to fine-tune the satellite in its final designated position.

NASA's highly successful space shuttle is now routinely used to put satellites into orbit. A separate rocket is not used. Instead, one or more satellites is loaded into the shuttle which is put into orbit. Astronauts deliver the satellites to their approximate location and then spin them out of the cargo bay into orbit. Ground stations make the final attitude and position adjustments.

Self-Test

Answer the following questions.

1. A satellite is kept in orbit by a balance between two forces: _____ .
2. A satellite that revolves in the same direction as the earth rotates is said to be in a _____ (posigrade, retrograde) orbit.
3. The geometric shape of a noncircular orbit is a(n) _____ .
4. The center of gravity of the earth is called the _____ .
5. The time for one orbit is called the _____ .
6. The angle of inclination of a satellite is _____ degrees if it orbits over the equator and is _____ degrees if it orbits over the north and south poles.
7. To prevent excessive signal attenuation and noise in the atmosphere, satellite angles of elevation of less than _____ degrees should be avoided.
8. A satellite that rotates around the equator 22,300 mi from the earth is said to be in _____ orbit.
9. Small jet thrusters on a satellite are fired to correct the satellite's _____ .
10. Name two ways satellites are stabilized in space.
11. The three axes of a satellite are _____ .
12. The point on the earth directly below a satellite is called the _____ .
13. Satellites are located by earth coordinates expressed in terms of _____ .
14. The two angles used to point a ground station antenna are _____ .
15. A satellite is put into final geosynchronous orbit from its transfer orbit by firing the _____ .

11-2 SATELLITE COMMUNICATIONS SYSTEMS

Communications satellites are not orginators of information to be transmitted. Although this is the case in some other types of satellites, it is not true for those used for communications purposes. Communications satellites are instead relay stations for other sources. If a transmitting station cannot communicate directly with one or more receiving stations because of line-of-sight restrictions, then a satellite can be used. The transmitting station sends the information to the satellite which, in turn, retransmits it to the receiving stations. The satellite in this application is what is generally known as a *repeater*.

Figure 11-15 shows the basic operation of a communications satellite. An earth station transmits information to the satellite. The satellite contains a receiver which picks up the transmitted signal, amplifies it, and translates it to another frequency. This new frequency is then retransmitted to the receiving stations back on earth. The original signal being transmitted from the earth station to the satellite is called the *up link,* and the retransmitted signal from the satellite to the receiving stations is called the *down link*. Usually the downlink frequency is lower than the up-link frequency. A typical up-link frequency is 6 GHz and a common down-link frequency is 4 GHz.

The transmitter-receiver combination in the satellite is known as a transponder. The basic function of a transponder is amplification and frequency translation. A simplified block diagram of a transponder is shown in Fig. 11-16. The reason for frequency translation is that the transponder cannot transmit and receive on the same frequency. The transmitter's strong signal would overload the receiver and block out the very small up-link signal, thereby prohib-

Fig. 11-16   A satellite transponder.

iting any communications. By using widely spaced transmit and receive frequencies, no interference is encountered.

Transponders are also wide-bandwidth units so that they may receive and retransmit more than one signal. Any earth station signal within the receiver's bandwidth will be amplified, translated, and retransmitted on a different frequency.

Even though the typical transponder has a wide bandwidth, it is only used with a single signal to minimize interference and improve communications reliability. However, a satellite would not be an economical repeater if it were capable of handling only a single channel. As a result, most satellites contain multiple transponders each operating at a different frequency. A typical communications satellite has 12, 24, or more transponders. Each transponder represents an individual communications channel. Various multiplexing schemes are used so that each channel may carry multiple information transmissions.

Most communications satellites operate in the microwave frequency spectrum. However, there are some exceptions to this. For example, many military satellites operate in the 200-to 400-MHz range which is generally considered to be in the UHF range. Also, the amateur radio OSCAR satellites operate in the VHF to UHF range.

The microwave spectrum is divided up into frequency bands which have been allocated to satellites as well as other communications services such as radar. These frequency bands are generally designated by a letter of the alphabet. The table in Fig. 11-17 shows the various frequency bands used in satellite communications.

The most widely used satellite communications band is the C band. The up-link frequencies are in the 5.925- to 6.425-GHz range. In any general discussion of the C band, the up-

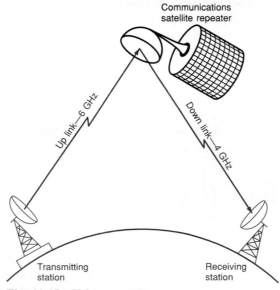

Fig. 11-15   **Using a satellite as a microwave relay link.**

Frequency	Band
225–390 MHz	P
350–530 MHz	J
1530–2700 MHz	L
2500–2700 MHz	S
3400–6425 MHz	C
7250–8400 MHz	X
10.95–14.5 GHz	Ku
17.7–21.2 GHz	Kc
27.5–31 GHz	K
36–46 GHz	Q
46–56 GHz	V
56–100 GHz	W

Fig. 11-17 **Frequency bands used in satellite communications.**

link frequency is generally said to be 6 GHz. The down-link frequencies are in the 3.7- to 4.2-GHz range, but again, in any general discussion of the C band, the down-link frequency is nominally said to be 4 GHz. Occasionally, you will see the C band referred to by the designation 6 GHz/4 GHz where the up-link frequency is given first.

Although most satellite communications activity takes place in the C band, there is a steady move toward the higher frequencies. In 1989 the Ku band received the most attention. The up-link frequencies are in the 14- to 14.5-GHz range, and the down-link frequencies are in the 11.7- to 12.2-GHz range. You will see the Ku band designated as 14/12 GHz.

Most new communications satellites will operate in this band. The reason for this shift upward in frequency is that the C band is becoming overcrowded. There are many communications satellites in orbit now, most of them operating in the C band. However, there is some difficulty with interference because of the heavy usage. The only way this interference will be minimized is to shift all future satellite communications to higher frequencies. Naturally, the electronic equipment needed to achieve these higher frequencies is more complex and expensive. Yet, the crowding and interference problems cannot be resolved in any other way. Further, for a given antenna size, the gain is higher in the Ku band than in the C band. This can improve communications reliability while decreasing antenna size and cost.

Two other bands of interest are the X and L bands. The military uses the X band for its satellites and radar, and the L band is used in marine and aeronautical communications and radar.

Recall the frequencies designated for the C

band up link and down link. These are 5925 to 6425 MHz and 3700 to 4200 MHz, respectively. You can see that the bandwidth between the upper and lower limits is 500 MHz. This is an incredibly wide band which is capable of carrying an enormous number of signals. In fact, 500 MHz covers all the radio spectrum we know so well from VLF through VHF and then some. Most communications satellites are designed to take advantage of this full bandwidth. This allows them to carry the maximum possible number of communications channels. Of course, this extremely wide bandwidth is one of the major reasons why microwave frequencies are so useful in communications. Not only can many communications channels be supported, but also very high speed digital data requiring wide bandwidth can also be dealt with.

The transponder receiver "looks at" the entire 500-MHz bandwidth and picks up any transmission there. However, the input is "channelized" because the earth stations operate on selected frequencies or channels. The 500-MHz bandwidth is typically divided into 12 separate transmit channels, each 36 MHz wide. There are 4-MHz guard bands between channels that are used to minimize adjacent-channel interference. These communications channels are illustrated in Fig. 11-18 on the next page. Note the center frequency for each channel. Remember that the up-link frequencies are translated by frequency conversion to the down-link channel. In both cases, the total bandwidth (500 MHz) and channel bandwidth (36 MHz) are the same. On board the satellite, a separate transponder is allocated to each of the 12 channels.

Although 36 MHz seems narrow compared to 500 MHz, each transponder bandwidth is still capable of carrying an enormous amount of information. For example, one typical transponder can handle up to 1000 one-way analog telephone conversations as well as one full-color TV channel. Each transponder channel can also carry high-speed digital data. Using certain types of modulation, a standard 36-MHz bandwidth transponder can handle digital data at rates up to 60 Mbits/s.

Although the transponders are quite capable, they nevertheless rapidly become overloaded with traffic. Further, there is at times more traffic than there are transponders to handle it. For that reason, numerous techniques have been developed to effectively increase the bandwidth and signal-carrying capacity of the sat-

Fig. 11-18 Receive and transmit bandwidths in a C band communications satellite.

ellite. Two of these techniques are known as frequency reuse and spatial isolation.

Frequency reuse effectively doubles the bandwidth and information-carrying capacity of a satellite. In this system, a communications satellite is provided with two identical sets of 12 transponders. The first channel in one transponder operates on the same channel as the first transponder in the other set, and so on. With this arrangement, the two sets of transponders transmit in the same frequency spectrum and, therefore, would appear to interfere with one another. However, this is not the case. The two systems, while operating on exactly the same frequencies, are isolated from one another by the use of special antenna techniques.

Recall that any radio signal is a combination of both electric and magnetic fields that exist in quadrature with one another. That is, the magnetic field is 90° offset from the electric field. Further, the electric field determines the so-called polarization of the signal. If the electric field is oriented vertically with respect to the earth, then the signal is said to be vertically polarized. If the electric field is horizontal to the earth, then the signal is said to be horizontally polarized. Of course, it is the positioning of the antenna that determines

whether the signal is either vertically or horizontally polarized.

In theory, a signal that is transmitted with vertical polarization will not be received on an antenna that is horizontally polarized. The same is true of a horizontally transmitted signal which will not be received by a vertically polarized antenna. In practice, perfect vertical or horizontal polarization can typically be received by antennas of either polarization. However, at microwave frequencies, the polarization can be more accurately controlled.

Circular polarization is also used in some types of antennas. There is left-hand and right-hand circular polarization. An antenna using one cannot work with the other.

By using transmitting and receiving antennas that are vertically or horizontally polarized or that use left- or right-hand polarization, two completely separate sets of transponders operating on the same frequency can be used simultaneously. One set of 12 transponders will have a vertically polarized or left-hand circular polarized antenna. The other set will use horizontal or right-hand circular polarization. By careful positioning and orientation of the antennas, one set of signals will not interfere with the other. For example, if two earth stations transmit on the same frequency but with different polarizations, one will be rejected by one transponder but picked up by another. Two transponders transmitting signals on the same frequency but with different polarizations will not interfere with one another as ground station antenna polarization will selectively sort them out. In this way, two 500-MHz bandwidths with signal-carrying capacity can be included in a satellite. The special antenna techniques that permit this will be discussed in more detail in another chapter.

Spatial isolation is another technique whereby additional information-carrying channels can be obtained. With this technique, extra sets of transponders are used. Special antenna techniques are used to isolate the inputs and outputs. In spatial isolation, very narrow beamwidth antennas are used to focus the down-link signals to specific areas of the earth. Such antennas are referred to as spot-beam antennas. By using such antennas on the spacecraft, the signals can be confined to a particular area. In this way, different earth stations can use the same frequencies. They do not interfere with one another because of the highly directional antennas. In this way, the total bandwidth or information-carrying capacity of the satellite can be doubled. For example, a satellite could contain up to four sets of 12 transponders each, all using the same frequencies. However, with frequency reuse techniques as well as spot beams, a total of 48 transponders can be used.

Self-Test

Supply the missing information in each statement.

16. The most common use of a satellite is _____ .
17. A communications satellite is basically a radio _____ .
18. A(n) _____ transmits an up-link signal to the satellite.
19. The signal path from a satellite to a ground station is called the _____ .
20. The basic communications electronics unit on a satellite is known as a(n) _____ .
21. Most communications satellites operate in the _____ frequency range.
22. The most popular satellite frequency range is 4 to 6 GHz and is called the _____ band.
23. Military satellites often operate in the _____ band.
24. The Ku band extends from _____ to _____ GHz.

25. The bandwidth of a typical satellite is _____ MHz.
26. A typical C band transponder can carry _____ channels, each with a bandwidth of _____ MHz.
27. The term used to refer to the technique of using antenna polarization to separate signals on the same frequency to double the number of channels is _____ .
28. Using very narrow beamwidth antennas to isolate signals on the same frequency is known as _____ .

11-3 SATELLITE SUBSYSTEMS

All satellite communications systems consist of two basic parts: the satellite or spacecraft and one or more earth stations. The satellite performs the function of a radio repeater or relay station. Two or more earth stations may communicate with one another through the satellite rather than directly point to point on the earth.

The heart of a communications satellite is the communications subsystem. This is a set of transponders that receive the up-link signals and retransmit them to earth. A transponder is a repeater that implements a wideband communications channel which can carry many simultaneous communications transmissions.

The transponders are supported by a variety of additional "housekeeping" subsystems. These include the power subsystem, the telemetry tracking and command subsystems, the antennas, and the propulsion and attitude stabilization subsystems. These are essential to the self-sustaining nature of the satellite.

Figure 11-19 on the next page shows a general block diagram of a satellite. All the major subsystems are illustrated. The solar panels supply the electrical power for the spacecraft. They drive regulators that distribute dc power to all other subsystems. And they charge the batteries that operate the satellite during eclipse periods. Both dc-to-dc converters and dc-to-ac inverters are used to supply special voltages to some subsystems.

The communications subsystem consists of multiple transponders. These receive the up-link signals, amplify them, translate them in frequency, and amplify them again for retransmission as down-link signals. The transponders share a common antenna subsystem for both reception and transmission.

The telemetry, tracking, and command subsystem monitors on-board conditions, such as

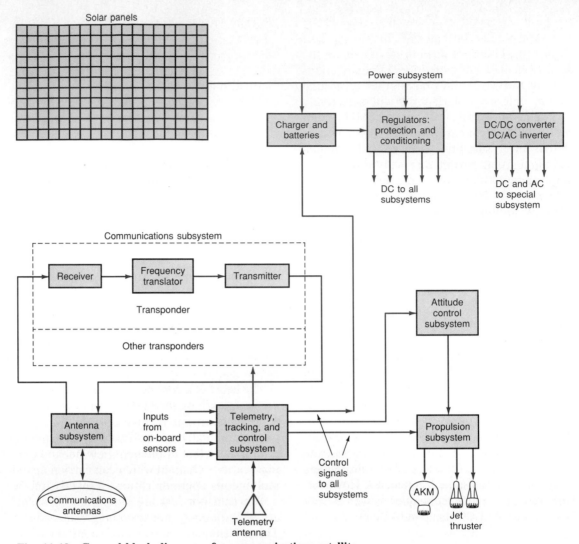

Fig. 11-19 **General block diagram of a communications satellite.**

temperature and battery voltage, and transmits this data back to a ground station for analysis. The ground station may then issue orders to the satellite by transmitting a signal to the command subsystem, which then is used to control many spacecraft functions, such as firing the jet thrusters.

The jet thrusters and the apogee kick motor (AKM) are part of the propulsion subsystem. They are controlled by commands from the ground.

The attitude-control subsystem provides stabilization in orbit and senses changes in orientation. It fires the jet thrusters to perform attitude adjustment and stationkeeping maneuvers that keep the satellite in its assigned orbital position.

The main payload on a communications satellite, of course, is the communications subsystem that performs the function of a repeater or relay station. An earth station takes the signals to be transmitted, known as baseband signals, and frequency modulates a microwave carrier. The most common baseband signals are voice, video, and computer data. These up-link signals are then amplified, translated in frequency, and retransmitted on the down link to one or more earth stations. The component that performs this function is known as a transponder. Most modern communications satellites contain at least 12 transponders. More advanced satellites contain many more. These transponders operate in the microwave frequency range.

The basic purpose of a transponder is simply to rejuvenate the up-link signal and retransmit it over the down link. In this role the transponder simply performs the function of an amplifier. By the time the up-link signal reaches the satellite, it is extremely weak.

Therefore, it must be amplified before it can be retransmitted to the receiving earth station.

However, transponders are more than just amplifiers. An amplifier is a circuit that does nothing more than increase the voltage, current, or power level of a signal without changing its frequency or content. Such a transponder then would literally consist of a receiver and a transmitter that would operate on the same frequency. Because of the close proximity of the transmitter and the receiver in the satellite, the high transmitter output power for the down link would be picked up by the satellite receiver. Naturally, the up-link signal would be totally obliterated. Further, the transmitter output fed back into the receiver input would cause oscillation.

To avoid this problem, the receiver and transmitter in the satellite transponder are designed to operate at separate frequencies. In this way, they will not interfere with one another. The frequency spacing is made as wide as practical to prevent oscillation and to minimize the effect of the transmitter desensitizing the receiver. In many repeaters, even though the receive and transmit frequencies are different, the high output power of the transmitter can still affect the sensitive receiver input circuits and, in effect, desensitize them, making them less sensitive in receiving the weak up-link signals. The wider the frequency spacing between transmitter and receiver, the less of a problem this desensitizing is.

In typical satellites, the input and output frequencies are separated by huge amounts. At C band frequencies, the up-link signal is around 6 GHz, and the down-link signal is around 4 GHz. This 2-GHz spacing is sufficient to eliminate most problems. However, to ensure maximum sensitivity and minimum interference between up-link and down-link signals, the transponder contains numerous filters, which not only provide channelization, but also help to eliminate interference from external signals, regardless of their source.

There are three basic transponder configurations used in communications satellites. They are all essentially minor variations of one another, but each has its own advantages and disadvantages. These are the single-conversion, double-conversion, and regenerative transponders. Let's take a look at each.

Figure 11-20 shows a basic block diagram of a *single-conversion transponder*. The term "single conversion" refers to the fact that only a single-frequency translation process from the received signal to the transmitted signal takes place within the satellite.

The up-link signal is picked up by the receiving antenna and is first routed to a low-noise amplifier (LNA), usually a GASFET circuit. The signal is very weak at this point even though it has been multiplied somewhat by the gain of the receive antenna. Amplifiers with an extremely low noise figure or noise temperature must be used to increase the level of the signal.

Once the up-link signal has been amplified, it is translated in frequency. This is done in a mixer circuit. Any original modulation is retained during the frequency translation process. The mixer output is then amplified again and fed to a bandpass filter. The purpose of the bandpass filter is to remove all but the desired down-link signal of 4 GHz. Another function of the bandpass filter is to channelize the output. This means that only input signals on a specific frequency will be accepted by the bandpass filter. The LNA will increase their level, the mixer will translate them into the down-link frequency, and the amplifier following the mixer will further increase their level. However, if the signal falls out of the range of the filter, it will be rejected. The bandpass filters simply set up portions of the spectrum designated as channels.

Finally, the down-link signal is amplified by a high-power amplifier (HPA), usually a traveling-wave tube (TWT). Its output is further filtered at the channel frequencies to eliminate harmonics and intermodulation products. Intermodulation products are spurious, unwanted signals that are generated as the result of nonlinearities in the TWT. The input signals

Fig. 11-20 A single-conversion transponder.

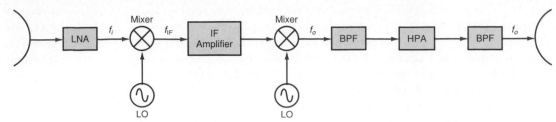

Fig. 11-21   A double-conversion transponder.

and any harmonics can be mixed together, producing many sum and difference frequencies unrelated to the desired signal. Filtering is used to eliminate them.

The resulting output is fed to the down-link antenna. In some cases, both the receive and transmit antennas are one and the same. A diplexer is used to keep the signals separate. The diplexer is a waveguide assembly that allows one antenna to be shared by a transmitter and a receiver.

Figure 11-21 shows a simplified block diagram of a *double-conversion transponder*. All the basic circuits of this system are similar in function to those of the single-conversion repeater. The only difference is that two frequency conversions are carried out.

Again, the up-link signal is picked up by the antenna, amplified by the LNA, and fed to the first mixer. The mixer translates the incoming signal into an IF. Typical IFs are 70 and 150 MHz. The IF signal out of the first mixer is fed to an IF amplifier where very high gain is achieved.

The output of the IF amplifier is fed to another mixer which translates the signal to the output frequency. Then the bandpass filter eliminates the unwanted mixer output and provides channelization. The HPA increases the signal level, and the output bandpass filter again provides for the elimination of harmonics and intermodulation components. Finally, the transmit antenna sends the signal on its way over the down link.

The main advantage of the double-conversion transponder is that greater flexibility in filtering and amplification can be achieved.

Amplification and selectivity at the lower IF level is far easier to attain.

In some satellites using multiple transponders, there are switching arrangements for taking the input of one channel and cross-connecting it to the output of another channel. When this is done, it is frequently done at the IF level since switching is simpler at these lower frequencies.

A third configuration of a satellite transponder is the *regenerative repeater*. Like the other configurations, it performs the basic amplification and frequency translation functions. But in this case, the received signal is actually demodulated. Therefore, remodulation is necessary to create the down-link signal.

The basic block diagram of a regenerative transponder is shown in Fig. 11-22. The up-link signal is amplified and frequency-translated as before. However, the output is then demodulated. The output of the demodulator, of course, is the baseband signal.

A baseband signal refers to the basic information content transmitted on the up-link carrier. The baseband signal may be a voice telephone conversation, a TV signal, or digital data. The output of the demodulator, therefore, in the satellite is this baseband signal. Some amplification may be provided.

Next, the baseband signal is fed to a modulator along with a carrier at the down-link frequency. The modulation is usually FM. The output of the modulator is a new carrier containing the same information. The signal is then amplified, filtered, and transmitted over the down link.

This regenerative configuration has several ba-

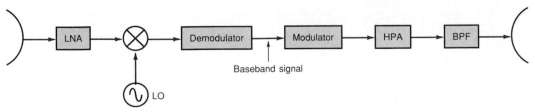

Fig. 11-22   A regenerative transponder.

sic advantages over the other configurations. First, it permits both the receiver and transmitter sections of the transponder to be optimized. In the other configurations, there is essentially no difference between the transmitter and the receiver. The transponder is a combination of both. You can't really tell where the receiver ends and the transmitter begins. In a regenerative repeater, the division line is clear since the receiver ends at the demodulator output. The transmitter begins at the modulator input. Clearly this separation permits each section to be fully optimized in performance without worrying about the interaction that normally occurs in more conventional configurations.

The amplification in a regenerative repeater is also easier to obtain at the very low baseband frequencies. The circuits are simpler and less expensive.

Further, the S/N ratio of the transponder overall can be improved using this technique. It has been shown that an S/N ratio increase of 2 to 3 dB is achievable over transponders with more conventional arrangements.

Finally, in multitransponder satellites with input and output switching, the regenerative arrangement is simpler and more flexible to implement. It is a lot easier to switch baseband signals than it is to switch higher-frequency signals. This flexibility allows a wider variety of switching options to be implemented.

Virtually all modern communications satellites contain multiple transponders. This permits many more signals to be received and transmitted. A typical commercial communi-

cations satellite contains 12 transponders, or 24 if frequency reuse is incorporated. Military satellites often contain fewer transponders, whereas the newer, larger commercial satellites have provisions for up to 50 channels. Each transponder operates on a separate frequency, but its bandwidth is wide enough to carry multiple channels of voice, video, and digital information.

There are two basic multichannel architectures in use in communications satellites. One is a broadband system, and the other is a fully channelized system.

As indicated earlier, a typical communications satellite spectrum is 500-MHz wide. This is usually divided into 12 separate channels, each with a bandwidth of 36 MHz. The center frequency spacing between adjacent channels is 40 MHz, thereby providing a 4-MHz spacing between channels to minimize adjacent-channel interference. Refer back to Fig. 11-18 for details. A wide-bandwidth repeater is designed to receive any signal transmitted within the 500-MHz total bandwidth. Such a transponder is shown in Fig. 11-23.

The receive antenna is connected to an LNA as in every transponder. Very wideband tuned circuits are used so that the entire 500-MHz bandwidth is received and amplified. A mixer translates all incoming signals into their equivalent lower down-link frequencies. In a C band communications satellite, the incoming signals are located between 5.925 and 6.425 MHz. An LO operating at the frequency of 2.225 GHz is used to translate the input signals down into

Fig. 11-23 A broadband multiple-channel repeater.

the 3.7- to 4.2-GHz range. A wideband amplifier following the mixer amplifies this entire spectrum.

The channelization process occurs in the remainder of the transponder. For example, in a 12-channel satellite, 12 bandpass filters, each centered on one of the 12 channels, is used to separate all the various received signals. Figure 11-18 shows the 12 basic channels with their center frequencies, each having a bandwidth of 36 MHz. The bandpass filters separate out the unwanted mixer output signals and retain only the difference signals. Then, individual HPAs are used to increase signal level. These are usually TWTs. The output of each TWT amplifier is again filtered to minimize harmonic and intermodulation distortion problems. These filters are usually part of a larger assembly known as a multiplexer or combiner. In any case, it is a waveguide–cavity-resonator assembly that filters and combines all the signals for application to a single antenna.

It is logical to assume that if the receive function can be accomplished by wideband amplifier and mixer circuits, why isn't it possible to provide the transmit function in the same way? The answer is that it is generally not possible to generate the very high output power over such a wide bandwidth. In other words, there are no components and circuits currently available that do this well. The HPAs in most transponders are TWTs which inherently have limited bandwidth. They operate well over a small range but cannot deal with the entire 500-MHz bandwidth allocated to a satellite. Therefore, in order to achieve the necessary high power levels, the channelization process is used.

Another transponder architecture used in communications satellites uses individual narrowband input channels instead of the single wideband input described previously. This arrangement is illustrated in Fig. 11-24. The receive antenna feeds a demultiplexer, a waveguide assembly, that separates the single wideband input into equal feeds for separate channels. Separate LNAs and bandpass filters centered on the designated channels are used at the inputs. The filters sort out the input signals, and the LNAs provide the desired amount of gain. Although each bandpass filter contributes to the noise level, this is offset by the narrowband operation. Keep in mind that the wider the bandwidth of a channel, the higher its noise content. By limiting the bandwidth, noise decreases. Therefore, with the individual channel arrangement, some improvement in noise temperature is achieved.

Each channel also has its own mixer, driven by a common LO whose frequency is 2.225 GHz in the C band. The mixer outputs drive bandpass filters which, in turn, drive the HPA or TWT output amplifiers. Again, bandpass filters are used at the output of the TWTs. The output bandpass filters are part of a multiplexer-combiner unit which creates one feed for the transmitting antenna.

Now let's look at the other subsystems that support the transponders. Refer to Fig. 11-19, which shows all the subsystems.

A key component of the satellite is its power subsystem. Figure 11-25 shows a block diagram of the power subsystem. Everything on board operates electrically or has electrical components in it. Most satellites, therefore, depend entirely upon their power supplies for

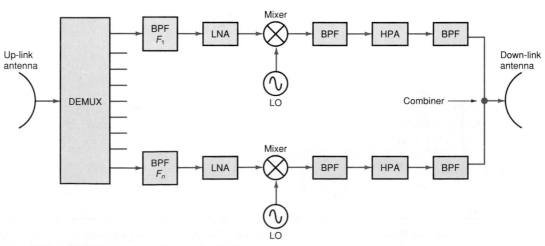

Fig. 11-24 **Multichannel receiver transponder.**

the power subsystem.

n-board
atteries
t be re-
pendent
es used
r drain.
lar pan-
lar pan-
cted in
reate a
y solar
atts of
able of
rement
ted to-
nfigu-
t page
haped
entire
em is
ilized
panels
to en-
th re-

rrent
... is used to operate the various components of the satellite. However, the dc power
is typically used to charge nickel-cadmium bat-

teries that act as a buffer. At times when the
satellite goes into an eclipse or when the solar
panels are not properly positioned, the batteries take over temporarily and keep the satellite operating. The batteries aren't large enough
to power the satellite for a long period of time;
they are simply used as a backup system for
eclipses, initial satellite orientation and stabilization, or emergency conditions.

The basic dc voltage from the solar panels
is then conditioned in various ways. For example, it is typically passed through voltage
regulator circuits before being used to power
individual electronic circuits. Most electronic
equipment works best with fixed, stable voltages; therefore, regulators are incorporated in
most satellite systems.

Occasionally, voltages higher than those produced by the solar panels must also be generated. For example, the TWT amplifiers in most
communications transponders require thousands of volts for proper operation. Special dc-
to-dc converters are used to translate the lower
dc voltage of the solar panels to the higher dc
voltage required by the TWTs.

All satellites have a telemetry, tracking, and
control (TTC) subsystem that allows a ground
station to monitor and control conditions in the
satellite. The telemetry system is used to re-

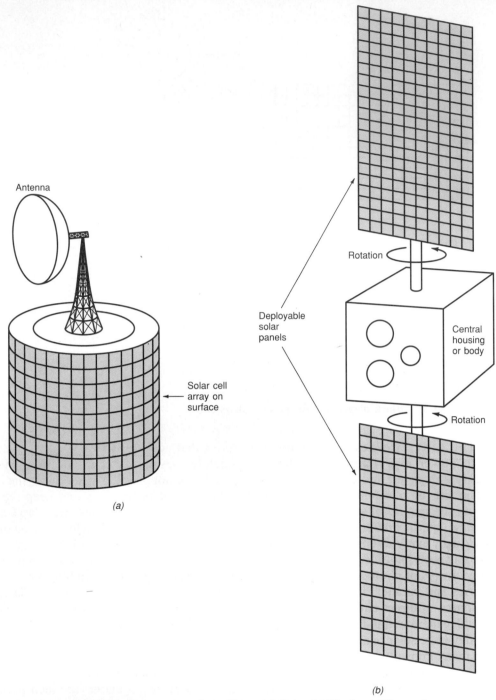

Antenna

Solar cell
array on
surface

(a)

Rotation

Deployable
solar
panels

Central
housing
or body

Rotation

(b)

Fig. 11-26   Satellite solar array configurations. (a) Spin-stabilized satellite, and (b) body-stabilized satellite.

port the status of the on-board subsystems to the ground station. A typical system is shown in Fig. 11-27. The telemetry system typically consists of various electronic sensors for measuring temperature, radiation level, power supply voltages, and other key operating characteristics. Both analog and digital sensors may be used. The sensors are selected by a multiplexer and are converted to a digital signal, which then modulates an internal transmitter. This transmitter sends the telemetry information back to the earth station where it is recorded and monitored. With this information, the ground station is then able to determine the operational status of the satellite at all times.

A command and control system permits the ground station to control the satellite. Refer to

Fig. 11-27   General block diagram of a satellite telemetry unit.

Fig. 11-28   The command receiver and controller.

Fig. 11-28. Typically the satellite contains a command receiver which receives control signals from an earth station transmitter. The control signals are typically made up of various digital codes that tell the satellite what to do. Various commands may initiate a telemetry sequence, activate thrusters for attitude correction, reorient an antenna, or perform other operations as required by the special equipment specific to the mission. Usually, the control signals are processed by an on-board computer.

Most satellites now contain a small digital computer, usually microprocessor-based, that acts as a central control unit for the entire sat-

ellite. The computer contains a built-in ROM with a master control program. This master control program operates the computer and causes all other subsystems to be operated as required. The command and control receiver typically passes on new command codes that it receives from the ground station to the computer which then carries out the desired action.

The computer may also be used to make necessary computations and decisions. Information collected from the telemetry system may be first processed by the computer before it is sent to the ground station. The memory of the computer may also be used to store data temporarily before the data is processed or transmitted back to earth. The computer may also serve as an event timer or clock. The computer is a versatile control element that can typically be reprogrammed via the command and control system to carry out any additional functions that may be required, particularly those that were not properly anticipated by those designing the mission.

All satellites contain an attitude-stabilization subsystem with built-in jet thrusters for making corrections to the attitude of the satellite. By way of the command and control subsystem and the on-board computer, the attitude-stabilization subsystem receives signals that fire the appropriate thrusters at the proper time to achieve an adjustment in attitude. This may be to reorient antennas, maximize exposure of the solar panels to the sun, or make corrections indicated by photo- or infrared sensors that provide orientation input for a three-axis stabilized satellite. The attitude-stabilization system may also consist of gyroscopes, flywheels for spin stabilization, and antenna de-spin mechanisms.

In addition to the jet thrusters for attitude stabilization, most satellites also contain a propulsion system. This propulsion system is usually the apogee kick motor used to put the satellite into final orbit, or it may be one or more liquid or solid propellant rockets that could be used to change the orbit of a satellite or remove the satellite from orbit. The propulsion system is also operated by the on-board computer in response to the command control subsystem.

All satellites have one or more antennas used for receiving signals from the ground station and transmitting information back to earth. Some antennas have a dual function. Numerous antennas are generally required because of the various satellite requirements. For example, one antenna may be used for communications relay purposes, while another antenna is dedicated to the telemetry, tracking, and control functions. Most of these are highly directional gain antennas that must be accurately pointed. However, most satellites contain an omnidirectional antenna that is used on the command receiver so that the satellite may pick up ground control signals during launch, orbit transfer, and other periods prior to full attitude stabilization.

All satellites have an applications subsystem made up of special components that make the satellite useful for the purpose intended. For a communications satellite, this subsystem is the transponders.

For a scientific satellite designed to gather information about space, a special instrumentation package will be included. This would be made up of various sensors, their signal-conditioning electronics, and additional telemetry equipment to transmit the information back to earth. The system may also include a tape recorder for temporarily storing information and transmitting it back to earth at a time to be determined later by the ground station. Nonsynchronous satellites often use recorders since data will be gathered during the times the ground station has no access to the satellite. The recorder captures data for later transmission.

An observation satellite such as those used for intelligence gathering or weather monitoring may use television cameras or infrared sensors to pick up various conditions on earth and in the atmosphere. This information is then transmitted back to earth by a special transmitter designed for this purpose. And, of course, there are many other variations of this subsystem, depending upon what it will be used for.

Self-Test

Answer the following questions.

29. The most common baseband signals handled by a satellite are _____ .
30. The input circuit to a transponder is the _____ .
31. List the three main functions of a transponder.
32. A(n) _____ circuit in the transponder performs the frequency conversion.
33. Power amplification in a transponder is usually provided by a(n) _____ .
34. A(n) _____ transponder provides amplification at an IF of 70 or 150 MHz.
35. A transponder that demodulates the baseband signals and then remodulates a carrier is known as a(n) _____ transponder.
36. The circuit that provides channelization in a transponder is the _____ .
37. The two basic architectures for multichannel transponders are _____ .
38. The main power supplies in a satellite are the _____ .
39. During an eclipse, the satellite is powered by _____ .

40. The TTC subsystem is used to _____ operations on a satellite.

41. Attitude correction is made by firing _____ .

11-4 EARTH STATIONS

The earth station or ground station is the terrestrial base of the system. The earth station communicates with the satellite to carry out the designated mission. The earth station may be located at the end user's facilities or may be located remotely with ground-based intercommunication links between the earth station and the end user. In the early days of satellite systems, earth stations were typically situated in remote country locations. Because of their enormous antennas and other critical requirements, it was not practical to locate them in downtown or suburban areas. Today earth stations are much less complex and the antennas are smaller. Many earth stations are now located on top of tall buildings or in other urban areas directly where the end user resides. This offers the advantage of eliminating complex intercommunications systems between the earth station and the end user.

Like the satellite, the earth station is made up of a number of different subsystems. The subsystems, in fact, generally correspond to those on board the satellite. However, most of the subsystems are larger and much more complex. Further, several additional subsystems exist at earth stations that are not applicable to the satellite itself.

An earth station consists of five major subsystems. These are the antenna subsystem, the receive subsystem, the transmit subsystem, the ground communications equipment (GCE) subsystem, and the power subsystem. A general block diagram of these is illustrated in Fig. 11-29 on the next page. Not shown here are the telemetry, control, and instrumentation subsystems.

The antenna subsystem, of course, consists of the parabolic reflector, the horn antennas, waveguide transmission line, and the related mounts. The receive subsystem consists of the LNA and filters as well as any power dividers and related circuitry. Many earth stations have multiple circuits for redundancy and reliability. These are switched on automatically if the main unit fails.

The transmit subsystem consists of the HPA and any intermediate driver stages plus related filters. The GCE consists of both receiving and transmitting circuits. The receiving portion consists of down converters, filters, demodulators, and demultiplexing equipment. The transmit portion of the GCE consists of multiplexers, modulators, up converters, and the related filters. Connections to the telephone system, terrestrial microwave relay links, computer interfaces, and so on are made through the GCE.

The power subsystem furnishes all the power to the other equipment. The primary sources of power are the standard ac power lines. The subsystem operates power supplies which distribute a variety of dc voltages to the other equipment. The power subsystem also consists of emergency power sources such as diesel generators, batteries, and inverters to ensure continuous operation during power failures.

Refer back to Fig. 11-29. The antenna subsystem usually includes a diplexer, a waveguide assembly that permits both the transmitter and receiver to use the same antenna. The diplexer feeds a bandpass filter in the receiver section that ensures that only the received frequencies pass through to the sensitive receiving circuits. This bandpass filter blocks the high-power transmit signal which can occur simultaneously with reception. This prevents overload and damage to the receiver.

The output of the bandpass filter feeds an LNA. This LNA drives a power divider. This is a waveguidelike assembly that splits the received signal into smaller but equal power signals. The power divider feeds several down converters. These are standard mixers fed by LOs that translate the received signals down to an IF, usually of 70 MHz. A bandpass filter ensures the selection of the proper sidebands out of the down converter.

The IF signal containing the data is then sent to the receive GCE where it is demodulated and fed to a demux where the original signals are finally obtained. The demux outputs are usually the baseband or original communications signals. In actual systems, several levels of demodulation and demultiplexing may have to take place to obtain the original signals.

In the GCE transmit subsystem, the baseband signals such as telephone conversations are applied to a multiplexer which permits multiple signals to be carried on a single channel. The multiplex signals are applied to a modulator along with the carrier oscillator. The modulator output is usually at the 70-MHz IF. Several levels of multiplexing and modulation

Fig. 11-29 General block diagram of an earth station.

may be used depending upon the application. This IF is fed to the transmit subsystem where it is translated to the final output frequency by way of an up-converter circuit. The up converter is a mixer with its LO tuned to generate the correct carrier channel. All the up-converted signals are then applied to a power combiner. This waveguidelike device equally combines all the RF signals into the final signal to be transmitted. This feeds a driver circuit which has sufficient power to operate the final amplifier or HPA. This is either a TWT or klystron. The output signal is applied to a bandpass filter and then to the diplexer and antenna.

All earth stations have a relatively large parabolic dish antenna that is used for sending and receiving signals to and from the satellite. Early satellites had very low-power transmitters, and, therefore, the signals received on earth were extremely small. Huge high-gain antennas were required to pick up minute signals from the satellite. The earth station "dishes" were 80 to 100 ft or more in diameter. Antennas of this size are still used in some satellite systems today, and even larger antennas have been used for deep-space probes.

Modern satellites now transmit with much more power. Advances have also been made in receiver components and circuitry. For that reason, smaller earth station antennas are now practical. In some applications, antennas as small as 2 or 3 ft in diameter can be used.

Typically the same antenna is used for both transmitting and receiving. A diplexer, a special coupling device for microwave signals, permits a single antenna to be used for multiple transmitters and/or receivers. In some applications, a separate antenna is used for telemetry and control functions.

The antenna in an earth station must also be steerable. That is, it must be possible to adjust the antenna's azimuth and elevation so that the antenna can be properly aligned with the satellite. Earth stations supporting geosynchronous satellites can generally be fixed in position. However, azimuth and elevation adjustments are necessary to initially pinpoint the satellite and to permit minor adjustments over the satellite's life.

Satellites in nonstationary orbits must be tracked. Earth station antennas for such systems incorporate large, sophisticated tracking systems in which both the azimuth and elevation of the antenna is adjusted by large motors that are part of a servo system. The servo system is driven by the tracking system. The tracking system may be nothing more than a timer which operates in synchronism with the satellite orbit. The servo motors are driven continuously through gear trains so that the antenna is always pointed at the satellite. A computer may be used for the controlling purposes if it knows the satellite orbit characteristics. The tracking system can also be based on signals received by the satellite. When the satellite comes into view, its signal is picked up and its amplitude is used to drive the tracking system. The system is optimized so that the antenna adjustments are made to continuously seek the maximum amplitude signal from the satellite. A beacon transmitter on the satellite is used for transmitting the tracking signal.

The down link is the receive subsystem of the earth station. It usually consists of very low noise preamplifiers that take the small signal received from the satellite and amplify it to a level suitable for further processing. The signal is then demodulated and sent on to other parts of the communications system.

The receive subsystem consists of the LNA, down converters, and related components. The purpose of the receive subsystem is to amplify the down-link satellite signal and translate it to a suitable IF. From that point, the IF signal is demodulated and demultiplexed as necessary to generate the original baseband signals.

Refer back to the general block diagram of the receive subsystem shown in Fig. 11-29. The output from the diplexer or circulator in the antenna subsystem feeds the RF signal to a bandpass filter (BPF). The purpose of this filter is to pass only the receive signal and to block any transmitted energy. Although the circulator or diplexer generally provides sufficient isolation between the received and transmitted

signals, the filter provides additional protection of the receiver circuits.

The filter output is applied to an LNA. This provides initial amplification of the down-link signal. Typically several stages of amplification are used to raise the tiny down-link signal to an amplitude suitable for use with the remaining circuits.

The output of the LNA is fed to a down-converter circuit consisting of a mixer and an LO. The down converter translates the RF signal to an IF where it will be dealt with further by the ground communications equipment.

The next stage in a satellite earth station receiver is a down converter which translates the microwave signal down to one or more IFs. The IF signal containing the original modulation is then sent to the GCE where it is demodulated and demultiplexed.

There are basically two types of down converters used in earth station receivers: single conversion and double conversion. A single-conversion circuit is shown in Fig. 11-30. The broadband down-link satellite signal is amplified by the LNA and is fed to a mixer circuit which is combined with an LO frequency. The resulting output is an IF of 70 MHz, the standard frequency used in most earth stations. The mixer itself may be a Schottky or a hot-carrier diode or in newer systems a GaAs MESFET. The LO is set to a frequency that will pick out one of the numerous channels in the wideband satellite signal. After the mixer is a bandpass filter which filters out any unwanted components from the mixer, including images. For example, to select 1 of the 12 channels in a standard 500-MHz bandwidth satellite signal, the LO would be tuned to a frequency of 70 MHz above or below the input channel carrier desired. The bandpass filter has a bandwidth of 36 MHz to pass the modulated and multiplexed signals on this channel.

In practice, single-conversion down converters are rarely used. The reason for this is image rejection problems and the difficulty of tuning or channel selection. To solve these

Fig. 11-30 Single-conversion down converter.

problems, dual-conversion down converters are normally used.

Figure 11-31(a) shows a diagram of a typical dual-conversion down converter. The input bandpass filter, of course, passes the entire 500-MHz satellite signal. This is fed to a mixer along with an LO. The output of the mixer is an IF signal, usually 770 MHz. This is passed through a bandpass filter at that frequency with a bandwidth of 36 MHz.

The signal is then applied to another mixer. When combined with the LO frequency, the mixer output is the standard 70-MHz IF value. An IF of 140 MHz is used in some systems. After the mixer is a 36-MHz-wide bandpass filter which passes the desired channel.

In dual-conversion down converters, two different tuning or channel selection arrangements are used. One is referred to as RF tuning, and the other is referred to as IF tuning. In RF tuning, shown in Fig. 11-31(a), the first LO is made adjustable. Typically a frequency synthesizer is used in this application. The frequency synthesizer generates a highly stable signal at selected frequencies whose increment can be changed. The synthesizer is set to a frequency that will select a desired channel. In RF tuning, the second LO has a fixed frequency to achieve the final conversion.

In IF tuning, the first LO is fixed in frequency, and the second is made tunable. Again, a frequency synthesizer is normally used for the tunable LO. See Fig. 11-31(b).

Figure 11-32 shows a full set of down converters as used in a standard earth station receiver. The amplified signal from the LNA is fed to a power divider circuit or branching network. This is either a waveguidelike device or in some cases a special stripline assembly that divides a single signal into multiple signals. The signals are identical to the received signal except that their amplitude is lower. The power dividers simply provide a means of allocating some of the amplified received signal to each of the independent down-converter circuits. It also provides necessary impedance matching.

The receiver GCE consists of one or more racks of equipment used for demodulating and demultiplexing the received signals. The down converters provide initial channelization by transponder, and the demodulators and demultiplexing equipment processes the 70-MHz IF signal into the original baseband signals. Other intermediate signals may be developed as required by the application.

The outputs from the down converters are usually made available on a patch panel of coax connectors. See Fig. 11-32. These are interconnected via coax to the demodulators. The demodulators are typically packaged in a thin, narrow vertical module that plugs into a chassis in a rack. Many of these demodulators are provided. They are all identical in that they demodulate the IF signal. In FDM systems, each demodulator is an FM detector. The most commonly used type is the PLL discriminator. Equalization and de-emphasis are also taken care of in the demodulator.

In systems using TDM, the demodulators are typically those used to detect four-phase or

(a)

(b)

Fig. 11-31 **Dual-conversion down converters.** (a) **RF tuning, and** (b) **IF tuning.**

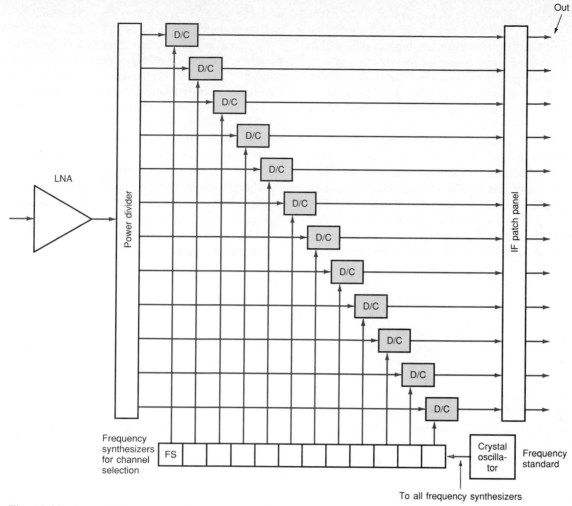

Fig. 11-32 **Typical down-converter arrangement.**

quadrature PSK at 60 or 120 Mbits/s. (*Note:* These circuits are covered in Chap. 12.) The IF is usually at 140 MHz. Again, a patch panel between the down converters and the demodulators permits flexible interconnection to provide any desired configuration. When video data is transmitted, the output of the frequency demodulator is the baseband video signal. It may then be transmitted by cable or used on premises as required.

If the received signals are telephone calls, then the demodulator outputs are sent to demultiplexing circuits. See Fig. 11-33. Again in many systems, a patch panel between the demodulator outputs and the demultiplexer inputs is provided to make various connections as may be required. In FDM systems, standard frequency division multiplexing is used. This consists of additional SSB demodulators and filters. Depending upon the number of signals multiplexed, several levels of channel filters and SSB demodulators may be required to generate

the original baseband voice signals. Once this is achieved, the signal may be transmitted over the standard telephone system network as necessary.

In TDM systems, time division demultiplexing equipment is used to reassemble the original transmitted data. The original baseband

Fig. 11-33 **The GCE for receiving.**

digital signals may be developed in some cases or in others these signals are used with modems as required for interconnecting the earth station with the computer that will process the data.

The up link is the transmitting subsystem of the earth station. It consists of all the electronic equipment that takes the signal to be transmitted, amplifies it, and sends it to the antenna. In a communications system, the signals to be sent to the satellite might be TV programs, multiple telephone calls, or digital data from a computer. These signals are used to modulate the carrier which is then amplified by a large TWT or klystron amplifier. Such amplifiers generally generate many hundreds of watts of output power. This is sent to the an-

tenna by way of microwave waveguides, combiners, and diplexers.

The transmit subsystem consists of two basic parts: the up converters and the power amplifiers. The up converters translate the baseband signals modulated onto carriers up to the final up-link microwave frequencies. The power amplifiers generate the high-power signals that are applied to the antenna. The modulated carriers are created in the transmit GCE.

The transmit subsystem begins with the baseband signals. These are first fed to a multiplexer if multiple signals are to be carried by a single transponder. Telephone calls are a good example. Frequency or time division multiplexers are used to assemble the composite

Fig. 11-34   Typical GCE transmit circuits.

signal. The multiplexer output is then fed to a modulator. In analog systems, a wideband frequency modulator is normally used. It operates at a carrier frequency of 70 MHz with a maximum deviation of ± 18 MHz. Video signals are fed directly to the modulator; they are not multiplexed.

In digital systems, analog signals are first digitized with PCM converters. The resulting serial digital output is then used to modulate a QPSK modulator. The carrier frequency for these modulators is usually 140 MHz. Figure 11-34 shows the general transmit GCE configuration.

Once the modulated IF signals have been generated, up conversion and amplification will take place prior to transmission. Individual up converters are connected to each modulator output. Each up converter is driven by a frequency synthesizer which allows selection of the final transmitting frequency. The frequency synthesizer selects the transponder it will use in the satellite. The synthesizers are ordinarily adjustable in 1-kHz increments so that any up converter can be set to any channel frequency or transponder.

As in down converters, most modern systems use dual conversion. Both RF and IF tuning are used. Refer to Fig. 11-35. With IF tuning, a tunable carrier from a frequency synthesizer is applied to the mixer to convert the modulated signal to an IF level, which is usually 700 MHz. Another mixer fed by a fixed-frequency LO then performs the final up conversion to the transmitted frequency.

In RF tuning, a mixer fed by a fixed-frequency LO performs an initial up conversion to 700 MHz. Then, a sophisticated RF synthesizer applied to a second mixer provides up conversion to the final microwave frequency.

In some systems, all the IF signals at 70 or 140 MHz are combined prior to the up conversion. In this system, different carrier frequencies are used on each of the modulators to provide the desired channelization. The carrier frequencies when translated by the up converter will translate to the individual transponder center frequencies. A special IF combiner circuit mixes all the signals linearly and applies them to a single up converter. This up converter creates the final microwave signal.

In most systems, however, individual up converters are used on each modulated channel. At the output of the up converters, all the signals are combined in a microwave combiner which produces a single-output signal which will be fed to the final amplifiers. This arrangement is illustrated in Fig. 11-36 on the next page.

The final combined signal to be transmitted to the satellite appears at the output of the RF combiner, but it must first be amplified considerably before being sent to the antenna. This is done by the power amplifier. The power amplifier usually includes an initial stage called the intermediate power amplifier (IPA). This provides sufficient drive to the final, the HPA. See Fig. 11-36. Note also in Fig. 11-36 that redundant amplifiers are used. The main IPA and

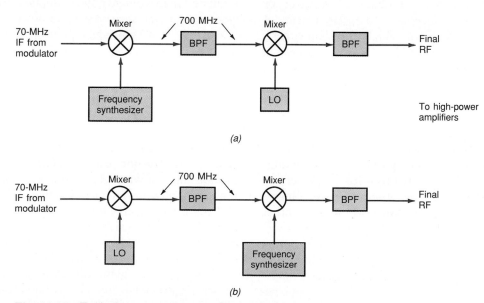

(a)

(b)

Fig. 11-35 **Typical up-converter circuits.** (*a*) **IF tuning, and** (*b*) **RF tuning.**

Fig. 11-36   An RF combiner and power amplifiers.

HPA are used until a failure occurs. Switches (SW) automatically disconnect the defective main amplifier and connect the spare to ensure continuous operation. The amplified signal is then sent to the antenna via a waveguide, the diplexer, and a filter.

There are three types of power amplifiers used in earth stations: transistors, TWTs, and klystrons. Let's take a brief look at each. Transistor power amplifiers are used in small- and medium-size earth stations with low power. Powers up to 10 W are common. Normally, power GaAs FETs are used in this application. In some instances, a chain of varactor or IMPATT diodes are used to create a string of frequency multipliers which translate an RF power signal generated by a bipolar transistor at a lower frequency to the desired final operating frequency. Again, powers of up to 10 W are achievable with these systems. The newer GaAs MESFET power devices when operated in parallel can achieve powers up to approximately 40 W. Further improvements are expected.

Most medium- and high-power earth stations use either TWTs or klystrons for the power amplifiers. There are two typical power ranges, one in the 200- to 400-W range and the other in the 2- to 3-kW range. The amount of power used depends upon the location of the station and its antenna size. Satellite transponder characteristics also influence the power required by the earth station. As improvements have been made in LNAs and satellites have been able to carry higher-power transponders, earth station transmitter power requirements have greatly decreased. This is also true for antenna sizes.

The majority of medium- and high-power earth stations use TWTs for power amplification. Traveling-wave tubes are available in sizes from a few watts to 20,000 W. Their greatest virtue is very wide bandwidth. A single TWT can transmit an entire 500-MHz bandwidth satellite signal.

On the down side, TWTs are extremely expensive. Further, they are very complex in operation. They require numerous expensive, regulated high-voltage power supplies. Finally, when they reach their maximum power saturation levels, TWTs produce considerable intermodulation products, spurious signals which affect the reliability of communications. When operating near maximum power, it is sometimes necessary to back off on input signal power to reduce the intermodulation signals.

The alternative to TWTs is klystrons. Klystrons are also available in a wide range of power sizes, from a fraction of a watt all the way to tens of thousands of watts. Klystrons are simpler in their construction and operation than TWTs and are, therefore, usually lower in cost. Less complex and less expensive power supplies are also required. Further, klystrons provide better intermodulation performance than TWTs.

However, the main disadvantage is that klystrons are basically narrowband devices. The maximum bandwidth for most high-power klystrons is approximately 150 MHz. This makes them suitable for transmitting only three adjacent 36-MHz-bandwidth channels. More klystrons must be used, therefore, to transmit all the desired signals. In some systems, narrowband klystrons are used one per channel. In such cases where multiple klystrons are used, the outputs are all combined in a single RF combiner to create a single signal that is fed to the antenna.

Regardless of whether klystrons or TWTs are used, if power levels in excess of 2 kW are required, liquid-cooling systems must typically be used. These are larger, more expensive devices and require the necessary hardware and plumbing to provide the liquid cooling.

The final HPA, whether it uses klystrons or TWTs, is usually driven by an IPA. This amplifier takes the low-level up-converted signals and amplifies them to a level capable of driving the final HPA devices. This IPA is typically a smaller TWT or klystron. The total power gain of the IPA and HPA combined is usually in the 60- to 70-dB range.

The output of the HPA located after any combiner is usually passed through a final filter to remove any undesirable harmonics and to provide final bandwidth shaping. The signal is then fed through a diplexer and the waveguide to the feed horns.

Most earth stations receive their power from normal ac sources. Standard power supplies convert the ac to the dc voltages required to operate all subsystems. However, most earth stations have backup power systems. Satellite systems, particularly those used for reliable communications of telephone conversations, TV programs, computer data, etc., must not "go down." The backup power system, which may consist of a diesel engine driving an ac generator, takes over if there is an ac power failure. When ac power fails, an automatic system starts the diesel engine. The generator creates the equivalent ac power which is automatically switched to the system. Smaller systems may use uninterruptable power supplies (UPS) which derive their main power from batteries. Large battery arrays drive dc-to-ac inverters which produce the ac voltages for the system. Uninterruptible power supplies are not suitable for long power failures and interruptions because the batteries quickly become exhausted. However, for short interruptions of power, less than an hour, they are adequate.

The telemetry equipment consists of a receiver and the recorders and indicators that display the telemetry signals. The signal may be received by the main antenna or by a separate telemetry antenna. A separate receiver on a frequency different from the communications channels is used for telemetry purposes. The telemetry signals from the various sensors and transducers in the satellite are multiplexed onto a single carrier and are sent to the earth station. The earth station receiver demodulates and demultiplexes the telemetry signals into the individual outputs. These are then recorded and are also sent to various indicators, such as strip chart recorders, meters, and digital displays. Signals may be in digital form or converted to digital. They can also be sent to a computer where they can be further processed and stored. Mathematical operations on the signals may be necessary. For example, the number received from the satellite representing temperature may have to be processed by some mathematical formula to obtain the actual output temperature in degrees Celsius or Fahrenheit. The computer can also generate and format printed reports as required.

The control subsystem permits the ground station to control the satellite. This system usually contains a computer for entering the commands which modulate a carrier that is amplified and fed to the main antenna. The command signals can make adjustments in the satellite attitude, turn transponders off and on, actuate redundant circuits if the circuits they are backing up fail, and so on.

In some satellite systems where communications is not the main function, some instrumentation may be a part of the ground station. Instrumentation is a general term for all the electronic equipment that is used to deal with the information transmitted back to the earth station. It may consist of demodulators and demultiplexers, amplifiers, filters, A/D converters, or signal processors. The instrumentation subsystem is in effect an extension of the telemetry system. Besides relaying information about the satellite itself, the telemetry system may also be used to send back information related to various scientific experiments being conducted on the satellite. In satellites used for surveillance, the instrumentation may be such that it can deal with television signals sent back from an on-board TV camera. The possibilities are extensive, depending upon the actual satellite mission.

At the ground station, additional equipment may be required to process this information. The ground station equipment almost always includes tape recorders for recording the information, and the various pieces of equipment used to display it. Again, a computer may be part of the instrumentation for analyzing, processing, and reformatting the data.

Not all earth stations have the telemetry, control, and instrumentation subsystems; only those designated as control stations do. Most earth stations just send or receive.

Self-Test

Answer the following questions.

42. The signals to be communicated by the earth station to the satellite are known as _____ signals.

43. The receive or up-link subsystem uses a _____ to translate the satellite signal to an IF before demodulation and demultiplexing.

44. A common IF in an earth station receiver is _____ MHz.

45. The functions performed by the receive GCE section are _____ .

46. The functions performed by the transmit GCE section are _____ .

47. The two types of modulation used in the transmit GCE are _____ .

48. A(n) _____ in the transmit subsection translates the modulated signal to the final transmission frequency.

49. _____ are often used to replace LOs for channel selection in earth station transmitters and receivers.

50. List the three main types of power amplifiers used in earth stations.

51. Most earth stations operate from the ac power line but have built-in _____ for emergency operation.

52. True or false. All earth stations contain telemetry, control, and instrumentation subsystems.

11-5 APPLICATIONS OVERVIEW

Every satellite is designed to perform some specific task. Its predetermined application specifies the kind of equipment it must have on board and its orbit. Although our emphasis on satellites in this chapter is communications, satellites are useful for many other purposes. We have already mentioned some of them.

The main applications for satellites today is communications. Satellites used for this purpose act as relay stations in the sky. They permit reliable long-distance communications worldwide, and they solve many of the grow-

ing communications needs of industry and government. Communications applications will continue to dominate this industry.

Another application of satellites is surveillance or observation. From their vantage point high in the sky, satellites can look at the earth and transmit what they see back to ground stations for a wide variety of purposes. For example, military satellites are used to perform reconnaissance. On-board film cameras take photographs which can later be ejected from the satellite and brought back to earth for recovery. Television cameras can take pictures and send them back to earth as electrical signals. Infrared sensors detect heat sources, and small radars can profile earth features.

Intelligence satellites collect information about enemies and potential enemies. They permit monitoring for the purpose of proving compliance with nuclear test ban and missile treaties and agreed-upon conditions with other countries.

There are many different kinds of observation satellites. One special type is the meteorological or weather satellite. These satellites photograph cloud cover, and the pictures which are sent back to earth are used for determining and predicting the weather. Geodetic satellites photograph the earth for the purpose of creating more-accurate and more-detailed maps.

Satellites can also monitor the status of the earth's resources such as our land and oceans. They can observe crops, forests, lakes, and rivers. Satellites can spot diseased crop areas, mineral resources, sources of pollution, and other characteristics that are difficult, if not impossible, to detect in other ways.

One of the newest and most useful satellite systems is the Global Positioning System (GPS). Its primary application is navigation. The system is a network of 24 low-earth-orbit satellites spaced equally around the world in overlapping patterns. Each satellite transmits a unique signal back to earth on low microwave frequencies such as 1.57542 and 1.2276 GHz. Receivers on the earth pick up transmissions from four satellites simultaneously. The receiver uses the signals in a microprocessor to compute the exact position of the receiver on earth. The receiver output is a display giving the latitude, longitude, and altitude of the receiver.

GPS was designed primarily as a navigation system for the military. It can also be used by individuals and industry for general navigational purposes. The military system has more accuracy than the civilian system. The military system error is less than 40 m whereas the civilian system accuracy is less than 100 m from the actual location. Such accuracies are more than suitable in most air and sea navigation situations. However, greater accuracy may be required in land applications. Regardless of the accuracy, the real value of the system is that you can know your position on earth at any time by using a low-cost hand-held microwave GPS receiver.

Eighteen of the satellites are in orbit as of this writing, but all 24 are expected to be operational by the end of 1994. As use increases, the cost of the receivers is expected to decline, making them easily affordable by everyone.

Each satellite is maintained in a *very* accurate orbit. Each satellite has a very accurate atomic clock. Time information is encoded on the transmitted signal. The receiver will receive a unique time reference from each satellite. The times are unique because of the different propagation times. For example, path 1 in Fig. 11-37 is the longest, and path 3 is the shortest. Signal 1 will be delayed the most. Since the velocity of propagation is known, the relative path lengths can be computed in the receiver microprocessor. Three (or more) simultaneous equations are then solved by the microprocessor to determine where the receiver is located.

For years, TV signals have been transmitted through satellites for redistribution. Because of the very high frequency signals involved in TV transmission, the standard telephone system cannot be used. Special coax as well as microwave relay links have been used to transmit TV signals from one place to another. However, with today's communications satellites, TV signals can be transmitted easily from one place to another. All the major TV networks and cable TV companies rely on communications satellites for TV signal distribution.

Consumers have discovered that with an appropriate home satellite antenna and receiver, they may pick up these "free" TV signals. Satellites carrying TV broadcasts have been identified, and an entire industry has grown up around receiving these network distribution TV signals at home. At the time this book is being written, there are over one million such home satellite TV receivers in use. Those who have them enjoy the broadest range of TV possible. These signals contain virtually everything that exists on the standard TV channels, much of which never reaches the public. The programs being transmitted are often uncut and unedited.

Self-Test

Answer the following questions.

53. The main use of satellites is for _____ .

54. Name four types of surveillance sensors.

55. What are the name and the purpose of the 24-satellite system developed by the military?

56. Name three uses for surveillance satellites.

57. Consumers use satellites for _____ .

Time codes leave all three satellites at the same time

Satellite

Satellite

Satellite

Path 2

Path 1
Longest propagation time

Shortest propagation time
Path 3

Receiver

Fig. 11-37 Satellite identification information relates each satellite location with respect to points on the earth's surface.

1. A satellite is a physical object that orbits a celestial body.

2. A communications satellite containing electronic equipment acts as a repeater or relay station between two earth stations.

3. The basic component of a communications satellite is a receiver-transmitter combination called a transponder.

4. A satellite stays in orbit because the gravitational pull of the earth is balanced by the centripetal force of the revolving satellite.

5. Satellite orbits about the earth are either circular or elliptical.

6. Satellites orbit the earth from heights of 100 to 22,300 mi and travel at speeds of 6800 to 17,500 mi/h.

7. A satellite that orbits directly over the equator 22,300 mi from earth is said to be in a geostationary orbit. It revolves in synchronism with the earth's rotation, so it appears to be stationary when seen from points on the earth.

8. A satellite is stabilized in orbit by spinning it on its axis or building in spinning flywheels for each major axis (roll, pitch, yaw).

9. Attitude adjustments on a satellite are made by firing small jet thrusters to change the satellite's position or speed.

10. The location of a satellite is determined with latitude and longitude measurements that designate a point on the earth [subsatellite point (SSP)] directly below the satellite.

11. Azimuth and elevation angles determine where to point an earth station so that it intercepts the satellite.

12. Satellites are launched into orbit by rockets that give them vertical as well as forward motion.

13. A geosynchronous satellite is initially put into an elliptical orbit where its apogee is 22,300 mi high. The apogee kick motor is then fired to put the satellite into its final circular geostationary orbit.

14. Many satellites are put into orbit by launching them from NASA's space shuttle.

15. Most satellites operate in the microwave region.

16. Microwave satellites operate on assigned frequency bands designated by a letter. Common communications satellite bands are the C (3.4 to 6.425 GHz) and Ku (10.95 to 14.5 GHz) bands.

17. Satellite bands are typically 500 MHz wide and are divided into 12 segments, each 36 MHz wide. A transponder is used to cover each segment.

18. Frequency reuse is a technique that allows two sets of transponders to operate on the same frequency, thus doubling channel capacity. The two sets of channels use antennas of different polarizations to prevent interference with one another.

19. Spatial isolation is another technique for frequency sharing. It uses highly directional spot-beam antennas to prevent interference between stations on the same frequency.

20. The main subsystems in a satellite are the communications; power; telemetry, tracking, and control (TTC); propulsion; attitude stabilization; and antenna subsystems.

21. A transponder consists of a low-noise amplifier (LNA) that receives and amplifies the up-link signal, a mixer that converts the signal to another (lower) frequency, and a high-power amplifier that retransmits the signal on its new down-link frequency.

22. Double-conversion transponders use two mixers, one to translate the up-link signal to an IF where it is amplified and filtered, and another to translate the signal to its final down-link frequency.

23. Regenerative transponders demodulate the up-link signal to recover the baseband signals and then use them to remodulate a down-link transmitter. This improves the S/N ratio.

24. In a broadband transponder, a single mixer converts all channels within the 500-MHz bandwidth simultaneously to their down-link frequencies. These are selected by channel bandpass filters and then amplified by individual power amplifiers.

25. In a channelized transponder, each channel has its own LNA, bandpass filters, mixer, and high-power amplifier.

26. The power subsystem consists of solar panels, batteries, dc-to-dc converters, and regulators. The solar panels convert sunlight into dc power to operate all satellite electronics and to charge the batteries that take over when sunlight is blocked.

27. The TTC subsystem contains a receiver that picks up commands from a ground station and translates them into control signals that initiate some action on board. The

telemetry system monitors physical conditions within the satellites and converts them into electrical signals that are transmitted back to earth.

28. The propulsion system consists of the apogee kick motor that puts the satellite into final orbit and the jet thrusters that are used for positioning and attitude control.

29. The stabilization subsystem for attitude control consists of spin components or three-axis flywheel gyros.

30. The antenna system consists of one or more highly directional horn or parabolic antennas and an omnidirectional TTC antenna.

31. Earth stations consist of transmit, receive, power, antenna, TTC, and ground control equipment (GCE) subsystems.

32. The transmit subsystem takes the baseband voice, video, or computer data signals; multiplexes them; and uses the composite signal to modulate a carrier. An up converter translates the signal to its final up-link frequency before it is amplified and transmitted.

33. The most common forms of modulation used are FM and QPSK.

34. Transistor power amplifiers are used in low-power earth stations; klystrons are used in high-power narrowband stations; and TWTs are used in high-power broadband stations.

35. Earth stations feature large parabolic dish antennas with high gain and directivity for receiving the weak satellite signal.

36. The receive subsystem in an earth station amplifies the signal with an LNA and then separates the channels with bandpass filters. Down converters translate the signals to a lower IF where they are demodulated and demultiplexed.

37. The GCE in an earth station interfaces the baseband signals to the transmit and receive subsystems. The receive GCE performs demodulation and demultiplexing. The transmit GCE performs modulation and multiplexing.

38. The most common application for satellites is communications.

39. Another major use of satellites is surveillance and reconnaissance.

40. Film cameras, TV cameras, infrared sensors, and radars are all used to observe a variety of conditions on earth from surveillance satellites.

41. Satellites play a major role in military and defense systems not only for communications but also for surveillance.

42. The 24-satellite Global Positioning System (GPS) makes accurate navigation possible anywhere on earth with a low-cost microwave receiver.

43. Consumers use satellite TV receivers to intercept TV signals transmitted by networks and cable TV companies.

CHAPTER REVIEW QUESTIONS

Choose the letter which best answers each question.

11-1. As the height of a satellite orbit gets lower, the speed of the satellite
  a. Increases.
  b. Decreases.
  c. Remains the same.
  d. None of the above.

11-2. The main function of a communications satellite is as a(n)
  a. Repeater.
  b. Reflector.
  c. Beacon.
  d. Observation platform.

11-3. The key electronic component in a communications satellite is the
  a. Telemetry.
  b. On-board computer.
  c. Command and control system.
  d. Transponder.

11-4. A circular orbit around the equator with a 24-h period is called a(n)
  a. Elliptical orbit.
  b. Geostationary orbit.
  c. Polar orbit.
  d. Transfer orbit.

11-5. A satellite stays in orbit because the following two factors are balanced.
    a. Satellite weight and speed
    b. Gravitational pull and inertia
    c. Centripetal force and speed
    d. Satellite weight and the pull of the moon and sun

11-6. The height of a satellite in a synchronous equatorial orbit is
    a. 100 mi.   *c.* 22,300 mi.
    b. 6800 mi.   *d.* 35,860 mi.

11-7. Most satellites operate in which frequency band?
    a. 30 to 300 MHz   *c.* 3 GHz to 30 GHz
    b. 300 MHz to 3 GHz   *d.* Above 300 GHz

11-8. The main power sources for a satellite are
    a. Batteries.
    b. Solar cells.
    c. Fuel cells.
    d. Thermoelectric generators.

11-9. The maximum height of an elliptical orbit is called the
    a. Perigee.   *c.* Zenith.
    b. Apex.   *d.* Apogee.

11-10. Batteries are used to power all satellite subsystems
    a. At all times.
    b. Only during emergencies.
    c. During eclipse periods.
    d. To give the solar arrays a rest.

11-11. The satellite subsystem that monitors and controls the satellite is the
    a. Propulsion subsystem.
    b. Power subsystem.
    c. Communications subsystem.
    d. Telemetry, tracking, and command subsystem.

11-12. The basic technique used to stabilize a satellite is
    a. Gravity-forward motion balance.
    b. Spin.
    c. Thruster control.
    d. Solar panel orientation.

11-13. The jet thrusters are usually fired to
    a. Maintain attitude.
    b. Put the satellite into the transfer orbit.
    c. Inject the satellite into the geosynchronous orbit.
    d. Bring the satellite back to earth.

11-14. Most commercial satellite activity occurs in which band(s)?
    a. L   *c.* X
    b. C and Ku   *d.* S and P

11-15. How can multiple earth stations share a satellite on the same frequencies?
    a. Frequency reuse   *c.* Mixing
    b. Multiplexing   *d.* They can't

11-16. The typical bandwidth of a satellite band is
    a. 36 MHz.   *c.* 70 MHz.
    b. 40 MHz.   *d.* 500 MHz.

11-17. Which of the following is *not* usually a part of a transponder?
    a. LNA   *c.* Modulator
    b. Mixer   *d.* HPA

11-18. The satellite communications channels in a transponder are defined by the
    a. LNA.   *c.* Mixer.
    b. Bandpass filter.   *d.* Input signals.

11-19. The HPAs in most satellites are
    a. TSTs.
    b. Klystrons.
    c. Vacuum tubes.
    d. Magnetrons.
11-20. The physical location of a satellite is determined by its
    a. Distance from the earth.
    b. Latitude and longitude.
    c. Reference to the stars.
    d. Position relative to the sun.
11-21. The receive GCE system in an earth station performs what function(s)?
    a. Modulation and multiplexing
    b. Up conversion
    c. Demodulation and demultiplexing
    d. Down conversion
11-22. Which of the following types of HPA is *not* used in earth stations?
    a. TWT
    b. Transistor
    c. Klystron
    d. Magnetron
11-23. A common up-converter and down-converter IF is
    a. 36 MHz.
    b. 40 MHz.
    c. 70 MHz.
    d. 500 MHz.
11-24. The type of modulation used on voice and video signals is
    a. AM.
    b. FM.
    c. SSB.
    d. QPSK.
11-25. The modulation normally used with digital data is
    a. AM.
    b. FM.
    c. SSB.
    d. QPSK.
11-26. Which of the following is *not* a typical output from a GPS receiver?
    a. Latitude
    b. Speed
    c. Altitude
    d. Longitude

CRITICAL THINKING QUESTIONS

11-1. What do you think are the primary factors that determine a satellite's useful life in space?
11-2. If you want an earth station to receive another satellite, what adjustments do you have to make and on what components?
11-3. What factors do you think determine an earth station's ability to select and distinguish one satellite from another closely adjacent satellite in orbit?
11-4. List as many civilian applications as you can think of for the Global Positioning System.

Answers to Self-Tests

1. centripetal force, gravitational pull
2. posigrade
3. ellipse
4. geocenter
5. period
6. 0, 90
7. 5
8. geosynchronous or geostationary
9. attitude
10. spin, three-axis stabilization
11. yaw, pitch, roll
12. subsatellite point
13. latitude, longitude
14. azimuth, elevation
15. apogee kick motor
16. communications
17. relay station
18. earth station
19. down link
20. transponder
21. microwave
22. C
23. X
24. 10.95, 14.5
25. 500
26. 12, 36
27. frequency reuse
28. spatial isolation
29. voice, video, computer or digital data
30. low-noise amplifier
31. signal amplification, frequency translation, power amplification
32. mixer
33. TWT
34. double-conversion
35. regenerative
36. bandpass filter
37. broadband, multichannel
38. solar panels
39. batteries
40. monitor, control
41. jet thrusters
42. baseband
43. down converter
44. 70
45. demodulation, demultiplexing
46. modulation, multiplexing
47. FM, QPSK
48. up converter
49. frequency synthesizers
50. transistor, TWT, klystron
51. backup power or uninterruptible power supply
52. false
53. communications
54. film cameras, TV cameras, infrared sensors, radar
55. Global Positioning System (GPS), navigation
56. weather forecasting, spying, earth resources, monitoring
57. TV reception

CHAPTER 12

Data Communications

CHAPTER OBJECTIVES

This chapter will help you to:

1. *Explain* the concepts of communications by serial binary data and *define* the terms ASCII, synchronous, asynchronous, and baud rate.
2. *State* the relationship between communications channel bandwidth and data rate in bits per second and *compute* one given the other.
3. *Define* the term modem and *explain* its operation and application.
4. *Explain* the processes of modulation and demodulation with frequency-shift keying (FSK) and phase-shift keying (PSK).
5. *Explain* the need for and types of communications protocols and error-detection and error-correction schemes.
6. *Name* the three basic configurations of local networks and *discuss* the pros and cons of each.
7. *Explain* the operation and benefits of frequency hopping and direct-sequence spread spectrum systems.

Data communications is the transmission of binary or digital information from one point to another. Data communications systems permit the transfer of information between computers and also permit the remote operation of a computer from a terminal. Further, since any type of signal can be digitized, the transmission of voice, video, or other traditionally analog information in binary form is also known as data communications. Because of the low cost and effectiveness of digital techniques, data communications is becoming more widely used. This chapter introduces the basic concepts and equipment used in data communications.

12-1 DIGITAL COMMUNICATIONS CONCEPTS

Digital signals are binary pulses that have two distinct states, each represented by a voltage level. The pulses switch rapidly between these two levels. One level is referred to as a binary 0 or low, and the other is referred to as a binary 1 or high. Figure 12-1 shows a typical binary or digital signal. The binary 0 level might be 0 V or ground, and the binary 1 level might be +5 V. Any other two voltages can also be used.

Binary signals are easy to generate and pro-

cess with electronic circuits. Binary pulses are usually produced by and manipulated with high-speed transistor switches. These switches are connected in a variety of configurations to form digital logic circuits. These digital logic circuits are available in IC form and can generate and process digital data at high speeds.

Binary 0 = 0 V (ground)
Binary 1 = +5 V

Fig. 12-1 **A binary signal.**

Usually, the binary signals represent data or information. Specifically, the binary signals are codes made up of groups or patterns of 0s and 1s. Each pattern represents a numerical value, a letter of the alphabet, or some special symbol, meaning, or message. The digital logic circuits process this data in various ways.

The earliest forms of data communications were binary in nature. The very first type of electronic communications in use was the telegraph, which was invented in 1832 and first demonstrated successfully in 1844 by Samuel F. B. Morse. A telegraph hand key acting as a switch connects to a remote receiver or sounder. The sounder is nothing more than a magnetic coil which attracts an armature that makes a clicking sound. Whenever the telegraph key is depressed, current flows through the coil producing a magnetic field that attracts the armature and makes a click. When the key is opened, the armature makes another click. The operation is strictly binary since there are only two states: off and on. For two-way communications, both stations have a key and a sounder so that either station may transmit or receive.

Morse created a special code using this off/on capability. Known as the Morse code, it is a series of dots and dashes that represent letters of the alphabet, numbers, and punctuation marks. These are summarized in Fig. 12-2. When the armature is closed for a short duration, a dot is produced. When the armature is closed for a longer time, a dash is produced. With special training, humans can easily send and receive messages at high speeds. Speeds of 15 to 20 words per minute are easily mastered with the upper limit being 70 to 80 words per minute.

In the earliest radio communications the Morse code was used to send messages. A key turned the carrier of a transmitter off and on to produce the dots and dashes. These were detected at the receiver and converted by an operator back into letters and numbers making up the message. This type of radio communications is known as continuous wave (CW).

Although telegraph and CW communications are still used, for the most part they have been replaced by voice communications. However, the communications industry has rediscovered the two-state or binary communications, and today more and more communications are taking place by that method. Even voice, video, and other analog signals are being converted into binary codes that can be transmitted by data communications techniques by using A/D and D/A conversion techniques.

One of the first binary codes was the Baudot code used in the early teletype machines (the Baudot code is shown in Fig. 12-3). A teletype machine is a typewriterlike device that is used to send and receive coded signals over a communications link. Whenever a key on the typewriter keyboard is pressed, a unique code is generated and transmitted to the receiving machine which recognizes and then prints the corresponding letter, number, or symbol. Teletype machines were designed to eliminate

Fig. 12-2 The Morse telegraph code. A dot (·) is a short click; a dash (—) is a long click.

Character Shift		Binary Code				
Letter	Figure	Bit: 4	3	2	1	0
A	—	1	1	0	0	0
B	?	1	0	0	1	1
C	:	0	1	1	1	0
D	$	1	0	0	1	0
E	3	1	0	0	0	0
F	!	1	0	1	1	0
G	&	0	1	0	1	1
H	#	0	0	1	0	1
I	8	0	1	1	0	0
J	'	1	1	0	1	0
K	(	1	1	1	1	0
L	)	0	1	0	0	1
M	.	0	0	1	1	1
N	,	0	0	1	1	0
O	9	0	0	0	1	1
P	0	0	1	1	0	1
Q	1	1	1	1	0	1
R	4	0	1	0	1	0
S	bel	1	0	1	0	0
T	5	0	0	0	0	1
U	7	1	1	1	0	0
V	;	0	1	1	1	1
W	2	1	1	0	0	1
X	/	1	0	1	1	1
Y	6	1	0	1	0	1
Z	"	1	0	0	0	1
Figure shift		1	1	1	1	1
Letter shift		1	1	0	1	1
Space		0	0	1	0	0
Line feed (LF)		0	1	0	0	0
Blank (null)		0	0	0	0	0

Fig. 12-3 **The Baudot code.**

the need for telegraph operators to learn the Morse code. Anyone who could type could send messages with a teletype machine.

The binary code used in teletype machines is a 5-bit code called the Baudot (pronounced "baw dough"). With 5 bits, $2^5 = 2 \times 2 \times 2 \times 2 \times 2 = 32$ different symbols can be represented. This, of course, is not enough to represent the 26 letters of the alphabet, 10 numbers, and various punctuation marks. The Baudot code gets around this by using two "shift codes." Refer to Fig. 12-3 and note the figure-shift and letter-shift codes. If the message is preceded by the letter-shift code (11011), all the following codes are interpreted as letters of the alphabet. Sending the figure-shift code (11111) causes all the following characters to be interpreted as numbers or punctuation marks. The Baudot code is still used in telegraph communications, but in most data communications, it has been replaced by codes that can represent more characters and symbols.

The most widely used data communications code is the American Standard Code for Information Interchange, abbreviated ASCII (pronounced "ass key"). This is a 7-bit binary code used for representing 128 numbers, letters, punctuation marks, and other symbols. Its big advantage over the Baudot code is that a sufficient number of code combinations is available to represent both upper- and lowercase letters of the alphabet. See Fig. 12-4.

Another code similar to ASCII is the Ex-

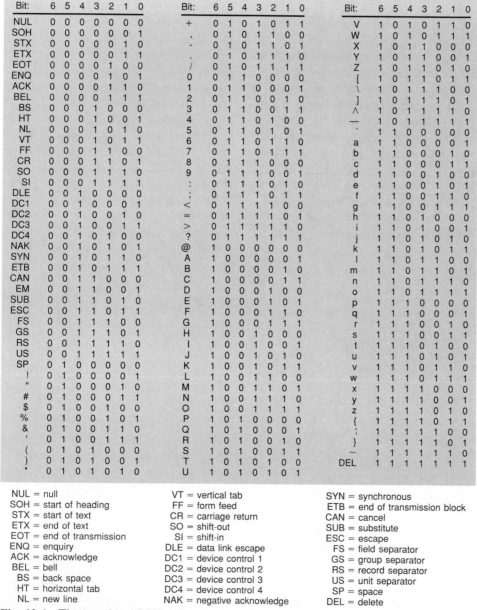

Bit:	6	5	4	3	2	1	0
NUL	0	0	0	0	0	0	0
SOH	0	0	0	0	0	0	1
STX	0	0	0	0	0	1	0
ETX	0	0	0	0	0	1	1
EOT	0	0	0	0	1	0	0
ENQ	0	0	0	0	1	0	1
ACK	0	0	0	0	1	1	0
BEL	0	0	0	0	1	1	1
BS	0	0	0	1	0	0	0
HT	0	0	0	1	0	0	1
NL	0	0	0	1	0	1	0
VT	0	0	0	1	0	1	1
FF	0	0	0	1	1	0	0
CR	0	0	0	1	1	0	1
SO	0	0	0	1	1	1	0
SI	0	0	0	1	1	1	1
DLE	0	0	1	0	0	0	0
DC1	0	0	1	0	0	0	1
DC2	0	0	1	0	0	1	0
DC3	0	0	1	0	0	1	1
DC4	0	0	1	0	1	0	0
NAK	0	0	1	0	1	0	1
SYN	0	0	1	0	1	1	0
ETB	0	0	1	0	1	1	1
CAN	0	0	1	1	0	0	0
EM	0	0	1	1	0	0	1
SUB	0	0	1	1	0	1	0
ESC	0	0	1	1	0	1	1
FS	0	0	1	1	1	0	0
GS	0	0	1	1	1	0	1
RS	0	0	1	1	1	1	0
US	0	0	1	1	1	1	1
SP	0	1	0	0	0	0	0
!	0	1	0	0	0	0	1
"	0	1	0	0	0	1	0
#	0	1	0	0	0	1	1
$	0	1	0	0	1	0	0
%	0	1	0	0	1	0	1
&	0	1	0	0	1	1	0
'	0	1	0	0	1	1	1
(	0	1	0	1	0	0	0
)	0	1	0	1	0	0	1
*	0	1	0	1	0	1	0

Bit:	6	5	4	3	2	1	0
+	0	1	0	1	0	1	1
,	0	1	0	1	1	0	0
-	0	1	0	1	1	0	1
.	0	1	0	1	1	1	0
/	0	1	0	1	1	1	1
0	0	1	1	0	0	0	0
1	0	1	1	0	0	0	1
2	0	1	1	0	0	1	0
3	0	1	1	0	0	1	1
4	0	1	1	0	1	0	0
5	0	1	1	0	1	0	1
6	0	1	1	0	1	1	0
7	0	1	1	0	1	1	1
8	0	1	1	1	0	0	0
9	0	1	1	1	0	0	1
:	0	1	1	1	0	1	0
;	0	1	1	1	0	1	1
<	0	1	1	1	1	0	0
=	0	1	1	1	1	0	1
>	0	1	1	1	1	1	0
?	0	1	1	1	1	1	1
@	1	0	0	0	0	0	0
A	1	0	0	0	0	0	1
B	1	0	0	0	0	1	0
C	1	0	0	0	0	1	1
D	1	0	0	0	1	0	0
E	1	0	0	0	1	0	1
F	1	0	0	0	1	1	0
G	1	0	0	0	1	1	1
H	1	0	0	1	0	0	0
I	1	0	0	1	0	0	1
J	1	0	0	1	0	1	0
K	1	0	0	1	0	1	1
L	1	0	0	1	1	0	0
M	1	0	0	1	1	0	1
N	1	0	0	1	1	1	0
O	1	0	0	1	1	1	1
P	1	0	1	0	0	0	0
Q	1	0	1	0	0	0	1
R	1	0	1	0	0	1	0
S	1	0	1	0	0	1	1
T	1	0	1	0	1	0	0
U	1	0	1	0	1	0	1

Bit:	6	5	4	3	2	1	0
V	1	0	1	0	1	1	0
W	1	0	1	0	1	1	1
X	1	0	1	1	0	0	0
Y	1	0	1	1	0	0	1
Z	1	0	1	1	0	1	0
[	1	0	1	1	0	1	1
\	1	0	1	1	1	0	0
]	1	0	1	1	1	0	1
^	1	0	1	1	1	1	0
—	1	0	1	1	1	1	1
`	1	1	0	0	0	0	0
a	1	1	0	0	0	0	1
b	1	1	0	0	0	1	0
c	1	1	0	0	0	1	1
d	1	1	0	0	1	0	0
e	1	1	0	0	1	0	1
f	1	1	0	0	1	1	0
g	1	1	0	0	1	1	1
h	1	1	0	1	0	0	0
i	1	1	0	1	0	0	1
j	1	1	0	1	0	1	0
k	1	1	0	1	0	1	1
l	1	1	0	1	1	0	0
m	1	1	0	1	1	0	1
n	1	1	0	1	1	1	0
o	1	1	0	1	1	1	1
p	1	1	1	0	0	0	0
q	1	1	1	0	0	0	1
r	1	1	1	0	0	1	0
s	1	1	1	0	0	1	1
t	1	1	1	0	1	0	0
u	1	1	1	0	1	0	1
v	1	1	1	0	1	1	0
w	1	1	1	0	1	1	1
x	1	1	1	1	0	0	0
y	1	1	1	1	0	0	1
z	1	1	1	1	0	1	0
{	1	1	1	1	0	1	1
|	1	1	1	1	1	0	0
}	1	1	1	1	1	0	1
~	1	1	1	1	1	1	0
DEL	1	1	1	1	1	1	1

NUL = null
SOH = start of heading
STX = start of text
ETX = end of text
EOT = end of transmission
ENQ = enquiry
ACK = acknowledge
BEL = bell
BS = back space
HT = horizontal tab
NL = new line

VT = vertical tab
FF = form feed
CR = carriage return
SO = shift-out
SI = shift-in
DLE = data link escape
DC1 = device control 1
DC2 = device control 2
DC3 = device control 3
DC4 = device control 4
NAK = negative acknowledge

SYN = synchronous
ETB = end of transmission block
CAN = cancel
SUB = substitute
ESC = escape
FS = field separator
GS = group separator
RS = record separator
US = unit separator
SP = space
DEL = delete

Fig. 12-4 The popular ASCII code.

tended Binary Coded Decimal Interchange Code, abbreviated EBCDIC (pronounced "ebb see dick") developed by IBM. It is an 8-bit code allowing a maximum of 256 characters to be represented. Although it is an improvement over ASCII, its primary use is in IBM and IBM-compatible computing systems and equipment.

In data communications, messages are composed using one of these codes. The numbers and letters to be represented are coded usually by way of a keyboard, and the binary word representing each character is stored in a computer memory. The message may also be stored on magnetic tape or disk. The purpose of data communications is to transmit that data or information from one place to another.

As you know, there are two basic ways to transfer binary information from one place to another: parallel and serial. In parallel data transfers, all the bits of a code word are transferred simultaneously. This concept is illustrated in Fig. 12-5. The binary word to be transmitted is usually loaded into a register containing one flip-flop for each bit. The flip-flop outputs are each connected to a wire to carry that bit to the receiving circuit. The receiving circuit is also usually a storage register. As you can see, in parallel data transmission, there is one wire for each bit of information to be transmitted. This means a multiwire cable must be used. Multiple parallel lines that carry binary data are usually referred to as a *data bus*.

Parallel data transmission is extremely fast since all the bits of the data word are transferred simultaneously. The transmit speed is limited to the speed of the logic circuits used in the data transfer. Such data transfers can occur in only a few nanoseconds in many applications.

Parallel data transmission is not practical for long-distance communications. To transfer an 8-bit data word from one place to another, eight separate communications channels are needed, one for each bit. Although multiwire cables can be used over limited distances, for practical long-distance data communications they are impractical because of cost and signal attenuation. And, of course, parallel data transmission by radio would be even more complex and expensive.

Data transfers in communications systems are made serially. Each bit of the word is transmitted one after another. Figure 12-6 shows the concept of serial transmission. The figure shows the ASCII code for the letter J (1001010) being transmitted one bit at a time. The least significant bit (LSB) is transmitted first, the most significant bit (MSB) last. Each bit is transmitted for a fixed interval of time (t). The voltage levels representing each bit appear on a single data line one after another until the entire word has been transmitted. For example, the bit interval may be 1 ms, meaning that the voltage level for each bit in the word will appear for 1 ms. A 7-bit ASCII word would, therefore, take 7 ms to transmit.

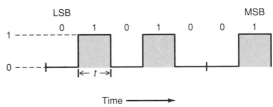

Time ⟶

Fig. 12-6   Serial data transmission for the ASCII character J.

Because of the sequential nature of serial data transmission, naturally it takes longer to send data this way than it does to transmit it by parallel means. However, with a high-speed logic circuit, even serial data transfers can take place at very high speeds.

In most computers and other digital equipment, the binary data is transferred between circuits in parallel format. When it is transmitted over long distances, it is sent in serial format. This means there must be techniques for converting between parallel and serial, and se-

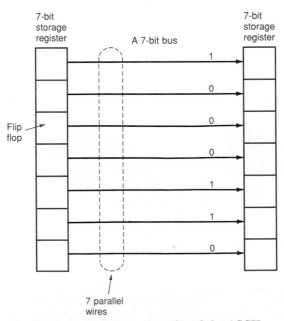

Fig. 12-5   A parallel data transfer of the ASCII character F.

rial and parallel. Such data conversions are usually taken care of by shift registers. A typical circuit is shown in Fig. 12-7. A shift register is a sequential logic circuit made up of a number of flip-flops connected in cascade. The flip-flops are capable of storing a multibit binary word. Usually this word is loaded in parallel into the transmitting register. When a clock pulse (CP) is applied to the flip-flops, the bits of the word are shifted from one flip-flop to another in sequence. The last (right-hand) flip-flop in the transmitting register will ultimately store each of the bits in sequence as it is shifted out.

The serial data word is then transmitted over the communications link and is received by another shift register. The bits of the word are shifted into the flip-flops one at a time until the entire word is contained within the register. The flip-flop outputs may then be observed and the data stored in them transferred in parallel to other circuits.

The speed of data transfer is usually indicated by the number of bits per second (bits/s). Most data rates take place at relatively slow speeds, usually several thousand bits per second. Personal computers, for example, communicate at 1200 and 2400 bits/s over the telephone lines. However, in some data communications systems, bit rates as high as several hundred million bits per second are used.

Another term used to express the data speed in digital communications systems is baud rate. *Baud rate* is the number of signaling elements or symbols that occur in a given unit of time, such as 1 s. A signaling element or symbol in many cases is just a binary logic level, either 0 or 1. In that case, the baud rate is simply equal to the data rate in bits per second. However, in other cases the signaling element or symbol may be one of several discrete signal amplitudes or phase shifts, each of which represents two or more data bits. Several unique modulation schemes have been developed which use each symbol or baud to represent multiple bits. The number of symbol changes per unit of time is no higher than the straight binary bit rate, but more bits per unit time are transmitted. As a result, higher bit rates can be transmitted over telephone lines or other severely bandwidth lim-

Fig. 12-7 **Parallel-to-serial and serial-to-parallel data transfers with shift registers.**

ited communications channels which cannot ordinarily handle them. Several of the modulation methods are discussed in Sec. 12-2.

An illustration will make this clearer. Assume a system that represents 2 bits of data as different voltage levels. There are four possible combinations of 2 bits, and we assign a discrete level to each.

| 00 | 0 V | 10 | 2 V |
| 01 | 1 V | 11 | 3 V |

Here we have a system with four symbols, in this case four different voltage levels. Instead of transmitting a binary signal with only two levels, we transmit one of the four symbols or voltage levels. Assume we wanted to transmit the decimal number 201. This can be represented in binary by the 8-bit number 11001001. We could transmit the number serially as a sequence of equal-time-interval pulses that are either on (1) or off (0). Refer to Fig. 12-8(a). If each bit interval is 1 ms, then the bit rate is 1000 bits/s. The baud rate is also 1000 bits/s or 1000 baud.

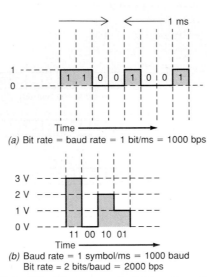

(a) Bit rate = baud rate = 1 bit/ms = 1000 bps

(b) Baud rate = 1 symbol/ms = 1000 baud
Bit rate = 2 bits/baud = 2000 bps

Fig. 12-8   The bit rate can be higher than the baud rate when each symbol represents two or more bits.

Using our new four-level system, we could divide the word into 2-bit groups and transmit the appropriate voltage level representing each. Our number 11001001 would be divided into groups of 11/00/10/01. Thus the transmitted signal would be voltage levels of 3, 0, 2, and 1 V, respectively, each occurring for a fixed interval of say 1 ms. This is illustrated in Fig. 12-8(b). The baud rate is still 1000 baud because there is only one symbol per time interval (1 ms). However, the bit rate is 2000 bits/s because although only four symbols were

transmitted, each symbol represents 2 bits. So, we have doubled the bit rate while keeping the baud rate constant. We have also shortened the total transmission time. It takes 8 ms to transmit the 8-bit binary word, but it only takes 4 ms to transmit the four-level signal.

There are two basic ways of transmitting serial binary data: asynchronous and synchronous. In *asynchronous* communications, each data word is accompanied by stop and start bits that identify the beginning and ending of the word. Each binary code word transmitted represents one character.

The use of start and stop bits to transmit an ASCII character is illustrated in Fig. 12-9. When no information is being transmitted, the communications line is usually high, or binary 1. In data communications terminology, this is referred to as a mark. To signal the beginning of a word, a start bit is transmitted. A start bit is a binary 0 or space as shown in Fig. 12-9. The start bit has the same duration as all other bits in the data word. The transition from mark to space indicates the beginning of the word and allows the receiving circuits to prepare themselves for the reception of the remainder of the bits.

Fig. 12-9   Asynchronous transmission with start and stop bits.

After the start bit, the individual bits of the word are transmitted. In this case, the 7-bit ASCII code for the letter U is transmitted. Once the last code bit has been transmitted, a stop bit is included. The stop bit is the same duration as all other bits and again is a binary 1 or mark. In some systems, two stop bits are transmitted, one after the other, to signal the end of the word.

Most low-speed data communications, up to 1200 or 2400 baud, take place using the asynchronous technique. It is extremely reliable, and the start and stop bits ensure that the sending and receiving circuits remain in step with one another.

The primary disadvantage of asynchronous communications is that the extra start and stop bits effectively slow down the data transmission. This is not a problem in low-data-volume, low-speed applications, but in those cases where huge volumes of information are to be transmitted, the start and stop bits represent a significant overhead; that is, they are a major percentage of the bits transmitted. For the transmission of a 7-bit ASCII character with start and stop bits, 9 bits are needed. Two of 9 is 22.2 percent. By removing the start and stop bits and stringing the ASCII characters end to end, many more data words can be transmitted per second. In other words, the speed of transmission can be significantly increased. When large volumes of data are to be transmitted between computers, high speed is important to ensure minimum transmission time on expensive long-distance circuits.

The technique of transmitting each data word one after another without start and stop bits is referred to as *synchronous* data communications. Data is usually transmitted in multiword blocks. To maintain synchronization between transmitter and receiver, a group of synchronization bits is placed at the beginning of the block and at the end of the block. Figure 12-10 shows one arrangement. Each block of data may represent hundreds or even thousands of characters. At the beginning of each block is a unique series of bits that identifies the beginning of the block. In Fig. 12-10, two 8-bit synchronization (SYN) codes signal the start of a transmission. The receiving equipment looks for these characters and then begins to receive the continuous data sent in sequential 8-bit words or bytes. At the end of the block is another special code (ETX) signaling the end of transmission. The receiving equipment will look for this code, detect it, and thereby recognize the end of the transmission. One or more error codes usually follow.

Although the special synchronization codes at the beginning and end do add extra bits to the transmission, overall they represent a very small increase in the total number of bits being transmitted. In any case, the number of bits used for synchronization purposes is small compared to the number of start and stop bits that would be transmitted if asynchronous transmission were used. Synchronous transmissions are much faster than asynchronous transmissions.

An important concept in data communica-

tions, and analog communications for that matter, is that the amount of information that can be transmitted is proportional to the bandwidth of the communications channel and the time of transmission. The greater the bandwidth, the more information that can be transmitted in a given time. To transmit voice so that it is intelligible and recognizable, a bandwidth of only about 3 kHz is required. However, to transmit music with full fidelity, a bandwidth of 15 to 20 kHz is required because of the high frequencies and harmonics produced by the various musical instruments. Music inherently contains more information than voice and thus requires greater bandwidth.

Television is another example. Video requires a bandwidth of 4 MHz because of all the light and color detail in the scene or picture transmitted. A picture signal contains more information than a voice or music signal. Therefore, greater bandwidth is required to transmit it. A typical TV signal contains both voice and picture and is usually allocated 6 MHz of spectrum space.

It is also possible to transmit a picture in less bandwidth. This is done by transmitting it over a longer period of time. A black and white photo can be transmitted by TV in about 33 ms which to the human eye is virtually instantaneous. That same photo can also be transmitted over a voice-grade telephone line using a facsimile (fax) machine. The difference is that because the bandwidth is so much narrower, the transmission requires more time. It would take the fax machine 10 to 15 s to transmit the photo over the telephone line.

Just remember that the greater the bandwidth of a channel, the greater the amount of information you can transmit in a given time. You can still transmit the same amount of information over a narrower channel, but it will simply take longer. This general concept is known as Hartley's law. It is one of the many concepts that together form what is called information theory. Information theory is the study of the most efficient ways to transmit information.

When dealing with binary data transmission, Hartley's law also applies. The greater the

Fig. 12-10 Synchronous data transmission.

number of bits transmitted in a given time, the greater the amount of information that is conveyed. But the higher the bit rate, the wider the bandwidth needed to pass the signal without distortion. Narrowing the bandwidth of a channel will cause the harmonics in the binary pulses to be filtered out.

Binary pulses are rectangular waves made up of a fundamental sine wave plus many harmonics. The channel bandwidth must be wide enough to pass all the harmonics to preserve the waveshape. On the other hand, most communications channels or media act like low-pass filters. Voice-grade telephone lines, for example, act as a low-pass filter with an upper cutoff frequency of about 3400 Hz. Naturally, harmonics higher in frequency than the cutoff will be filtered out. This produces signal distortion. Eliminating the harmonics rounds the signal off.

This filtering essentially converts the binary signal into its fundamental sine wave. The data is distorted but not lost. The information is still transmitted reliably but in the minimum possible bandwidth. The sine wave signal shape can easily be restored to a rectangular wave at the receiver with a Schmitt trigger, comparator, or other waveshaping circuit.

The upper cutoff frequency of the channel is approximately equal to the channel bandwidth. It is this bandwidth that determines the information capacity of the channel. The channel capacity C expressed in bits per second is twice the channel bandwidth B in hertz.

$$C = 2B$$

The bandwidth B is usually the same as the upper cutoff (3-dB-down) frequency of the channel. This is the maximum theoretical limit, and it assumes that no noise is present.

For example, assume a channel bandwidth of 10 kHz. The maximum theoretical bit capacity then is

$$C = 2B$$
$$= 2(10,000)$$
$$= 20,000 \text{ bits/s}$$

The channel capacity expression can be modified by taking into consideration multiple-level encoding schemes that permit more bits per symbol or baud to be transmitted. The new expression becomes

$$C = 2B \log_2 N$$

Here N is the number of different encoding lev-

els per time interval. What this expression says is that for a given bandwidth, the channel capacity in bits per second can be higher if we use more than two levels or other symbols per time interval.

In our previous example, we used two levels per millisecond (0 or 1) in the binary signal of Fig. 12-8(a). The bit rate (and baud rate) is 1000 bits/s. The bandwidth needed to transmit this signal can be computed from the original expression.

$$C = 2B$$

Therefore,

$$B = \frac{C}{2}$$

With a 1000 bits/s channel rate, we will need an absolute minimum bandwidth of

$$B = \frac{1000 \text{ bits/s}}{2}$$
$$= 500 \text{ Hz}$$

The same result is obtained with the new expression. If $C = 2B \log_2 N$, then

$$B = \frac{C}{2 \log_2 N}$$

The logarithm of a number to the base 2 can be computed with the expression

$$\log_2 N = \frac{\log_{10} N}{\log_{10} 2}$$
$$= \frac{\log_{10} N}{0.301}$$
$$= 3.32 \log_{10} N$$

where N is the number whose logarithm is to be calculated. The base 10 or common logarithm can be computed on any scientific calculator. With two coding levels (binary 0 and 1), the bandwidth is

$$B = \frac{C}{2 \log_2 N}$$
$$= \frac{1000 \text{ bits/s}}{2(1)}$$
$$= 500 \text{ Hz}$$

Note that $\log_2 2$ for a binary signal is simply 1. That's how the basic expression is derived, or

$$C = 2B \log_2 N$$

For binary, $\log_2 N = \log_2 2 = 1$. Therefore,

$$C = 2B(1)$$
$$= 2B$$

Now, in Fig. 12-8(b), let's see what our multilevel coding scheme does.

$$B = \frac{C}{2 \log_2 N}$$

The channel capacity is still 1000 bits/s or 1000 baud as shown in Fig. 12-8(b) because each symbol (level) interval is 1 ms long. However, each symbol represents 2 bits. Therefore, the channel capacity is 2000 bits/s. The number of symbol levels $N = 4$. The bandwidth then is

$$B = \frac{2000 \text{ bits/s}}{2 \log_2 4}$$

and $\log_2 4 = 2$.

$$B = \frac{2000}{2(2)}$$

$$= \frac{2000}{4}$$

$$= 500 \text{ Hz}$$

Look what happened here. With a multilevel coding scheme, we can transmit at twice the speed in the same bandwidth.

Another important aspect of information theory is the impact of noise on the signal. As it turns out, the greater the amount of noise, the lower the channel capacity. The presence of noise reduces the amount of information that can be transmitted in a given bandwidth. Increasing the bandwidth increases the information rate. On the other hand, as you learned in a previous chapter, a wider bandwidth also allows more noise to pass. As you can see then, the choice of a bandwidth is a compromise between information transmission rate and the amount of noise.

The relationship between channel capacity, bandwidth, and noise is summarized in what is known as the *Shannon-Hartley theorem.*

$$C = B \log_2 (1 + S/N)$$

where C = channel capacity, bits/s
     B = bandwidth, Hz
    S/N = signal-to-noise power ratio

To see how the Shannon-Hartley theorem works, assume we wish to calculate the maximum channel capacity of a voice-grade telephone line with its bandwidth of 3100 Hz. Assume an S/N ratio of 30 dB.

First, we must convert 30 dB into a power ratio. If

$$dB = 10 \log P$$

where P is the power ratio, then

$$P = \text{antilog } \frac{dB}{10}$$

Antilogs are easily computed on a scientific cal-

culator. A 30-dB S/N ratio translates to a power ratio of

$$P = \text{antilog } \frac{30}{10}$$

$$= \text{antilog } 3$$

$$= 1000$$

We can now calculate the channel capacity.
$$C = B \log_2 (1 + S/N)$$

$$= 3100 \log_2 (1 + 1000)$$

$$= 3100 \log_2 1001$$

The base 2 logarithm of 1001 is

$$\log_2 1001 = 3.32 \log_{10} 1001$$

$$= 3.32(3)$$

$$= 9.97$$

$$\approx 10$$

Therefore, the channel capacity is
$$C = 3100(10)$$

$$= 31,000 \text{ bits/s}$$

That is a surprisingly high bit rate for such a narrow bandwidth. In fact, it appears to conflict with what we learned earlier. That is, the maximum channel capacity is twice the channel bandwidth. If the bandwidth of the voice-grade line is 3100 Hz, then the channel capacity is only

$$C = 2B$$

$$= 2(3100)$$

$$= 6200 \text{ bits/s}$$

This is for a binary (two-level) system only, and it assumes no noise. How then can the Shannon-Hartley theorem predict that the channel capacity is 31,000 bits/s when noise is present?

The Shannon-Hartley expression says that we can theoretically achieve a 31,000 bits/s channel capacity on our 3100-Hz bandwidth line. What it doesn't say is that multilevel encoding is needed to do so. Let's go back to the basic channel capacity expression of

$$C = 2B \log_2 N$$

We know $C = 31,000$ bits/s and B is 3100 Hz. What we don't know is the number of coding or symbol levels N. Rearranging the formula, we get

$$\log_2 N = \frac{C}{2B}$$

$$= \frac{31,000}{2(3100)}$$

$$= \frac{31,000}{6200}$$

$$= 5$$

Therefore,
$$N = \text{antilog}_2\, 5$$

The antilog of a number is simply the value we get when we raise the base (2) to the number, in this case 5.
$$N = 2^5 = 32$$

What these calculations say is that we can achieve the channel capacity of 31,000 bits/s if we use a multilevel encoding scheme that uses 32 different levels or symbols per interval rather than pure two-level binary. The baud rate of the channel C is still 6200 or twice the bandwidth.
$$C = 2B$$
$$= 2(3100)$$
$$= 6200 \text{ baud}$$

But because we use a 32-level encoding scheme, the bit rate is 31,000 bits/s. As it turns out, we rarely attempt to use the maximum channel capacity. In practice, it is very difficult to achieve. Typical systems limit the channel capacity to one-third to one-half the maximum to ensure more reliable transmission in the presence of noise.

Self-Test

Answer the following questions.

1. Data communications is the transmission of _____ information.
2. True or false. Audio and video signals may be transmitted using data communications techniques.
3. The earliest form of data communications was the _____ .
4. The earliest form of radio data communications is known as _____ .
5. The Morse code is made up of combinations of _____ and _____ to represent characters.
6. Give the Morse code for the following characters (see Fig. 12-2): K, 8, ,.
7. The 5-bit code used in teletype systems is called the _____ code.
8. A system of data communications using typewriterlike devices to send and receive coded messages is known as _____ .
9. If in the Baudot code the character 11011 is sent and then the code 00111, the character _____ is received.
10. The most widely used binary data code is _____ which uses 7 bits to represent characters.

11. An 8-bit data code used in IBM systems is known as _____ .
12. Write the ASCII codes for the following characters: B, 3, ?.
13. What ASCII character is transmitted to ring a bell? BEL = _____
14. The two types of data transmission are _____ .
15. True or false. Serial data transfers are faster than parallel transfers.
16. The digital circuit often used to perform parallel-to-serial and serial-to-parallel data conversions is the _____ .
17. In asynchronous transmissions, _____ and _____ bits are used to signal the beginning and end of a character.
18. The speed of a serial data transmission is usually expressed in _____ .
19. The number of symbols occurring per second in a data transmission is called the _____ .
20. True or false. There can be more than 1 bit per baud in data transmissions.
21. True or false. Asynchronous transmission is faster than synchronous transmission.
22. In synchronous data transmission, data is sent in _____ of characters with special codes to signal the beginning and end.
23. In serial data transmission, a binary 0 is called a(n) _____ and a binary 1 is called a(n) _____ .
24. The study of efficient information transfer is known as _____ .
25. The amount of information transmitted is proportional to the system _____ and the _____ allowed for transmission.
26. The channel capacity of a 5-kHz bandwidth binary system is _____ bits/s assuming no noise.
27. If an 8-level encoding scheme is used in a 10-kHz bandwidth system, the channel capacity is _____ bits/s.
28. True or false. Multilevel or multisymbol binary encoding schemes permit more data to be transmitted in less time, assuming a constant symbol interval.
29. By using a multisymbol encoding scheme, a higher bit rate can be achieved with _____ (more, less) bandwidth.
30. The channel capacity of a 6-MHz channel with an S/N ratio of 25 dB is _____ Mbits/s.
31. The minimum allowable bandwidth for a binary signal with a bit rate of 200 kbits/s is _____ kHz.

12-2 MODEMS

What is the fastest and easiest way to communicate electronically with anyone else in the United States or in the world? Radio communications is common and simple, but few people or organizations have radio transmitters and receivers. If they did, they would have to operate on your frequency, use your type of modulation, and be available at the appropriate time. Other conditions such as radio wave propagation also enter into the picture. So radio communications is simply not the most convenient.

Without belaboring the point further, the answer to the question is simple. The fastest and easiest way to communicate with anyone else in the world is by the telephone. The telephone network is a vast electronic communications system that is in place, and virtually everyone uses it. You can talk to anyone else in the world by simply dialing their number. Your voice signal then travels over hundreds or even thousands of miles of wire and even over microwave relay and satellite communications links. In newer long-distance telephone systems, your voice may be carried on a light beam in fiber-optic cable.

Although voice is the primary signal carried by the telephone system, this network is now widely used to carry digital information as well. When it is necessary for computers to communicate or for a personal computer or terminal to have access to a large mainframe computer, the telephone network serves as the communication medium.

There are two primary problems in transmitting digital data over the telephone network. The first problem is that binary signals are usually switched dc pulses. That is, the binary 1s and 0s are represented by pulses of a single polarity, usually positive. The telephone line is designed to carry only ac analog signals. Voice and other analog signals are processed by numerous components, circuits, and equipment along the way. Usually only ac signals of a specific frequency range are allowed. Voice frequencies in the 300- to 3000-Hz range are the most common. If a binary signal were applied directly to the telephone network, it simply would not pass. The transformers, capacitive coupling, and other ac circuitry virtually ensure that no dc signals get through. Further, binary data is usually transmitted at high speeds. This high-speed data would essentially be filtered out by the system with its limited bandwidth. The question then is: "Just how does digital data get transmitted over the telephone network?" The answer is by using modems.

A *modem* is a device containing both a *mod*ulator and a *dem*odulator. A modem is used to convert binary signals into analog signals capable of being transmitted over the telephone lines and to demodulate such analog signals and reconstruct the equivalent binary output. Figure 12-11 shows two ways that modems are commonly used in digital data transmission. In Fig. 12-11(*a*), two mainframe computers exchange data by speaking through modems. While one modem is transmitting, the other is receiving. Full duplex operation is also possible. In Fig. 12-11(*b*), a remote video terminal or personal computer uses a modem to communicate with a large mainframe computer.

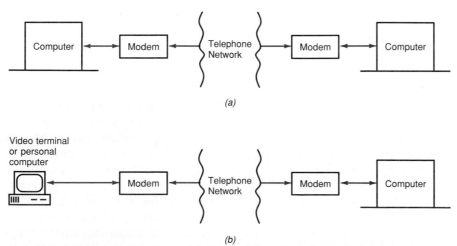

(a)

(b)

Fig. 12-11 How modems are used to permit digital data transmission on the telephone network.

Again, modems are used at each end of the communications system. The modem is the interface between these units.

A modem causes the binary dc pulses to modulate an analog sine wave carrier that is compatible with the telephone line. The modem makes the binary signals compatible with the analog telephone network. And even though the bandwidth of the telephone network is essentially restricted to 3 kHz, a variety of tricky modulation techniques permit serial data transmission rates as high as about 15,000 bits/s to be transmitted. In this section we take a look at the various modulation and demodulation techniques used in modern data communications modems. These include frequency-shift keying, phase-shift keying, and quadrature amplitude modulation.

The oldest and simplest form of modulation used in modems is frequency-shift keying (FSK). In FSK, two sine wave frequencies are used to represent binary 0s and 1s. For example, a binary 0, usually called a space in data communications jargon, has a frequency of 1070 Hz. A binary 1, referred to as a mark, is 1270 Hz. These two frequencies are alternately transmitted to create the serial binary data. The resulting signal looks something like that shown in Fig. 12-12. Note that both of the frequencies are well within the 300- to 3000-kHz bandwidth normally associated with the telephone system. This is illustrated in Fig. 12-13.

To permit simultaneous transmit and receive operations with a modem, known as full-duplex operation, another set of frequencies are defined. These are also indicated in Fig. 12-13. A binary 0 or space is 2025 Hz, while a binary 1 or mark is 2225 Hz. These tones are

(a)

(b)

Fig. 12-12   Frequency-shift keying. (a) Binary signal, and (b) FSK signal.

Fig. 12-13   The FSK signals within the telephone audio bandpass.

also within the telephone bandwidth but are spaced far enough from the other frequencies so that selective filters can be used to distinguish between the two. The 1070- and 1270-Hz tones are used for transmitting (originate) and the 2025- and 2225-Hz tones are used for receiving (answer).

A general block diagram of an FSK modem is illustrated in Fig. 12-14. Each modem contains an FSK modulator and an FSK demodulator so that both send and receive operations can be achieved. Bandpass filters at the inputs to each modem separate the two tones. For example, in the upper modem, a bandpass filter allows frequencies between 1950 and 2300 Hz to pass. This means that the 2025- and 2225-Hz tones will be passed but the 1070- and 1270-Hz tones generated by the internal modulator will be rejected. The lower modem has a bandpass filter that accepts the lower-frequency tones while rejecting the upper-frequency tones generated internally.

Note also that both systems contain a device called a UART which stands for *u*niversal *a*synchronous *r*eceiver *t*ransmitter. This is a digital IC used to perform parallel-to-serial and serial-to-parallel data transfers. Most data transfers and operations inside the computer are parallel, but as you know, data communications uses only serial binary data. The UART performs the necessary conversions.

Figure 12-15 on page 322 shows a general block diagram of a UART. All this circuitry is typically contained within a single large-scale IC. Parallel data from the computer data bus is fed into and out of a bidirectional data buffer which is usually a storage register with appropriate level-shifting circuits. The data, usually parallel 8-bit words, is put on an internal data bus. Before being transmitted, the data is stored first in a buffer storage register and then sent to a shift register. A clock signal shifts the data out serially 1 bit at a time. Note here that the internal circuitry adds start and stop bits and a parity bit. The start and stop bits signal the beginning and end of the word, and the par-

Fig. 12-14 Block diagram of an FSK modem.

ity bit is used to detect error. The resulting serial data word is transmitted 1 bit at a time.

The receive section of the UART is at the bottom of the diagram in Fig. 12-15. Serial data is shifted into a shift register. There the start, stop, and parity bits are stripped off. The remaining data is transferred to a buffer register, to the internal data bus, and through the bidirectional data buffer to the computer in parallel form. The clock and control logic circuits in the UART control all internal shifting and data transfer operations under the direction of control signals from the computer.

A wide variety of modulator and demodulator circuits are used to produce and recover FSK. Virtually all the circuits described in the chapter on FM can or have been used. A typical FSK modulator is simply an oscillator whose frequency can be switched between two frequencies, usually by switching in different capacitor values. Both *RC* and *LC* oscillators are used. At the lower audio frequencies, *RC* oscillators are preferred because of their simplicity. A variety of demodulators are also used. These include the PLL, pulse-averaging discriminator, and others.

Most of the newer modems use digital techniques because they are simpler and more adaptable to IC implementation. One type of digital FSK modulator is shown in Fig. 12-16 on the next page. A clock oscillator generates

a clock at a frequency of 271,780 Hz. It is applied to a binary frequency divider. This divider is usually some form of binary counter with various feedback logic gates to set the divide ratio. This frequency divider is set up so that it will divide by two different integer values. One divide ratio will produce the mark frequency, while the other will produce the space frequency.

To transmit a space at 1070 Hz, the divide ratio logic in Fig. 12-16 produces frequency division by 127. That is, when the serial binary input is zero, the frequency divider output will be $\frac{1}{127}$ of its input. This means the output frequency will be 2140 Hz. This is fed to a single flip-flop which divides the frequency by 2, producing the desired 1070-Hz output. The flip-flop produces a 50 percent duty cycle square wave; that is, its on and off times are equal in length. The reason for this is that if the duty cycle of the frequency divider is other than 50 percent, the square wave when converted to a sine wave will produce excessive distortion. The flip-flop output is passed through a low-pass filter which removes the higher odd harmonics producing a 1070-Hz sine wave tone.

When a binary 1 is applied to the divide ratio logic, the frequency divider will divide by 107. This produces an output frequency of 2540 Hz which when divided by 2 in the flip-flop produces the desired 1270-Hz output. The

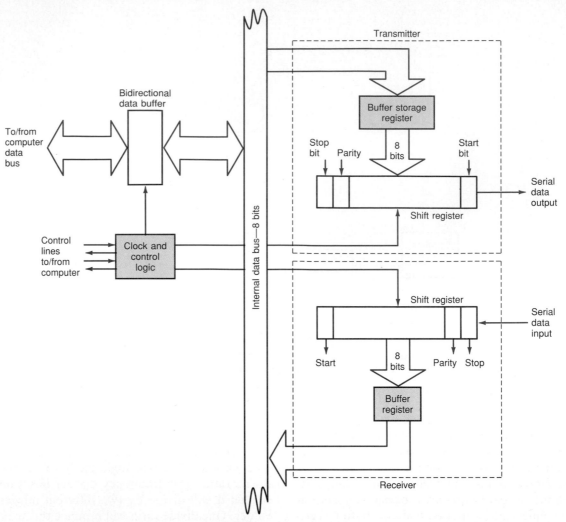

Fig. 12-15   General block diagram of a UART.

low-pass filter removes the higher-frequency harmonics, producing a sine wave output.

A digital method of FSK demodulation is shown in Fig. 12-17. The sine wave FSK signal is applied to a limiter which removes any amplitude variations and which shapes the signal into a square wave. The square wave is then applied to a gate circuit. The gate is used to turn a 1-MHz clock signal off and on. A binary counter counts or accumulates the 1-MHz clock pulses.

When a low-frequency space signal is applied, the period of the signal will be long, thereby allowing the gate to open and the bi-

Fig. 12-16   A digital FSK modulator.

Fig. 12-17 A digital FSK demodulator.

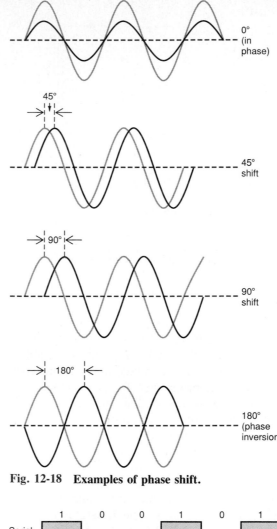

Fig. 12-18 Examples of phase shift.

nary counter to accumulate the clock pulses. The detection logic which is a set of binary gates determines whether the number in the binary counter is above or below some predetermined value. For the low-frequency tone input, the number in the counter will be higher than the counter value when the high-frequency or mark signal is applied to the input. The detection logic produces a binary 0 or binary 1 output depending upon whether the number in the counter is above or below a specific value between the two limits.

Frequency-shift keying is used primarily in low-speed modems capable of transmitting data up to a speed of 300 baud or bits/s. This is a relatively slow data rate and is rarely used today. Higher speeds are used to shorten the transmission time for long messages or sets of data. Most communications data rates are 1200 baud or higher. Rates of 2400, 4800, and 9600 baud are common. To achieve these higher speeds, other forms of modulation are used. These include phase-shift keying and quadrature amplitude modulation.

In *phase-shift keying (PSK)*, the binary signal to be transmitted changes the phase shift of a sine wave character depending upon whether a binary 0 or binary 1 is to be transmitted. Recall that phase shift is a time difference between two sine waves of the same frequency. Figure 12-18 illustrates several examples of phase shift. Note that a phase shift of 180° represents the maximum difference and is also known as a phase reversal.

Figure 12-19 illustrates the simplest form of PSK known as binary PSK (BPSK). During the time that a binary 0 occurs, the carrier signal is transmitted with one phase, but when a binary 1 occurs, the carrier is transmitted with a 180° phase shift.

Fig. 12-19 Binary phase-shift keying (BPSK).

Figure 12-20 on the next page shows one kind of circuit used for generating BPSK. It is a standard lattice ring modulator or balanced modulator used for generating DSB signals. The carrier sine wave, usually 1600 or 1700 Hz, is applied to the input transformer T_1, and the binary signal is applied to the transformer center taps. The binary signal provides a switching signal for the diodes. When a binary 0 appears at the input, A is positive and B is negative, so diodes D_1 and D_4 conduct. They act as closed switches,

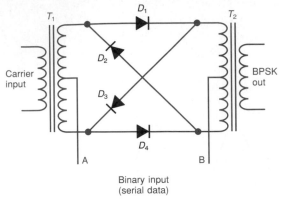

Fig. 12-20   A BPSK modulator.

thereby connecting the secondary of T_1 to the primary of T_2. The windings are phased so that the BPSK output is in phase with the carrier input.

When a binary 1 appears at the input, A is negative and B is positive so diodes D_1 and D_4 are cut off while diodes D_2 and D_3 conduct. This causes the secondary of T_1 to be connected to the primary of T_2 but with the interconnections reversed. This introduces a 180° phase-shift carrier at the output.

Demodulation of a BPSK signal is also done with a balanced modulator. A version of the diode ring or lattice modulator can be used as shown in Fig. 12-21. This is actually the same circuit as in Fig. 12-20, but the output is taken from the center taps. The BPSK and carrier signals are applied to the transformers. Integrated-circuit balanced modulators can also be used at the lower frequencies.

The key to demodulating BPSK is that a carrier with the correct frequency and phase relationship must be applied to the balanced

Fig. 12-21   A BPSK demodulator.

modulator along with the BPSK signal. Typically the carrier is derived from the BPSK signal itself. This is done with a carrier recovery circuit shown in Fig. 12-22. A bandpass filter ensures that only the desired BPSK signal is passed. The signal is then squared or multiplied by itself by a balanced modulator or analog multiplier which applies the same signal to both inputs. Squaring removes all the 180° phase shifts. The result is an output that is twice the input signal frequency or $2f$. A bandpass filter set at twice the carrier frequency passes only this signal. The resulting signal is applied to the phase detector of a PLL. Note that a ×2 frequency multiplier is used between the VCO and phase detector. This ensures that the VCO frequency is at the carrier frequency. Because the PLL is used, the VCO will track any carrier frequency shifts. The result is a signal with the correct frequency and phase relationship for proper demodulation. The carrier is applied to the balanced modulator-demodulator along with the BPSK signal. The output is the recovered binary data stream.

In order to correctly recover the binary data transmitted by BPSK, the received signal must be fed to a balanced modulator along with a carrier whose frequency and phase is identical to that used on the original modulator. This ensures that the phase of the received signal is compared to the proper phase of the carrier. The carrier recovery circuit just described produces a carrier of the correct phase at the receiver. However, the required circuitry is complex and expensive. To eliminate this problem and simplify the demodulation process, a version of BPSK called differential PSK (DPSK) can be used. In DPSK, there is no absolute carrier phase reference. Instead, the transmitted signal itself becomes the phase reference. In demodulating DPSK, the phase of the received bit is compared to the phase of the previously received bit.

In order for DPSK to work, the original binary bit stream must undergo what is known as differential phase coding. This amounts to processing the serial bit stream through an inverted exclusive OR (XNOR) circuit as shown in Fig. 12-23(a). Note that the output of the XNOR circuit is applied to a 1-bit delay circuit before being applied back to the input. The 1-bit delay may simply be a clocked flip-flop or a delay line. The resulting bit pattern will permit the signal to be recovered by comparing the present bit phase with the previously received bit phase.

Fig. 12-22 A BPSK carrier recovery circuit.

In Fig. 12-23(b), the input binary word to be transmitted is shown along with the output of the XNOR circuit. An XNOR circuit is simply a 1-bit comparator that produces a binary 1 output when both inputs are alike and a binary 0 output when the two bits are different. The output of the circuit is delayed for a 1-bit interval by storing it in a flip-flop. Therefore, the XNOR inputs are the current bit plus the previous bit. The XNOR signal is then applied to the balanced modulator along with the carrier to produce a BPSK signal.

Demodulation is extremely simple with the circuit shown in Fig. 12-24. The DPSK signal is applied to one input of the balanced modulator and a 1-bit delay circuit, either a flip-flop or a delay line. The output of the delay circuit is used as the carrier. The resulting output is filtered by a low-pass filter to recover the binary data. Typically the low-pass filter output is shaped with a Schmitt trigger or comparator to produce clean, high-speed binary levels.

The main problem with BPSK and DPSK is that the speed of data transmission is limited in a given bandwidth. One way to increase the binary data rate while not increasing the bandwidth required for the signal transmission is to encode more than 1 bit per phase change. There is a symbol change for each bit change so the baud (symbol) rate is the same as the bit rate. In BPSK and DPSK, each binary bit produces a specific phase change. It is also possible to use combinations of two or more bits to specify a particular phase shift. Therefore, a symbol change (phase shift) can represent multiple bits. We can encode more bits per baud. The bit rate of data transfer can be higher than the baud rate, while the signal occupies no more bandwidth.

One commonly used system for doing this is known as quadrature, quaternary, or quadraphase PSK (QPSK or 4-PSK). In QPSK, each pair of successive digital bits in the transmitted word are assigned a particular phase as indicated in Fig. 12-25(a). Each pair of bits called a *dibit* is represented by a spe-

(a) DPSK modulator

(b) Differential phase encoding

Fig. 12-23 The DPSK process.

Fig. 12-24 A DPSK demodulator.

Dibit		Phase shift
0	0	45°
0	1	135°
1	1	225°
1	0	315°

(a)

(b)

Fig. 12-25 Quadrature PSK modulation.
(a) Phase angle of carrier for different pairs of bits.
(b) Phasor representation of carrier sine wave.

CHAPTER 12 DATA COMMUNICATIONS   325

cific phase. A 90° phase shift exists between each pair of bits. Other phase angles can also be used as long as they have a 90° separation. For example, it is common to use phase shifts of 45°, 135°, 225°, and 315° as shown in Fig. 12-25(b).

A modulator circuit for producing QPSK is shown in Fig. 12-26. It consists of a 2-bit shift register implemented with flip-flops. This shift register is more commonly known as a bit splitter. The serial binary data train is shifted through this register. The bits from the two flip-flops are applied to balanced modulators. The carrier oscillator is applied to balanced modulator 1 and through a 90° phase shifter to balanced modulator 2. The outputs of the balanced modulators are linearly mixed to produce the QPSK signal.

The output from each balanced modulator is a BPSK signal as described before. With a binary 0 input, one phase of the carrier will be produced, and with a binary 1 input, the carrier phase will be shifted 180°. The output of balanced modulator 2 will also have two phase states 180° out of phase with one another. Just keep in mind that the 90° carrier phase shift at the input causes the outputs from balanced modulator 2 to be shifted 90° from those of balanced modulator 1. The result is four different carrier phases which are combined two at a time in the linear mixer. The result is four unique output phase states.

Figure 12-27 shows the outputs of one possible set of phase shifts. Note that the carrier

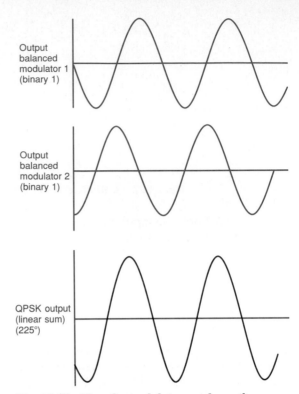

Fig. 12-27 How the modulator produces the correct phase by adding two signals.

outputs from the two balanced modulators are shifted 90°. When the two carriers are algebraically summed in the mixer, the result is an output sine wave that has a phase shift of 225° which falls between the two balanced modulator signals.

A demodulator for QPSK is illustrated in Fig. 12-28. The carrier recovery circuit is sim-

Fig. 12-26 A QPSK modulator.

Fig. 12-28   A QPSK demodulator.

ilar to the one described previously. The carrier is applied to balanced modulator 1 and is shifted 90° before being applied to balanced modulator 2. The outputs of the two balanced modulators are filtered and shaped into the dibit. The two bits are combined in a shift register and shifted out to produce the originally transmitted binary signal.

More bits per phase change can also be encoded to produce even higher data rates. For example, three serial bits can be used to produce a total of 8 different phase changes. This produces 8-PSK. Four serial input bits can produce 16 different phase changes. This results in 16-PSK which produces an even higher data rate. Again, the basic idea here is to encode more bits per carrier phase change or more bits per baud in order to produce a higher data rate within a restricted bandwidth.

One of the most popular modulation techniques used in modems for encoding more bits per baud is *quadrature amplitude modulation (QAM)*. It uses both AM and PM of a carrier. In addition to producing different phase shifts, the amplitude of the carrier is also varied.

A popular version of QAM is known as 8-QAM. This system uses a total of four different phase shifts as in QPSK and two carrier amplitudes. With four possible phase shifts and two different carrier amplitudes, a total of eight different states can be transmitted. With eight states, 3 bits can be encoded for each baud or symbol transmitted. Each 3-bit binary word transmitted uses a different PM/AM combination.

One way to illustrate an 8-QAM signal is to use what is known as a constellation diagram, which shows all possible phase and amplitude combinations. An 8-QAM constellation diagram is shown in Fig. 12-29. The points in the diagram indicate the eight possible PM/AM combinations. Note that there are two amplitude levels for each phase position. Point A shows a low carrier amplitude with a phase shift of 135°. It represents 100. Point B shows a higher amplitude and a phase shift of 315°. This sine wave represents 011.

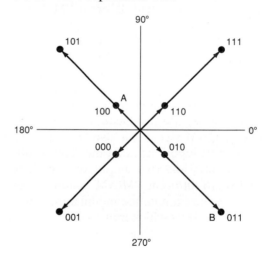

Note: Each vector has a specific amplitude and phase shift and represents one 3-bit word.

Fig. 12-29   A constellation diagram of QAM signal.

A block diagram of an 8-QAM modulator is shown in Fig. 12-30 on the next page. The binary data to be transmitted is shifted serially into the 3-bit shift register. These bits are applied in pairs to two 2-to-4 level converters. A 2-to-4 level converter is a circuit that translates a pair of binary inputs into one of four possible dc output voltage levels. A 2-to-4 level converter is basically a simple D/A converter. The idea is to produce four voltage levels corresponding to the different combina-

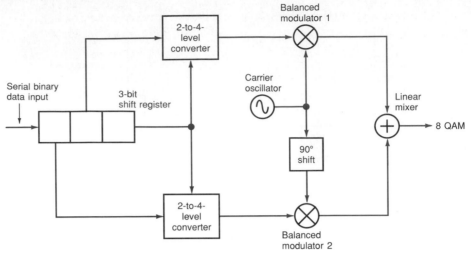

Fig. 12-30 **An 8-QAM modulator.**

tions of two input bits. The result is four equally spaced voltage levels. These are applied to the two balanced modulators fed by the carrier oscillator and a 90° phase shifter as in a QPSK modulator. Each balanced modulator produces four different output PM/AM combinations. When these are combined in the linear mixer, eight different PM/AM combinations are produced. The result is 8-QAM as described previously. The most critical part of the circuit is the 2-to-4 level converters which must have very precise output amplitudes so that when they are combined in the linear summer, the correct output and phase combinations are produced.

A 16-QAM signal can also be generated by encoding 4 input bits at a time. The result is 8 phase shifts and two amplitude levels, for a total of 16 different PM/AM combinations.

Quadrature amplitude modulation is widely used in modems which transmit computer data over the telephone lines. Recall that the telephone system bandwidth is restricted to approximately 3 kHz. For that reason, it is extremely difficult to transmit binary data at speeds higher than approximately 1200 baud or bits/s. However, by using QPSK, QAM, or similar techniques, very high data rates can be achieved. For example, using QAM, data rates of 9600 bits/s are regularly achieved over the standard telephone network.

The BPSK, QPSK, QAM, and other techniques are also widely used to transmit digital data in microwave and satellite radio communications. With very high frequency carriers, bit rates of millions of bits per second can easily be achieved with minimum transmission error in a noisy environment.

Self-Test

Supply the missing information in each statement.

32. A modem contains both _____ and _____ circuits.
33. Modems convert _____ signals to _____ signals and vice versa.
34. The transmission medium most widely used with modems is the _____ .
35. Modems are the interface between _____ .
36. _____ modulation is used in low-speed modems.
37. In FSK, binary 0 and 1 levels are represented by different _____ .
38. In an FSKs modem, a high frequency represents a(n) _____ and a low frequency represents a(n) _____ .
39. The typical frequencies used in a full-duplex FSK modem are as follows: originate 0 _____ Hz and 1 _____ Hz; answer 0 _____ Hz and 1 _____ Hz.
40. The IC used to perform serial-to-parallel and parallel-to-serial conversions and other operations in a modem is known as a(n) _____ .
41. The modulation method that represents bits as different phase shifts of a carrier is known as _____ .
42. In BPSK, phase shifts of _____ degrees and _____ degrees used to represent binary 0 and binary 1, respectively.
43. The circuit used to produce BPSK is a(n) _____ .

44. The circuit used to demodulate BPSK is a(n) _____ .
45. A(n) _____ circuit is used to generate the carrier to be used in demodulating a BPSK signal.
46. A carrier recovery circuit is not needed with _____ PSK modulation.
47. The key circuit used in a DPSK modulator is the _____ .
48. The number of different phase shifts used in QPSK is _____ .
49. In QPSK, how many bits are represented by each phase shift?
50. The number of bits represented by each phase shift in 16-PSK is _____ .
51. The circuit used to create a dibit is known as a(n) _____ .
52. A circuit that converts a 2-bit binary code into one of four dc voltage levels is known as a _____ .
53. Quadrature amplitude modulation is a combination of _____ modulation and _____ modulation.

Determine whether each statement is true or false.

54. Modems must transmit signals in the 300-Hz to 3-kHz range.
55. Carrier recovery is not required in a QPSK demodulator.
56. When QPSK is used, the bit rate is faster than the baud (symbol) rate.
57. Each phase and/or amplitude change can represent only one bit.
58. With QAM, a 9600 bits/s signal can be transmitted within a 3000-Hz bandwidth.

12-3 PROTOCOLS AND ERROR DETECTION AND CORRECTION

To communicate successfully using serial digital data, some guidelines must be established between the sender and the receiver. That is, specific rules and procedures must be agreed upon at the transmitting and receiving ends of the communications system. These rules and procedures are called *protocols*. A variety of different protocols are used in data communications.

When transmitting high-speed binary data over a communications link, be it a cable or radio, errors will occur. These errors are changes in the bit pattern caused by interference, noise, or equipment malfunctions. Such errors will cause incorrect data to be received.

To ensure reliable communications, schemes have been developed to detect such bit errors so that corrections can be made. In this section we will review some of the basic error-detection and -correction schemes used in data communications.

The simplest form of protocol is the use of stop and start bits in the transmission of individual characters using asynchronous methods. Refer back to Fig. 12-9. When no information is being transmitted, the line between the transmitter and receiver stays at the binary 1 or mark state. To begin transmission of a single 7- or 8-bit ASCII character, a start bit is generated. The start bit is a binary 0 or space transmitted for a 1-bit time. This signals the receiver that a character is about to be sent so that the receiver can then prepare itself for the forthcoming serial bits. The character bits are then transmitted. The character is followed by a stop bit which is a binary 1 or mark transmitted for 1- or 2-bit intervals. This signals the receiver that the character has been transmitted.

Even though this protocol is extremely simple, it illustrates the concept used in the transmission of all digital data. The minimum protocol then is a start bit and a stop bit framing a single character. But what happens when you want to transmit multiple characters in synchronous form?

In data communications, a message is more than one character. It is many characters composed of letters of the alphabet, numbers, punctuation marks, and other symbols in the desired sequence. A group of characters forms a block. When a message is sent, it is usually broken up into blocks, a block being a fixed number of characters. In synchronous data communications applications, the block is the basic transmission unit.

To identify a block, one or more special characters are transmitted prior to the block and after the block. These additional characters added to the beginning and end of the block perform a number of functions. These special characters are usually represented by 7- or 8-bit codes. Like the start and stop bits for a character, they signal the beginning and end of the transmission, but they are also used to identify a specific block of data and also provide means for error checking and detection.

Some of the characters at the beginning and end of each block are used for "handshaking" purposes. These are characters that give the transmitter and receiver status information.

Fig. 12-31   The handshaking process in data communications.

Figure 12-31 illustrates the basic handshaking process. For example, the transmitter may send a character indicating that it is ready to send data to the receiver. The receiver identifies that character and indicates its status. If the receiver is busy, it may indicate this by sending a character representing ''busy'' back to the transmitter. The transmitter will continue to send its ready signal until the receiver signals back that it is not busy or is ready to receive.

At that point, data transmission takes place. Once the transmission is complete, some additional handshaking takes place. The receiver acknowledges that it has received the information. The transmitter then sends a character indicating that the transmission is complete, and the receiver may acknowledge this.

To give you a more specific example, let's discuss the Xmodem protocol which is widely used in data transmission between personal computers. Such data transmissions use asynchronous methods and the ASCII code.

The data transmission procedure begins with the receiving computer transmitting a negative acknowledge (NAK) character to the transmitter. The NAK is an ASCII character code of 7 bits that is transmitted serially back to the transmitter every 10 s until the transmitter recognizes it. Once the transmitter recognizes the NAK character, it will begin to send a 128-byte block of data. This block begins with a start of header (SOH) character, which is another ASCII character meaning that the transmission is beginning. A header is usually two or more characters preceding the actual data block which give auxiliary information. In the Xmodem protocol, the header consists of 2 bytes designating the block number. In most messages, several blocks of data are transmit-

ted and each is numbered sequentially. The first byte is the block number in binary code. The second byte is the complement of the block number; that is, all bits have been inverted. Then, the 128-byte block is transmitted. At the end of the block, the transmitting computer sends a check sum byte. The check sum is a method of error detection which will be described later. It is the binary sum of all the binary information sent in the block.

The receiving computer looks at the block of data and also computes the check sum. If the check sum of the received block is the same as that transmitted, then it is assumed that the block was received correctly. If the block was received correctly, it sends an acknowledge (ACK) character back to the transmitter. Again, ACK is simply another ASCII code. Assuming ACK is received by the transmitter, then the next block of data is sent. However, if the block was not received correctly due to interference or equipment problems, the check sums will not match. As a result, the receiving computer will send the NAK code back to the transmitter. If the transmitter receives NAK, it will automatically respond by sending the block again. This process is repeated until the entire block and the entire message is sent without errors.

When the entire message has been sent, the transmitting computer will send an end of transmission (EOT) character. The receiving computer replies with an ACK character, thus terminating the communication. Keep in mind that each character is sent along with its start and stop bits since this is an asynchronous protocol.

In synchronous data communications, the protocols become even more complex. Like the protocols used in the asynchronous Xmodem system, various control characters are used for signaling purposes at the beginning and ending of the block of data to be transmitted.

An example of a typical synchronous protocol is IBM's Bisync protocol widely used in computer communications. The protocol usually begins with the transmission of two or more sync (SYN) characters. See Fig. 12-32.

Fig. 12-32   Bisync synchronous protocol.

These characters signal the beginning of the transmission and are also used to initialize the clock timing circuits in the receiving modem. This will ensure proper synchronization of the data transmitted 1 bit at a time.

After the SYN characters, a start of header (SOH) character is transmitted. The header is a group of characters that usually identifies the type of message to be sent, a block number, a priority code, or some specific routing destination. The end of the header is signaled by a start of text (STX) character. At this point, the desired message is transmitted, 1 byte at a time. Remember that in synchronous transmission no start and stop bits are sent. The 7- or 8-bit words are simply strung together one after another. The receiver must sort them out into individual binary words which will be dealt with on a parallel basis later in the computer. A block usually consists of no more than 256 characters.

At the end of the block an end of block (ETB) character is transmitted. If this is the last block in a complete message, than an end of text (ETX) character is transmitted. Finally, an end of transmission (EOT) character is transmitted signaling the end of the transmission. At the very end of the transmission, an error-detection code is appended. This is usually 1 or 2 bytes known as the binary check code (BCC) which will be discussed later.

As you have seen, all the protocols contain some form of error-checking method. A parity bit is used in asynchronous character transmission. A BCC is used in both the Xmodem asynchronous and the Bisync synchronous protocols. Let's take a look at the error-detection and -correction process in more detail.

The main objective in error detection and correction is to ensure a fully correct transmission of the message. When high data rates are used in a noisy environment, bit errors are inevitable. If the S/N ratio is good, the number of errors will be extremely small. The number of bit errors that occur for a given number of bits transmitted is referred to as the *bit error rate (BER)*. The BER is similar to a probability in that it is the ratio of the number of bit errors to the total number of bits transmitted. If there is one error for 100,000 bits transmitted, the BER is $1/100,000 = 10^{-5}$. The bit rate may be higher or lower depending upon the equipment, the environment, and other considerations.

Many different methods have been used to ensure 100 percent accuracy in transmission.

The simplest, of course, is simply to transmit each character or each message multiple times until it is properly received. This is known as redundancy. For example, one protocol may specify transmitting each character twice in succession. Entire blocks or messages may be treated in the same way.

Another approach to reliable transmission is to use special codes which permit automatic checking for accuracy. One such code is the ARQ exact-count code. This is a special 7-bit binary code used for representing letters of the alphabet, numbers, and other symbols. Refer to Fig. 12-33. Each 7-bit word contains exactly three binary 1s. One way to determine if a character is received properly is to count the number of binary 1s in each character received. If the count is 3, then the character was most likely sent correctly. Should noise occur and cause one of the bits to change, the number of 1s would be a different value and an error would be indicated. This might signal the system to repeat the character or the entire block.

Binary Code							Character	
Bit: 1	2	3	4	5	6	7	Letter	Figure
0	0	0	1	1	1	0	Letter shift	
0	1	0	0	1	1	0	Figure shift	
0	0	1	1	0	1	0	A	–
0	0	1	1	0	0	1	B	?
1	0	0	1	1	0	0	C	:
0	0	1	1	1	0	0	D	(WRU)
0	1	1	1	0	0	0	E	3
0	0	1	0	0	1	1	F	%
1	1	0	0	0	0	1	G	@
1	0	1	0	0	1	0	H	£
1	1	1	0	0	0	0	I	8
0	1	0	0	0	1	1	J	(bell)
0	0	0	1	0	1	1	K	(
1	1	0	0	0	1	0	L	)
1	0	1	0	0	0	1	M	.
1	0	1	0	1	0	0	N	,
1	0	0	0	1	1	0	O	9
1	0	0	1	0	1	0	P	0
0	0	0	1	1	0	1	Q	1
1	1	0	0	1	0	0	R	4
0	1	0	1	0	1	0	S	'
1	0	0	0	1	0	1	T	5
0	1	1	0	0	1	0	U	7
1	0	0	1	0	0	1	V	=
0	1	0	0	1	0	1	W	2
0	0	1	0	1	1	0	X	/
0	0	1	0	1	0	1	Y	6
0	1	1	0	0	0	1	Z	+
0	0	0	0	1	1	1		(blank)
1	1	0	1	0	0	0		(space)
1	0	1	1	0	0	0		(line feed)
1	0	0	0	0	1	1		(carriage return)

Fig. 12-33   **The ARQ exact-count code.**

One of the most widely used systems of error detection is known as parity. *Parity* is a system where each character transmitted contains one additional bit. That bit is known as a parity bit. The bit may be a binary 0 or binary 1 depending upon the number of 1s and 0s in the character itself.

Two systems of parity are normally used: odd and even. Odd parity means that the total number of binary 1 bits in the character, including the parity bit, is odd. Even parity means that the number of binary 1 bits in the character, including the parity bit, is even. Examples of odd and even parity are indicated below. The seven left-hand bits are the ASCII character, and the right-hand bit is the parity bit.

Odd parity:

  10110011

  00101001

Even parity:

  10110010

  00101000

The parity of each character to be transmitted is generated by a parity generator circuit. The parity generator is made up of several levels of exclusive OR (XOR) circuits as shown in Fig. 12-34. Normally the parity generator circuit monitors the shift register in the computer or modem UART. Just prior to transmitting the data in the register by shifting it out, the parity generator circuit generates the correct parity value and inserts it as the last character in the data stream. In an asynchronous system, the start bit comes first, followed by the character bits, the parity bit, and finally one or more stop bits.

At the receiving modem or computer, the serial data word is transferred into a shift register in the UART. A parity generator in the receiving UART produces the parity on the re-ceived character. If it matches the transmitted and received parity bit, then it is assumed that the character was transmitted correctly. If a parity error is detected, the system signals this to the computer. Then, depending upon the protocol or other procedures, the character may be retransmitted, the block may be transmitted, or the error may simply be ignored.

Although parity checking is simple and effective, it is useful only for detecting single bit errors. If two or more bit errors occur, the parity circuit may not detect it. If an even number of bit changes occurs, the parity circuit will not detect the error. In situations requiring more reliable transmission, more thorough schemes can be used.

Sometimes you will hear the individual character parity method of error detection referred to as the vertical redundancy check (VRC). In displaying characters transmitted in a data communications system, the bits are sometimes written vertically as illustrated in Fig. 12-35. The bit at the bottom is the parity or VRC bit as described earlier.

The term "vertical" also distinguishes the parity check from another method of error detection known as the horizontal or longitudinal redundancy check (LRC). An LRC is the process of exclusive-ORing all the characters in a particular block of transmitted data. For example, the top bit of the first word is exclusive-ORed with the top bit of the second word. The result of this operation is exclusive-ORed with the top bit of the third word, and so on until all the bits in a particular horizontal row have been exclusive-ORed. The final bit value then becomes 1 bit in a character known as the block check character (BCC) or the block check sequence (BCS). Each row of

Fig. 12-34 **A parity generator circuit.**

Character	D	A	T	A		C	O	M	LRC or BCC
(LSB)	0	1	0	1	0	1	1	1	1
ASCII	0	0	0	0	0	1	1	0	1
	1	0	1	0	0	0	1	1	0
	0	0	0	0	0	0	1	1	0
Code	0	0	1	0	0	0	0	0	1
	0	0	0	0	1	0	0	0	1
(MSB)	1	1	1	1	0	1	1	1	1
Parity → or VRC (odd)	1	1	0	1	0	0	0	1	0

Fig. 12-35 Vertical and horizontal redundancy checks.

bits is manipulated in the same way to produce the BCC. All the characters transmitted in the text as well as STX, ETX, ETB, EOT, and so on are included as part of the BCC. Typically, SYN and SOH characters are not included as part of the BCC.

The BCC is computed by circuits in the computer or modem as the data is transmitted, and it is then appended to the message. At the receiving end, the computer computes its own version of the BCC on the received data and compares it to the received BCC. Again, the two should be the same.

As it turns out, by knowing the parity on each character as well as the BCC, the exact location of a faulty bit can be determined. The individual character parity bits and the BCC bits provide a form of coordinate system that allows a particular bit error in one character to be identified. Once it is identified, the bit is simply complemented to correct it. The VRC identifies which character contains the bit error, while the LRC identifies which bit contains the error.

Another more reliable error-detection scheme is the cyclic redundancy check (CRC).

The CRC is a mathematical technique which is applied to the transmitted data. It is effective in catching 99.9 percent or more of transmission errors. The mathematical process is essentially a division. The entire string of bits in a block of data are considered to be one giant binary number which is divided by some preselected constant. The quotient resulting from the division is discarded, and any remainder is retained. This remainder is known as the CRC character.

Although the CRC mathematical process can actually be programmed using the computer's instruction set, in most applications it is computed by a special CRC hardware circuit. A typical circuit is shown in Fig. 12-36. It consists of several shift registers into which XOR gates have been inserted at specific points. The data to be checked is fed into the registers serially. The data is simply shifted in a bit at a time, and at the conclusion of the data, the contents will be the remainder of the division or the desired CRC value. Since 16 flip-flops total are used in the shift register, the CRC can be transmitted as two sequential 8-bit bytes. The CRC is computed as the data is transmitted, and the resulting CRC is appended to the block.

At the receiving end, the CRC is computed by the receiving computer and compared to the received CRC characters. Assuming the two are alike, the message is correctly received. Any difference indicates an error which then initiates retransmission or some other form of corrective action.

A number of error-correction schemes have also been devised to complement the error-detection methods just described. Typically these involve various mathematical processes on the characters to be transmitted which re-

Fig. 12-36 A CRC error-detection circuit made with a 16-bit shift register and XOR gates.

sult in the addition of several bits of information to each character transmitted. The character and these extra bits can then be processed to determine if an error occurs. If a bit error does occur, these error-correction codes will automatically identify it, making it possible for it to be corrected. The techniques for error correction involve complex computer and digital circuitry and are beyond the scope of this chapter.

Self-Test

Answer the following questions.

59. Rules and procedures that describe how data will be transmitted and received are referred to as _____ .
60. List five methods of error detection and correction.
61. The process of exchanging signals between transmitter and receiver to indicate status or availability is called _____ .
62. A popular protocol used in personal computer data communications is known as _____ .
63. A string of characters making up a message or part of a message is referred to as a(n) _____ of data.
64. Stop and start bits are used with _____ (asynchronous, synchronous) data.
65. Synchronous protocols usually begin with the _____ character.
66. The characters that indicate message number, destination, or other facts are collectively referred to as the _____ .
67. The _____ character designates the beginning of a message, while the _____ character indicates its end.
68. One simple but time-consuming way to ensure an error-free transmission is to send the data multiple times. This is called _____ .
69. One type of special self checking error code is the _____ code.
70. Errors in data transmission are usually caused by _____ .
71. The ratio of the number of bit errors to the total number of bits transmitted is known as the _____ .
72. A bit added to the transmitted character to help indicate an error is called the _____ bit.
73. A number added to the end of a data block to assist in detecting errors is known as the _____ .

74. Write the correct parity bits for each number.
    a. Odd: 0011000____
    b. Even: 1111101____
    c. Odd: 0101101____
    d. Even: 1000110____
75. Another name for parity is _____ .
76. The basic building block of a parity generator circuit is the _____ gate.
77. Performing an exclusive OR on corresponding bits in successive data words produces a(n) _____ word.
78. An error-detection system that uses a BCC at the end of a block is known as a(n) _____ .
79. Dividing the data block by a constant produces a quotient that is discarded and a remainder called the _____ character.
80. The CRC circuit is basically a(n) _____ .
81. True or false. If a bit error can be identified, it can be corrected.

12-4 INTRODUCTION TO NETWORKS

Up to now we have discussed communications as primarily a two-way channel where both parties can transmit and receive. Although this one-to-one mode of communications is relatively common, in many communications systems a user may communicate with many more than one additional station. In order for a station to communicate with other stations, some form of connection or link must be established. If wired communications is used, each station must be connected to all the other stations with which communication is desired. This interconnection of one station to many others is referred to as networking. A *network* is any interconnection of two or more stations that wish to communicate. Each station in a communications network is called a *node*. These nodes are interconnected with the desired wiring configuration. One possible arrangement of a network is shown in Fig. 12-37. Each node is connected to all other nodes. Many other organized interconnections are possible as you will see.

Today, electronic communications networks are very common. Many of our most familiar communications systems use network techniques. In this section, we will introduce you briefly to communications networks.

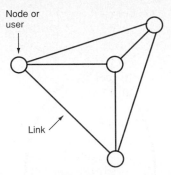

Fig. 12-37 A simple communications network.

Perhaps the most familiar of all communications networks is the telephone system. The telephone system may indeed be the largest and most sophisticated network of all. Each person has a transmitter and receiver known as a telephone, and through many different networks that individual can communicate by telephone with virtually any other person on this planet. Large networks such as those that cover a complete country or state are generally referred to as *wide area networks (WANs)*. Most WANs are telephone systems. The long-distance systems of AT&T, MCI, GTE, and others are examples of WANs. These networks use cables of various types for their interconnection, and, in many cases, radio links using microwave relay stations or satellites also form part of the network. Most telephone networks, of course, carry voice communications, but as you have seen, when a modem is used, digital data can be converted into analog signals which are fully compatible with the voice-grade telephone system.

Smaller networks are also common. A typical example is a local cable television system. A cable TV company picks up programming from sources available on a satellite and rebroadcasts them via a system of coaxes to subscribers. In this case, the communications is only one-way, but the total interconnection is a network where each home is a node in the network. Such medium-sized networks are sometimes referred to as *metropolitan area networks (MANs)*.

Even smaller networks are used. Known as *local area networks (LANs)*, these networks interconnect multiple stations over a very small area. Most LANs have 10 to 100 users but never more than 1000. For example, LANs are widely used in companies to connect several offices within the same building or to connect different floors of the building. Somewhat larger LANs may interconnect several buildings within a complex, such as networks used on college campuses and in military installations. In a LAN, the nodes are usually computers or terminals. Local area networks made up of PCs are the most common type. Data transmitted over a LAN is typically digital rather than voice or video.

One of the most common types of LANs is a large mainframe or minicomputer connected to many remote terminals. Multiple users time-share the system. The remote nodes are dumb terminals or PCs set up to emulate a terminal. Users have access to all the system's capabilities and may communicate with one another by an electronic mail system.

Initially LANs were developed so that multiple PC users could share expensive or limited-use peripherals such as printers, hard disks, and modems. One computer in the network is set up as a controller through which other users have access to the peripherals. Such units are called servers, disk servers, printer servers, or modem servers. A variation of the disk server is a file server that combines a disk server with sophisticated software that lets users share data in a large data base. The LANs also permit software and data sharing.

In addition to sharing peripherals, LANs permit users to talk with one another directly or through an electronic mail system. In the electronic mail system, one computer has a large disk storage unit in which users are assigned "mailboxes" where messages from other users are stored.

Although many LANs stand alone, others are interconnected to other systems. Frequently a LAN is connected to a larger mainframe or minicomputer through one of the PCs in the network. This PC with its software and often special controller hardware is called a gateway. A LAN is also sometimes connected to other LANs. A PC in each LAN acts as the link and controller and is called a bridge.

There are many different ways that stations or nodes can be interconnected to form a network. However, there are several popular configurations which are used again and again. These different network configurations or arrangements are known as *network topologies*. The more popular topologies are the star, the ring, the bus, and the tree. Let's take a look at each in more detail.

The *star* network is shown in Fig. 12-38 on the next page. It consists of a central controller node and each individual station. In order for one station to communicate with another,

Wide area network (WAN)

Metropolitan area network (MAN)

Local area network (LAN)

Network topologies

Star network

Private branch exchange (PBX)

Ring network

Bus network

the communication must take place through the central controller station. It is the central controller that allows any one user to be linked up with any other user.

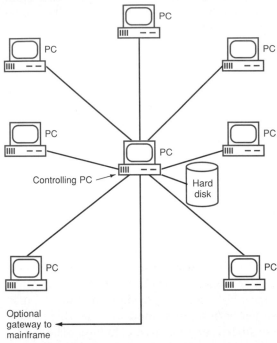

Fig. 12-38   A star LAN configuration.

An example of a star network is a mainframe or minicomputer which serves as the central controller for multiple video terminals with keyboards and CRT displays or PCs. Each station can communicate with the computer but must go through the computer to communicate with any other station.

Another example of a star network is a PBX system. The acronym PBX stands for *private branch exchange*, which is the terminology used to describe a small telephone system within an office or building. The PBX allows any telephone in the building to be linked to any other telephone. The PBX is nothing more than a large master switching circuit which automatically connects one station to another when the correct number is dialed. Although most PBXs are for voice applications, they are being more widely used for digital data.

A *ring* network configuration is shown in Fig. 12-39. There is no central controller station since each station performs some control functions. All the stations are connected end to end to form a continuous loop, so each station is effectively connected to two other stations. Each small box in the diagram is the interface circuitry.

Fig. 12-39   A ring LAN configuration.

Stations in the ring can both send and receive. All nodes have the ability to recognize their assigned code or address and the capability of retransmitting a received signal. The message used to communicate between stations begins with the address of the station to whom the message is to be sent. The desired message then follows. The station originating the message sends it to the next station in line. In a ring, the direction of communication is usually unidirectional. If the message is not for the next station in line, the address will not be recognized, so that station will simply repeat the message and send it on to the next node. Each station in the ring performs that operation until the desired station is reached, at which time it recognizes the address and receives the message.

Another widely used network configuration is the *bus* as shown in Fig. 12-40. Here all the stations effectively share a common cable called the bus. Communications is bidirectional on the bus by allowing any one station to talk to any other station. In a bus configuration, the sending station effectively broadcasts its message on the bus which all stations can receive.

The main advantage of the bus system is that it is much faster than any of the other topologies. In both the star and ring configurations, other devices become involved in the transmission. In the bus system, the two stations desiring to communicate do so directly without

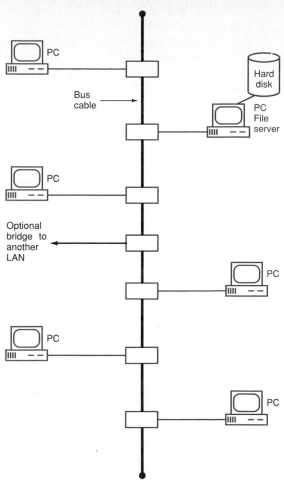

Fig. 12-40   A bus LAN configuration.

Type	Advantages	Disadvantages
Star	1. Excellent control 2. Facilities sharing 3. Failure of one node does not disable system	1. Slow speed response during heavy use 2. Limited number of nodes possible
Ring	1. Lowest cost of all topologies 2. Ease of expansion	1. Slow speed since nodes are accessed sequentially 2. Failure of one node disables the system
Bus	1. Fastest speed due to direct access 2. Failure of one node will not disable the system 3. Ease of expansion	

Fig. 12-41   Comparison of LAN topologies.

interference from or assistance from any other station. Bus systems are fast and simple as a result.

The *tree* topology is essentially a variation of the bus system. The bus is extended into several branches so that more stations can be interconnected. In some large tree systems, a controller node is used.

The three basic LAN topologies are compared and contrasted in the table of Fig. 12-41.

There are three basic transmission media used in interconnecting LANs: twisted pair, coax, and fiber-optic cable. Twisted pair is a cable made of two #22 or #24 gage insulated solid-copper wires twisted loosely together. It is very inexpensive and easy to work with. It is excellent for short distances, no more than about 2 mi, usually less. It can accommodate data rates up to about 1 Mbits/s. Twisted pair is popular in many simple, low-speed LANs. The biggest disadvantage is that because it is unshielded, it can pick up noise which can lead to bit errors.

Coax is also widely used in LANs. Because it is shielded, noise is less of a problem. Coax can also be used over longer distances, up to about 10 mi. Its wider bandwidth permits data rates up to 10 Mbits/s. It is by far the most widely used LAN medium.

Fiber-optic cable is gaining in popularity as its cost and complexity come down. Its main advantages are superior noise immunity and very high data rates, up to 50 Mbits/s. Distances up to about 5 mi can be accommodated.

The LANs can also be categorized according to whether baseband or broadband operation is chosen. Baseband refers to the data or information signal whether it is binary or analog voice or video. In baseband systems, the digital data is applied directly to the medium. The entire bandwidth of the medium is used by the signal. Such systems are simple, inexpensive, and easy to work with, but only one signal can be carried on the medium at a time. Further, the cable attenuation and distortion greatly limit the transmission distance, usually up to 1 mi with twisted pair or 5 mi with coax. Despite these disadvantages, most LANs are of the baseband type.

Broadband means that modulation techniques are used. Analog methods are used to transmit digital data. In this way, data signals can be translated up in frequency to specific channels. Frequency division multiplexing techniques are used to create multiple channels for binary signals. Broadband systems use coax because it has a bandwidth of up to 300 to 450 MHz. As a result, an enormous number of high-speed channels can be created. In addition, transmission over longer distances (up to 10 mi) on coax can be achieved. Of course,

CHAPTER 12   DATA COMMUNICATIONS   **337**

broadband systems are more complex and expensive because modems are required at each node.

Self-Test

Answer the following questions.

82. The telephone system is an example of a _____ .

83. True or false. A cable TV system is considered a network.

84. Most networks are smaller and cover shorter distances and are referred to as _____ .

85. The acronym PBX means _____ .

86. True or false. PBXs cannot be used with digital data.

87. The three main types of LAN topologies are _____ .

88. A PBX is a type of _____ topology LAN.

89. An LAN that uses a central controller for multiple stations is the _____ topology.

90. One type of LAN connects multiple terminals and PCs to a large central _____ computer.

91. A LAN in which the message is passed from one station to the next until the destination is reached is called a(n) _____ .

92. The stations or users in a network are referred to as _____ .

93. A network in which all stations attach to a common cable is called a(n) _____ .

94. The fastest network configuration is the _____ .

95. In a(n) _____ LAN configuration, if one station fails, the whole system fails.

96. A connection from a LAN to a mainframe computer is known as a(n) _____ .

97. A connection from one LAN to another is called a(n) _____ .

98. The three transmission media used in LANs are _____ .

99. An extended variation of the bus topology is called the _____ .

100. List the three main disadvantages of twisted pair cable.

101. _____ is the most widely used medium in LANs.

102. The medium that permits the highest data rates is _____ .

103. The longest transmission distances can be achieved with _____ medium.

104. Give the average maximum data rates in Mbits/s for each medium: twisted pair, coax, fiber-optic cable.

105. The digital, voice, or video voltages are referred to as _____ signals.

106. When the basic data signal is sent directly over the medium, the type of operation is said to be _____ .

107. _____ operation uses modulation techniques and FDM techniques to increase speed and the number of operational channels.

108. Broadband operation requires the use of a(n) _____ at each node.

12-5 SPREAD SPECTRUM

Spread spectrum (SS) is a modulation and multiplexing technique that distributes the signal and its sidebands over a very wide bandwidth. Traditionally, we evaluate the efficiency of modulation and multiplexing methods by how little bandwidth they consume. The continued increase in all types of radio communications, the resulting crowding, and the finite bounds of usable spectrum space have made everyone sensitive to how much bandwidth a given signal occupies. And designers of communications systems and equipment typically do all in their power to minimize the amount of bandwidth a signal takes. If that is the case, then how can a scheme that spreads the signal over a very wide piece of the spectrum be of value? The answer to this question and a discussion of the benefits and concepts of spread spectrum are the subjects of this section.

Spread spectrum is not new. It was developed primarily by the military after World War II. Its use has been restricted to military communications applications because it is a secure communications technique that is essentially immune to jamming. In the mid-1980s, the FCC authorized use of SS in civilian applications. Currently, unlicensed operation is permitted in the 902- to 928-MHz, 2.4- to 2.483-GHz, and 5.725- to 5.85-GHz ranges with 1 W of power. Spread spectrum on these frequencies is now being widely incorporated into a variety of commercial communications systems. One of the most popular new applications is wireless data communications. A variety of new wireless local area networks

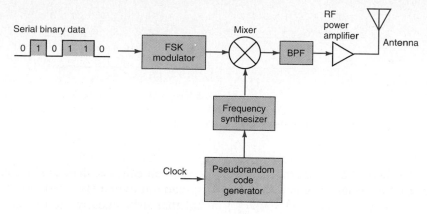

Fig. 12-42   A frequency-hopping SS transmitter.

(LANs) and portable personal computer modems use SS techniques. The use of SS is expected to grow because of the numerous benefits it offers.

There are two basic types of SS: *frequency hopping (FH)* and *direct sequence (DS)*. In frequency hopping, the frequency of the carrier of the transmitter is changed randomly at a rate higher than that of the serial binary data modulating the carrier. In direct sequence, the serial binary data is mixed with a higher-frequency pseudorandom binary code at a faster rate and the result is used to phase modulate a carrier. Let's examine each technique in more detail.

Figure 12-42 shows a block diagram of a frequency-hopping SS transmitter. The serial binary data to be transmitted is applied to a conventional FSK modulator. The two-tone FSK modulator output is applied to a mixer. Also driving the mixer is a frequency synthesizer. The output signal from the bandpass filter (BPF) after the mixer is the difference between one of the two FSK sine waves and the frequency of the frequency synthesizer. Note that the synthesizer is driven by a pseudorandom code generator. This is either a special

digital circuit or the output of a specially programmed microprocessor.

The pseudorandom code is a pattern of binary 0s and 1s that changes randomly. For example, an 8-bit code word can make the synthesizer hop to 256 different frequencies. The randomness of the 1s and 0s makes the serial output of this circuit seem like digital noise. Sometimes the output of this generator is called *pseudorandom noise (PSN)*. The binary sequence is not totally random, because it does repeat after many bit changes. However, the randomness is sufficient to minimize the possibility of someone accidentally duplicating the code.

Pseudorandom noise sequences are usually generated by a shift register circuit similar to that shown in Fig. 12-43. Here eight flip-flops in the shift register are clocked by an external clock oscillator. The input to the shift register is derived by X-ORing two or more of the flip-flop outputs. It is this connection that produces the pseudorandom sequence. The output is taken from the last flip-flop in the register. By changing the number of flip-flops in the register and/or changing the outputs which are X-ORed and fed back, the code sequence can

Fig. 12-43   A typical PSN code generator.

Fig. 12-44 Serial data and the PSN code rate.

be changed. The serial PSN sequence can be converted to parallel codes for controlling the divide-by-N counter in the feedback loop of a phase-locked loop synthesizer. Or, the parallel codes can be converted to corresponding frequencies by means of direct digital synthesis. Alternatively, a microprocessor can be programmed to generate pseudorandom sequences.

In a frequency-hopping SS system, the rate of synthesizer frequency change is higher than the data rate. This means that although the data bit and the FSK tone it produces remain constant for one data interval, the frequency synthesizer will switch frequencies many times during this period. This is shown in Fig. 12-44. Here, the frequency synthesizer changes frequencies four times for each bit time of the serial binary data. The time that the synthesizer remains on a single frequency is called the *dwell time*. The frequency synthesizer puts out a pseudorandom sine wave frequency to the mixer. The mixer creates a new carrier frequency for each dwell interval. The resulting signal, the frequency of which jumps around, effectively scatters pieces of the signal all over the band. This wideband signal is amplified and fed to the antenna.

This is a signal that rapidly jumps around all over the place. Specifically, the carrier randomly switches between dozens or even hundreds of frequencies over a given bandwidth. The actual dwell time on any frequency varies with the application and data rate, but it may be as short as 10 milliseconds (ms). Currently, the FCC regulations specify that there must be a minimum of 50 hopping frequencies and that the dwell time must be no longer than 400 ms. Figure 12-45 shows what one frequency-hopping sequence might be. The horizontal axis is time divided into increments of the dwell time. The vertical axis is the transmitter output frequency divided into step increments of the phase-locked loop (PLL) frequency synthesizer.

As you can see, the signal is divided up and spread out over a very wide bandwidth. A signal that may occupy only a few kilohertz of spectrum is spread out over a range that is 10 to 10,000 times that wide. However, the signal does not remain on any one frequency for a long period of time. Because it jumps around randomly, the SS signal will not interfere with a traditional signal on any of the hopping frequencies. The SS signal actually appears to be more like noise to a conventional narrow bandwidth receiver. A conventional receiver picking up such a signal will not respond even to a signal with a duration of tens of milliseconds. In addition, the conventional receiver cannot receive the SS signal because it does not have a wide enough bandwidth and it cannot follow, or track, its random frequency changes. Thus the SS signal is as secure as if it were scrambled.

Also, consider the situation when there are two or more SS transmitters operating over the same bandwidth but with different pseudorandom codes. Since the transmitters will be hopping to different frequencies at different

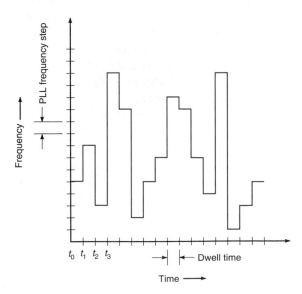

Fig. 12-45 A random frequency hop sequence.

times, they will not typically occupy a given frequency simultaneously with any of the other signals. For this reason, the signals do not interfere with one another. Because of this unique operation, SS is also a kind of multiplexing, for it permits two or more signals to use a given bandwidth concurrently without interference. Therefore, this method makes it possible to pack more signals in a given band than would be possible with other methods.

Another distinct advantage of SS is its ability to mostly eliminate selective fading or multipath distortion and cancellation. Consider an FM telemetry transmitter operating at a narrow band of frequencies. Reflected and direct signals will cause nulls or dropouts as two waves arrive at the receiver out of phase. Spread spectrum is redundant (because of dwell time), and although a given receiver location might cancel one frequency, the others will get through.

As more and more signals use a band, the background noise produced by many signals switching increases, but not enough to prevent highly reliable communications.

A frequency-hopping receiver is shown in Fig. 12-46. The very-wide-band signal picked up by the antenna is applied to a broadband RF amplifier and then to a conventional mixer. The mixer, like that in any superheterodyne receiver, is driven by a local oscillator which in this case is a frequency synthesizer like the one used at the transmitter. To receive the signal, the receiver local oscillator must have the same pseudorandom code sequence so that it is tuned to the correct frequency to receive the signal on the correct frequency when it occurs. The signal is thus reconstructed as an IF signal which contains the original FSK data. The signal is then applied to a FSK demodulator which reproduces the original binary data train.

One of the most important parts of the receiver is the circuit that is used to acquire and synchronize the transmitted signal with the internally generated pseudorandom code. Even though the codes must be the same, it is still a problem to get the two codes into step with one another. This problem is solved by a preamble signal and code at the beginning of the transmission. Once synchronism has been established, the code sequences occur in step. Any receiver not having the correct code cannot receive the signal. You can see how secure an SS is because of this characteristic.

One way different stations are distinguished from one another is by their code sequence. Many stations can share a common band by using different codes. Instead of assigning each station a single frequency on which to operate, each is given a different pseudorandom code within the same band. This permits a transmitter to transmit selectively to a single receiver without other receivers in the band being able to pick up the signal.

Another type of SS is direct sequence (DS). A block diagram of a DS transmitter is shown in Fig. 12-47 on the next page. The serial binary data is applied to an X-OR gate along with a serial pseudorandom code that occurs faster than the binary data. Figure 12-48 on the next page shows typical waveforms. One bit time for the pseudorandom code is called a *chip,* and the rate of the code is called the *chipping rate.* The chipping rate is faster than the data rate.

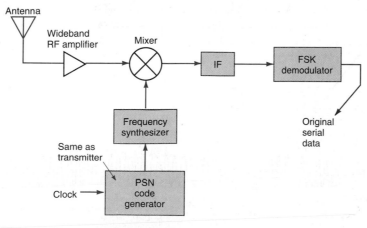

Fig. 12-46   A frequency-hopping receiver.

Fig. 12-47 **A direct-sequence SS transmitter.**

The signal developed at the output of the X-OR is then applied to a PSK modulator. Binary phase-shift keying (BPSK) is commonly used. The carrier phase is switched between 0° and 180° by the 1s and 0s of the X-OR output. Quadrature phase-shift keying (QPSK), quadrature amplitude modulation (QAM), and other forms of PSK may also be used. The PSK modulator is generally some form of balanced mixer as described earlier in this chapter. The signal phase modulating the carrier, being much higher in frequency than the data signal, causes the modulator to produce multiple widely spaced sidebands whose strength is such that the complete signal takes up a great deal of the spectrum. Thus the signal is spread. And because of its randomness, the resulting signal appears to be nothing more than wideband noise to a conventional narrowband receiver.

One type of direct-sequence receiver is shown in Fig. 12-49. The broadband SS signal

is amplified and then translated down to a lower IF frequency by mixing it in mixer 1 with a local oscillator signal. For example, the SS signal at an original carrier of 902 MHz might be translated down to an IF of 70 MHz. This IF signal is then compared to an IF signal that is produced in mixer 3 with the PSN sequence that is similar to that transmitted. This comparison takes place in mixer 2. The comparison process, called *correlation,* tells how alike the two signals being compared are. If the two signals are identical, the correlation is 100 percent. If the two signals are not at all alike, the correlation is 0 percent. The correlation process in the mixer produces a signal that is averaged in the low-pass filter at the output of mixer 2. The output signal will be a high average value if the transmitted and received PSN codes are alike.

The signal out of mixer 2 is fed to a synchronization circuit. The sync circuit has the toughest job of any in the receiver. It must

Fig. 12-48 **Data signals in direct-sequence SS.**

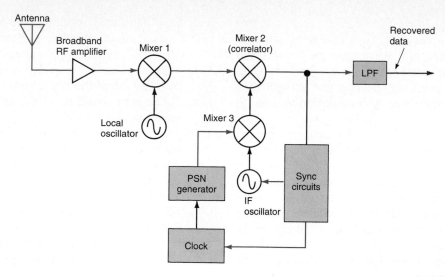

Fig. 12-49 **A direct-sequence receiver.**

Despread

Code division multiple access (CDMA)

Chirp

Time-hopping

re-create the carrier in the exact frequency and phase so that demodulation can occur. The synchronizing circuit varies the clock frequency so that the PSN code output frequency varies in order to find the same chip rate as the incoming signal. The clock drives a PSN code generator containing the exact code used at the transmitter. The PSN code in the receiver is the same as that of the received signal, but the two are out of sync with one another. Adjusting the clock by speeding it up or slowing it down will eventually cause the two to come into synchronization.

The PSN code produced in the receiver is used to phase-modulate a carrier at the IF in mixer 3. Like all the other mixers, this one is usually of the double-balanced diode ring type. The output of mixer 3 is a BPSK signal similar to that being received. It is compared to the received signal in mixer 2, which acts as a correlator. The output of mixer 2 is then filtered to recover the original serial binary data. The received signal is said to be *despread*.

Direct sequence is also called *code division multiple access (CDMA)* or *spread spectrum multiple access*. The term *multiple access* refers to any technique that is used for multiplexing many signals on a single communications channel. CDMA is used in satellite systems so that many signals can use a transponder, thus making it more efficient and generating more income for the owner. CDMA is also being considered as a method for future cellular telephone service because it permits more users to occupy a given band than other methods.

Although frequency hopping and direct sequence are the two most widely used types of SS, there are other methods. One example is called *chirp*. With this method, the signal modulates a carrier that is linearly swept over the band being used. This technique is used primarily in radar systems and in distance-ranging equipment.

Another type of SS is called *time-hopping*. In this system, the carrier is keyed on and off by the pseudorandom code sequence. The data signal modulates the carrier in the usual way.

Hybrid systems can also be created by combining two types of SS. Depending upon the application, such a combination will produce some additional benefits. The most common hybrid is a combination of frequency hopping and direct sequence.

What are the benefits of SS? First, the communications are secure; unauthorized listening is prevented. Unless the receiver has a very wide bandwidth and the exact pseudorandom code and type of modulation, the signal will not be received.

Second, SS signals are highly resistant to jamming. Jamming signals typically are restricted to a single frequency; jamming one frequency will not interfere with an SS signal.

In the same way, unintentional interference from a signal occupying the same band is greatly minimized and in most cases virtually eliminated.

A big advantage is that many users can share a single band with little or no interference. Because the signals are spread out and random, they will not interfere with one another. With very high usage, a slight increase in the background noise level may result, but in general it is not enough to create a problem. With SS, more signals can use a band than with any other type of modulation and multiplexing.

Another benefit is a resistance to fading. Frequency-selective fading occurs during signal propagation because signals of different frequency arrive at the receiver at slightly different times. Wide variations of signal strength due to reflections and other phenomena during propagation are virtually eliminated by SS.

Spread spectrum also provides a way for the duration of transmission of a signal to be precisely timed. The pseudorandom code makes it possible to accurately determine the start and end of a transmission. Thus SS is a superior method for radar and other ranging applications that rely upon the time of transmission to determine distance.

Spread spectrum is a rapidly growing technique in communications. It is finding more and more applications in data communications as its benefits are discovered and as new components and equipment become available to implement it.

Self-Test

Answer the following questions.

109. Spread spectrum is both a _____ and a _____ technique.
110. The two main types of spread spectrum are _____ and _____ .
111. In a frequency-hopping SS system, a _____ circuit generates the transmitter frequency.
112. In a frequency-hopping SS system, a _____ circuit selects the frequency produced by the synthesizer.
113. True or false. The hop rate is slower than the bit rate of the digital data.
114. To a narrowband receiver, an SS signal appears as a type of _____ .
115. Two or more stations using SS and sharing a common band are identified and distinguished from one another by their unique _____ .
116. The length of time a frequency-hopping SS transmitter stays on one frequency is called the _____ .
117. In a direct-sequence SS transmitter, the data signal is mixed with a PSN signal in an _____ circuit.
118. True or false. In a direct-sequence SS system, the chip rate is faster than the data rate.
119. The type of modulation used with direct-sequence SS is usually _____ .
120. The most difficult part of SS communications is initiating and maintaining _____ prior to and during transmission.
121. The process of comparing one direct-sequence SS signal to another in an effort to obtain a match is called _____ .
122. The two main benefits of SS are _____ and _____ .
123. The main circuit used in making a PSN code generator is a _____ .
124. The frequencies allocated to spread spectrum by the FCC for civilian applications are _____ , _____ , and _____ MHz.
125. Two growing applications for SS are _____ and _____ .

SUMMARY

1. Data communications is the transmission and reception of binary data between computers and other digital equipment.
2. The earliest form of electronic communications, the telegraph, was a type of data communications.
3. Turning a carrier off and on in a code of dots and dashes is a kind of data communications known as continuous wave (CW).
4. Teletype is a form of telegraph that uses the 5-bit Baudot code to transmit between typewriterlike units.
5. The most widely used binary data communications code is the 7-bit American Standard Code for Information Interchange (ASCII).
6. Another popular code is the 8-bit Extended Binary Coded Decimal Inter-

change Code (EBCDIC) used mainly in IBM systems.

7. The two main methods of data transmission are serial and parallel. In serial transmission, each bit is transmitted sequentially. In parallel transmission, all bits are transmitted simultaneously.

8. Serial transfers are slower than parallel transfers but require only a single line or channel. Parallel transfers require multiple channels of lines called a bus.

9. The speed of data transmission is designated in terms of bits per second (bits/s) or baud.

10. Baud rate is the number of symbol changes per second. A symbol is an amplitude, frequency, or phase change.

11. The channel- or bit-rate capacity of a channel is directly proportional to the channel bandwidth and the time of transmission.

12. The channel capacity or binary signal transmission speed C in bits per second is equal to twice the channel bandwidth B when no noise is present ($C = 2B$).

13. When multiple levels or symbols are used to encode the data, the channel capacity C is greater for a given bandwidth B, or $C = 2B \log_2 N$ where N is the number of symbols used.

14. The channel capacity C in bits per second is proportional to the channel bandwidth B and the power S/N ratio, or

$$C = B \log_2 (1 + S/N)$$

15. The bit rate is higher than the baud (symbol) rate if multiple-level (symbol) encoding is used.

16. The two methods of data transmission are asynchronous and synchronous. In asynchronous transmission, data is sent one character at a time with start and stop bits. In synchronous transmission, data is sent as a continuous block of multiple characters framed with synchronization characters.

17. Synchronous transmission is faster than asynchronous transmission.

18. In data communications, a binary 1 is referred to as a mark and a binary 0 as a space.

19. Signals, whether voice, video, or binary, transmitted directly over a cable are known as baseband signals.

20. Voice and video signals are analog but may be converted to digital for data communications transmission.

21. Signals that involve a modulated carrier are called broadband signals.

22. Communications of binary data signals over the telephone network which is designed for analog signals is made possible by using a modem.

23. A modem is a modulator-demodulator unit that converts digital signals to analog and vice versa.

24. The most commonly used modulation techniques in modems are frequency-shift keying (FSK), phase-shift keying (PSK), and quadrature amplitude modulation (QAM).

25. Frequency-shift keying uses two frequencies for binary 0 and 1 (1070 and 1270 Hz or 2025 and 2225 kHz). It operates at speeds of 300 baud or less.

26. Modems capable of transmitting at standard higher rates of 1200, 2400, 4800, and 9600 bits/s use PSK and/or QAM.

27. Binary PSK (BPSK) uses a carrier of 1600 or 1700 Hz where a phase of 0° represents a binary 0 and a 180° phase shift represents a binary 1, or vice versa.

28. Binary PSK is generated by a balanced modulator.

29. Binary PSK is demodulated by a balanced modulator.

30. To properly demodulate BPSK, the carrier at the demodulator must have exactly the same phase as the transmitting carrier.

31. A special carrier recovery circuit in the receiver produces the correct phase carrier from the BPSK signal.

32. Differential PSK eliminates the need for a special reference phase carrier by using a coding technique where the phase of each bit is referenced to the previous bit.

33. Quadrature PSK uses four equally spaced phase shifts of the carrier to represent two bits (dibit). For example, 00 = 45°, 01 = 135°, 11 = 225°, 10 = 315°.

34. In 8-PSK, 3 bits are coded per phase change. In 16-PSK, 4 bits are coded per phase change. Thus the bit rate is 3 or 4 times the symbol rate change or baud rate.

35. Quadrature amplitude modulation uses a combination of QPSK and two-level AM to code 3 bits per baud. Each of the eight possible 3-bit combinations is represented by a unique phase and amplitude signal.

36. A protocol is a rule or procedure that defines how data is sent and received.

37. Protocols include "handshaking" signals between the transmitter and receiver that indicate the status of each.

38. The Xmodem protocol is widely used in personal computers.

39. In synchronous communications, a variety of special characters are sent before and after the block of data to ensure that the data is correctly received.

40. Bit errors that occur during transmission are caused primarily by noise.

41. The ratio of the number of bit errors that occur for a given number of bits transmitted is known as the bit error rate (BER).

42. Error-detection and -correction schemes have been devised to reduce bit errors and increase data accuracy.

43. One of the most widely used error-detection schemes adds a parity bit to each character transmitted, making the total number of binary 1s transmitted odd or even. If a bit error occurs, the parity bit derived at the receiver will differ from the one transmitted.

44. Parity generator circuits are made up of multiple levels of exclusive OR (XOR) gates.

45. Another name for parity is vertical redundancy check (VRC).

46. The longitudinal redundancy check (LRC) is another way to test for errors. Corresponding bits in adjacent data words are exclusive-ORed to generate a block check character (BCC) that is appended to the transmitted block.

47. The VRC and LRC provide a coordinate system that will identify the exact location of a bit error so that it may be corrected.

48. A widely used error-detection scheme is the cyclical redundancy check (CRC) where the data block is divided by a constant to produce a quotient and remainder. The remainder is the CRC character which is attached to the transmitted data block and compared to the CRC computed at the receiver.

49. A network is any interconnection of two or more stations that can communicate with one another.

50. An example of a wide area network (WAN) is the telephone system. An example of a metropolitan area network (MAN) is a cable TV system.

51. A local area network (LAN) is an interconnection of stations in a small area over short distances such as in an office building, on a military base, or on a college campus.

52. Local area networks were conceived to allow PC users to share expensive peripherals such as hard disks and printers but are now used for general communication and the sharing of software and data.

53. The most common physical configurations or topologies of LANs are the star, ring, and bus.

54. The bus is the fastest. The ring is the least expensive but is disabled if one station fails.

55. The star configuration is commonly used to connect many terminals and PCs to a larger mainframe or minicomputer.

56. The three most commonly used transmission media in networks are twisted pair, coax, and fiber-optic cable.

57. Twisted pair is inexpensive and easy to work with and is widely used for short-distance, low-speed LANs. It is not shielded, so it is susceptible to noise.

58. Coax is the most widely used medium because of its high-speed capability and its shielding against noise.

59. Fiber-optic cable is growing in usage as its price declines. It has the highest speed capability of any medium.

60. Local area networks use both baseband and broadband techniques. Baseband refers to transmitting the information signal directly on the medium.

61. Broadband refers to using modulation techniques to transmit the data on carriers that can be assigned specific channels over a wide frequency range on a common medium.

62. Broadband systems require the use of modems at each node.

63. Spread spectrum (SS) is a modulation and multiplexing technique used primarily in data communications that deliberately spreads the signal out over a wide bandwidth rather than trying to restrict it to a narrow band.

64. The two most widely used types of spread spectrum are frequency hopping (FH) and direct sequence (DS).

65. Spread spectrum offers the benefits of privacy or security of communications, immunity to jamming, and lower sensitivity to frequency-selective fading.

66. In frequency-hopping SS, the serial binary data usually modulates a carrier by FSK.

The FSK signal is mixed with a sine wave from a frequency synthesizer to form the final RF signal. The frequency synthesizer is switched at a rate of speed higher than the rate of the data signal, dwelling only briefly on each of many channel frequencies. Thus, the signal is broken up into small pieces and spread over a wide frequency range.

67. The frequency-hopping scheme is controlled by a pseudorandom binary code that switches at random from binary 0 to 1 and vice versa. The random nature of the signal causes the frequency to jump all over the band, distributing pieces of the signal hither and yon. The pseudorandom code acts like digital noise and thus is called a pseudorandom noise (PSN) code.

68. In direct-sequence SS, the serial data is mixed with a higher-frequency PSN code in an X-OR circuit. The resulting higher-frequency binary signal produces more higher-frequency sidebands, thereby spreading the signal out over a wider bandwidth. The X-OR output usually phase-modulates the final carrier.

69. Because an SS signal is spread out randomly over a wide bandwidth, many signals can share a band without interference. A narrowband receiver will not respond to an SS signal except to interpret the random signals as a form of low-level noise.

70. The main problem in receiving SS signals is in acquiring the signal and synchronizing the transmitted PSN code to the same internally generated PSN code in the receiver. In direct-sequence SS, an electronic correlator circuit is responsible for achieving synchronism.

71. Different stations sharing a given band are defined and identified by their unique PSN code.

72. Spread spectrum was originally developed for and used in military equipment. In 1985, the FCC authorized SS for commercial or civilian use in the 902- to 928-, 2400- to 2483-, and 5725- to 5850-MHz bands.

73. The new applications for spread spectrum include wireless LANs, wireless computer modems, and telemetry systems.

CHAPTER REVIEW QUESTIONS

Choose the letter which best answers each question.

12-1. Data communications refers to the transmission of
    a. Voice.
    b. Video.
    c. Computer data.
    d. All the above.

12-2. Data communications uses
    a. Analog methods.
    b. Digital methods.
    c. Either of the above.
    d. Neither of the above

12-3. Which of the following is *not* primarily a type of data communications?
    a. Telephone
    b. Teletype
    c. Telegraph
    d. CW

12-4. The main reason that serial transmission is preferred to parallel transmission is that
    a. Serial is faster.
    b. Serial requires only a single channel.
    c. Serial requires multiple channels.
    d. Parallel is too expensive.

12-5. Mark and space refer respectively to
    a. Dot and dash.
    b. Message and interval.
    c. Binary 1 and binary 0.
    d. On and off.

12-6. The number of amplitude, frequency, or phase changes that take place per second is known as the
    a. Data rate in bits per second.
    b. Frequency of operation.
    c. Speed limit.
    d. Baud rate.

12-7. Data transmission of one character at a time with start and stop bits is known as what type of transmission?
- *a.* Asynchronous
- *b.* Serial
- *c.* Synchronous
- *d.* Parallel

12-8. The most widely used data communications code is
- *a.* Morse.
- *b.* ASCII.
- *c.* Baudot.
- *d.* EBCDIC.

12-9. The ASCII code has
- *a.* 4 bits.
- *b.* 5 bits.
- *c.* 7 bits.
- *d.* 8 bits.

12-10. Digital signals may be transmitted over the telephone network if
- *a.* Their speed is low enough.
- *b.* They are converted to analog first.
- *c.* They are ac instead of dc.
- *d.* They are digital only.

12-11. Start and stop bits, respectively, are
- *a.* Mark, space.
- *b.* Space, mark.
- *c.* Space, space.
- *d.* Mark, mark.

12-12. Which of the following is correct?
- *a.* The bit rate may be greater than the baud rate.
- *b.* The baud rate may be greater than the bit rate.
- *c.* The bit and baud rates are always the same.
- *d.* The bit and baud rates are not related.

12-13. A modem converts
- *a.* Analog signals to digital.
- *b.* Digital signals to analog.
- *c.* Both *a* and *b*.
- *d.* None of the above.

12-14. Slow-speed modems use
- *a.* FSK.
- *b.* BPSK.
- *c.* QPSK.
- *d.* QAM.

12-15. A carrier recovery circuit is *not* needed with
- *a.* BPSK.
- *c.* QPSK.
- *c.* DPSK.
- *d.* QAM.

12-16. The basic modulator and demodulator circuits in PSK are
- *a.* PLLs.
- *b.* Balanced modulators.
- *c.* Shift registers.
- *d.* Linear summers.

12-17. The carrier used with a BPSK demodulator is
- *a.* Generated by an oscillator.
- *b.* The BPSK signal itself.
- *c.* Twice the frequency of the transmitted carrier.
- *d.* Recovered from the BPSK signal.

12-18. A 9600-baud-rate signal can pass over the voice-grade telephone line if which kind of modulation is used?
- *a.* BPSK
- *b.* QPSK
- *c.* DPSK
- *d.* QAM

12-19. Quadrature amplitude modulation is
- *a.* Amplitude modulation only.
- *b.* QPSK only.
- *c.* AM plus QPSK.
- *d.* AM plus FSK.

12-20. A QAM modulator does *not* use a(n)
- *a.* XNOR.
- *b.* Bit splitter.
- *c.* Balanced modulator.
- *d.* 2-to-4 level converter.

12-21. A rule or procedure that defines how data is to be transmitted is called a(n)
- *a.* Handshake.
- *b.* Error-detection scheme.
- *c.* Data specification.
- *d.* Protocol.

12-22. A popular PC protocol is
    a. Parity.   *c.* CRC.
    b. Xmodem.   *d.* LRC.

12-23. A synchronous transmission usually begins with which character?
    a. SYN   *c.* SOH
    b. STX   *d.* ETB

12-24. The characters making up the message in a synchronous transmission are collectively referred to as a data
    a. Set.   *c.* Block.
    b. Sequence.   *d.* Collection.

12-25. Bit errors in data transmission are usually caused by
    a. Equipment failures.   *c.* Noise.
    b. Typing mistakes.   *d.* Poor S/N ratio at the receiver.

12-26. Which of the following is *not* a commonly used method of error detection?
    a. Parity   *c.* CRC
    b. BCC   *d.* Redundancy

12-27. Which of the following words has the correct parity bit? Assume odd parity. The last bit is the parity bit.
    a. 1111111 1   *c.* 0011010 1
    b. 1100110 1   *d.* 0000000 0

12-28. Another name for parity is
    a. Vertical redundancy check.   *c.* Longitudinal redundancy check.
    b. Block check character.   *d.* Cyclical redundancy check.

12-29. Ten bit errors occur in two million transmitted. The bit error rate is
    a. 2×10^{-5} *c.* 5×10^{-6}
    b. 5×10^{-5} *d.* 2×10^{-6}

12-30. The building block of a parity or BCC generator is a(n)
    a. Shift register.   *c.* 2-to-4 level converter.
    b. XOR.   *d.* UART.

12-31. A longitudinal redundancy check produces a(n)
    a. Block check character.   *c.* CRC.
    b. Parity bit.   *d.* Error correction.

12-32. Dividing the data block by a constant produces a remainder that is used for error detection. It is called the
    a. Vertical redundancy check.   *c.* Block check character.
    b. Horizontal redundancy check. *d.* Cyclical redundancy check.

12-33. A CRC generator uses which components?
    a. Balanced modulator   *c.* Binary adder
    b. Shift register   *d.* Multiplexer

12-34. Which of the following is *not* a LAN?
    a. PBX system   *c.* Office building system
    b. Hospital system   *d.* Cable TV system

12-35. The fastest LAN topology is the
    a. Ring.   *c.* Star.
    b. Bus.   *d.* Square.

12-36. Which is *not* a common LAN medium?
    a. Twin lead   *c.* Fiber-optic cable
    b. Twisted pair   *d.* Coax

12-37. A mainframe computer connected to multiple terminals and PCs usually uses which configuration?
    a. Bus   *c.* Star
    b. Ring   *d.* Tree

12-38. A small telephone switching system that can be used as a LAN is called a
    a. Ring.   *c.* UART.
    b. WAN.   *d.* PBX.

12-39. Which medium is the least susceptible to noise?

    a. Twin lead

    b. Twisted pair

    c. Fiber-optic cable

    d. Coax

12-40. Which medium is the most widely used in LANs?

    a. Twin lead

    b. Twisted pair

    c. Fiber-optic cable

    d. Coax

12-41. Transmitting the data signal directly over the medium is referred to as

    a. Baseband.

    b. Broadband.

    c. Ring.

    d. Bus.

12-42. The technique of using modulation and FDM to transmit multiple data channels of a common medium is known as

    a. Baseband.

    b. Broadband.

    c. Ring.

    d. Bus.

12-43. What is the minimum bandwidth required to transmit a 56 kbits/s binary signal with no noise?

    a. 14 kHz

    b. 28 kHz

    c. 56 kHz

    d. 112 kHz

12-44. Sixteen different levels (symbols) are used to encode binary data. The channel bandwidth is 36 MHz. The maximum channel capacity is

    a. 18 Mbits/s.

    b. 72 Mbits/s.

    c. 288 Mbits/s.

    d. 2.176 Gbits/s.

12-45. What is the bandwidth required to transmit at a rate of 10 Mbits/s in the presence of a 28-dB S/N radio?

    a. 1.075 MHz

    b. 5 MHz

    c. 10 MHz

    d. 10.75 MHz

12-46. Which circuit is common to both frequency-hopping and direct-sequence SS transmitters?

    a. Correlator

    b. PSN code generator

    c. Frequency synthesizer

    d. Sweep generator

12-47. Spread spectrum stations sharing a band are identified by and distinguished from one another by

    a. PSN code.

    b. Frequency of operation.

    c. Clock rate.

    d. Modulation type.

12-48. The type of modulation most often used with direct-sequence SS is

    a. QAM.

    b. SSB.

    c. FSK.

    d. PSK.

12-49. The main circuit in a PSN generator is a(n)

    a. X-OR.

    b. Multiplexer.

    c. Shift register.

    d. Mixer.

12-50. To a conventional narrowband receiver, an SS signal appears to be like

    a. Noise.

    b. Fading.

    c. A jamming signal.

    d. An intermittent connection.

12-51. Which of the following is *not* a benefit of SS?

    a. Jam-proof.

    b. Security.

    c. Immunity to fading.

    d. Noise proof.

12-52. Spread spectrum is a form of multiplexing.

    a. True.

    b. False.

12-53. The most critical and difficult part of receiving a direct-sequence SS signal is

    a. Frequency synthesis.

    b. Synchronism.

    c. PSN code generation.

    d. Carrier recovery.

CRITICAL THINKING QUESTIONS

12-1. How is voice intelligence transmitted via SS?

12-2. Could SS be used for satellite communications? Explain how and describe any benefits.

12-3. What are the three basic types of symbol changes that can be used to represent more bits per baud?

12-4. Explain how a wireless LAN works.

Answers to Self-Tests

1. binary or digital
2. true
3. telegraph
4. continuous wave (CW)
5. dots, dashes
6. K = $-\cdot-$; 8 $---\cdot\cdot$; , = $--\cdot\cdot--$
7. Baudot
8. teletype
9. M
10. ASCII
11. EBCDIC
12. B = 1000010; 3 = 0110011; ? = 0111111
13. 0000111

14. serial, parallel
15. false
16. shift register
17. start, stop
18. bits per second (bits/s)
19. baud rate
20. true
21. false
22. blocks
23. space, mark
24. information theory
25. bandwidth, time
26. 10,000
27. 60,000
28. true

29. less
30. 50
31. 100
32. modulator, demodulator
33. analog, digital (or binary)
34. telephone network
35. computers
36. frequency-shift keying
37. frequencies
38. mark (1), space (0)
39. 1070, 1270, 2025, 2225
40. universal asynchronous receiver transmitter (UART)
41. phase-shift keying

42. 0, 180
43. balanced modulator
44. balanced modulator
45. carrier recovery
46. differential
47. XNOR
48. four
49. 2 bits
50. four
51. bit splitter (or shift register)
52. 2-to-4 level converter
53. amplitude, PSK
54. true
55. false
56. true
57. false
58. true
59. protocols
60. redundancy, special error-checking codes, parity, block check code, cyclical redundancy check
61. handshaking
62. Xmodem
63. block
64. asynchronous
65. SYN or sync
66. header
67. STX, ETX or ETB
68. redundancy
69. ARQ exact-count
70. noise

71. bit error rate
72. parity
73. block check character
74. a. 1
    b. 0
    c. 1
    d. 1
75. vertical redundancy check
76. exclusive OR
77. block check character
78. longitudinal redundancy check
79. cyclical redundancy check
80. shift register
81. true
82. wide area network
83. true
84. local area networks
85. private branch exchange
86. false
87. star, ring, bus
88. star
89. star
90. mainframe (or mini)
91. ring
92. nodes
93. bus
94. bus
95. ring
96. gateway
97. bridge
98. twisted pair, coax, fiber-optic cable

99. tree
100. noise susceptibility, short distances, low data rates
101. coax
102. fiber-optic cable
103. coax
104. 1, 10, 50
105. baseband
106. baseband
107. broadband
108. modem
109. modulation, multiplexing
110. frequency hopping, direct sequence
111. frequency synthesizer
112. pseudorandom noise code generator
113. false
114. noise
115. PSN code
116. dwell time
117. X-OR
118. true
119. PSK
120. synchronism
121. correlation
122. security, jam-proof
123. shift register
124. 902- to 928-, 2400- to 2483-, 5725- to 5850-MHz
125. wireless LANs, wireless computer modems

CHAPTER 13

Fiber-Optic Communications

■

CHAPTER OBJECTIVES

This chapter will help you to:

1. *Draw* a basic block diagram of a fiber-optic communications system and *tell* what each part of it does.
2. *State* eight benefits of fiber-optic cables over electrical cables for communications.
3. *Name* six typical communications applications for fiber-optic cable.
4. *Explain* how light is propagated through a fiber-optic cable.
5. *Name* the three basic types of fiber-optic cable and *state* the two materials from which they are made.
6. *Calculate* the transmission loss in decibels (dB) of fiber-optic cable over a distance.
7. *Name* the two types of optical transmitter components and the main operating frequency range.
8. *Explain* the operation of an optical detector and receiver.

───────────

The communications medium in most electronic communications systems is either a wire conductor cable or free space. But now, a new medium is growing in popularity, the fiberoptic cable. A fiberoptic cable is essentially a light pipe that is used to carry a light beam from one place to another. Light is an electromagnetic signal like a radio wave. It can be modulated by information and sent over the fiber-optic cable. Because the frequency of light is extremely high, it can accommodate very wide bandwidths of information and/or extremely high data rates can be achieved with excellent reliability. This chapter introduces you to the basic concepts and circuits used in lightwave communications.

■

13-1 LIGHT-WAVE COMMUNICATIONS SYSTEMS

One of the main limitations of communications systems is their restricted information-carrying capabilities. In more specific terms what this means is that the communications medium can only carry so many messages. And, as you have seen, this information-handling ability is directly proportional to the bandwidth of the communications channel. In telephone systems, the bandwidth is limited by the charac-

teristics of the cable used to carry the signals. As the demand for telephones has increased, better cables and wiring systems have been developed. Further, multiplexing techniques have been developed to transmit multiple telephone conversations over a single cable. These techniques have the same effect as if the number of cables or channels of communications were greatly multiplied.

In radio communications systems, the information modulates a high-frequency carrier. The modulation produces sidebands, and therefore, the signal occupies a narrow portion of the RF spectrum which we call a channel. However, the RF spectrum is finite. There is only so much space for radio signals. To increase the information capacity of a channel, the bandwidth of the channel must be increased. This reduces available spectrum space. Multiplexing techniques are used to send more signals in a given channel bandwidth, and methods have been developed to transmit more information in less bandwidth.

The information-carrying capacity of a radio signal can be increased tremendously if higher carrier frequencies are used. As the demand for increased communications capacity has gone up over the years, higher and higher RFs are being used. Today, microwaves are the preferred radio channels for this reason, but it is more complex and expensive to use these higher frequencies because of the special equipment required. Further, transmission distances are limited to the line of sight.

Today the use of microwave frequencies for communications has been greatly expanded. Although the microwave frequency spectrum is not totally occupied at this point, it is well used. A great deal of communication takes place by microwave signals transmitted through satellite relay stations. Satellites can receive and retransmit thousands of telephone calls to any place on earth or can carry very high speed digital data. Slowly but surely, however, we are reaching the limits of our radio spectrum.

One way to expand communications capability further is to use light as the transmission medium. Instead of using an electrical signal traveling over a cable or electromagnetic waves traveling through space, the information is put on a light beam and transmitted through space or through a special cable. In the late nineteenth century, Alexander Graham Bell, the inventor of the telephone, demonstrated that information could be transmitted by light. A special transmitter made up of a microphone attached to a mirror was used to transmit voice over a distance of several hundred feet using a light beam from the sun. The mirror was used to reflect the light from the sun to the receiving site. The microphone, mechanically attached to the mirror, caused the mirror to vibrate as words were spoken into it. The mirror vibrated at the voice frequency, thereby changing the amplitude of the reflected sunlight. At the receiving site, a special photocell picked up the reflected light rays. This photocell was a small batterylike device that produced a dc output voltage proportional to the amount of light received. As the reflected light varied, the voltage output from the photocell varied in the same way. The voltage could be connected to an earphone so that the voice would be accurately reproduced.

Communications by light beam in free space is impractical over very long distances. The main disadvantage is the great attenuation of the light due to atmospheric effects. Fog, haze, smog, rain, snow, and other conditions absorb, reflect, and refract the light, greatly attenuating it and thereby limiting the transmission distance. Artificial light beams used to carry information are also virtually obliterated during daylight hours by the sun.

Light-beam communication was made more practical with the invention of the laser. The laser is a special high-intensity, single-frequency light source. It produces a very narrow beam of brilliant light of a specific wavelength (color). Because of its great intensity, the laser beam can penetrate atmospheric obstacles better than other types of light, thereby making light-beam communication more reliable over longer distances. The primary problem with such free-space light-beam communication is that the transmitter and receiver must be perfectly aligned with one another.

Instead of using free space, some type of light-carrying cable can also be used. For centuries it has been known that light is easily transmitted through various types of transparent media such as glass and water, but it wasn't until the early 1900s that scientists were able to develop practical light-carrying media. By the mid-1950s, glass fibers were developed that permitted long light-carrying cables to be constructed. Over the years, these glass fibers have been perfected. Further, low-cost plastic fiber cable was also developed. Developments in these cables permitted them to be made longer with less attenuation of the light.

Today, these fiber-optic cables have been highly refined. Cables many miles long can be constructed and interconnected for the purpose of transmitting information on a light beam over very long distances. Thanks to these highly refined fiber-optic cables, a new transmission medium is now available. Its great advantage is that light beams have an incredible information-carrying capacity. Whereas hundreds of telephone conversations may be transmitted simultaneously at microwave frequencies, many thousands of signals can be carried on a light beam through a fiber-optic cable. Using multiplexing techniques similar to

Fig. 13-1 **Basic elements of a fiber-optic communications system.**

those used in telephone and radio systems, fiber-optic communications systems have an almost limitless capacity for information transfer.

The components of a typical fiber-optic communications system are illustrated in Fig. 13-1. The information signal to be transmitted may be voice, video, or computer data. The first step is to convert the information into a form compatible with the communications medium. This is usually done by converting continuous analog signals such as voice and video (TV) signals into a series of digital pulses. An A/D converter is used for this purpose. Computer data is already in digital form.

These digital pulses are then used to flash a powerful light source off and on very rapidly. In simple low-cost systems that transmit over short distances, the light source is usually a light-emitting diode (LED). This is a semiconductor device that puts out a low-intensity red light beam. Other colors are also used. Infrared beams like those used in TV remote controls are also used in transmission. Another commonly used light source is the solid-state laser. This is also a semiconductor device that generates an extremely intense single-frequency light beam.

The light-beam pulses are then fed into a fiber-optic cable where they are transmitted over long distances. At the receiving end, a light-sensitive device known as a photocell or light detector is used to detect the light pulses. This photocell or photodetector converts the light pulses into an electrical signal. The electrical pulses are amplified and reshaped back

into digital form. They are fed to a decoder, such as a D/A converter, where the original voice or video is recovered.

In very long transmission systems, repeater units must be used along the way. Since the light is greatly attenuated when it travels over long distances, at some point it may be too weak to be received reliably. To overcome this problem, special relay stations are used to pick up the light beam, convert it back into electrical pulses that are amplified, and then retransmit the pulses on another light beam. Several stages of repeaters may be needed over very long distances. But despite the attenuation problem, the loss is less than the loss that occurs with the electric cables.

Fiber-optic communications systems are being used more and more each day. Their primary use is in long-distance telephone systems. The fiber-optic cables are no more expensive or complex to install than standard electric cables, yet their information-carrying capacity is many times greater. Most new long-distance telephone systems being installed use fiber-optic cables.

Fiber-optic communications systems are also being used in other applications. For example, they are being used to interconnect computers in networks within a large building, and they carry control signals in airplanes and ships. Many TV systems also use fiber-optic cable because of its wide bandwidth. In all cases, the fiber-optic cables replace conventional coax or twisted pair cables. The table in Fig. 13-2 (on the next page) lists some of the applications in which fiber-optic cables are being used.

Fig. 13-2 **Applications of fiber-optic cables.**

We have already mentioned the main benefit of fiber-optic cables, that is, their enormous information-carrying capability. This is dependent upon the bandwidth of the cable. Bandwidth refers to the range of frequencies which a cable will carry. Electric cables such as coax do have a wide bandwidth, but the bandwidth of fiber-optic cable is much greater. Therefore, more signals can be carried.

But that's not all. Fiber-optic cable has many other benefits. These are summarized briefly in Fig. 13-3. Is it any wonder that fiber-optic cables are becoming more widely used?

There are some disadvantages to fiber-optic cable. For example, its small size and brittleness make it much more difficult to work with. Special, expensive tools and techniques are re-

Fig. 13-3 **Benefits of fiber-optic cables over conventional electric cables.**

quired. It is also more costly, at least for now. In time, with the continuously rising price of copper, fiber-optic cable will no doubt cost less than electric cable. Part of the high cost of fiber-optic cable is due to the difficulty of manufacturing it. As methods improve, costs will decline.

Self-Test

Answer the following questions.

1. Fiber-optic cables carry _____ rather than electrical signals.
2. The three main types of information carried by fiber optic-cables are _____ , _____ , and _____ .
3. The major use of fiber-optic cables is in _____ .
4. Fiber-optic cables are made of _____ or _____ .
5. The main benefit of fiber-optic cable over electric cable is its _____ .
6. True or false. Fiber-optic cable has more loss than electric cable over long distances.
7. True or false. Fiber-optic cable is smaller, lighter, and stronger than electric cable.
8. List the two main disadvantages of fiber-optic cable.
9. The two most commonly used light sources in fiber-optic systems are _____ .
10. Voice and video signals are converted into _____ before being transmitted by a light beam.
11. The device that converts the light pulses into an electrical signal is a(n) _____ .
12. Regenerative units called _____ are often used to compensate for signal attenuation over long distances.

13-2 HOW FIBER-OPTIC CABLES WORK

A fiber-optic cable is nothing more than a light pipe. It is not really a hollow tube carrying light, but it is a long thin strand of glass or plastic fiber. Most fiber cables have a circular cross section with a diameter of only a fraction of an inch. Some fiber-optic cables only have a diameter the size of a human hair. A light source is placed at the end of the fiber, and light passes through it and exits at the other end of the cable. How the light actually propagates through the fiber depends upon the laws of optics. In this section, we will review the basic fundamentals of light and those theories of optics that explain how light is contained within the light pipe.

Light is a kind of electromagnetic radiation. Another form of electromagnetic radiation that you are familiar with is a radio wave. Any electromagnetic wave is made up of both electric and magnetic fields that travel through space from one place to another. The basic characteristic of electromagnetic radiation is its frequency or wavelength. To put light waves into perspective with the lower radio and other frequencies, refer to the spectrum chart in Fig. 13-4. At the left end of the spectrum are very low frequency electrical signals such as 60-cycle ac power and audio or sound frequencies in the 20-Hz to 20-kHz range. These are low frequencies which have long wavelengths. Higher up the scale to the right is a wide range of radio frequencies. Radio frequencies range from approximately 10 kHz to 300 GHz. Microwaves extend from 1 GHz to 300 GHz.

Even further up the scale we find visible light. The frequency of the optical spectrum is in the range of 3×10^{11} to 3×10^{16} Hz. This includes both the infrared and ultraviolet as well as the visible parts of the spectrum. The visible spectrum is from 4.3×10^{14} to 7.5×10^{14} Hz, but we rarely refer to the frequency of light. Instead, we state light in terms of wavelength. Recall that wavelength is a distance measured in meters between peaks of a wave. Light waves are very short and are usually expressed in nanometers (one billionth of a meter) or micrometers or microns (one millionth of a meter). Visible light is in the range of 400 to 700 nanometers (nm) 0.4 to 0.7 micrometers (μm), depending upon the color of the light. Short-wavelength light is violet (400 nm), and red (700 nm) is long-wavelength light.

Right below visible light is a region known as infrared. Infrared rays cannot be seen. However, they have the properties of light and, therefore, can be manipulated in the same way with mirrors, lenses, and other devices normally used to manipulate light. The source of most infrared is heat. Right above the visible spectrum is the ultraviolet range. Infrared and ultraviolet are included in what we call the optical spectrum. Beyond that are the x-rays, gamma rays, and cosmic rays which we are not concerned with here.

Light waves travel in a straight line like microwaves do. Light rays emitted by a candle, light bulb, or other light source move out in a straight line in all directions. Like electricity, these light rays travel at the speed of light, which is generally considered to be 300,000,000 m/s or about 186,000 mi/s in free space.

The speed of light depends upon the medium through which the light passes. The figures given above are correct for free space, meaning for light traveling in air or a vacuum. When light passes through another material such as glass, its speed is slower.

Light can be manipulated in many ways. For example, lenses are widely used to focus, enlarge, or decrease the size of light waves from some source. Lenses permit useful devices such as cameras, microscopes, binoculars, and telescopes to be constructed. Another simple way of manipulating light is to reflect it. When light rays hit a reflective surface, such as a mirror, the light waves are thrown back or deflected. By using mirrors, the direction of a light beam can be changed.

The *reflection* of light from a mirror follows a simple physical law. That is, the direction of the reflected light wave can be easily predicted if the angle of the light beam striking the mirror is known. Refer to Fig. 13-5. Assume an

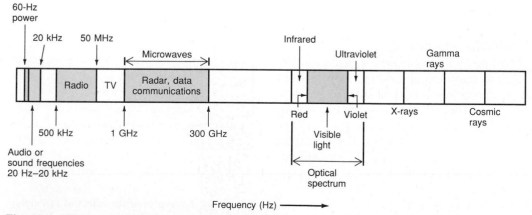

Fig. 13-4 Electromagnetic frequency spectrum.

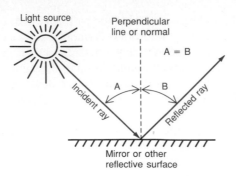

Fig. 13-5 **The law of light reflection. The angle of reflection equals the angle of incidence.**

imaginary line that is perpendicular to the flat mirror surface. A perpendicular line, of course, makes a right angle with the surface as shown. This imaginary perpendicular line is referred to as the *normal*. The normal is usually drawn at the point where the mirror reflects the light beam.

If the light beam follows the normal, then the reflection will simply be back along the same path. The reflected light ray will exactly coincide with the original light ray.

If the light ray strikes the mirror at some angle A from the normal, then the reflected light ray will leave the mirror at the same angle B to the normal. This principle is known as the law of reflection. It is usually expressed in the following form: *The angle of incidence is equal to the angle of reflection.*

The light ray from the light source is normally called the incident ray. It makes an angle A with the normal at the reflecting surface. This is called the angle of incidence. The reflected ray is the light wave that leaves the mirror surface. Its direction is determined by the angle of reflection B which is exactly equal to the angle of incidence.

Another way that the direction of the light ray can be changed is by refraction. *Refraction* is the bending of a light ray that occurs when the light rays pass from one medium to another. In reflection, the light ray bounces away from the reflecting surface rather than being absorbed by or passing through the mirror. Refraction only occurs when light passes through transparent material such as air, water, or glass. This refraction or the bending of the light rays takes place at the point where two different substances meet. For example, where air and water meet, refraction will occur. The dividing line between the two different substances or media is known as the boundary or interface.

A simple example of refraction is to place a spoon or straw into a glass of water as shown in Fig. 13-6(*a*). If you observe the glass of water from the side, it will appear as if the spoon or straw is bent or offset at the surface of the water.

Another phenomenon caused by refraction is the effect that occurs whenever you observe an object under water. For example, you may be standing in a clear stream and observing a stone at the bottom. The stone is actually in a different position than it appears to be from your observation. See Fig. 13-6(*b*).

The refraction or bending of light occurs simply because light travels at different speeds in different materials. The speed of light in free space is typically much higher than the speed of light in water, glass, or other materials. The

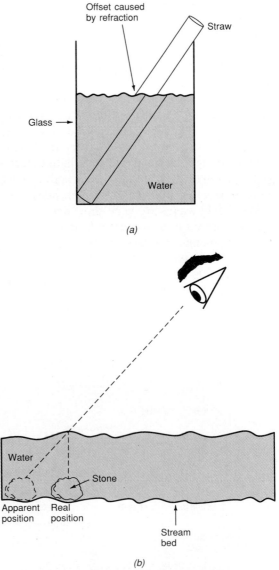

Fig. 13-6 **Examples of the effect of refraction.**

amount of refraction or bending of the light in a material is usually expressed in terms of the index of refraction n. This is simply the ratio of the speed of light in air to the speed of light in the substance.

Index of refraction (n)

$$= \frac{\text{speed of light in air}}{\text{speed of light in substance}}$$

Naturally, the index of refraction of air is 1, simply because 1 divided by itself is 1. The refractive index of water is approximately 1.3, whereas that of glass is 1.5.

To get a better understanding of this idea, let's consider a piece of glass with a refractive index of 1.5. What this means is that light will travel through 1.5 ft of air, but during that same time, the light will only travel 1 ft through the glass. The glass slows down the light wave considerably. The index of refraction is important because it tells exactly how much a light wave will be bent in various substances.

Whenever a light ray passes from one medium to the next, the light wave will be bent depending upon the index of refraction. In Fig. 13-7, you see a light ray passing through air. It makes an angle A with the normal. At the interface between the air and the glass, the direction of the light ray is changed. The speed of light is slower; therefore, the angle that the light beam makes to the normal B is different from the incident angle. By knowing the index of refraction, the exact angle can be determined.

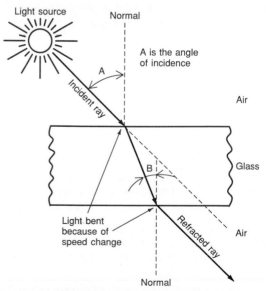

Fig. 13-7 How light rays are bent when passing from one medium to another.

If the light ray should then pass from the glass back into air, it will again change directions as Fig. 13-7 shows. The important point to note is that the angle of the refracted ray B is not equal to the angle of incidence A.

If you keep increasing the angle of incidence, at some point the angle of refraction will equal 90° to the normal as shown in Fig. 13-8(a). When this happens, the refracted light ray travels along the interface between the air and glass and the angle of incidence A is said to be the critical angle. The critical angle value depends upon the index of refraction of the glass.

(a)

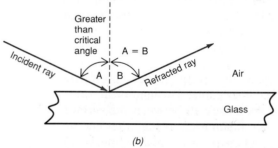

(b)

Fig. 13-8 Special cases of refraction. (a) Along the surface, and (b) reflection.

Now if you make the angle of incidence greater than the critical angle, the light ray will actually be reflected from the interface. See Fig. 13-8(b). When the light ray strikes the interface at an angle greater than the critical angle, the light ray does not pass through the interface into the glass. The effect is as if a mirror existed at the interface. When this occurs, the angle of reflection B is equal to the angle of incidence A as if a real mirror were used. This particular action is known as *total internal reflection*. Total internal reflection occurs only in those materials where the velocity of light is slower than in air. It is this basic principle that allows a fiber-optic cable to work.

Now let's see how this principle applies to the fiber-optic cable. Figure 13-9 shows a thin

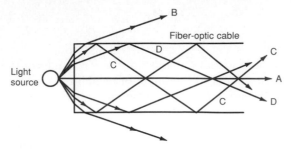

Fig. 13-9   Light rays in a fiber-optic cable.

fiber-optic cable. A beam of light is focused on the end of the cable as shown. This beam of light can be positioned a number of different ways so that the light enters the fiber at different angles. For example, light ray A enters the cable perpendicularly to the end surface. Therefore, the light beam simply travels straight down the fiber and exits at the other end.

The angle of light beam B is such that its angle of incidence is less than the critical angle, and, therefore, refraction takes place. The light wave passes through the fiber and exits the edge into the air at a different angle.

The angle of incidence of light beams C and D is greater than the critical angle. Therefore, total internal reflection takes place and the light beams are simply reflected off the surface of the fiber cable. The light beam simply bounces back and forth between the surfaces until the light beam exits at the other end of the cable.

When the light beam reflects off the inner surface, the angle of incidence is equal to the angle of reflection. Because of this principle, light rays entering at different angles will take different paths through the cable. Since some paths will be longer than others, some light rays will exit sooner and some later.

In practice, the light source is placed so that the angle is such that the light beam passes directly down the center axis of the cable or so that the reflection angles are great. This prevents the light from being lost due to refraction at the interface. Because of total internal reflection, the light beam will continue to propagate through the fiber even though it is bent. Care must be taken in bending the cable so as not to break it and so that the bend angle is not great enough to cause the angle of incidence to change which, in turn, will cause the light to be lost due to refraction. With long slow bends, the light will stay within the cable.

Self-Test

Answer the following questions.

13. Light is a type of _____ radiation.
14. A fiber-optic cable can be viewed as a light _____ .
15. Light travels in a
    a. Circle.
    b. Straight line.
    c. Curve.
    d. Random way.
16. The wavelength of light is usually expressed in _____ or _____ .
17. The lowest-frequency visible light is
    a. Red
    b. Violet
18. A micrometer, or micron, is a length of _____ of a _____ .
19. The wavelength range of visible light is _____ to _____ nm.
20. Infrared light has a wavelength that is
    a. Less than 400 nm.
    b. More than 700 nm.
    c. Less than 700 nm.
21. The optical spectrum is made up of what three parts?
22. The speed of light in air is _____ m/s or _____ mi/s.
23. True or false. The speed of light is slower in glass or water than it is in air.
24. The number that tells how fast light travels in a medium compared to air is the _____ of _____ .
25. Light beams can be bounced or their direction can be changed by _____ with a(n) _____ .
26. The bending of light rays due to speed changes when moving from one medium to another is called _____ .
27. If a light ray strikes a mirror at an angle of 30° from the normal, it is reflected at an angle of _____ ° from the normal.
28. When the angle of refraction is 90° to the normal, the ray travels along the _____ between the two media. Therefore, the incident ray strikes the surface at the _____ angle.
29. When the incident ray strikes the interface at an angle greater than the critical angle, _____ occurs.
30. The critical angle depends upon the value of the _____ of _____ .

13-3 FIBER-OPTIC CABLES

Just as standard electric cables come in a variety of sizes, shapes, and types, fiber-optic cables are available in different configurations. The simplest cable is just a single strand of fiber, whereas complex cables are made up of multiple fibers with different layers and other elements. In this section, we will introduce you to the various types of fiber-optic cables, show you how they are constructed, and explain the most popular kinds that you will encounter.

As indicated earlier, the portion of a fiber-optic cable that carries the light is made from either glass or plastic. Another name for glass is silica. Special techniques have been developed to create nearly perfect optical glass or plastic which is transparent to light. Such materials can carry light over a long distance. The glass or plastic is melted and pulled through a form to produce a fine threadlike fiber.

Glass has superior optical characteristics over plastic. However, glass is far more expensive and more fragile than plastic. Although the plastic is less expensive and more flexible, its attenuation of light is greater. For a given intensity, light will travel a greater distance in glass than in plastic. For very long distance transmissions, glass is certainly preferred. For shorter distances, plastic is much more practical.

In an electric cable, the current-carrying element is a copper wire. However, copper wires are usually not used alone. Typically, most wires are covered with an insulation not only to protect the wire but to prevent electrical shorts between adjacent wires.

In the same way, a fiber-optic cable is rarely used alone. The fiber, which is called the core, is usually surrounded by a protective cladding as illustrated in Fig. 13-10. The cladding is also made of glass or plastic but has a lower index

of refraction. This ensures that the proper interface is achieved so that the light waves remain within the core. In addition to protecting the fiber core from nicks and scratches, the cladding adds strength. Some fiber-optic cables have a glass core with a glass cladding. Others have a plastic core with a plastic cladding. Another common arrangement is a glass core with a plastic cladding. It is called plastic-clad silica (PCS) cable.

In observing a fiber-optic cable, you typically cannot tell the division between the core and the cladding. Since the two are usually made of the same types of material, to the naked eye it is not possible to see the difference. A plastic jacket similar to the outer insulation on an electric cable is usually put over the cladding.

There are two basic ways of classifying fiber-optic cables. The first way is an indication of how the index of refraction varies across the cross section of the cable. The second way of classification is by mode. Mode refers to the various paths that the light rays can take in passing through the fiber. Usually these two methods of classification are combined to define the types of cable.

There are two basic ways of defining the index of refraction variation across a cable. These are step index and graded index. *Step index* refers to the fact that there is a sharply defined step in the index of refraction where the fiber core and the cladding interface. It means that the core has one constant index of refraction N_1, while the cladding has another constant index of refraction N_2. Where the two come together, there is a distinct step as illustrated in Fig. 13-11 on the next page. If you were to plot a curve showing how the index of refraction varies vertically as you move from left to right across the cross section of the cable, there would be a sharp increase in the index of refraction as the core is encountered and then a sharp decline in the index of refraction as the cladding is encountered.

The other type of cable has a *graded index*. In this type of cable, the index of refraction of the core is not constant. Instead, the index of refraction varies smoothly and continuously over the diameter of the core as shown in Fig. 13-12 on the next page. As you get closer to the center of the core, the index of refraction gradually increases, reaching a peak at the center and then declining as the other outer edge of the core is reached. The index of refraction of the cladding is constant.

Fig. 13-10 Basic construction of a fiber-optic cable.

Core

Cladding

Step index

Graded index

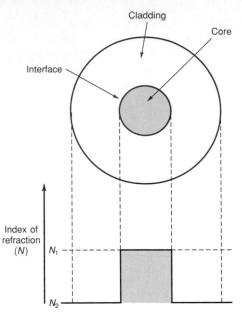

Fig. 13-11 A step-index cable cross section.

Mode refers to the number of paths for the light rays in the cable. There are two classifications: *single mode* and *multimode*. In single mode, light follows a single path through the core. In multimode, the light takes many paths through the core.

Each type of fiber-optic cable is classified by one of these methods of rating the index or mode. In practice, there are three commonly used types of fiber-optic cable. These are multimode step index, single-mode step index, and multimode graded index. Let's take a look at each of these types in more detail.

The *multimode step-index* fiber cable is probably the most common and widely used type.

It is also the easiest to make and, therefore, the least expensive. It is widely used for short to medium distances at relatively low pulse frequencies.

The main advantage of a multimode step-index fiber is the large size. Typical core diameters are in the 50- to 1000-μm range. Such large-diameter cores are excellent at gathering light and transmitting it efficiently. This means that an inexpensive light source such as an LED can be used to produce the light pulses. The light takes many hundreds or even thousands of paths through the core before exiting. You saw this earlier in Fig. 13-9. Because of the different lengths of these paths, some of the light rays take longer to reach the end of the cable than others. The problem with this is that it stretches the light pulses.

For example, in Fig. 13-13, a short light pulse is applied to the end of the cable by the source. Light rays from the source will travel in multiple paths. At the end of the cable, those rays which travel the shortest distance reach the end first. Other rays begin to reach the end of the cable later in time until the light ray with the longest path finally reaches the end, concluding the pulse. In Fig. 13-13 ray A reaches the end first, then B, then C. The result is a pulse at the other end of the cable that is lower in amplitude due to the attenuation of the light in the cable and increased in duration due to the different arrival times of the various light rays. This stretching of the pulse is referred to as modal dispersion.

Because the pulse has been stretched, input pulses cannot occur at a rate faster than the output pulse duration permits. Otherwise the pulses will essentially merge together as shown

Fig. 13-13 A multimode step-index cable.

Fig. 13-12 Graded-index cable cross section.

in Fig. 13-14. At the output, one long pulse will occur and will be indistinguishable from the three separate pulses originally transmitted. This means that incorrect information will be received. The only cure for this problem is to reduce the pulse repetition rate or the frequency of the pulses. When this is done, proper operation occurs. But with pulses at a lower frequency, less information can be handled.

Fig. 13-14 The effect of modal dispersion on pulses occurring too rapidly in a multimode step-index cable.

In a *single-mode*, or *mono-mode, step-index* fiber cable the core is so small that the total number of modes or paths through the core are minimized and modal dispersion is essentially eliminated. See Fig. 13-15. Typical core sizes are 2 to 15 μm. Virtually the only path through the core is down the center. With minimum refraction, no pulse stretching occurs. The output pulse has essentially the same duration as the input pulse.

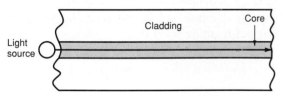

Fig. 13-15 Single-mode step-index cable.

The single-mode step-index fibers are by far the best since the pulse repetition rate can be high and the maximum amount of information can be carried. For very long distance transmission and maximum information content, single-mode step-index fiber cables should be used.

The main problem with this type of cable is that because of its extremely small size, it is difficult to make and is, therefore, very expensive. Handling, splicing, and making interconnections are also more difficult. Finally, for proper operation an expensive, superintense light source such as a laser must be used. For

long distances, however, this is the type of cable preferred.

Multimode graded-index fiber cables have several modes or paths of transmission through the cable, but they are much more orderly and predictable. Figure 13-16 shows the typical paths of the light beams. Because of the continuously varying index of refraction across the core, the light rays are bent smoothly and converge repeatedly at points along the cable. The light rays near the edge of the core take a longer path but travel faster since the index of refraction is lower. All the modes or light paths tend to arrive at one point simultaneously. The result is that there is less modal dispersion. It is not eliminated entirely, but the output pulse is not nearly as stretched as in multimode step-index cable. The output pulse is only slightly elongated. As a result, this cable can be used at very high pulse rates and, therefore, a considerable amount of information can be carried on it.

Fig. 13-16 A multimode graded-index cable.

This type of cable is also much wider in diameter with core sizes in the 50- to 100-μm range. Therefore, it is easier to splice and interconnect, and cheaper, less-intense light sources may be used.

Most fiber-optic cables consist of more than the core, cladding, and a jacket. They usually contain one or more additional elements. The simplest cable is the core with its cladding surrounded by a protective jacket. As in electric cables, this outer jacket or insulation is made of some type of plastic, typically polyethylene, polyurethane, or polyvinyl chloride (PVC). The main purpose of this outer jacket is to protect the core and cladding from damage. Usually fiber-optic cables are buried underground or are strung between supports. Therefore, the fibers themselves have to be protected from moisture, dirt, temperature variations, and other conditions. The outer jacket also helps minimize physical damage such as cuts, nicks, and crushing. The more complex cables may contain two or more fiber-optic elements. Typical cables are available with 2, 6, 12, 18, and 24 fiber-optic cores.

There are many different types of cable configurations. Many have several layers of protective jackets. Some cables incorporate a flexible strength or tension element which helps minimize damage to the fiber-optic elements when the cable is being pulled or when it must support its own weight. Typically, this strength element is made of stranded steel or a special yarn known as Kevlar. Kevlar is strong and is preferred over steel because it is an insulator. In some cables, the Kevlar forms a protective sleeve or jacket over the cladding. Most claddings are covered with a clear protective coating for added strength and resistance to moisture and damage. Figure 13-17 shows one type of cable.

Fig. 13-17   Typical layers in a fiber-optic cable.

Fiber-optic cables are also available in a flat ribbon form as shown in Fig. 13-18. When working with multiple fibers, flat ribbon cable is generally easier to use because handling and identification of individual fibers is far easier. The flat cable is also more space-efficient for some applications.

Fig. 13-18   Multicore flat ribbon cable.

The main specification of a fiber-optic cable is its attenuation. Attenuation refers to the loss of light energy as the light pulse travels from one end of the cable to the other. The light pulse of a specific amplitude or brilliance is applied to one end of the cable, but the light pulse output at the other end of the cable will be much lower in amplitude. The intensity of the light at the output is lower due to various losses in the cable. The main reason for the loss in light intensity over the length of the cable is due to light absorption, scattering, and dispersion.

Absorption refers to how the light waves are actually "soaked up" in the core material due to the impurity of the glass or plastic. Scattering refers to the light lost due to light waves entering at the wrong angle and being lost in the cladding due to refraction. Dispersion, as you have already learned, refers to the pulse stretching caused by the many different paths through the cable. Although no light is lost as such in dispersion, the output is still lower in amplitude than the input but the length of the light pulse has increased in duration.

The amount of attenuation, of course, varies with the type of cable and its size. Glass has less attenuation than plastic. Wider cores have less attenuation than narrower cores. But more importantly, the attenuation is directly proportional to the length of the cable. It is obvious that the longer the distance the light has to travel, the greater the loss due to absorption, scattering, and dispersion. Doubling the length of a cable doubles the attenuation, and so on.

The attenuation of a fiber-optic cable is expressed in decibels per unit of length. The standard is decibels per kilometer. The standard decibel formula used is

$$dB = 10 \log \frac{P_o}{P_i}$$

where P_o is the power out and P_i is the power in.

Figure 13-19 is a table that shows the percentage of output power for various decibel losses. For example, 3 dB represents half power. In other words, a 3-dB loss means that only 50 percent of the input appears at the output. The other 50 percent of the power is lost in the cable. The higher the decibel figure, the greater the attenuation and loss. A 30-dB loss means that only one-thousandth of the input power appears at the end.

The standard specification for fiber-optic cable is the attenuation expressed in terms of decibels per kilometer. A kilometer is a smaller unit of distance than a mile. A mile is 5280 ft, while a kilometer is only 3281 ft. In other words,

Loss (dB)	Power output (%)
1	79
2	63
3	50
4	40
5	31
6	25
7	20
8	14
9	12
10	10
20	1
30	.1
40	.01
50	.001

Fig. 13-19 Decibel loss table.

a kilometer is approximately 0.62 mi. That means that there are about 1.6 km per mile.

The attenuation ratings of fiber-optic cables vary over a considerable range. The finest single-mode step-index cables have an attenuation of only 1 dB/km. However, very large core plastic-fiber cables can have an attenuation of several thousand decibels per kilometer. A typical cable might have a standard loss of 10 to 20 dB/km. Typically, those fibers with an attenuation of less than 10 dB/km are called low-loss fibers, while those with an attenuation of between 10 and 100 dB/km are medium-loss fibers. High-loss fibers are those with over 100 dB/km ratings. Naturally, the smaller the decibel number, the less the attenuation and the better the cable.

You can easily determine the total amount of attenuation for a particular cable if you know the attenuation rating. For example, if a cable has an attenuation of 15 dB/km, then a 5-km length cable has a total attenuation of 5(15) = 75 dB. If two cables are spliced together and one has an attenuation of 17 dB and the other 24 dB, the total attenuation is simply the sum, or 17 + 24 = 41 dB.

When long fiber-optic cables are needed, two or more cables may be spliced together. The ends of the cable are perfectly aligned and then glued together with a special, clear, low-loss epoxy. Connectors are also used. A variety of connectors provide a convenient way to splice cables and attach them to transmitters, receivers, and repeaters.

Connectors are special mechanical assemblies that allow fiber-optic cables to be connected to one another. You might say they are a form of temporary splice that can be conveniently disconnected. Fiber-optic connectors are the optical equivalent of electrical plugs and sockets. They are mechanical assemblies that hold the ends of the cable and cause them to be accurately aligned with another cable. Most of them either snap together or have threads which allow the two pieces to be screwed together.

Connectors force better alignment of the cables. The two ends of the cables must be aligned with precision so that excessive light is not lost. Otherwise, a splice or connection will introduce excessive attenuation. Figure 13-20 shows several ways that the cores can be misaligned. A connector corrects these problems.

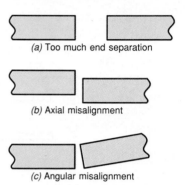

(a) Too much end separation

(b) Axial misalignment

(c) Angular misalignment

Fig. 13-20 Misalignment of fiber-optic cores causes excessive light loss.

A typical fiber-optic connector is shown in Fig. 13-21(a) on the next page. One end of the connector called the ferrule holds the fiber securely in place. A matching fitting holds the other fiber securely in place. When the two are screwed together, the ends of the fibers touch, providing a low-loss coupling. Figure 13-21(b) shows how the connector aligns the fibers.

Dozens of different kinds of connectors are available for different applications. These connectors are used in those parts of the system where it may be desirable to occasionally disconnect the fiber-optic cable for making tests or repairs. Connectors are normally used at the end of the cable applied to the light source or at the end of the cable connected to the photodetector. Connectors are also used at the repeater units where the light is picked up, converted into an electrical pulse, amplified and reshaped, and then used to create a new pulse to continue the transmission over a long line.

The first step in installing a connector is to cleave the fiber so it is perfectly square. Then the manufacturer's directions should be followed when attaching the connector to the fiber. Many manufacturers supply rolls of cable in different lengths with connectors already attached. In working with fiber-optic cable, connectors, and other accessories, you should always follow the manufacturer's directions.

(a)

(b)

Fig. 13-21 Details of a fiber cable connector.

Self-Test

Answer the following questions.

31. Which material has the best optical characteristics and lowest loss?
    a. Plastic
    b. Glass
    c. they are equal
32. The core is protected by the _____ .
33. In PCS cable, the core is _____ and the cladding is _____ .
34. The index of refraction is highest in the
    a. Core.
    b. Cladding.
35. List the three main types of fiber-optic cable.
36. Stretching of the light pulse is called _____ .
37. List the two types of cable in which light-pulse stretching occurs.
38. High-frequency pulses can be best transmitted over _____ cable.
39. Pulse stretching causes the information capacity of a cable to
    a. Increase.
    b. Decrease.
40. Graded index means that the _____ of _____ of the core varies over its cross section.

41. Single-mode step-index cable has a core diameter in the range of
    a. 100 to 1000 μm
    b. 50 to 100 μm
    c. 2 to 15 μm
42. A _____ is applied over the cladding to protect against moisture, damage, etc.
43. A common protective layer in a cable is made of _____ mesh.
44. Fiber-optic cables are available with the following number of cores: _____ .
45. Light loss in a cable is called _____ .
46. Light loss is caused by _____ .
47. Light loss is measured in _____ per _____ .
48. A cable with a loss of 5 dB will have _____ percent of the input appear at the output (see Fig. 13-19).
49. True or false. A kilometer is longer than a mile.
50. Which cable length will have the least attenuation?
    a. 40 ft
    b. 120 ft
    c. 1780 ft
    d. 1 km
51. Three cables with attenuations of 9, 22, and 45 dB are spliced together. The total attenuation is _____ dB.
52. True or false. Fiber-optic cables may be spliced.
53. To conveniently link and attach fiber-optic cables to one another and related equipment, _____ are used.

13-4 OPTICAL TRANSMITTERS AND RECEIVERS

In an optical communications system, transmission begins with the transmitter, which consists of a modulator and the circuitry that generates the carrier. In this case, the carrier is a light beam that is modulated by digital pulses which turn it on and off. The basic transmitter is nothing more than a light source.

Conventional light sources such as incandescent lamps cannot be used in fiber-optic systems. The reason for this is that they are simply too slow. An incandescent light source consists of a filament which heats up and emits light. Such a light source cannot be turned off and on fast enough because of the thermal delay in the filament. In order to transmit high-speed digital pulses, a very fast light source must be used. The two most commonly used light sources are LEDs and semiconductor lasers.

An LED is a PN junction semiconductor device that emits light when forward-biased. When a free electron encounters a hole in the semiconductor structure, the two combine, and in the process they give up energy in the form of light. Semiconductors such as gallium arsenide (GaAs) are superior to silicon in light emission. Most LEDs are GaAs devices optimized for producing red light. They are used for displays indicating whether a circuit is off or on, or for displaying decimal and binary data. However, because an LED is a fast semiconductor device, it can be turned off and on very quickly. Therefore, it is capable of transmitting the narrow light pulses required in a digital fiber-optic system.

Light-emitting diodes can be designed to emit virtually any color of light desired. Red LEDs are the most common, but yellow, green, and blue LEDs are also available. The LEDs used for fiber-optic transmission are usually in the red and low-infrared ranges. Typical wavelengths of LED light commonly used are 0.82, 0.94, 1.3, and 1.55 μm. These wavelengths are all in the near-infrared range just below red light. The light is not visible to the naked eye. These wavelengths have been chosen primarily because most fiber-optic cables have the lowest losses in these wavelength ranges. The most commonly used wavelength is 1.3 μm because many fiber-optic cables have minimum loss at that frequency.

Special LEDs are made just for fiber-optic applications. These units are made of GaAs indium phosphide (GaAsInP) and emit light at 1.3 μm. They come with a fiber-optic "pigtail" already attached for optimum coupling of light. The pigtail usually has a connector that attaches to the main cable.

There are two basic ways that digital data is formatted in fiber-optic systems. These are the return-to-zero (RZ) and non–return-to-zero (NRZ) formats. Both of these are illustrated in Fig. 13-22. The NRZ format at (a) is already generally familiar to you as we have discussed it previously in the chapter on data communications. Each bit occupies a separate time slot and is either a binary 1 or binary 0 during that time period. In the RZ format, the same time period is allotted for each bit but each bit is transmitted as a very narrow pulse (usually 50 percent of the bit time) or as an absence of a pulse as shown in Fig. 13-22(b). The digital data to be transmitted is converted into a serial

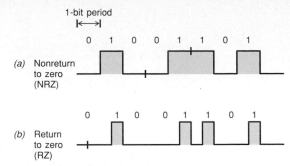

Fig. 13-22 Binary data formats.

pulse train and then into the desired RZ or NRZ format. These pulses are then applied to a light transmitter.

The light transmitter consists of the LED and its associated driving circuitry. A typical circuit is shown in Fig. 13-23 on the next page. The binary pulses are applied to a logic gate which, in turn, operates a transistor switch Q_1 that turns the LED off and on. A positive pulse at the NAND gate input causes the NAND output to go to zero. This turns off Q_1, so the LED is forward-biased through R_2 and turns on. With zero input, the NAND output is 1, so Q_1 turns on and shunts current away from the LED. Very high current pulses are used to ensure a brilliant high-intensity light. High intensity is required if data is to be transmitted reliably over long distances. Most LEDs are capable of generating power levels up to approximately several hundred microwatts. With such low intensity, LED transmitters are good for only short distances. Further, the speed of the LED is limited. Turn-off and turn-on times are no faster than several nanoseconds, and, therefore, transmission rates are limited. Most LED-like transmitters are used for short-distance, low-speed digital fiber-optic systems.

The other commonly used light transmitter is a laser. A laser is a light source that emits coherent monochromatic light. Monochromatic means a single light frequency. Although an LED emits red light, that light covers a narrow spectrum of red frequencies. Monochromatic light has a pure single frequency. Coherent refers to the fact that all the light waves emitted are in phase with one another. Coherent light waves are focused into a narrow beam, which, as a result, is extremely intense. The effect is somewhat similar to that of using a highly directional antenna to focus radio waves into a narrow beam which also increases the intensity of the signal.

Fig. 13-23 Optical transmitter circuit using an LED.

There are many different ways to make lasers. Some lasers are made from a solid rod of special material which is capable of emitting light when properly stimulated. Certain types of gases are also used. However, lasers used in fiber-optic systems are usually nothing more than specially made LEDs which are capable of operating as lasers.

The most widely used light source in fiber-optic systems is an injection laser diode (ILD). Like the LED, it is a PN junction diode usually made of GaAs. See Fig. 13-24. At some current level, it will emit a brilliant light. The physical structure of the ILD is such that the semiconductor structure is cut squarely at the ends to form internal reflecting surfaces. One of the surfaces is usually coated with a reflecting material such as gold. The other surface is only partially reflective. When the diode is properly biased, the light will be emitted and will bounce back and forth internally between the reflecting surfaces. The distance between the reflecting surfaces has been carefully measured so that it is some multiple of a half wave at the light frequency. The bouncing back and forth of the light waves causes their intensity

to reinforce and build up. The structure is like a cavity resonator for light. The result is an incredibly high brilliance, single-frequency light beam that is emitted from the partially reflecting surface.

Injection laser diodes are capable of developing light power up to several watts. They are far more powerful than LEDs and, therefore, are capable of transmitting over much longer distances. Another advantage ILDs have over LEDs is their ability to turn off and on at a faster rate. High-speed laser diodes are capable of gigabit per second digital data rates.

Injection lasers dissipate a tremendous amount of heat and, therefore, must be connected to a heat sink for proper operation. Because their operation is heat-sensitive, most injection lasers are used in a circuit that provides some feedback for temperature control. This not only protects the laser, but also ensures proper light intensity and frequency.

A typical injection laser transmitter circuit is shown in Fig. 13-25. When the input is zero, the AND gate output is zero, so Q_1 is off and so is the laser. Capacitor C_2 charges through R_3 to the high voltage. When a binary 1 input

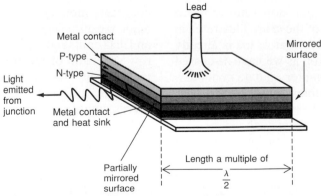

Fig. 13-24 An injection laser diode (ILD).

Fig. 13-25 An ILD transmitter.

occurs, Q_1 conducts, connecting C_2 to the ILD. Then C_2 discharges a very high current pulse into the laser, turning it on briefly and creating an intense light pulse.

The receiver part of the optical communications system is relatively simple. It consists of a detector that will sense the light pulses and convert them into an electrical signal. This signal is then amplified and shaped into the original serial digital data. The most critical component, of course, is the light sensor.

The most widely used light sensor is a photodiode. This is a silicon PN junction diode which is sensitive to light. This diode is normally reverse-biased as shown in Fig. 13-26. The only current that flows through it is an extremely small reverse leakage current. Whenever light strikes the diode, this leakage current will increase significantly. It will flow through a resistor and develop a voltage drop across it. The result is an output voltage pulse.

The reverse current in a diode is extremely small even when the diode is exposed to light. The resulting voltage pulse is very small, so it must be amplified. This can be done by using

Fig. 13-26 How a photodiode converts light into voltage pulses.

a phototransistor. The base-collector junction is exposed to light. The base leakage current produced causes a larger emitter-to-collector current to flow. Thus the transistor amplifies the small leakage current into a larger, more useful output.

The sensitivity and response time of a photodiode can be increased by adding an undoped or intrinsic (I) layer between the P and N semiconductors to form a PIN diode. The I layer is exposed to the light. The diode is reverse-biased as previously discussed.

A more widely used photosensor is the avalanche photodiode (APD). It is the fastest and most sensitive photodiode available, but it is expensive and its circuitry is complex. Like the standard photodiode, the APD is reverse-biased. However, the operation is different. The APD uses the reverse breakdown mode of operation that is commonly found in zener and IMPATT microwave diodes. When a sufficient amount of reverse voltage is applied, an extremely high current will flow due to the avalanche effect. Normally, several hundred volts of reverse bias, just below the avalanche threshold, is applied. When light strikes the junction, breakdown occurs and a large current flows. This high reverse current requires less amplification than the small current in a standard photodiode. The APDs are also significantly faster and are capable of handling the very high gigabit per second data rates possible in some systems.

Figure 13-27 on the next page shows the basic circuit used in most receivers. The current through the photodiode generated when light is sensed produced a current which is then amplified in an op amp. Following the amplifier is an op-amp comparator used as a

Fig. 13-27   A photodiode fiber-optic receiver.

shaping circuit that squares the pulses to ensure fast rise and fall times. The output is passed through a logic gate so that the correct binary voltage levels are produced.

The most important specification in a fiber-optic communications system is the data rate, that is, the speed of the optical pulses. For the very best systems using high-power ILDs and APD detectors, data rates of several billion bits per second are possible. This is known as a gigabit rate. Most systems have a data rate of 10 to 500 Mbits/s.

A system's performance is usually indicated by the product of the bit rate and the distance. This rating tells the fastest bit rate that can be produced over a 1-km cable. Assume a system with a 100 Mbits·km/s rating. This is a constant figure. If the distance increases, the bit rate decreases in proportion. In the above system, at 2 km the rate would drop to 50 Mbits/s. At 4 km, the rate is 25 Mbits/s, and so on.

Another important consideration is the maximum distance between repeaters. Obviously the fewer the repeaters, the better. The average distance between repeaters is now in the 10- to 30-km range. The more components and systems improve, the greater this distance will become.

Self-Test

Answer the following questions.

54. The two most common light sources used in fiber-optic transmitters are _____ .

55. The most popular light wavelength is _____ μm because fiber-optic cable attenuation is lowest at that wavelength.
56. True or false. The light from a 1.55-μm LED is visible.
57. Usually LEDs are made of _____ .
58. Single-frequency light is called _____ .
59. The condition of all emitted light waves being in phase is known as _____ .
60. A single-frequency intense light source is known as a(n) _____ .
61. The reflective surfaces on a laser diode structure form a(n) _____ that produces in-phase light waves.
62. For normal operation, LED and ILDs are _____ (reverse, forward)-biased.
63. Which is faster, an LED or ILD?
64. Which produces the brightest light, an LED or ILD?
65. During normal operation, all photodiodes are _____ (reverse, forward)-biased.
66. Light falling on the PN junction of a photodiode causes the diode's _____ current to increase.
67. The most sensitive and fastest light detector is the _____ .
68. The two main circuits in a fiber-optic receiver are _____ .
69. The product of the bit rate and distance of a system is 90 Mbits·km/s. The rating at 3 km is _____ Mbits/s.
70. In today's systems, the average maximum distance between repeaters is between _____ and _____ mi.

SUMMARY

1. The information-carrying capacity of a cable or radio channel is directly proportional to its bandwidth.

2. The RF spectrum is heavily used and occupied. Only in the microwave region is there room for expansion.

3. Light is an electromagnetic signal like a radio wave but is much higher in frequency. It can be used as a carrier for information signals.

4. Because of the very high frequency of light compared to typical information signals, tremendous bandwidth is easily available.

5. Light waves carrying data signals can be transmitted in free space but are greatly attenuated by atmospheric effects and require pinpoint alignment.

6. Most light-wave communication is by way of a glass or plastic fiber cable that acts as a "light pipe" to carry light modulated by information signals.

7. The main components of a light-wave communications system are an A/D converter, a light source transmitter, a fiber-optic cable, a photo- or light detector with amplifier and shaper, and a D/A converter.

8. Because of the great attenuation of light in a fiber-optic cable, repeater units are used to amplify and regenerate the signals over long distances.

9. The primary application of fiber-optic communications is in long-distance telephone systems.

10. The primary advantages of fiber-optic cables over conventional cables and radio are wider bandwidth, lower loss, light weight, small size, strength, security, interference immunity, and safety.

11. The main disadvantage of fiber-optic cable is that its small size and brittleness make it more difficult to work with.

12. Light waves, like radio waves, are a kind of electromagnetic radiation.

13. Light waves occur at very high frequencies in the range of 3×10^{11} to 3×16^{16} Hz.

14. Wavelength rather than frequency is used to express the place of light in the spectrum.

15. The wavelength of light is expressed in terms of nanometers (1 nm $= 10^{-9}$ m) or micrometers (1 μm $= 10^{-6}$ m). Micrometers are also called microns.

16. The visible light spectrum is from 700 nm (red) to 400 nm (violet).

17. The optical spectrum is made up of visible light, infrared at lower frequencies and ultraviolet at higher frequencies.

18. Infrared rays cannot be seen, but they act like light waves and can be manipulated in similar ways as with a lens or mirrors.

19. Light waves, like microwaves, travel in a straight line.

20. The angle at which light strikes a surface is called the angle of incidence. The angle at which light is reflected from a surface is called the angle of reflection. The angle of incidence is equal to the angle of reflection.

21. When a light ray passes from one medium to another, it is bent. This is called refraction.

22. The amount of refraction is called the index of refraction n and is the ratio of the speed of light in air to the speed of light in another medium, such as water, glass, or plastic ($n = 1$ in air, $n = 1.3$ in water, $n = 1.5$ in glass).

23. The angle of the incident light ray determines whether the ray will be reflected or refracted.

24. The critical angle is the angle of incidence that causes the refracted light to travel along the interface between two different media.

25. If the angle of incidence is made greater than the critical angle, reflection occurs instead of refraction.

26. Light entering a fiber-optic cable has an angle of incidence such that the light is reflected or bounced off the boundary between the fiber and the external media. This is called total internal reflection.

27. Fiber-optic cables are made from glass and plastic. Glass has the lowest loss but is brittle. Plastic is cheaper and more flexible, but has high attenuation.

28. A popular fiber-optic cable with a glass core and plastic cladding is called plastic-clad silica (PCS).

29. The cladding surrounding the core protects the core and provides an interface with a controlled index of refraction.

30. Step index means there is a sharp difference in the index of refraction between the core and cladding.

31. Graded index means that the index of refraction of the core varies over its cross section, highest in the center and lowest at the edges.

32. A single-mode cable is very small in diameter and essentially provides only a single path for light.

33. Multimode cores are large and provide multiple paths for the light.

34. Multiple light paths through a step-index core cause a light pulse to be stretched and

attenuated. This is called modal dispersion, and it limits the upper pulse repetition rate and thus the information bandwidth.

35. Multiple light paths in a graded-index core are controlled so that they converge at multiple points along the cable. Modal dispersion does occur, but it is not as severe as that caused by a step-index core.

36. Modal dispersion does not occur in single-mode cores.

37. The three most widely used types of fiber-optic cables are multimode step-index, single-mode step-index, and multimode graded-index.

38. The primary specification of a fiber-optic cable is attenuation, which is usually expressed as the loss in decibels per kilometer.

39. Light loss in a fiber-optic cable is caused by absorption, scattering, and dispersion.

40. Cable attenuation is directly proportional to its length.

41. Cable losses range from 1 dB/km in glass single-mode step-index cable to 100 dB/km for plastic multimode step-index cable.

42. Fiber-optic cables can be spliced by gluing.

43. Special connectors are used to connect cables to one another and to the equipment.

44. Fiber-optic systems use light-emitting diodes (LEDs) and semiconductor lasers as the main light sources.

45. Light-emitting diodes are used in short-distance, low-speed systems. Injection laser diodes (ILDs) are used in long-distance, high-speed systems.

46. Most LEDs and ILDs emit light in the invisible near-infrared range (0.82 to 1.55 μm).

47. A popular operating frequency is 1.3 μm because fiber-optic cable has an attenuation null at this wavelength.

48. Laser diodes emit monochromatic or single-frequency light. The light waves are coherent, so they reinforce one another to create an intense and finely focused beam.

49. Intense laser light is produced by an ILD because reflecting surfaces in the structure form a cavity resonator for the light waves.

50. The most commonly used light sensor is a photodiode.

51. A photodiode is a PN junction that is reverse-biased and exposed to light. Light increases the leakage current across the junction. This current is converted into a voltage pulse.

52. PIN junction diodes are faster and more sensitive than conventional photodiodes.

53. The fastest and most sensitive light detector is the avalanche photodiode (APD).

54. The APD operates with a high reverse bias so that when light is applied, breakdown occurs and produces a fast, high-current pulse.

55. The receiver portion of a fiber-optic system is made up of a photodiode, amplifier, and shaper.

56. Fiber-optic systems are rated by the speed and the product of the bit rate and the distance.

57. A measure of the quality of a fiber-optic system is the maximum distance between repeaters.

CHAPTER REVIEW QUESTIONS

Choose the letter which best answers each question.

13-1. Which of the following is *not* a common application of fiber-optic cable?
  a. Computer networks
  b. Long-distance telephone systems
  c. Closed circuit TV
  d. Consumer TV

13-2. Total internal reflection takes place if the light ray strikes the interface at an angle with what relationship to the critical angle?
  a. Less than
  b. Greater than
  c. Equal to
  d. Zero

13-3. The operation of a fiber-optic cable is based on the principle of
  a. Refraction.
  b. Reflection.
  c. Dispersion.
  d. Absorption.

13-4. Which of the following is *not* a common type of fiber-optic cable?
  a. Single-mode step-index   *c.* Single-mode graded-index
  b. Multimode graded-index   *d.* Multimode step-index

13-5. Cable attenuation is usually expressed in terms of
  a. Loss per foot.   *c.* Intensity per mile.
  b. dB/km.   *d.* Voltage drop per inch.

13-6. Which cable length has the highest attenuation?
  a. 1 km   *c.* 95 ft
  b. 2 km   *d.* 500 ft

13-7. The upper pulse rate and information-carrying capacity of a cable is limited by
  a. Pulse shortening.   *c.* Light leakage.
  b. Attenuation.   *d.* Modal dispersion.

13-8. The core of a fiber-optic cable is made of
  a. Air.   *c.* Diamond.
  b. Glass.   *d.* Quartz.

13-9. The core of a fiber-optic cable is surrounded by
  a. Wire braid shield.   *c.* Cladding.
  b. Kevlar.   *d.* Plastic insulation.

13-10. The speed of light in plastic compared to the speed of light in air is
  a. Less.   *c.* The same.
  b. More.   *d.* Zero.

13-11. Which of the following is *not* a major benefit of fiber-optic cable?
  a. Immunity from interference *c.* Excellent data security
  b. No electrical safety problems *d.* Lower cost

13-12. The main benefit of light-wave communications over microwaves or any other communications media are
  a. Lower cost.   *c.* Wider bandwidth.
  b. Better security.   *d.* Freedom from interference.

13-13. Which of the following is *not* part of the optical spectrum?
  a. Infrared   *c.* Visible color
  b. Ultraviolet   *d.* X-rays

13-14. The wavelength of visible light extends from
  a. 0.8 to 1.6 μm.   *c.* 200 to 660 nm.
  b. 400 to 750 nm.   *d.* 700 to 1200 nm.

13-15. The speed of light is
  a. 186,000 mi/h.   *c.* 300,000 m/s.
  b. 300,000 mi/h.   *d.* 300,000,000 m/s.

13-16. Refraction is the
  a. Bending of light waves.   *c.* Distortion of light waves.
  b. Reflection of light waves. *d.* Diffusion of light waves.

13-17. The ratio of the speed of light in air to the speed of light in another substance is called the
  a. Speed factor.   *c.* Index of refraction.
  b. Index of reflection.   *d.* Speed gain.

13-18. A popular light wavelength in fiber-optic cable is
  a. 0.7 μm.   *c.* 1.5 μm.
  b. 1.3 μm.   *d.* 1.8 μm.

13-19. Which type of fiber-optic cable is the most widely used?
  a. Single-mode step-index   *c.* Single-mode graded-index
  b. Multimode step-index   *d.* Multimode graded-index

13-20. Which type of fiber-optic cable is best for very high speed data?
  a. Single-mode step-index   *c.* Single-mode graded-index
  b. Multimode step-index   *d.* Multimode graded-index

13-21. Which type of fiber-optic cable has the least modal dispersion?
  a. Single mode step-index   *c.* Single-mode graded-index
  b. Multimode step-index   *d.* Multimode graded-index

13-22. Which of the following is *not* a factor in cable light loss?
    a. Reflection   *c.* Scattering
    b. Absorption   *d.* Dispersion

13-23. A distance of 8 km is the same as
    a. 2.5 mi.   *c.* 8 mi.
    b. 5 mi.   *d.* 12.9 mi.

13-24. A fiber-optic cable has a loss of 15 dB/km. The attenuation in a cable 1000 ft long is
    a. 4.57 dB.   *c.* 24 dB.
    b. 9.3 dB.   *d.* 49.2 dB.

13-25. Fiber-optic cables with attenuations of 1.8, 3.4, 5.9, and 18 dB are linked together. The total loss is
    a. 7.5 dB.   *c.* 29.1 dB.
    b. 19.8 dB.   *d.* 650 dB.

13-26. Which light emitter is preferred for high-speed data in a fiber-optic system?
    a. Incandescent   *c.* Neon
    b. LED   *d.* Laser

13-27. Most fiber-optic light sources emit light in which spectrum?
    a. Visible
    b. Infrared
    c. Ultraviolet
    d. X-ray

13-28. Both LEDs and ILDs operate correctly with
    a. Forward bias.   *c.* Neither *a* or *b*.
    b. Reverse bias.   *d.* Either *a* or *b*.

13-29. Single-frequency light is called
    a. Pure.   *c.* Coherent.
    b. Intense.   *d.* Monochromatic.

13-30. Laser light is very bright because it is
    a. Pure.   *c.* Coherent.
    b. White.   *d.* Monochromatic.

13-31. Which of the following is *not* a common light detector?
    a. PIN photodiode
    b. Photovoltaic diode
    c. Photodiode
    d. Avalanche photodiode

13-32. Which of the following is the fastest light sensor?
    a. PIN photodiode
    b. Photovoltaic diode
    c. Phototransistor
    d. Avalanche photodiode

13-33. Photodiodes operate properly with
    a. Forward bias.   *c.* Neither *a* or *b*.
    b. Reverse bias.   *d.* Either *a* or *b*

13-34. The product of the bit rate and distance of a fiber-optic system is 2 Gbits·km/s. What is the maximum rate at 5 km?
    a. 100 Mbits/s
    b. 200 Mbits/s
    c. 400 Mbits/s
    d. 1000 Gbits/s

13-35. Which fiber-optic system is better?
    a. 3 repeaters
    b. 8 repeaters
    c. 11 repeaters
    d. 20 repeaters

CRITICAL THINKING QUESTIONS

13-1. Explain how you would transmit analog signals over a fiber-optic cable.

13-2. Name three potential new applications for fiber-optic communications that are not on the list in Fig. 13-2 but take advantage of the benefits listed in Fig. 13-3.

13-3. Explain how a single fiber-optic cable can handle two-way communications, both half- and full-duplex.

13-4. Could an incandescent light be used for a fiber-optic transmitter? Explain its possible benefits and disadvantages.

Answers to Self-Tests

1. light
2. voice, video, computer data
3. telephone systems
4. glass, plastic
5. wide bandwidth
6. false
7. true
8. brittleness, difficult to work with
9. LEDs, semiconductor lasers
10. binary or digital pulses
11. photocell or light detector
12. repeaters
13. electromagnetic
14. pipe
15. *b*
16. micrometers (microns), nanometers
17. *a*
18. one-millionth, meter
19. 400, 700
20. *b*
21. visible, infrared, ultraviolet
22. 300,000,000; 186,000

23. true
24. index, refraction
25. reflection, mirror
26. refraction
27. 30
28. critical
29. reflection
30. index, refraction
31. *b*
32. cladding
33. glass, plastic
34. *a*
35. multimode step-index, single-mode step-index, multimode graded-index
36. modal dispersion
37. multimode step-index, multimode graded-index
38. single-mode step-index
39. *b*
40. index, refraction
41. *c*
42. protective coating
43. Kevlar
44. 2, 6, 12, 18, 24
45. attenuation
46. dispersion, absorption, scattering

47. decibels, kilometer
48. 31
49. false
50. *a*
51. 76
52. true
53. connectors
54. LED, laser
55. 1.3
56. false
57. gallium arsenide
58. monochromatic
59. coherence
60. laser
61. resonant cavity
62. forward
63. ILD
64. ILD
65. reverse
66. leakage
67. avalanche photodiode
68. amplifier, comparator or shaper
69. 30
70. 6.17, 18.5 (10 and 30 km)

CHAPTER 14

Modern Communications Applications

∎

CHAPTER OBJECTIVES

This chapter will help you to:

1. *Draw* a block diagram of a facsimile transceiver and *explain* its send and receive operations.
2. *Explain* the concept behind the modern cellular telephone system, *name* the major parts of the system, and *state* the operation of each during a typical telephone call.
3. *State* the concept of radar and *name* its major uses.
4. *Calculate* the distance from radar station to target given the pulse return time.
5. *Draw* the frequency spectrum of a TV signal and *name* the types of modulation and multiplexing used.
6. *Draw* a block diagram of a TV receiver and briefly *explain* the function of each major circuit.

───────

To give you a feel for the extent of electronic communications, this chapter will introduce you to several important communications technologies. These include facsimile, cellular telephones, radar, and television. All of them employ circuits and techniques discussed in previous chapters. As you will discover, electronic communications methods can be combined in a variety of ways to achieve unique results tailored to contemporary needs.

∎

14-1 FACSIMILE

Facsimile or *fax* is an electronic system for transmitting graphical information by wire or radio. Facsimile is used to send printed material by scanning it and converting it into electronic signals that modulate a carrier to be transmitted over the telephone lines or by radio communications. With facsimile, documents such as letters, photographs, line drawings, or any printed information can be converted into an electrical signal and transmitted by means of conventional communications techniques.

The concept of a fax machine is illustrated in Fig. 14-1.

Although facsimile is used to transmit pictures, it is not television since it does not transmit sound messages or live scenes and motion. However, it does use scanning techniques which are in some way generally similar to those used in television.

Like most other communications technologies, facsimile is not new. It was invented in 1842 by a Scotsman named Alexander Bain. The techniques have been in continuous use since then. The biggest users of fax in the past

Fig. 14-1   Concepts of a fascimile machine.

have been newspapers and wire service companies who used fax for transmitting news information and photos. It has also been used by telegram companies, the government and military, and in many cases, commercial business organizations for transmitting information from one point to another. Fax is also widely used to transmit weather photos from satellites to ground stations.

Early fax machines were very slow. It took many minutes to transmit one page. Further, low resolution produced poor-quality documents. The disadvantages kept fax machines from becoming more popular. With new electronic techniques, fax has been reborn in the 1980s. With new, higher-resolution scanning techniques and high-speed modems, graphical information can now be transmitted and received by fax machines over the telephone lines with greater quality and shorter times. Transmission times of less than 15 s per page are typical for the newer machines. With fast scanning and high baud rates, fax machines have become *the* way to transmit letters, reports, and other important business information. Fax has become the preferred business communications method and has, in many cases, quickly replaced overnight mail services and even electronic computer mail. The out-

put quality compares favorably to that of the average photocopy machine.

The early facsimile machines were electro-mechanical devices consisting of a rotary scanning drum and some form of printing mechanism. The typical facsimile machine was capable of both sending and receiving. Figure 14-2 shows the basic scanning mechanism. The document to be transmitted was wrapped around and tightly affixed to the drum. A motor rotated the drum through a gear train mechanism. The motor was designed to operate at an accurate speed so that the scanning rate was precise. A precision tuning-fork oscillator was used to control the scanning drum motor for precise speed value.

Scanning of the document was done with a light source and photocell arrangement. A lead screw driven by the gear train moved a scanning head consisting of the light source and photocell. An incandescent light source focused to a tiny point with a lens system was used to scan the document. The lens was also used to focus the reflected light from the document onto the photocell. As the light scanned the letters and numbers in a typed or printed document or the gray scale in a photograph, the photocell produced a varying electronic signal whose output amplitude was proportional to the amount of reflected light. This baseband signal was then used to amplitude or frequency modulate a carrier in the audio frequency range. This permitted the signal to be transmitted over the telephone lines.

Figure 14-3 on the next page shows how a printed letter might be scanned. Assume the letter F is black on a white background. The output of a photodetector as it scans across line A is shown at A. The output voltage is high for white and low for black. The output of the photodetector is also shown for scan lines B and C.

The resolution of the transmission is determined by the number of scan lines per vertical inch. The greater the number of lines scanned,

Fig. 14-2   Drum scanner used in early fax machines.

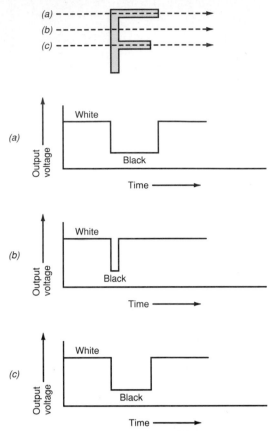

Fig. 14-3 Output of a photosensitive detector during different scans.

the finer the detail transmitted and the higher the quality of reproduction. The older systems had a resolution of 96 lines per inch, while the new systems use 200 lines per inch.

On the receiving end, a demodulator recovered the original signal information which was then applied to a stylus. The purpose of the stylus was to redraw the original information on a blank sheet of paper. A typical stylus converted the electrical signal into heat variations that would burn the image into heat-sensitive paper. Other types of printing mechanisms were also used.

In the printing process, a blank sheet of paper was wrapped around a cylindrical drum and rotated at the same speed as the transmitting drum. The drum rotation was precisely controlled by a precision tuning-fork oscillator. The drum rotations in the transmitter and receiver were synchronized to start simultaneously so that the information would be reproduced simultaneously with transmission. Because of the frequency spectrum limitations of the telephone lines, scanning speed was naturally slow to ensure signals that were within the available bandwidth. Transmission times of many minutes per page were typical.

Today's modern fax machine is a high-tech electro-optical machine. Scanning is done electronically, and the scanned signal is converted into a binary signal. Then digital transmission with standard modem techniques is used.

Most fax machines use charge coupled devices (CCDs) for scanning. A CCD is a light-sensitive semiconductor device that converts varying light amplitudes into an electrical signal. The typical CCD device is made up of many tiny capacitorlike devices manufactured in a matrix on a silicon chip. The capacitors are actually tiny reverse-biased silicon photodiodes. A typical CCD assembly is shown in Fig. 14-4. The base forms one large plate of a capacitor which is electrically separated by a dielectric from many thousands of tiny capacitor plates as shown. When the CCD is exposed to light, the CCD capacitors charge to a value proportional to the light intensity. The capacitors are then scanned or sampled electronically to determine their charge. This creates an analog output signal that accurately depicts the image focused on the CCD.

One way of looking at a CCD device is as a method of breaking any scene or picture up into individual picture elements or pixels. The greater the number of CCD capacitors or pixels, the higher the resolution and the more faithfully a scene, photograph, or document can be reproduced. The CCD devices are available with a matrix of many thousands of pixels, thereby permitting very high resolution picture transmission. They are widely used in modern video cameras in place of the more delicate and more expensive vidicon tubes. In the video camera (camcorders), the lens focuses the entire scene on a CCD matrix. This same approach is used in some fax machines. In the bed-type fax machine, the document to be transmitted is placed face down as it might be in a copy machine. The document is then illuminated with brilliant light from a xenon or fluorescent bulb. A lens system focuses the reflected light on a CCD. The CCD is then scanned, and the resulting output is an analog signal whose amplitude is proportional to the amplitude of the reflected light.

In most desktop fax machines, the entire document is not focused on a single CCD. Instead, only a narrow portion of the document is lighted and examined as it is moved through the fax machine with rollers. A complex system of mirrors is used to focus the lighted area on the CCD. See Fig. 14-5 on page 380.

A number of different modulation schemes are used in fax systems. Analog fax systems

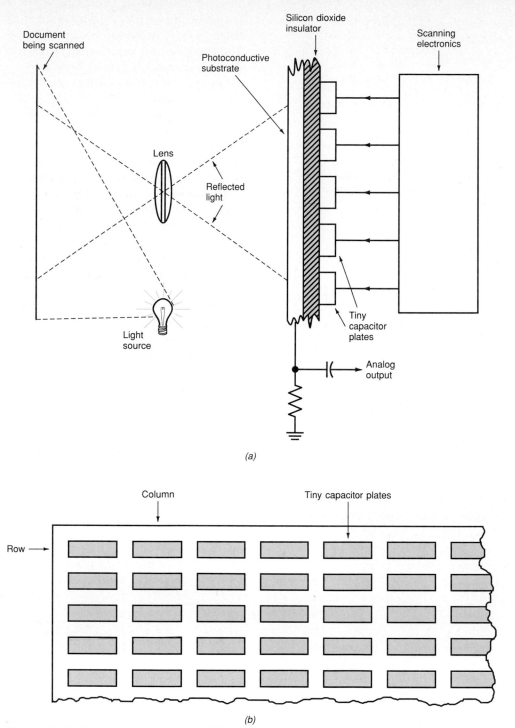

(a)

(b)

Fig. 14-4 A charge coupled device (CCD) is used to scan documents in modern fax machines. (*a*) Cross section and (*b*) detail of capacitor matrix.

use AM or FM. Digital fax systems use PSK or QAM. To ensure compatibility between fax machines of different manufacturers, standards have been developed for speed, modulation methods, and resolution. These were established by the International Telegraph and Telephone Consultative Committee, better known by its French abbreviation CCITT. The CCITT fax standards are divided into four groups.

These are summarized below:

Group 1 (G1 or GI) uses analog transmission by FM where white is 1300 Hz and black is 2100 Hz. Most North American equipment uses 1500 Hz for white and 2300 Hz for black. The scanning resolution is 96 lines per inch. The average transmission speed is 6 min per page (8½ × 11 in. or A4 metric size, which is slightly longer than 11 in.).

Fig. 14-5 Scanning mechanism in a fax machine.

Group 2 (G2 or GII) uses analog transmission by using FM or vestigial sideband AM. The vestigial sideband AM uses a 2100-Hz carrier. The lower sideband plus part of the upper sideband is transmitted. The resolution is 96 lines per inch. The transmission speed is 3 min or less for an 8½ × 11 in. or A4 page.

Group 3 (G3 or GIII) uses digital transmission by PCM with black and white only or up to 32 shades of gray. Either PSK or QAM is

used to achieve transmission speeds of up to 9600 baud. The resolution is 200 lines per inch. The transmission speed is less than 1 min per page with 15 to 30 s being typical.

Group 4 (G4 or GIV) uses digital transmission at 56 kbits/s. The resolution is up to 400 lines per inch. The speed of transmission is less than 5 s per page.

The older G1 machines are no longer used. Many G2 machines are still in operation but most are being phased out. The most common configuration is Group 3. Most G3 machines can also read the G2 format.

Group 4 machines are not yet widely used. They are designed to use digital transmission only with no modem over very wide band dedicated digital-grade telephone lines. These machines will become popular when the new Integrated Services Digital Network (ISDN) telephone system goes into use in the future.

Both G3 and G4 formats also employ digital data compression methods that shorten the binary data stream considerably thereby speeding up page transmission. This is important since shorter transmission times cut long-distance telephone charges and reduce operating costs. A typical 8½ × 11-in. page represents about 40,000 bytes of data. This can be shortened by a factor of 10 or more with data compression techniques.

Figure 14-6 shows a simplified block diagram of the transmitting circuits in a modern G3 fax

Fig. 14-6 Block diagram of a fascimile transceiver.

transceiver. The analog output from the CCD is fed to an A/D converter which translates the continuously varying light intensity into a stream of binary numbers. Sixteen gray scale values between white and black are typical. The binary data is sent to a VLSI digital data compression circuit. The compression circuit reduces the redundancy in the converted image so that fewer bits are needed to represent the image. The result is less storage memory and higher data transmission rates. The binary output in serial data format is then used to modulate a carrier which is then transmitted over the telephone lines. The techniques used are similar to those employed in modems. Speeds of 2400, 4800, 7200, and 9600 baud are common. Most systems use some form of PSK or QAM to achieve very high data rates on voice-grade lines.

In the receiving portion of the fax machine, the received signal is demodulated and is then sent to VLSI circuits where the data compression is removed and the binary signals are restored to their original form. The signal is then applied to a printing mechanism. In most fax machines, thermal printers are used. Most thermal printers use a special heat-sensitive paper. The print head, or writing stylus as it is sometimes called, contains tiny heating elements that are turned off and on by the received signal. The heat turns the paper black at the appropriate point to re-create the original image. Some newer thermal printers use a heating-element print head to melt ink on a special ribbon onto plain 20-lb bond paper.

In the high-priced machines, laser scanning of an electrosensitive drum as used in laser printers produces output copies using the proven techniques of xerography.

In Fig. 14-6, the control logic is used for handshaking between the two machines that will communicate. This ensures compatibility. Handshaking is usually carried out by exchanging different audio tones. The called machine responds with tones designating its capability. The calling machine compares this to its own standards and then either initiates the transmission or terminates it because of incompatibility. If the transmission proceeds, the calling machine sends sync signals to ensure that both machines start at the same time. The called machine acknowledges the receipt of the sync signals, and transmission begins. Transmission is, of course, half duplex. Simultaneous transmission or full duplex is not allowed.

As improvements have been made in picture resolution quality, transmission speed, and cost, facsimile machines have become much more popular. Today a good fax machine can be purchased for approximately the same price as a good photocopy machine or computer printer. The units can be easily attached with standard connectors to any telephone system. In most business applications, the fax machine is typically dedicated to a single line. Most fax machines feature fully automatic operation with microprocessor-based control. A document can be sent to a fax machine automatically. The sending machine simply dials the receiving machine and initiates the transmission. The receiving machine answers the initial call and then reproduces the document before hanging up.

Other fax machines are available with a built-in telephone and are designed to share a single line with conventional voice transmission. The built-in telephone usually features touch-tone dialing and number memory plus automatic redial and other modern telephone features. Most fax machines also feature automatic send and receive features for fully unattended operation. Smaller portable fax machines are also available for those users who travel. Fax machines may also be used with standard cellular telephone systems in automobiles.

Self-Test

Supply the missing information in each statement.

1. Images are converted into electrical signals by a process of _____ with a light source and photosensitive detector.
2. The most common transmission medium for fax signals is _____ .
3. Standards for fax transmission are set by the _____ .
4. The modulation method used in group 1 machines was FM with black = _____ Hz and white = _____ Hz.
5. Vestigial sideband AM is used in group _____ machines.
6. The photosensitive device used in most modern fax machines to convert a scanned line into an analog signal is a(n) _____ .
7. Most modern fax machines are of the group _____ type.
8. To ensure compatibility between sending and receiving fax machines, control logic carries out the process of _____ by audio tones prior to transmission.

9. The most commonly used type of printer in a fax machine is a(n) _____ printer.
10. The upper speed limit of a G3 fax machine over telephone lines is _____ baud.
11. The resolution of a G3 fax machine is typically _____ lines per inch.
12. The two modulation techniques used in G3 machines are _____ .
13. A fax transmission is usually _____ (full, half) duplex.
14. The speed and resolution of group 4 fax transmissions are _____ kbits/s and _____ lines per inch.
15. True or false. Fax can transmit photos and drawings as well as printed text.
16. True or false. Fax signals may not be transmitted by radio.
17. True or false. Group 4 transmission do not use the standard telephone lines.

14-2 CELLULAR RADIO SYSTEMS

A cellular radio system provides standard telephone operation by two-way radio at remote locations. Cellular radios are usually installed in cars or trucks but are also available in hand-held and briefcase models. Each cellular radio allows the user to link up with the standard telephone system which permits calls to any part of the world.

By far the most widely used two-way electronic communications system is the telephone. The greatest percentage of personal, business, and other communications is conducted by telephone. But there are times when the telephone system is inadequate in that it does not provide communications services in remote locations. Telephone communications from a car or truck has always been a desirable feature. In most cases, communications from a mobile unit has been taken care of by other two-way radio systems, but today, many of those systems are quickly being replaced by cellular radio telephones.

Radio telephone service has actually been available to consumers and businesses since the late 1940s. Over the years it developed into a large and rather sophisticated service. Known as the mobile telephone system (MTS), this system functioned reasonably well but had numerous disadvantages. The area that it served and the number of RF channels available to users were limited. In large metropolitan areas with many users, there were invariably more individuals wanting to use the service than there were channels available to accommodate them. Most users had waits of 20 to 30 min or more before a free channel could be found. Further, thousands of other potential users were on waiting lists for MTS service.

A typical MTS used a large single-radio repeater station connected to the telephone system. The repeater station contained a receiver for picking up transmissions from the mobile telephone units. A transmitter with an output as high as several hundred watts transmitted a signal to the users over an omnidirectional antenna on a large tower. The typical service range was a radius of 20 to 30 mi, sometimes more depending upon the antenna height and the transmitter power. These systems used FM and typically operated in the 450-MHz frequency range.

To help overcome the basic limitations of the MTS, the Bell Telephone Company division of AT&T developed what is known as the cellular radio system. The system was designed and fully tested in the late 1970s and was fully implemented in the early 1980s. Today, cellular radio telephone service is available nationwide in most medium and large cities.

The basic concept behind the cellular radio system is that, rather than serving a given geographical area with a single transmitter and receiver, it divides the area into many smaller areas known as cells. The typical cell covers only several square miles and contains its own receiver and low-power transmitter. The cell site is designed to reliably serve only vehicles in its small cell area. The desired service area is divided into many multiple cells as shown in Fig. 14-7.

Each of the cells is connected by telephone lines to a master control center known as the mobile telephone switching office (MTSO). The MTSO controls all the cells and provides the interface between each cell and the main telephone office. As the vehicle containing the telephone passes through a cell, it will be served by the cell transceiver. The telephone call will be routed through the MTSO and through the standard telephone system. As the vehicle moves from one cell to the next, the system will automatically switch from one cell to the next. The receiver in each cell station continuously monitors the signal strength of the mobile unit. When the signal strength drops below a desired level, it automatically seeks a cell where the signal from the mobile unit is stron-

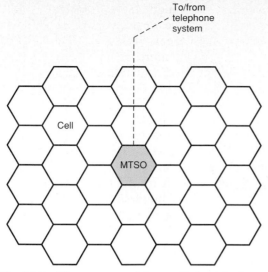

To/from telephone system

Cell

MTSO

Fig. 14-7 The area served by a cellular telephone system is divided up into small areas called cells.

ger. The computer at the MTSO causes the transmission from the vehicle to be switched from the weaker cell to the stronger cell. All of this takes place in a very short period of time, and it is completely unnoticeable to the user. The result is that optimum transmission and reception are obtained.

The cellular telephone system overcomes the limitations of the older MTS in a number of ways. First, it operates on much higher frequencies where more spectrum space is available. The cellular system operates in the 800- to 900-MHz range. Six hundred and sixty-six 30-KHz-wide full-duplex channels are available for communications. Second, the cellular system uses what is known as frequency reuse. This allows cells within the system to use the same frequency channel. Because the cells are physically small and low-power transmitters are used, the signal does not stray beyond the cell boundaries. This allows the same frequency channel to be used in other cells within the system without there being interference between the different signals. Frequency reuse tremendously multiplies the number of available channels. The same channel is not used in adjacent cells so as to prevent interference. Third, cells of different sizes can be accommodated. In low-usage areas, the cells can be larger. However, as the number of users increases, additional smaller cells can be added. Effectively, larger cells are divided into many smaller cells. This provides ease of expansion as the number of users grows. A typical system can serve up to about 50,000 subscribers.

Ten thousand of them may use the system simultaneously.

Figure 14-8 shows a general block diagram of a cellular mobile radio unit. It consists of five major sections: transmitter, receiver, synthesizer, logic unit, and control unit. Mobile radios derive their operating power from the car battery, and portable units contain built-in rechargeable batteries. The transmitter and receiver share a single antenna. Let's examine each section in greater detail.

The transmitter block diagram is given in Fig. 14-9 on the next page. It is a low-power FM unit operating in the frequency range of 825- to 845-MHz. There are 666 transmit channels spaced 30 kHz apart. Channel 1 is 825.03 MHz, channel 2 is 825.06 MHz, and so on up to channel 666 on 844.98 MHz. The carrier furnished by a frequency synthesizer is phase modulated by the voice signal. The phase modulator produces a deviation of $\pm$ 12 kHz. Preemphasis is used to help minimize noise. The modulator output is translated up to the final transmitter frequency by a mixer whose second input also comes from the frequency synthesizer. The mixer output is fed to class C power amplifier stages where the output signal is developed. The final amplifier stage is designed to supply 3 W to the antenna.

A unique feature of the transmitter is that its output power is controllable by the cell site

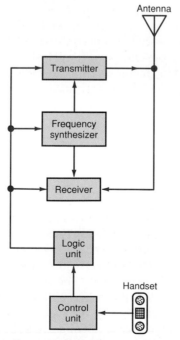

Fig. 14-8 General block diagram of a cellular radio.

Fig. 14-9 Cellular transmitter.

and MTSO. Special control signals picked up by the receiver are sent to an automatic power control (APC) circuit that sets the transmitter to one of eight power output levels. The APC circuit can introduce power attenuation in steps of 4 dB from 0 dB (3 W) to −28 dB (4.75 mW). This is done by controlling the supply voltage to one of the intermediate power amplifier stages.

The output power of the transmitter is monitored internally by built-in circuits. A microstrip directional coupler taps off an accurate sample of the transmitter output power and rectifies it into a proportional dc signal. This signal is used in the APC circuit and is transmitted back to the cell site permitting the MTSO to know the present power level.

This APC feature permits optimum cell site reception with minimal power. This helps to minimize interference with other stations in the same or adjacent cells.

The transmitter output is fed to a duplexer circuit or isolator that allows the transmitter and receiver to share the same antenna. Since cellular telephone units use full-duplex operation, the transmitter and receiver will be operating simultaneously. The transmit and receive frequencies are spaced 45 MHz apart to minimize interference. However, an isolator is still needed to keep transmitter power out of the sensitive receiver. The duplexer consists of two very sharp bandpass filters, one for the transmitter and one for the receiver. The transmitter output passes through this filter to the antenna.

The receiver is a dual-conversion superheterodyne. Refer to Fig. 14-10. An RF amplifier boosts the level of the received cell site signal. The receiver frequency range is 870.03 to 889.98 MHz. There are 666 receive channels spaced 30 kHz apart. The first mixer translates the incoming signal down to a first IF of 82.2 MHz. Some receivers use a 45-MHz first IF. The LO signal for the mixer is derived from the frequency synthesizer. The LO frequency sets the receive channel. The signal passes through IF amplifiers and filters to the second mixer. It is driven by a crystal-controlled LO. The second IF is usually either 10.7 MHz or 455 kHz. The signal is then demodulated, de-emphasized, filtered, and amplified before being applied to the output speaker in the handset.

The output of the demodulator is also fed to other filter circuits that select out the control audio tones and digital control data stream sent by the cell site to set and control both the transmitter and receiver. The demodulator output is also filtered into a dc level whose amplitude is proportional to the strength of the received signal. This is the receive signal strength indicator (RSSI) signal that is sent back to the cell site so that the MTSO can monitor the received signal from the cell and make decisions about switching to another cell.

The frequency synthesizer section develops all the signals used by the transmitter and receiver. A simplified block diagram is shown in Fig. 14-11. It uses standard PLL circuits and a mixer. A crystal-controlled oscillator provides the reference for the PLLs. One PLL incorporates a VCO (VCO 2) whose output fre-

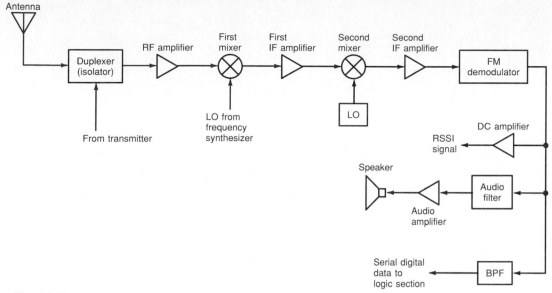

Fig. 14-10   Cellular receiver.

quency is used as the LO for the first mixer in the receiver. This signal is mixed with the output of a second PLL VCO to derive the transmitter output frequency.

As in other PLL circuits, the output VCO frequency is determined by the frequency division ratio of the divider in the feedback path between the VCO and the phase detector. In a cellular radio, this frequency division ratio is supplied by the MTSO via the cell site. When a mobile unit initiates or is to receive a call, the MTSO computer selects an unused channel. It then transmits a digitally coded signal

to the receiver containing the frequency division ratios for the transmitter and receiver PLLs. This sets the transmit and receive channel frequencies.

The logic unit shown in Fig. 14-12 on the next page contains the master control circuitry for the cellular radio. It contains a microprocessor with both random-access memory (RAM) and read-only memory (ROM) plus additional circuitry used for interpreting signals from the MTSO or cell site and generating control signals for the transmitter and receiver.

All cellular radios contain a programmable

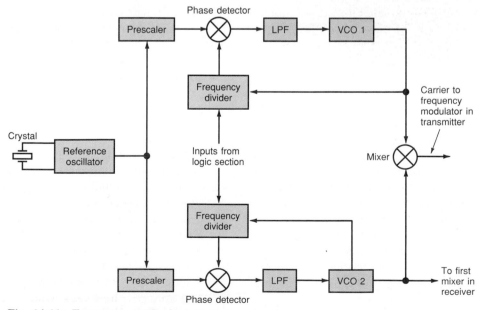

Fig. 14-11   Frequency synthesizer.

Number assignment module

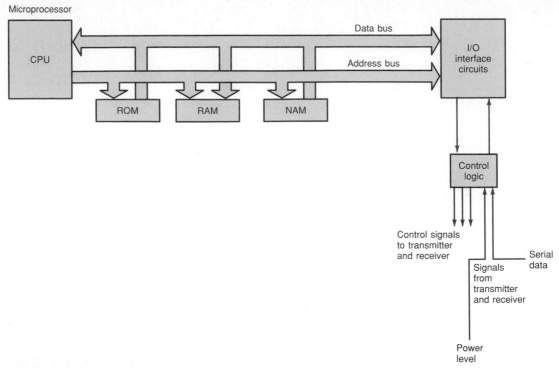

Fig. 14-12 Logic control circuits in a cellular radio.

read-only memory (PROM) chip called the number assignment module (NAM). The NAM contains the mobile identification number (MIN) which is the telephone number assigned to the unit. The NAM PROM is "burned" when the cellular radio is purchased and the MIN is assigned. This chip allows the radio to identify itself when a call is initiated or when interrogated by the MTSO.

All cellular mobile radios are fully under the control of the MTSO through the cell site. The MTSO sends a serial digital data stream at 10 kbits/s through the cell site to the radio to control transmit and receive frequencies and transmitter power. The MTSO monitors the received cell signal strength at the cellular radio by way of the RSSI signal, and it monitors transmitter power level. These are transmitted back to the cell site and MTSO. Audio tones are also used for signaling purposes.

The control unit contains the handset with speaker and microphone. It also contains a complete touch-tone dialing circuit. A simplified block diagram is given in Fig. 14-13. The control unit is operated by a separate microprocessor that drives the LCD display and other indicators. It also implements all manual control functions. The microprocessor memory permits storage of often-called numbers and an auto-dial feature.

The transmitter, receiver, synthesizer, and logic sections are housed in a cabinet which is

mounted in the trunk of the car. The control unit is mounted on the floor or dashboard near the user and communicates with the main unit via a cable.

Here is the sequence of operation for when a mobile unit initiates a call.

1. The operator enters the telephone number to be called via the keypad.
2. The operator picks up the handset, turning on the transmitter and receiver.
3. The receiver seeks an open control channel. There are 21 control or paging channels that are used to establish initial contact with a cell and the MTSO. When contact is made, the cell site reads the NAM data while the MTSO computer verifies that it is a valid number.
4. The operator sends the telephone number to be called.
5. The cell and MTSO search for an open channel and send frequency data to the cellular transceiver.
6. The RSSI signal is read to determine optimum cell selection. Transmitter power is adjusted.
7. Handshake signals are exchanged signifying that contact has been established.
8. The MTSO calls the designated number. The conversation takes place.
9. If the mobile unit passes from one cell to another, the MTSO senses the RSSI sig-

Fig. 14-13 Control unit with handset.

nal and "hands off" the mobile unit from one cell to another to maintain maximum signal strength.

10. The call is terminated.

When a mobile unit is to receive a call, the MTSO and cell site transmit a call signal containing the MIN over a control channel. The transceiver monitors the control channels, identifies the MIN, and turns on. From that point on, the sequence beginning at step 5 above is similar.

The modern cellular telephone system has grown so fast that in many areas of the country the cellular frequency bands are overcrowded. New cellular systems using digital modulation and multiplexing techniques (that is, spread spectrum) are under development. In the coming years they will not only expand and improve the current system but will also permit data transmission from computers as well as voice.

Self-Test

Answer the following questions.

18. The area to be served by a cellular telephone system is divided into adjacent _____ .

19. The master control station for a group of cells is called the _____ .

20. Name two reasons why there are more channels available in cellular systems than in older mobile telephone systems.

21. The number of channels available is _____ .

22. True or false. Cellular telephone radios operate full duplex.

23. The type of modulation used in a cellular radio is _____ .

24. The maximum power output of a cellular transmitter is _____ W.

25. The transmit frequency range is _____ to _____ MHz.

26. The receive frequency range is _____ to _____ MHz.

27. Channel spacing is _____ kHz.

28. Separation between send and receive frequencies is _____ MHz.

29. Transmit and receive frequencies are determined by a(n) _____ .

30. Common first IFs are _____ and _____ MHz. Common second IFs are _____ and _____ MHz.

31. The APC circuit causes the transmitter power to _____ based on an input control signal from the MTSO.

32. A _____ allows the transmitter and receiver to share an antenna.

33. True or false. The transmit and receive channel is selected by the operator.

34. The source of the frequency divider ratios in the frequency synthesizer PLLs is the _____ .

35. Name two signals that are transmitted back to the cell site and are monitored by the MTSO.
36. Name three conditions in the transceiver that are controlled by the MTSO.
37. A dc signal derived from the demodulator and used to tell the MTSO about received signal strength is called the _____ .
38. A receiver has a first IF of 82.2 MHz. The second IF is 456 kHz. The LO frequency on the second mixer is _____ MHz.
39. The section of the cellular transceiver that interprets the serial digital data from the cell site and MTSO is the _____ unit.
40. How many microprocessors do most cellular transceivers contain?
41. The handset is part of the _____ section of the transceiver.

14-3 RADAR

Radar is an acronym for the expression *r*adio *d*etection *and* *r*anging. It is an electronic communications system used to detect objects at distances that cannot be observed visually. The range, direction, and elevation of a remote object can be precisely determined with radar.

Radar is based on the principle that high-frequency RF signals are reflected by conductive targets. The most usual targets are airplanes, missiles, ships, and automobiles. In a radar system, a signal is transmitted toward the target. The reflected signal is picked up by a receiver in the radar unit. This reflected or returned radio signal is called an echo. The radar unit can then determine the distance to the target, its direction or azimuth, and in some cases, its elevation or distance above the ground. See Fig. 14-14.

Fig. 14-14   Concept of radar.

The ability of radar to determine the distance between a remote object and the radar unit is dependent upon knowing the exact speed of radio signal transmission. As you know, the speed of radio signals and light is 300,000,000 m/s. This is the familiar speed of 186,000 mi/s. In most applications where radar is used, the nautical mile is used instead of statute miles for measurement. Recall that a mile is equal to 5280 ft. A nautical mile (nmi) is equal to 6076 ft. The speed of a radio signal is 162,000 nmi/s. Sometimes, a special unit known as a radar mile is used. A radar mile is equal to 6000 ft.

Knowing the speed of a signal in miles per second, we can actually determine the length of time that it takes for the signal to travel 1 mi. For example, it takes a radio signal 5.375 μs to travel 1 mi. The time it takes for a signal to travel 1 nmi is 6.18 μs.

A radar signal must travel twice the distance between the radar unit and the remote object. The signal is transmitted, and a finite time passes before it reaches the remote object. The signal is then reflected and travels an equal distance back to the radar station. If an object is exactly 1 nmi away, the signal takes 6.18 μs to reach the target, and then 6.18 μs for the return trip. The total elapsed time from the instant of initial transmission to the reception of the echo is 12.36 μs.

The distance to the remote target, therefore, can be calculated using the simple expression

$$D = \frac{T}{12.36}$$

Here, D is the distance between the radar unit and the remote object in nautical miles, and T is the total time that occurs between the transmission and reception of the signal in microseconds. For example, if the time is 33.7 μs, the distance is

$$D = \frac{33.7}{12.36}$$
$$= 2.73 \text{ nmi}$$

It is the job of the radar unit to measure this time and compute the distance for display.

In short-distance applications, the yard (yd) is the common unit of distance measurement. A radio signal travels 328 yd/μs. Knowing this fact, the distance to an object can be computed using the expression

$$D = \frac{328T}{2}$$
$$= 164T$$

A measured time of 4.8 μs corresponds to a distance of

$$D = 164(4.8)$$
$$= 787.2 \text{ yd}$$

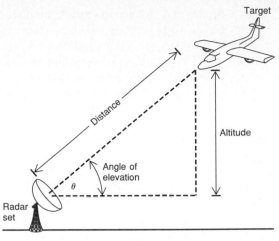

In order to obtain a strong reflection or echo from a distant object, the wavelength of the radar signal should be small compared to the size of the object being observed. If the wavelength of the radar signal is long with respect to the distant object, only a small amount of energy will be reflected. At higher frequencies, the wavelength is shorter and, therefore, the reflected energy is greater. Ideally, the size of the target should be one-quarter wavelength or more at the transmitted frequency for the optimum reflected signal.

The shorter the wavelength of the signal compared to the observed object, the greater the resolution or definition of the remote object. In most cases, it is only necessary to detect the presence of a remote object, but if very short wavelengths are used, the actual shape of the object can often be clearly determined.

The earliest radars operated in the VHF and UHF ranges primarily because technology did not permit reliable higher-frequency operation. However, as microwave components and techniques were developed, they were quickly adopted for radar applications. Virtually all radar systems use microwave signals. Many systems use frequencies as high as millimeter waves for precision detection.

Since radar uses microwave frequencies, line-of-sight communications results. In other words, radar cannot detect objects which are beyond the horizon. Objects need not be physically visible, but they must be within line-of-sight radio distance in order for detection to occur.

In locating a distant object by radar, there are three factors that must be determined. These are distance or range, azimuth or bearing, and elevation or altitude. These can be expressed by a right triangle as shown in Fig. 14-15. Assume that the radar is land-based and used to detect aircraft. The distance between the radar unit and the remote airplane is the hypotenuse of the right triangle. The angle of elevation is the angle between the hypotenuse and the base line which is a line tangent to the surface of the earth at the radar location. The altitude is defined by the angle of elevation. The greater the angle of elevation, the greater the altitude. By knowing the range and the angle of elevation, the altitude can be computed using standard trigonometric techniques.

Another important factor in locating a distant object is knowing its direction with respect to the radar set. If the radar station is fixed and land-based, the direction or azimuth is usually

Altitude = distance × sin θ

Fig. 14-15 The trigonometry of radar.

given as a compass direction. Recall that true north is 0° or 360°, east is 90°, south is 180°, and west is 270°. The direction or bearing of the remote object is usually given in degrees.

If the radar unit is located in a moving vehicle such as an airplane or ship, the azimuth or bearing is given as a relative bearing with respect to the forward direction of the vehicle. Straight ahead is 0° or 360°, directly to the right is 90°, directly behind is 180°, and directly to the left is 270°.

The ability of a radar unit to determine the direction of a remote object is strictly dependent upon a highly directional antenna. By using an antenna with an extremely narrow beam width, the radar unit will receive signals from only over a narrow angle. The narrower the beam width of the antenna, the more precisely the actual azimuth or bearing can be determined.

Since most radars operate in the microwave region, highly directional antennas can be easily obtained. Horn antennas with parabolic reflectors are the most common, and beam widths of less than 1° are readily attainable. These highly directional antennas are rotated continuously 360°. The same antenna is used for transmitting the original signal and receiving the reflected signal.

Circuits within the radar unit are calibrated so that the direction in which the antenna is pointing is accurately known. Then when the echo is received, it can be compared to the calibrated values and the precise direction can be determined.

The ability of a radar unit to determine the altitude of a remote target depends upon the vertical beam width of the radar antenna. The radar antenna may scan vertically while measuring the distance of the object during the scan. When the object is detected, the vertical elevation of the antenna is noted and the actual altitude can then be computed.

There are two basic types of radar systems: pulsed and continuous wave (CW). There are also numerous variations of each. By far the most commonly used radar system is the pulsed type. Signals are transmitted in short bursts or pulses as shown in Fig. 14-16. The duration or width *w* of the pulse is very short and, depending upon the application, can be anywhere from less than a microsecond to several microseconds. The time between transmitted pulses is known as the pulse repetition time (PRT). From this can be determined the pulse repetition frequency (PRF).

$$PRF = \frac{1}{PRT}$$

If the PRT is 120 μs, the PRF is

$$PRF = \frac{1}{120 \times 10^{-6}}$$
$$= 833.3 \text{ kHz}$$

The ratio of the pulse width *w* to the PRT is known as the duty cycle. The duty cycle is normally expressed as a percentage.

$$\text{Duty cycle} = \frac{w(100)}{PRT}$$

A pulse width of 9 μs with a PRT of 120 μs produces a duty cycle of

$$\text{Duty cycle} = \frac{9(100)}{120}$$
$$= 7.5\%$$

Note in Fig. 14-16 the time between the end of the transmitted pulse and the begin-

ning of the next pulse in sequence. It is during this interval that the echo will be received.

The duration of the transmitted pulse and the PRT are extremely critical in determining the performance of a radar and its application. Very short range radars will use narrow pulses and short PRTs. If the object is only a short distance away, then the time for the echo will be relatively short. In short-range radars, the pulse width is made narrow to ensure that the pulse is terminated prior to the receipt of the echo of the target. If the pulse is too long, the return signal may be masked or blanked by the transmitted pulse. On the other hand, long-range radars will typically use a longer PRT since it will take a greater time for the echo to return. This also permits a longer burst of energy to be transmitted, thereby ensuring a stronger return.

If the PRT is too short compared to the distance of the target, it is possible that the echo will not return during the time interval between two successive pulses. The return may occur after the second transmitted pulse. This is known as a double range or second return echo. Naturally, it will lead to ambiguous distance measurements.

Of course, the distance to the remote object is determined simply by measuring the time delay between the transmitted pulse and the received echo.

In CW radar, a constant-amplitude continuous microwave sine wave is transmitted. The echo, therefore, is also a constant-amplitude microwave sine wave of the same frequency but of lower amplitude. The question is then, "How can such a signal be useful in determining the range or other characteristics of a target?" The answer lies in the object itself which, in most cases, is moving with respect to the radar unit. When detecting a moving airplane, ship, missile, or automobile, the reflected signal undergoes a frequency change. It is this frequency change between the transmitted signal and the returned signal that is used to determine the speed of the target.

The frequency shift that occurs when there is relative motion between the transmitting station and a remote object is known as the *Doppler effect*. You have experienced the Doppler effect yourself with sound waves. For example, the horn on a car emits a tone at a fixed frequency. If the car is stationary, you hear this frequency. However, if the car is

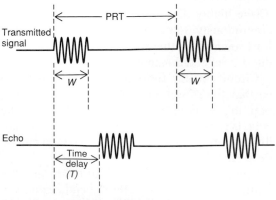

Fig. 14-16   Pulsed radar signals.

moving toward you while the horn is on, you will experience a tone of increased frequency. As the car moves closer to you, the sound waves are compressed, thereby giving the effect of a higher frequency. If the car is moving away from you with its horn on, you will experience a decreased frequency. As the car moves away, the sound waves are stretched out giving the effect of a lower frequency. This same effect works on both radio and light waves.

By measuring the amount of the frequency difference between the transmitted signal and the reflected signal, it is possible to determine the relative speed between the radar unit and the observed object.

$$V = 1.1 f \lambda$$

where f = frequency difference between transmitted and reflected signals, Hz
λ = wavelength of transmitted signal, m
V = relative speed between the two objects, mi/h

Assume a frequency shift of 1000 Hz at a frequency of 10 GHz. A frequency of 10 GHz represents a wavelength of

$$\lambda = \frac{300}{f \text{ (in MHz)}}$$
$$= \frac{300}{10,000}$$
$$= 0.03 \text{ m}$$

Therefore the speed is

$$V = 1.1(1000)(0.03)$$
$$= 33 \text{ mi/h}$$

In CW radar, it is the Doppler effect that provides frequency modulation of the carrier. In order for there to be a frequency change, the observed object must be moving toward or away from the radar unit. If the observed object moves parallel to the radar unit, there is no relative motion between the two objects, and no frequency modulation will occur. The greatest value of CW radar is its ability to measure the speed of distant objects. The familiar police radar units use CW Doppler radar for measuring the speed of cars and trucks.

A popular variation of CW radar is FM CW radar. In these systems, the carrier is continuous but is frequency modulated by a sawtooth or triangular wave. Since the frequency of the transmitted signal varies over time, the frequency of the echo signal can be compared to the frequency of the transmitted signal. The frequency difference will determine the time and, therefore, the distance between the transmitting station and the target. The frequency difference between the transmitted and echo signals is directly proportional to the range.

Some radar systems combine both pulse and Doppler techniques to improve performance and measurement capabilities. One such system evaluates successive echos to determine phase shifts that indicate when a target is moving. Such radars are said to incorporate moving target indication (MTI). Through a variety of special signal-processing techniques, multiple moving targets can be distinguished not only from one another but from fixed targets as well.

Figure 14-17 on the next page shows a comprehensive block diagram of a typical radar unit. There are four basic subsystems: the antenna, the transmitter, the receiver, and the display unit.

The transmitter in a pulsed radar system invariably uses a magnetron. Recall that a magnetron is a special high-power vacuum tube oscillator that operates in the microwave region. The cavity size of the magnetron sets the operating frequency. A master timing generator develops the basic pulses used for triggering the magnetron. The timing generator sets the pulse duration, PRT, and the duty cycle. The pulses from the timing network trigger the magnetron into oscillation, and the magnetron emits short bursts of microwave energy. Magnetrons are capable of extremely high powers, especially when operated on a pulse basis. Continuous average power may be low, but when pulsed, magnetrons can produce many megawatts of power for the short duration required by the application.

Although magnetrons are the most commonly used radar transmitter tube, both klystrons and TWTs are also used in some types of radars. These are more commonly found in CW Doppler radars. Low-power radars such as those used by the police for speed detection use Gunn diodes.

Referring to the block diagram, you can see that the transmitter output is then passed through a duplexer or circulator where it is then applied to the antenna. The duplexer is a special device that allows the transmitter and receiver to share a single antenna. The duplexer contains devices that prevent the

Fig. 14-17   General block diagram of a pulsed radar.

high-power transmitted signal from getting into the receiver and damaging it.

A duplexer is a waveguide assembly containing special devices that prevent interference between the transmitter and receiver. The most commonly used device is a spark-gap tube. Different types of spark-gap tubes are used and are referred to as transmit-receive (TR) and anti–transmit-receive (ATR) tubes. The purpose of the TR tube is to prevent transmitter power from reaching the receiver. When RF energy from the transmitter is detected, the spark gap breaks down creating a short circuit for the RF energy. The purpose of the ATR tube is effectively to disconnect the transmitter from the circuit during the receive interval. The TR and ATR tubes when combined with the appropriate quarter- and half-wavelength waveguides provide effective isolation between transmitter and receiver. In lower-power radar systems, a standard circulator or isolator can be used for the same purpose.

The antenna system is typically a horn antenna with a parabolic reflector used to produce a very narrow beam width. A special waveguide assembly with a rotating joint allows the waveguide and horn antenna to be rotated continuously over a 360° angle.

The same antenna is also used for reception. During the pulse off time, the received signal passes through the antenna, the associated waveguide, and either the duplexer or circulator to the receiver. The receiver is a standard high-gain superheterodyne type. The signal is usually fed directly to a mixer in most systems, although some radars use an RF amplifier. The mixer is typically a single diode or diode bridge assembly. The LO feeds a signal to the mixer at the appropriate frequency so that an IF is developed. Both klystrons and Gunn diodes are used in LO applications. The IF amplifiers provide very high gain prior to demodulation. Some radar receivers may use double conversion.

The demodulator in a pulse radar system is typically just a diode detector since only the pulses must be detected. In Doppler and other more complex radars, some form of frequency- and phase-sensitive demodulator is used. Phase-locked loops are common in this application. The output of the demodulator is fed to a video amplifier which creates signals that will ultimately be displayed.

The display in most radar systems is a cathode-ray tube (CRT). Various display formats can be used. The simplest form of display known as an A scan simply displays the transmitted and received pulses. The horizontal sweep on the oscilloscope is calibrated in yards or miles. See Fig. 14-18(a).

The most common type of CRT display is known as type P or plan position indicator

(a) A scan

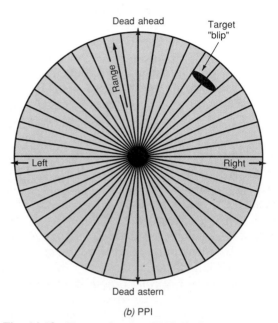

(b) PPI

Fig. 14-18   Types of radar CRT displays.

direction of the antenna. At the same time, the horizontal and vertical deflection coils (called the yoke) around the neck of the CRT rotate in synchronism with the antenna. The electron beam in the CRT is swept from the center of the screen out to the edge. The sweep begins at the instant of pulse transmission and moves outward from the center. The beginning of each transmitted pulse begins another sweep of the electron beam. As the deflection yoke rotates, the beam moves in such a way that it appears that a radius from the center to the edge is continuously rotating. If a target reflection occurs, a lighted blip will appear on the screen. Calibrations on the screen in the form of graticule markings or superimposed electron beam patterns permit the distance and azimuth reading to be read directly.

An important newer type of radar provides greater flexibility in scanning narrow sectors and tracking multiple targets. Known as phased array radars, these high-tech units operate on the same principles but use a unique antenna. Instead of a single horn and parabolic reflector, multiple dipoles are used. Slots in a waveguide are also used. A half-wave dipole at microwave frequencies is very short. Therefore, many of them can be mounted together in a matrix or array. The result is a special collinear or end-fire array on a reflecting surface with very high gain. By using a system of separate feed lines and a variable phase shifter for each dipole, the beam width and directivity can be controlled electronically. This permits rapid scanning and on-the-fly adjustment of directivity. Phased array radars eliminate the mechanical systems needed for conventional radars.

As for applications of radar, there are many. Their greatest use is in defense weapons systems and in safety and navigation applications. Search radars are used to locate enemy missiles, planes, and ships. Tracking radars are used on missiles and planes to acquire a target and zero in on them. Radars are widely used on planes and ships for navigating "blind" in fog or bad weather. Radars help ground controllers locate and identify nearby planes. Special radars assist planes in landing in bad weather when visibility is near zero.

Radars are also used as altimeters to measure height. High-frequency radars can actually be used to plot or map the terrain in an area. Special terrain-following radars allow high-speed jets to fly very close to the ground to avoid detection by enemy radar.

(PPI). See Fig. 14-18(*b*). The advantage of the PPI display is that it shows both the range and azimuth of the target. The center of the display is assumed to be the location of the radar unit. Concentric circles indicate the range. The azimuth or direction is indicated by the position of the reflected target on the screen with respect to the vertical radius line. The targets show up as lighted blips on the screen.

The PPI display is developed by a sophisticated scanning system tied in to antenna rotation. As the antenna is rotated by a motor, an encoder mechanism sends signals to the PPI control circuits that designate the azimuth or

Self-Test

Answer the following questions.

42. When the speed of radio signals is known, radar can be used to determine the _____ of a target.
43. An echo is the _____ from a target.
44. The directivity of the radar antenna determines the _____ of a target.
45. The speed of a radar signal is _____ μs/nmi.
46. The elapsed time between radiating a radar signal and receiving its echo is 43.2 μs. The target distance in nautical miles is _____ .
47. If a target is 1000 yd away, the time period between the transmission of the signal and receipt of the echo is _____ μs.
48. One nautical mile is equal to _____ statute miles.
49. For optimum reflection from a target, the wavelength of the radar signal should be _____ compared to the size of the target.
50. Radars operate in the _____ frequency range.
51. True or false. Both the transmitter and receiver in a radar share the same antenna.
52. The two main types of radar are _____ .
53. A frequency shift of sound, radio, or light waves that occurs as the result of the relative motion between objects is called the _____ effect.
54. A CW radar operates at a frequency of 14 GHz. A frequency shift of 25 kHz is produced by a moving target. Its speed is _____ mi/h.
55. True or false. Distance can be measured with constant-frequency CW radar.
56. A waveguide assembly that lets the transmitter and receiver share an antenna is called a(n) _____ .
57. The most commonly used component in a radar transmitter is a(n) _____ tube.
58. Both TR and ATR tubes incorporate a(n) _____ .
59. The output display in a radar is usually a(n) _____ .
60. A display that sweeps outward from the center of the screen while rotating is known as a(n) _____ .
61. A radar using multiple antennas to shape the beam width and automatically adjust directivity is known as a(n) _____ radar.

14-4 TELEVISION

Television is radio communications with both pictures and sound. Most radio communications are voice communications in which a microphone develops an audio signal that it can use to modulate a transmitter. This occurs in television too, but in addition, a television camera converts a visual scene into a voltage known as the *video signal*. This video signal, which represents the picture information, is used to modulate another transmitter. Both the picture and the sound signals are transmitted to the receiver, which is used to demodulate them and present the information to the user. The television receiver is a special superheterodyne that recovers both the sound and the picture information. The picture is displayed on a cathode-ray tube (CRT) better known as a *picture tube*.

A considerable amount of intelligence is contained in a television signal. As a result, it takes up a significant amount of spectrum space. As indicated earlier, the television signal consists of two main parts: the sound and the picture. The complete signal bandwidth of a television signal is illustrated in Fig. 14-19. The entire television signal occupies a channel in the spectrum whose bandwidth is 6 MHz. Because a TV signal uses so much spectrum space, TV stations must operate in the VHF and UHF frequency ranges from 50 to 806 MHz, or higher. The frequency ranges for each of the currently assigned 68 channels are given in Fig. 14-20.

Referring to Fig. 14-19, note that there are two signal carriers. The sound carrier is at the upper end of the spectrum. Frequency modulation is used to impress the sound signal on the carrier. High-fidelity sound is available with an audio bandwidth of 50 Hz to 15 kHz. The maximum permitted frequency deviation is 25 kHz, which is considerably less than the deviation permitted by conventional FM broadcasting. As a result, a TV sound signal occupies less bandwidth in the spectrum than a standard FM broadcast station, in this case only about 80 kHz, calculated using Carson's rule 2(25 kHz + 15 kHz) = 80 kHz.

Stereo sound is also available in television. The frequency division multiplexing method used to transmit two channels of sound information is virtually identical to that used in stereo transmission for FM broadcasting discussed in a previous chapter.

The picture information is transmitted on a separate carrier that is located 4.5 MHz lower

Fig. 14-19 **Spectrum of a broadcast TV signal.**

Channel	Frequency, MHz	Channel	Frequency, MHz
Low-band VHF		UHF (cont.)	
2	54–60	32	578–584
3	60–66	33	584–590
4	66–72	34	590–596
5	76–82	35	596–602
6	82–88	36	602–608
FM broadcast	88–108	37	608–614
Aircraft	118–135	38	614–620
Ham radio	144–148	39	620–626
Mobile or marine	150–173	40	626–632
		41	632–638
High-band VHF		42	638–644
		43	644–650
7	174–180	44	650–656
8	180–186	45	656–662
9	186–192	46	662–668
10	192–198	47	668–674
11	198–204	48	674–680
12	204–210	49	680–686
13	210–216	50	686–692
		51	692–698
UHF		52	698–704
14	470–476	53	704–710
15	476–482	54	710–716
16	482–488	55	716–722
17	488–494	56	722–728
18	494–500	57	728–734
19	500–506	58	734–740
20	506–512	59	740–746
21	512–518	60	746–752
22	518–524	61	752–758
23	524–530	62	758–764
24	530–536	63	764–770
25	536–542	64	770–776
26	542–548	65	776–782
27	548–554	66	782–788
28	554–560	67	788–794
29	560–566	68	794–800
30	566–572	69	800–806
31	572–578	Cellular telephone	806–902

Fig. 14-20 **VHF and UHF TV channel frequency assignments.**

in frequency than the sound carrier. The video signal derived from the camera is used to amplitude-modulate the picture carrier. Different methods of modulation are used for both sound and picture information so that there is less interference between the picture and sound signals. Further, amplitude modulation of the carrier takes up less bandwidth in the spectrum, and this is important for such a high-frequency modulating signal such as video.

Note in Fig. 14-19 that vestigial sideband AM is used. This was discussed in Chap. 2. The full upper sideband of the picture information is transmitted, but a major portion of the lower sideband is suppressed to conserve spectrum space. The video signal itself can contain frequency components up to 4 MHz. Therefore, if both sidebands were transmitted simultaneously, the picture signal alone would occupy 8 MHz.

The job of the television camera is to convert the scene or object you wish to view into an electrical signal which can be used to modulate a carrier. All visible scenes and objects are simply light that has been reflected by objects and transmitted to our eyes. It is the purpose of the camera to take the light and color details in a scene and convert them into an electrical signal.

To do this, the scene to be transmitted is collected and focused by a lens upon a light-sensitive imaging device. Both vacuum tube and semiconductor devices are used for converting the light information in the scene into an electrical signal. Common examples of imaging devices are the vidicon camera tube and the popular semiconductor charge-coupled device (CCD) so widely used in camcorders.

Because a scene contains so much light information, it is not possible for an imaging device to perform a simultaneous conversion of all this information into a composite signal. Instead, the scene is divided up into smaller segments which can be transmitted serially over a period of time. Again, it is the job of the camera to subdivide the scene in an orderly manner so that an acceptable signal is developed. This process, which is somewhat like sampling and time division multiplexing, is known as *scanning*.

Scanning is a technique that divides a rectangular scene into individual horizontal lines. The standard TV picture dimensions have an aspect ratio of 4:3. This means that the scene is 4 units wide for every 3 units of height. To create a picture, the scene is subdivided into many fine horizontal lines called *scan lines*. Each line represents a very narrow width of light variations in the scene. The greater the number of lines used, the higher the resolution and the greater the detail that can be observed. U.S. television standards, which were established in the 1950s by the National Television Standards Committee (NTSC), call for the scene to be divided into a maximum of 525 horizontal lines.

Figure 14-21 shows a simplified drawing of the scanning process. The scene is simply a large black letter F on a white background. The purpose of the television camera is to convert this scene into an electrical signal. This is done by transmitting a voltage of 1 V for black and 0 V for white. Note that the scene is divided into a total of 15 scan lines numbered 0 through 14. The scene is focused on the light-sensitive area of a vidicon tube or CCD imaging device, which scans the scene one line at a time, transmitting the light variations along that line as voltage levels. Figure 14-21 shows the light variations along several of the lines. When the white background is being scanned, a 0-V signal occurs. When a black picture ele-

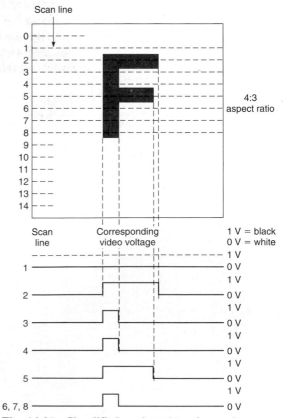

Fig. 14-21 Simplified explanation of scanning.

Fig. 14-22 **The scan line voltages are transmitted serially.**

ment is encountered, a 1-V level is transmitted. The electrical signals derived from each scan line are referred to as the *video signal.* They are transmitted serially one after the other until the entire scene has been sent (see Fig. 14-22). This is how a standard television picture is developed and transmitted.

Most scenes are more complicated than a black letter on a white background. Typically, the picture is transmitted as different shades of gray between black and white. Shades of gray are represented by some voltage level between the 0- and 1-V extremes represented by white and black. The resulting signal is known as the *brightness* or *luminance signal* and is usually designated by the letter *Y.*

A more detailed illustration of the NTSC scanning process is given in Fig. 14-23. In fact, the scene is scanned twice. One complete scanning of the scene is called a *field* and contains 262½ lines. The entire field is scanned in 0.01667 s for a 60-Hz field rate. In color TV the field rate is 59.94 Hz. Then the scene is scanned a second time again using 262½ lines. The second field is scanned so that its scan lines fall between those of the first field. This produces what is known as *interlaced scanning,* a total of 525 lines (2 × 262½). In prac-

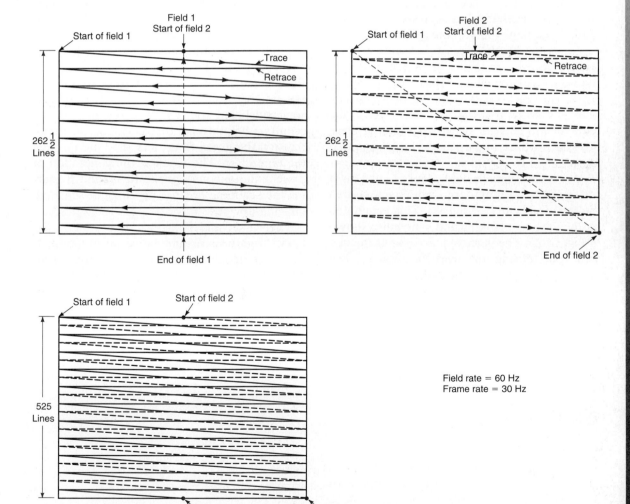

Fig. 14-23 **Interlaced scanning is used to minimize flicker.**

tice, only about 480 lines appear on the picture tube screen. Two interlaced fields produce a complete frame of video. With the field rate being 0.01667 s, two fields produce a frame rate of 0.03333 s, or 30 Hz. The frame rate for color TV is one-half the field rate, or 29.97 Hz. Interlaced scanning is used to reduce flicker, which is annoying to the eye. The rate is also fast enough that the human eye cannot detect individual scan lines, so the result is a stable picture.

The rate of occurrence of the horizontal scan lines is 15,750 Hz for monochrome or black-and-white TV and 15,734 Hz for color TV. This means that it takes about $1/15{,}734 = 63.56$ μs to trace out one horizontal scan line.

At the TV receiver, the picture tube is scanned in step with the camera to accurately reproduce the picture. To ensure that the receiver stays exactly in synchronization with the camera, special horizontal and vertical sync pulses are added to and transmitted with the video signal. This is illustrated in Fig. 14-24. After each line has been scanned, a horizontal blanking pulse comes along. At the receiver, the blanking pulse is used to cut off the electron beam in the picture tube during the time the beam must retrace from right to left to get ready for the next left to right scan line. The horizontal sync pulse is used at the receiver to keep the sweep circuits that drive the picture tube in step with the transmitted signal.

At the end of each field, the scanning must retrace from bottom to top of the scene so that the next field can be scanned. This is initiated by the vertical blanking and sync pulse shown in Fig. 14-25. The vertical pulse blanks the picture tube during the vertical retrace. The pulses on top of the vertical blanking pulse are the horizontal sync pulses which must con-

tinue to keep the horizontal sweep in sync during the vertical retrace. The equalizing pulses help synchronize the half scan lines in each field.

In color TV, the color information along each scan line must also be detected and transmitted. This is done by dividing the light in each scan line into three separate signals, each representing one of the three basic colors—red, green, and blue. It is a principle of physics that any color can be made by using some combination of the three primary light colors. This concept is illustrated in Fig. 14-26 on page 400. However, Fig. 14-26 does not show how colors such as orange are produced. They result from the combination of the primary colors (red, green, and blue) in the proper ratio. Orange requires more red and less green when compared to yellow. In theory, an infinite number of colors can be created by using different ratios of the primary colors.

In the same way, the light in any scene may be divided into the three basic colors by passing the light through red, green, and blue filters. This is done in a color TV camera, which is really three cameras in one (see Fig. 14-27 on page 400). The lens focuses the scene on three separate light-sensitive devices such as the vidicon or CCD by way of a series of mirrors and beam splitters. The red light in the scene passes through the red filter, the green through the green filter, and the blue through the blue filter. The result is the generation of three simultaneous signals (R, G, and B) during the scanning process by the light-sensitive imaging devices.

The R, G, and B signals also contain the basic brightness or luminance information. If the color signals are mixed in the correct proportion, the result is the *standard luminance, or Y, signal*. The Y signal is generated by scal-

Fig. 14-24 Sync pulses are used to keep the receiver in step with the transmitter.

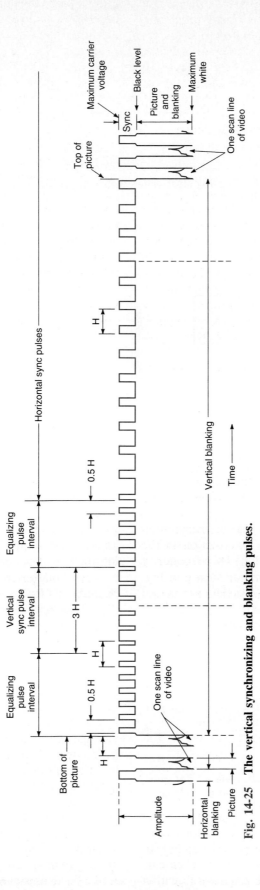

Fig. 14-25 The vertical synchronizing and blanking pulses.

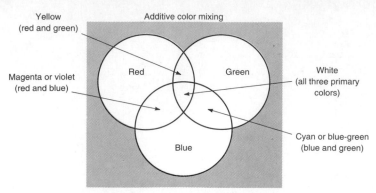

Fig. 14-26 Creating other colors with red, green, and blue light.

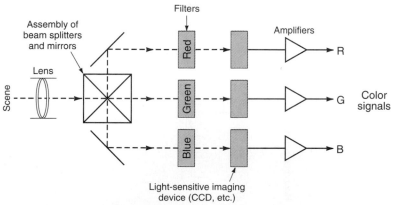

Fig. 14-27 How the camera generates the color signals.

ing each color signal with a potentiometer and then adding them together as shown in Fig. 14-28(a).

The color signals must also be transmitted along with the luminance information in the same bandwidth. This is done by a frequency division multiplexing technique shown in Fig. 14-28(a). The R, G, and B signals are mixed together to get the Y signal.

The Y signal is the only signal needed for monochrome reception. It is proportioned to be 30 percent red, 59 percent green, and 11 percent blue. Color reception requires additional information. The two additional signals prepared at the transmitter are called I (for in-phase) and Q (for quadrature, 90° shift). These chrominance signals are proportioned as follows:

I = 60% red, −28% green, and −32% blue
Q = 21% red, −52% green, and 31% blue

The minus signs mean that the signal is phase-inverted before linear mixing (addition).

Chrominance signals are phase-encoded. In the receiver, the sidebands around the subcarrier frequency of 3.579 MHz are mixed with a phase-locked oscillator to decode I and Q, or R–Y and B–Y signals. As can be seen in Fig. 14-28(b), the I–Q signals are rotated by 57°. Receiver design is simplified by using a 57° phase delay (compared to burst) for demodulation. This produces R–Y and B–Y; then it is a simple matter to produce the green signal because Y, R–Y, and B–Y are all available. (The G signal is contained in all of them.)

So why bother with I and Q in the first place? Original research showed that human vision is more sensitive to fine details in the color range around orange. This is where the I signal is phased. So, the I signal (approximately orange) is broadcast with more *bandwidth* than Q. Most receivers do not take advantage of this extra detail and simply decode R–Y and B–Y by phase-delaying 57° from the burst signal phase.

In order to demodulate the color signals and reproduce them faithfully, the subcarrier must be reinserted with the correct phase at the receiver. A 3.58-MHz oscillator in the receiver generates the subcarrier for the balanced modulator demod circuits. A sample of the 3.58-MHz subcarrier signal developed at the

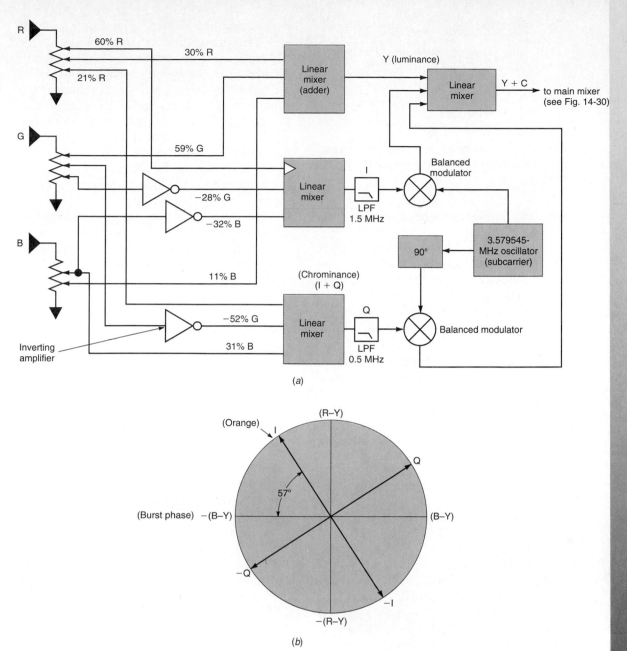

Fig. 14-28   (a) How the NTSC composite video signal is generated. (b) The chrominance signals are phase-encoded.

transmitter is added to the composite video signal. This is done by gating 8 to 12 cycles of the 3.58-MHz subcarrier and adding it to the horizontal sync and blanking pulse as shown in Fig. 14-29 on the next page. This is called the *color burst,* and it rides on what is called the *back porch* of the horizontal sync pulse. The receiver uses this signal to phase-synchronize the internally generated subcarrier before it is used in the demodulation process.

A complete block diagram of a TV transmitter is shown in Fig. 14-30 on the next page. Note the sweep and sync circuits that create the scanning signals for the vidicons or CCDs as well as generate the sync pulses that are transmitted along with the video and color signals. The sync signals (including the color burst), luminance Y, and the color signals are all added together to form the final video signal that is used to modulate the carrier. Low-level AM is used. The final AM signal is then amplified by very-high-power linear amplifiers and sent to the antenna via a diplexer. The diplexer is a filter circuit that passes the transmitter signal to the antenna but prevents signals from another transmitter (the sound) on

Fig. 14-29 The 3.58-MHz color subcarrier burst used to synchronize color demodulation at the receiver.

an adjacent frequency from getting back into the other transmitter.

At the same time, the voice or sound signals frequency-modulate a carrier that is then amplified by class C amplifiers and fed to the same antenna by way of the diplexer. The resulting VHF or UHF TV signal travels by line of sight propagation to the antenna and receiver.

A block diagram of a TV receiver is shown in Fig. 14-31(*a*). Although it is nothing more than a superheterodyne receiver, it is one of the most sophisticated and complex electronic devices ever developed. Today, most of the circuitry is incorporated in large-scale inte-

grated circuits. Yet, the typical TV receiver still uses many discrete component circuits.

The signal from the antenna or the cable is connected to the tuner, which consists of an RF amplifier, mixer, and local oscillator. The tuner is used to select which TV channel is to be viewed and convert the picture and sound carriers plus their modulation to an intermediate frequency (IF). As in most superheterodyne receivers, the local oscillator frequency is set higher than the incoming signal by the IF value.

Most TV set tuners are prepackaged in sealed and shielded enclosures. They are actually two tuners in one: one for the VHF signals and another for the UHF signals. The VHF tuner usually uses low-noise FETs for the RF amplifier and the mixer. UHF tuners use a FET, RF amplifier, and diode mixer. The local oscillators are phase-locked loop (PLL) frequency synthesizers set to frequencies that will convert the TV signals to the IF.

Tuning of the local oscillator is typically done digitally. The PLL synthesizer is tuned by setting the feedback frequency division ratio. In a TV set this is changed by a microprocessor which is part of the master control

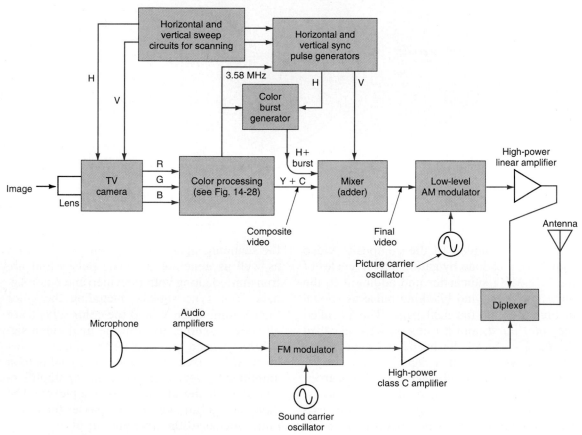

Fig. 14-30 Complete TV transmitter.

Fig. 14-31 (*a*) **Block diagram of TV receiver. The area outlined with a dashed line shows one way of generating horizontal and vertical sync pulses (see Fig. 14-31*b* on the next page).**

Phase detector

15.734-kHz
horizontal sync
pulses from
sync separator

LPF

VCO

31.468 kHz

Frequency
divider
÷ 525

Input
sample of
horizontal
output stage
signal

Flip-flop
÷ 2

Sync IC

15.734 kHz
to horizontal
output stage

59.94 Hz
to vertical
output stage

(b)

Fig. 14-31 (*b*) **An alternative way of generating horizontal and vertical sync pulses.**

system. The interstage LC resonant circuits in the tuner are controlled by varactor diodes. By varying the dc bias on the varactors, their capacitance is changed, thereby changing the resonant frequency of the tuned circuits. The bias control signals also come from the control microprocessor.

Most TV sets are also tuned by remote control. A handheld remote control is used to generate digital codes that indicate channel change, volume control, and other information. These codes normally gate a high-frequency carrier off and on. This carrier modulates an infrared (IR) LED light source. The receiver contains an infrared (IR) detector that picks up the digital codes, demodulates them, and feeds them to the control microprocessor for interpretation. The control micro then outputs signals that change channel, volume, picture brightness, or other functions.

The standard TV receiver IFs are 41.25 MHz for the sound carrier and 45.75 MHz for the picture carrier. Assume that the receiver is tuned to channel 4. The picture carrier is 67.25 MHz, and the sound carrier is 71.75 MHz. Note that their difference is 4.5 MHz. The synthesizer local oscillator is set to 113 MHz. The tuner produces an output that is the difference between the incoming signal and local oscillator frequencies, or 113 − 67.25 = 45.75

MHz for the picture and 113 − 71.75 = 41.25 MHz for the sound. Because the local oscillator frequency is above the incoming signals, their relationship is reversed at the intermediate frequencies, the picture carrier being 4.5 MHz above the sound carrier.

The IF signals are then sent to the video IF amplifiers. Selectivity is usually obtained with a special filter known as a *surface acoustic wave (SAW) filter*. This fixed tuned filter is designed to provide just the exact selectivity required to pass both of the IF signals. Figure 14-32(*a*) shows the concept of the filter. It is made on a piezoelectric ceramic substrate such as lithium niobate. A pattern of interdigital fingers on the surface convert the IF signals into acoustic waves that travel across the filter surface. Controlling the shapes, sizes, and spacings of the interdigital filters makes it possible to tailor the response to any application. Interdigital fingers at the output convert the acoustic waves back into electrical signals at the IF.

The response of the SAW IF filter is shown in Fig. 14-32(*b*). Note that the filter greatly attenuates the sound IF. The maximum response occurs in the 43- to 44-MHz range. The picture carrier IF is down 50 percent on the curve. The lower sideband contains frequencies up to 1.25 MHz away from the carrier fre-

(a) SAW filter

41.25 MHz (sound)   45.75 MHz (picture)

43–44 MHz

(b) Typical response curve

Fig. 14-32 (a) **Surface acoustic wave filter and** (b) **IF response curve.**

quency. With no compensation, these frequencies would be emphasized after detection in the receiver, since they are found in both the LSB and the USB of the transmitted TV signal. However, the response curve of Fig. 14-32(b) does compensate for this effect. Frequencies near the picture carrier (45.75 MHz) are attenuated because of the shape of the response curve. The overall effect is that vestigial LSB does not produce extra output, after detection, for the lower video frequencies.

The IF signals are next amplified by IC amplifiers. The video (luminance or Y) signal is then recovered by an AM demodulator. In older sets, a simple diode detector was used for video detection. In most modern sets a synchronous balanced modulator-type demodulator is used. It is part of the IF amplifier IC.

The output of the video detector is the Y signal and the composite color signals. These signals are amplified by the video amps. The Y signal is used to create an AGC voltage for controlling the gain of the IF amplifiers and the tuner amplifiers and mixers.

The color signals are selected from the video amp output by a filter and fed to color-balanced demodulator circuits. The color

burst signal is also picked up by a gating circuit and used to synchronize an oscillator that produces a 3.58-MHz subcarrier signal of the correct frequency and phase. The output of this oscillator is fed to two balanced demodulators that recover the I and Q signals or the R–Y and B–Y signals. These are combined in matrix with the Y signal and out comes the three color signals (R, G, and B). These are amplified and sent to the picture tube, which reproduces the picture. More about the picture tube later.

To recover the sound part of the TV signal, a separate sound IF and detector section are used. Continuing to refer to Fig. 14-31, note that the video IF signal is fed to the sound detector circuit. This video IF signal contains both the picture carrier at 45.75 MHz and the sound carrier at 41.25 MHz. The sound detector is a nonlinear circuit that heterodynes the two carriers and generates the sum and difference frequencies. The result is a 4.5-MHz difference signal which contains both the AM picture and the FM sound modulation. This is the sound IF signal. It is passed to the sound IF amplifiers, which are tuned to 4.5 MHz and which also perform a clipping and limiting function to remove the AM and leave only the

FM sound. The audio is recovered with a quadrature detector or differential peak detector as described in Chap. 5. The audio is amplified by one or more audio stages and sent to the speaker. If stereo is used, the appropriate demultiplexing is done by an IC and the left and right channel audio signals are amplified.

A major part of the TV receiver is concerned with the sweep and synchronizing functions which are unique to TV receivers. In other words, the receiver's job does not end with demodulation and recovery of the picture and sound. For the picture to be displayed on a picture tube, special sweep circuits are needed to generate the voltages and currents to operate the picture tube, and sync circuits are needed to keep the sweep in step with the transmitted signal.

The sweep and sync operations begin in the video amplifier. The demodulated video includes the horizontal blanking and sync pulses. The sync pulses are stripped off the video signal with a sync separator circuit and fed to the sweep circuits. The horizontal sync pulses are used to synchronize the horizontal oscillator to 15,734 Hz. This oscillator drives a horizontal output stage that develops a sawtooth of current that drives magnetic deflection coils in the picture tube yoke; these coils sweep the electron beams in the picture tube. The *yoke* is a coil assembly that fits around the picture tube's neck (the narrow part at the back of the tube).

The horizontal output stage, which is a high-power switch, is also used to create a switching power supply. The horizontal output transistor drives a step-up/step-down transformer called the *flyback*. The 15.734-kHz pulses are stepped up, rectified, and filtered to develop the 30- to 35-kV high-voltage DC required to operate the picture tube. Step-down windings on the flyback produce lower-voltage pulses that are rectified and filtered into low voltages used as power supplies for most of the circuits in the receiver.

The sync pulses are also fed to an integrating circuit that takes the horizontal sync pulses during the vertical blanking interval and integrates them into a 59.94-Hz sync pulse that is used to synchronize a vertical sweep oscillator. The output from this oscillator is a sawtooth sweep voltage at the field rate of 59.94 Hz. This is amplified and converted into a linear sweep current that drives the magnetic coils in the picture tube yoke that produce vertical deflection of the electron beams in the picture tube.

In most modern TV sets, the horizontal and vertical oscillators are replaced by digital sync circuits. See Fig. 14-31(*b*). The horizontal sync pulses from the sync separator are normally used to phase-lock a 31.468-kHz oscillator that is two times the normal horizontal rate of 15.734 kHz. Dividing this by two gives the horizontal pulses that are amplified and shaped in the horizontal output stage to drive the deflection coils on the picture tube. A digital frequency divider divides the 31.468-kHz signal by 525 to get a 59.94-Hz signal for vertical sync. Then this signal is shaped into a current sawtooth and is amplified by the vertical output stage which drives the deflection coils on the picture tube.

The component in the TV set that makes color TV pictures possible is the picture tube. The picture tube is a vacuum tube called a cathode-ray tube (CRT). The basic operation of a CRT is illustrated with a black-and-white or monochrome tube as shown in Fig. 14-33(*a*). The tube is housed in a bell-shaped glass enclosure. A filament heats a cathode which emits electrons. The negatively charged electrons are attracted and accelerated by positive-bias voltages on the elements in an electron gun assembly. The electron gun also focuses the electrons into a very narrow beam. A control grid that is made negative with respect to the cathode controls the intensity of the electron beam and the brightness of the spot it makes. The beam is accelerated forward by a very high voltage applied to an internal metallic coating called *aquadag*. The face or front of the picture tube is coated internally with a phosphor that glows and produces white light when it is struck by the electron beam.

Around the neck of the picture is a structure of magnetic coils called the *deflection yoke*. The horizontal and vertical current linear sawtooth waves are applied to the yoke coils. These produce magnetic fields inside the tube that influence the electron beam. Remember that when electrons flow, a magnetic field is produced around the conductor through which the current flows. The magnetic field that occurs around the electron beam is moved or deflected by the magnetic field produced by the deflection coils in the yoke. Thus the electron beam is swept across the face of the picture tube in the interlaced manner described earlier.

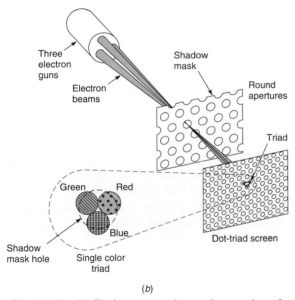

Fig. 14-33 (*a*) Basic construction and operation of a black-and-white (monochrome) cathode-ray tube (CRT). (*b*) Details of color picture tube.

As the beam is being swept across the face of the tube to trace out the scene, the intensity of the electron beam is varied by the luminance or Y signal. This signal is applied to the cathode or in some cases the control grid. The control grid is an element in the electron gun that is negatively biased with respect to the cathode. When the grid voltage is varied, the beam becomes stronger or weaker, thereby

varying the intensity of the light spot produced by the beam when it strikes the phosphor. Any shade of gray from white to black can be reproduced in this way.

The color picture tube works in much the same way. To produce color, the inside of the picture tube is coated with many tiny red, green, and blue phosphor dots arranged in groups of three called *triads*. Some tubes use a pattern of red, green, and blue stripes. These dots or stripes are energized by three separate cathodes and electron guns driven by the red, green, and blue color signals. Figure 14-33(*b*) shows how the three electron guns are focused so that they strike only the red, green, and blue dots as they are swept across the screen. A metallic plate has holes for each dot triad (called a *shadow mask*) and is located between the guns and the phosphor dots to ensure that the correct beam strikes the correct color dot. By varying the intensity of the color beams, the dot triads can be made to produce any color. The dots are small enough so that the eye cannot see them individually. What the eye sees is a color picture scanned out on the face of the tube.

Figure 14-34 on the next page shows how all the signals finally come together at the picture tube to produce the color picture. The R, G, and B signals are mixed with the Y signal to control the cathodes of the CRT. Thus the beams are properly modulated to reproduce

Fig. 14-34   Color picture tube circuits.

the color picture. Note the various controls associated with the picture tube. The R, G, and B screen, brightness, focus, and centering controls vary dc voltages that set the levels as desired. The convergence controls and assembly are used to control the positioning of the three electron beams so that they are centered on the holes in the shadow mask causing the electron beams to strike the color dots dead center.

The latest development in television is *high-definition television (HDTV)*. HDTV is a new system of television that greatly improves the resolution and definition of a TV picture by transmitting more horizontal scan lines. New systems under consideration provide from 700 to 1050 lines and have a wider screen with a 16:9 aspect ratio (which matches the film industry standard).

While both analog and digital techniques have been investigated for the new HDTV system, the new standards are expected to define an all-digital TV signal. The analog audio and video will be digitized, compressed, and transmitted serially using data communications techniques. Compression of the data is necessary to permit the HDTV signal to be transmitted within the existing 6-MHz TV channel bandwidth.

As of this writing, the FCC has not completed the HDTV standards. HDTV sets are not expected to appear on the market until 1995 or later.

Now, let's look at *cable television (CATV)*. Cable TV is a system of delivering the TV signal to home receivers by way of a coaxial cable rather than over-the-air by radio wave propagation. A cable TV company collects all the available signals and programs and multiplexes them on a single coaxial cable that is fed to the homes of subscribers. A special cable decoder box is used to receive the cable signals, select the desired channel, and feed a signal to the TV set.

Cable TV had its beginning in the 1950s and 1960s, when there were few TV stations and many viewers or potential viewers were located too far from the station to have decent reception. Companies emerged to offer TV signals by cable. These companies put up very tall high-gain TV antennas to permit reception. The resulting signals were amplified and fed to the subscribers by cable. Similar systems were developed by apartments and condos. A single master antenna system was installed at the building and the signals were amplified and distributed to each apartment or unit by cable. A complete cable TV system is illustrated in Fig. 14-35.

Today, cable TV companies collect signals and programs from many sources. Their main building or facility is called the *headend*. Their antennas receive local TV stations and other nearby stations plus the special cable channel signals distributed by satellite. The cable companies use parabolic dishes to pick up the so-called premium cable channels (CNN, HBO, Showtime, etc.). These signals are heterodyned to a lower intermediate frequency, amplified, and mixed with a local oscillator to translate them to one of many new cable TV channel frequencies. Some of the cable channels are the same as the standard TV channels; for others, special channel frequencies are used. All the resulting signals are frequency-multiplexed onto a single cable.

The main output cable is called the *trunk cable*. The cable is a high-quality solid-metal jacket coax with a characteristic impedance of 75 Ω. Often, it is pressurized with nitrogen to

Fig. 14-35   The modern cable TV system.

keep out water and other contaminants. It is buried and distributed to surrounding areas. A junction box containing amplifiers takes the signal and redistributes it to smaller cables called *feeders,* which go to specific areas and neighborhoods. From those points the signals are again rejuvenated with amplifiers and then sent to individual homes by coaxial cables called *drops*.

The coaxial cable used for drops (usually 75-Ω RG-59/U) enters a residence and is connected to a cable decoder box. This is essentially a heterodyne frequency converter. The desired cable channel is heterodyned to TV channel 3 or 4 and then fed to the TV set antenna jack. The TV receiver is tuned to channel 3 or 4 when a decoder box is used. Most sets today are "cable ready" and do not require a decoder box unless certain premium channels are desired.

Cable TV is a popular and widely used service in the United States. More than 70 percent of U.S. homes have cable TV service. It eliminates the need for a TV antenna. And because of the direct connection of amplified signals, there are fewer problems with weak, noisy, or snowy signals and increased reliability of good reception. In addition, many TV programs are available only on cable, the premium movie channels being an example. The only down side to cable TV is extra cost.

Self-Test

62. The bandwidth of a standard TV signal is _____ MHz.
63. The video carrier is _____ modulated and the sound carrier is _____ modulated.
64. The spacing between the sound and picture carriers is _____ MHz.
65. TV stations operate in the _____ and _____ regions of the spectrum.
66. The brightness signal produced by a video camera is called the _____ or _____ signal.
67. Two electronic imaging devices used in TV cameras to produce video signals are the _____ and the _____ .
68. The process that breaks up a picture or scene into serially transmitted signals is called _____ .
69. The three basic colors of light which can be used to produce any other color are _____ , _____ , and _____ .

70. A NTSC TV picture or one frame of video is composed of _____ scan lines.
71. A field consists of _____ scan lines.
72. The field and frame rates are _____ and _____ Hz respectively.
73. The time needed to scan one horizontal line in a color TV set is _____ .
74. To keep the picture at the receiver in step with the transmitted signal, _____ are added to the video.
75. The circuit that lets the picture and sound transmitters use the same antenna is the _____ .
76. The color in a scene is generated by a camera that develops three signals that represent the amount of _____ , _____ , and _____ light present along every scan line.
77. The red, green, and blue color signals are combined in a resistive matrix to produce the two color signals called _____ and _____ .
78. The two color signals are fed to _____ circuits that produce _____ modulation.
79. The color signals modulate a subcarrier with a frequency of _____ .
80. The characteristic of the composite color signal that tells the receiver the transmitted color is its _____ .
81. The _____ signal can be formed by adding the color signals in the following proportion: 0.11B + 0.59G + 0.3R.
82. Channel selection in the tuner is accomplished by using a(n) _____ for a local oscillator.
83. The _____ portion of a modern TV set uses digitally coded infrared signals for channel, volume, and other changes.
84. A channel 33 UHF TV station has a picture carrier frequency of 585.25 MHz. The sound carrier frequency is _____ .
85. The frequency of the video signal representing fine detail along a scan line can have a frequency up to about _____ .
86. The TV receiver sound IF is _____ MHz, and the picture IF is _____ MHz.
87. The name of the special filter that provides most of the adjacent channel selectivity for the TV receiver is _____ .
88. Spectrum space is conserved by using _____ modulation for the video in a TV signal.

89. In the TV receiver, the picture and sound IF beat together to form the sound IF of _____ MHz.

90. Two common sound demodulators in TV receivers are the _____ and the _____ .

91. The composite color signal is demodulated in two _____ demodulator circuits fed with quadrature 3.58-MHz signals from a subcarrier oscillator.

92. The 3.58-MHz subcarrier oscillator in the receiver is phase- and frequency-synchronized to the _____ signal transmitted along with the horizontal sync pulses.

93. The _____ circuit strips the horizontal sync pulses from the video detector output.

94. The horizontal sync pulses synchronize in internal sweep oscillator to a frequency of _____ Hz.

95. The vertical sweep oscillator is synchronized to a frequency of _____ Hz by a sync pulse derived by _____ the horizontal sync pulses occurring during the vertical blanking interval.

96. The horizontal and vertical sweep signals are currents with the shape of a(n) _____ .

97. The sweep signals are applied to the _____ assembly around the neck of the picture tube.

98. The electron beam is generated by the _____ element in the picture tube and focused into a narrow beam by the _____ .

99. The electron beam is deflected and swept across the face of the picture tube by _____ generated by the _____ .

100. A color CRT contains _____ electron guns to excite the color dot triads on the face of the tube.

101. The _____ circuits ensure that the electron beams strike the correct color dots.

102. The _____ stage is used as switching power supply to develop the HV required to operate the picture tube.

103. The transformer used to step up and step down the horizontal sync pulses is called the _____ .

104. The name given to the cable TV station that collects and distributes the cable signals is _____ .

105. The main cables used to distribute the TV signals to subscribers are called the _____ and _____ .

106. The coaxial cables that feed individual houses are called _____ and typically have a characteristic impedance of _____ ohms.

107. The cable TV station may change the TV signal frequency by translating it to another frequency by using a _____ .

108. Cable attenuation occurring during distribution is overcome by using _____ .

109. The cable box used by the subscriber contains two main circuits, the _____ and the _____ .

110. The output of the cable box is a TV signal on channel _____ or _____ that is connected to the _____ of the TV set.

111. Four advantages of cable TV are _____ , _____ , _____ , and _____ .

SUMMARY

1. Facsimile or fax is an electronic communication technique for transmitting printed documents including text, line drawings, photos, and other graphical information via telephone lines or radio.

2. Fax machines scan the document to be transmitted using photo-optical techniques to create a baseband signal that modulates a carrier prior to transmission.

3. The receiving fax machine demodulates the carrier to recover the baseband signal which is sent to a printer where the original document is faithfully reproduced.

4. The quality of reproduction is a function of the scanning resolution in lines per inch. The greater the number of scan lines, the finer the definition.

5. The most widely used device to convert the scanned lines into an electrical signal is a light-sensitive semiconductor component known as a charge coupled device (CCD).

6. The CCD contains thousands of tiny capacitors that charge to a value proportional to the light intensity. The capacitors are sequentially sampled. Their charges are read out, creating an analog signal corresponding to the lines scanned.

7. The baseband signal modulates a carrier prior to transmission; AM, FM, and PSK are commonly used.

8. Facsimile standards for modulation, transmission speed, and other factors are set by the International Telegraph and Telephone Consultative Committee (CCITT). There are four basic standards designated group 1 through 4.

9. Group 1 fax machines are analog and use FM where black is 1500 Hz (or 1300 Hz) and white is 2300 Hz (or 2100 Hz). Resolution is 96 lines per inch, and transmission speed is 6 min per page. Group 1 machines are no longer used.

10. Group 2 machines are analog and use FM or vestigial sideband AM with a 2100-Hz carrier. Resolution is 96 lines per inch, and transmission speed is 3 min per page or less.

11. Group 3 machines use digital techniques and PSK or QAM to achieve speeds of up to 9600 baud over the telephone lines. Resolution is 200 lines per inch, and transmission times are less than 1 min, with less than 30 s being common. Most modern fax machines are of the group 3 type.

12. Group 4 fax machines are digital and are designed for wideband telephone lines. Digital transmission rates are 56 kbits/s. Resolution is 400 lines per inch. Few group 4 machines are in use yet.

13. Most fax printers are of the thermal type and use special heat-sensitive paper, although laser printers using xerography are used in some higher-priced machines.

14. Fax is widely used by newspapers, business, and the military. Radio fax is used for transmitting weather satellite photos to earth.

15. Cellular radios provide telephone service in vehicles and are used as portable telephone units.

16. The area served by cellular telephones is divided into small zones called cells.

17. Each cell is served by a repeater containing a low-power transmitter and a receiver.

18. The cells are connected by wire to a computer-controlled master station called the mobile telephone switching office (MTSO). The MTSO links to the telephone system.

19. By operating at high frequencies in the spectrum and through the application of frequency reuse, many channels are available to users.

20. There are 666 full-duplex telephone channels in most service areas.

21. The receive frequencies are in the 870- to 890-MHz range.

22. The transmit frequencies are in the 825- to 845-MHz range.

23. Channel spacing is 30 kHz.

24. Spacing between the simultaneously used transmit and receive frequencies is 45 MHz.

25. Cellular radios use FM with a maximum deviation of $\pm$ 12 kHz.

26. Mobile cellular transmitters have a maximum output power of 3 W which can be decreased in steps under the control of the MTSO to minimize adjacent cell interference.

27. Mobile cellular receivers are of the dual-conversion type with a 45- or 82.2-MHz first IF and a 10.7-MHz or 455-kHz second IF.

28. Both transmit and receive frequencies are determined by PLL frequency synthesizers that are set by the MTSO to a clear channel.

29. The MTSO monitors received cell signal strength [received signal strength indicator (RSSI)] and transmitter power output and makes decisions about when to "hand off" the mobile unit to another cell to maintain optimum signal strength.

30. The MTSO controls transmitter power as well as transmit and receive frequencies. Serial digital data containing this information is transmitted to the mobile unit whose logic section interprets it and effects the changes.

31. Each cellular radio contains a PROM called the number assignment module (NAM) which stores the unit's telephone number referred to as the mobile identification number (MIN).

32. Most cellular radios contain two microprocessors, one for controlling the logic section and another for operating the displays and dialing circuits in the handset and control section.

33. Radar is the acronym for *ra*dio *d*etection *a*nd *r*anging.

34. Radar uses a reflected radio signal from a target to determine its distance, azimuth, elevation, and speed.

35. Radio waves travel at a speed of 186,000 mi/s or 162,000 nmi/s. Knowing the speed of radio waves permits the distance (range) of a remote target to be determined.

36. Radio signals travel at a speed of 5.375 μs/nmi or 6.18 μs/nmi.

37. The distance D in nautical miles to a target can be computed by knowing the total delay time T in microseconds for a signal to reach the target and return ($D = T/12.36$).

38. The strength of the reflection is a function of the wavelength of the radar signal and its relationship to the size of the target. Optimum reflection is obtained when the size of the target is one-quarter wavelength or larger of the signal frequency. Most radars operate in the microwave part of the spectrum.

39. The bearing or azimuth of a target with respect to the radar set is determined by a highly directional antenna. A narrow-beam-width antenna is rotated continuously over 360° and the detection of a reflection from a target at a given bearing gives the target's direction.

40. There are two basic types of radar: pulse and continuous wave (CW).

41. In pulse radar, the transmitter emits microwave energy in the form of repetitive sine wave bursts or pulses.

42. The pulse repetition time (PRT) is the time interval between the beginning of successive pulses.

43. The pulse repetition frequency (PRF) or the pulse repetition rate in pulses per second is the reciprocal of PRT (PRF = 1/PRT).

44. The ratio of the pulse burst duration to the PRT is known as the duty cycle and is usually expressed as a percentage.

45. The PRT and duty cycle set the range of a radar: a short PRT and pulse width for short range, and a long PRT and pulse width for long range.

46. The reflection from the target occurs in the interval between successively radiated pulses.

47. In CW radar, a constant-amplitude–constant-frequency signal is continuously radiated. If the target is stationary, the reflected signal contains no distance information.

48. Continuous wave radar relies on the Doppler effect to produce frequency modulation of the carrier.

49. The Doppler effect is the change in frequency that occurs as the result of relative motion between the transmitter and a target. The Doppler effect occurs with sound, radio, and light signals.

50. The relative speed between transmitter and target is directly proportional to the amount of Doppler frequency shift.

51. The CW Doppler radar is used for speed measurement. Police radars are an example.

52. By frequency modulating a CW radar with a sawtooth or triangular wave, the frequency difference between the transmitted and received signals can be used to compute the distance or range to a target.

53. Radar sets consist of a transmitter, a receiver, an antenna, a master timing section, and a display.

54. Most radar transmitters use a magnetron oscillator, although some high-power radars use klystrons or TWTs. Low-power radars use a Gunn diode oscillator.

55. Radar receivers are superheterodynes with diode mixers.

56. The radar antenna is usually a horn with a parabolic reflector that rotates over a 360° angle.

57. A duplexer is a waveguide assembly that allows both transmitter and receiver to share the same antenna.

58. Spark-gap tubes called transmit-receive (TR) and anti–transmit-receive (ATR) tubes prevent high-power energy from the transmitter from getting into and damaging or desensitizing the receiver.

59. The radar display is normally a cathode-ray tube (CRT) that is calibrated to read out the range, bearing, and other data.

60. The most common CRT readout is the plan position indicator (PPI) where the radar is at the center and a radius rotates to reveal reflected targets as blips.

61. Phased array radars use a matrix of dipoles or slot antennas with variable phase shifters to permit automatic, high-speed, electronic beam switching; beam width changes; and sweeping or scanning.

62. Television is the radio transmission of sound and pictures in the VHF and UHF ranges. The voice signal from a microphone frequency modulates a sound transmitter. A camera converts a picture or scene into an electrical signal called the

video or luminance Y signal, which amplitude-modulates a separate video transmitter. Vestigial sideband AM is used to conserve spectrum space. The picture and sound transmitter frequencies are spaced 4.5 MHz apart, with the sound frequency being the higher.

63. TV cameras use either a vacuum tube imaging device such as a vidicon or a solid-state imaging device such as the charge coupled device (CCD) to convert a scene into a video signal.

63. A scene is scanned by the imaging device to break it up into segments that can be transmitted serially. The National Television Standards Committee (NTSC) standards call for scanning the scene in two 262½ line fields which are interlaced to form a single 525-line picture called a frame. Interlaced scanning reduces flicker. The field rate is 59.94 Hz, and the frame or picture rate is 29.97 Hz. The horizontal line scan rate is 15,734 Hz or 63.6 μs per line.

64. The color in a scene is captured by three imaging devices which break a picture down into its three basic colors of red, green, and blue using color light filters. Three color signals are developed (R, G, B). These are combined in a resistive matrix to form the Y signal and are combined in other ways to form the I and Q signals. The I and Q signals amplitude-modulate 3.58-MHz subcarriers shifted 90° from one another in balanced modulators producing quadrature DSB suppressed signals that are added to form a carrier composite color signal. This color signal is then used to modulate the AM picture transmitter along with the Y signal.

65. A TV receiver is a standard superheterodyne receiver with separate sections for processing and recovering the sound and picture. The tuner section consists of RF amplifiers, mixers, and a frequency-synthesized local oscillator for channel selection. Digital infrared remote control is used to change channels in the synthesizer via a control microprocessor.

66. The tuner converts the TV signals to intermediate frequencies of 41.25 MHz for the sound and 45.75 MHz for the picture. These signals are amplified in IC IF amplifiers. Selectivity is usually provided by a surface acoustic wave (SAW) filter. The sound and picture IF signals are placed in a sound detector to form a 4.5 MHz sound IF signal. This is demodulated by a quadrature detector or other FM demodulator to recover the sound. Frequency multiplexing techniques similar to those used in FM radio are used for stereo TV sound. The picture IF is demodulated by a diode detector or other AM demodulator to recover the Y signal.

67. The color signals are demodulated by two balanced modulators fed with 3.58-MHz subcarriers in quadrature. The subcarrier is frequency- and phase-locked to the subcarrier in the transmitter by phase-locking to the color subcarrier burst transmitted on the horizontal blanking pulse.

68. To keep the receiver in step with the scanning process at the transmitter, sync pulses are transmitted along with the scanned lines of video. These sync pulses are stripped off the video detector and used to synchronize horizontal and vertical oscillators in the receiver. These oscillators generate deflection currents that sweep the electron beam in the picture tube to reproduce the picture.

69. The color picture tube contains three electron guns that generate narrow electron beams aimed at the phosphor coating on the inside of the face of the picture tube. The phosphor is arranged in millions of tiny red, green, and blue color dot triads or stripes. The electron beams excite the color dots or stripes in proportion to their intensity and generate light of any color depending upon the amplitude of the red, green, and blue signals. The electron beam is scanned or deflected horizontally and vertically in step with the transmitted video signals. Deflection signals from the internal sweep circuits drive coils in a deflection yoke around the neck of the picture creating magnetic fields that sweep the three electron beams.

70. The horizontal output stage, which provides horizontal sweep, is also used to operate a flyback transformer that steps up the horizontal sync pulses to a very high voltage. These are rectified and filtered into a 30- to 35-kV voltage to operate the picture tube. The flyback also steps down the horizontal pulses and rectifies and filters them into low-voltage dc supplies that are used to operate most of the circuits in the TV set.

71. Cable television is the transmission of

multiple TV signals which are frequency-multiplexed on a single cable to be used in lieu of over-the-air signal transmission. The headend of a cable TV station collects TV signals from local stations and from other sources by satellite and then modulates and multiplexes these signals on a cable that is sent to subscribers.

72. The main cable from the headend, called the trunk, takes the signals to other distribution points and to other cables called feeders, which transmit the signals to neighborhoods. The feeders are then tied to cable drops connected to each house. The cable signals are amplified at several points along the distribution path to maintain strong signals. A cable TV decoder box or tuner selects the desired channel and converts it into a channel 3 or 4 TV signal that connects to the TV receiver for presentation.

CHAPTER REVIEW QUESTIONS

Choose the letter which best answers each question.

14-1. Printed documents to be transmitted by fax are converted into a baseband electrical signal by the process of
    a. Reflection.
    b. Scanning.
    c. Modulation.
    d. Light variation.

14-2. The most commonly used light sensor in a modern fax machine is a
    a. Phototube.
    b. Phototransistor.
    c. Liquid-crystal display.
    d. Charge coupled device.

14-3. In FM fax, the frequencies for black and white are
    a. 1500 and 2300 Hz.
    b. 2300 and 1500 Hz.
    c. 1300 and 2400 Hz.
    d. 1070 and 1270 Hz.

14-4. Which resolution produces the best quality fax?
    a. 96 lines per inch
    b. 150 lines per inch
    c. 200 lines per inch
    d. 400 lines per inch

14-5. Group 2 fax uses which modulation?
    a. SSB
    b. FSK
    c. Vestigial sideband AM
    d. PSK

14-6. The most widely used fax standard is
    a. Group 1.
    b. Group 2.
    c. Group 3.
    d. Group 4.

14-7. Group 3 fax uses which modulation?
    a. QAM
    b. FSK
    c. Vestigial sideband AM
    d. FM

14-8. Most fax printers are of which type?
    a. Impact
    b. Thermal
    c. Electrosensitive
    d. Laser xerographic

14-9. Facsimile standards are set by the
    a. FCC.
    b. DOD.
    c. CCITT.
    d. IEEE.

14-10. What type of graphics are commonly transmitted by radio fax?
    a. Newspaper text
    b. Architectural drawings
    c. Cable movies
    d. Satellite weather photos

14-11. The transmission speed of group 4 fax is
    a. 4800 baud.
    b. 9600 baud.
    c. 56 kbits/s.
    d. 192 kbits/s.

14-12. The master control center for a cellular telephone system is the
    a. Cell site.
    b. Mobile telephone switching office.
    c. Central office.
    d. Branch office.

14-13. Each cell site contains a
a. Repeater.
c. Direct link to a branch exchange.
b. Control computer.
d. Touch-tone processor.
14-14. Multiple cells within an area may use the same channel frequencies.
a. True
b. False
14-15. Cellular telephones use which type of operation?
a. Simplex
c. Full-duplex
b. Half-duplex
d. Triplex
14-16. The maximum frequency deviation of an FM cellular transmitter is
a. 6 kHz.
c. 30 kHz.
b. 12 kHz.
d. 45 kHz.
14-17. The maximum output power of a cellular transmitter is
a. 4.75 mW.
c. 3 W.
b. 1.5 W.
d. 5 W.
14-18. Receive channel 22 is 870.66 MHz. Receive channel 23 is
a. 870.36 MHz.
c. 870.96 MHz.
b. 870.63 MHz.
d. 870.69 MHz.
14-19. A transmit channel has a frequency of 837.6 MHz. The receive channel frequency is
a. 729.6 MHz.
c. 867.6 MHz.
b. 837.6 MHz.
d. 882.6 MHz.
14-20. A receive channel frequency is 872.4 MHz. To develop an 82.2 MHz IF, the frequency synthesizer must supply an LO signal of
a. 790.2 MHz.
c. 954.6 MHz.
b. 827.4 MHz.
d. 967.4 MHz.
Note: The LO frequency is usually higher than the receive frequency.
14-21. The output power of a cellular radio is controlled by the
a. User or caller.
c. Called party.
b. Cell site.
d. MTSO.
14-22. When the signal from a mobile cellular unit drops below a certain level, what action occurs?
a. The unit is "handed off" to a closer cell.
b. The call is terminated.
c. The MTSO increases power level.
d. The cell site switches antennas.
14-23. In a cellular radio, the duplexer is a
a. Ferrite isolator.
b. Waveguide assembly.
c. Pair of TR/ATR tubes.
d. Pair of sharp bandpass filters.
14-24. The time from the transmission of a radar pulse to its reception is 0.12 ms. The distance to the target is how many nautical miles?
a. 4.85 nmi
c. 11.2 nmi
b. 9.7 nmi
d. 18.4 nmi
14-25. The ability of a radar to determine the bearing to a target depends upon the
a. Antenna directivity.
c. Speed of the target.
b. Speed of light.
d. Frequency of the signal.
14-26. The pulse duration of a radar signal is 600 ns. The PRF is 185 pulses per second. The duty cycle is
a. 1.1 percent.
c. 31 percent.
b. 5.5 percent.
d. 47 percent.
14-27. The Doppler effect is used to produce modulation of which type of radar signal?
a. Pulse
b. CW

14-28. The Doppler effect allows which characteristic of a target to be measured?
a. Distance   *c*. Altitude
b. Azimuth   *d*. Speed

14-29. The Doppler effect is a change in what signal characteristic produced by relative motion between the radar set and a target?
a. Amplitude   *c*. Frequency
b. Phase   *d*. Duty cycle

14-30. The most widely used radar transmitter component is a
a. Klystron.   *c*. TWT.
b. Magnetron.   *d*. Power transistor.

14-31. Low-power radar transmitters and receiver LOs use which component?
a. GaAs FET   *c*. Gunn diode
b. Magnetron   *d*. Klystron

14-32. What component in a duplexer protects the receiver from the high-power transmitter output?
a. Waveguide   *c*. Notch filter
b. Bandpass filter   *d*. Spark gap

14-33. Most radar antennas use a
a. Dipole.
b. Broadside array.
c. Horn and parabolic reflector.
d. Collinear array.

14-34. The most common radar display is the
a. A scan.   *c*. Liquid-crystal display.
b. Color CRT.   *d*. Plan position indicator.

14-35. A radar antenna using multiple dipoles or slot antennas in a matrix with variable phase shifters is called a(n)
a. A scan.   *c*. Broadside.
b. Phased array.   *d*. Circular polarized array.

14-36. Police radars use which technique?
a. Pulse
b. CW

14-37. Which of the following is a typical radar operating frequency?
a. 60 MHz   *c*. 900 MHz
b. 450 MHz   *d*. 10 GHz

14-38. The TV signal uses which types of modulation for picture and sound respectively?
a. AM, FM   *c*. FM, AM
b. DSB, FM   *d*. AM, DSB

14-39. If a TV sound transmitter has a carrier frequency of 197.75 MHz, the picture carrier is
a. 191.75 MHz.   *c*. 202.25 MHz.
b. 193.25 MHz.   *d*. 203.75 MHz.

14-40. The total bandwidth of an NTSC TV signal is
a. 3.58 MHz.   *c*. 6 MHz.
b. 4.5 MHz.   *d*. 10.7 MHz.

14-41. What is the total number of interlaced scan lines in one complete frame of a NTSC U.S. TV signal?
a. 262½   *c*. 480
b. 525   *d*. 625

14-42. What keeps the scanning process at the receiver in step with the scanning in the picture tube at the receiver?
a. Nothing
b. Color burst
c. Sync pulses
d. Deflection oscillators

14-43. What is the black-and-white or monochrome brightness signal in TV called?
- *a.* RGB
- *b.* Color subcarrier
- *c.* Q and I
- *d.* Luminance Y

14-44. What is the name of the solid-state imaging device used in TV cameras that converts the light in a scene into an electrical signal?
- *a.* CCD
- *b.* Photodiode matrix
- *c.* Vidicon
- *d.* MOSFET array

14-45. The I and Q composite color signals are multiplexed onto the picture carrier by modulating a 3.58-MHz subcarrier using
- *a.* FM.
- *b.* PM.
- *c.* DSB AM.
- *d.* Vestigial sideband AM.

14-46. The assembly around the neck of a picture tube that produces the magnetic fields that deflect and scan the electron beams is called the
- *a.* Shadow mask.
- *b.* Phosphor.
- *c.* Electron gun.
- *d.* Yoke.

14-47. The picture and sound carrier frequencies in a TV receiver IF are respectively
- *a.* 41.25 and 45.75 MHz.
- *b.* 45.75 and 41.25 MHz.
- *c.* 41.75 and 45.25 MHz.
- *d.* 45.25 and 41.75 MHz.

14-48. The sound IF in a TV receiver is
- *a.* 4.5 MHz.
- *b.* 10.7 MHz.
- *c.* 41.25 MHz.
- *d.* 45.75 MHz.

14-49. What type of circuit is used to modulate and demodulate the color signals?
- *a.* Phase-locked loop
- *b.* Differential peak detector
- *c.* Quadrature detector
- *d.* Balanced demodulator

14-50. What circuit in the TV receiver is used to develop the high voltage needed to operate the picture tube?
- *a.* Low-voltage power supply
- *b.* Horizontal output
- *c.* Vertical sweep
- *d.* Sync separator

14-51. What ensures proper color synchronization at the receiver?
- *a.* Sync pulses
- *b.* Quadrature modulation
- *c.* 4.5-MHz carrier spacing
- *d.* 3.58-MHz color burst

14-52. Which of the following is not a benefit of cable TV?
- *a.* Lower-cost reception
- *b.* Greater reliability
- *c.* Less noise, stronger signals
- *d.* Premium cable channels

14-53. What technique is used to permit hundreds of TV signals to share a common cable?
- *a.* Frequency modulation
- *b.* Mixing
- *c.* Frequency division multiplexing
- *d.* Time division multiplexing

CRITICAL THINKING QUESTIONS

14-1. Could fax be used to transmit and receive color pictures? Explain.

14-2. In electronic warfare between fighter planes, how does a plane's radar help and hurt in a battle?

14-3. Can data signals between computers be transmitted via cellular telephone? Explain the possible problems and limitations.

14-4. By using high-definition television (HDTV) with 1050 scan lines, a picture's resolution and quality are improved. What are the possible disadvantages of such a system?

Answers to Self-Tests

1. scanning
2. telephone lines
3. CITT
4. 2300, 1500
5. 2
6. charge coupled device
7. 3
8. handshaking
9. thermal
10. 9600
11. 200
12. PSK, QAM
13. half
14. 56K, 400
15. true
16. false
17. true
18. cells
19. mobile telephone switching office
20. more spectrum space at the higher frequencies, frequency reuse of channels.
21. 666
22. true
23. FM
24. 3
25. 825, 845
26. 870, 890
27. 30
28. 45
29. frequency synthesizer
30. 45, 82.2, 10.7, 0.455
31. change or decrease
32. duplexer or isolator
33. false
34. MTSO
35. transmitter power level, received signal level
36. transmit frequency, receive frequency, transmitter power.
37. receive signal strength indicator (RSSI)
38. 82.656
39. logic
40. two
41. control
42. distance or range
43. reflection
44. bearing or azimuth
45. 6.18
46. 3.5
47. 6.1
48. 1.15
49. short
50. microwave
51. true
52. pulse, continuous wave
53. Doppler
54. 589
55. false
56. duplexer
57. magnetron
58. spark gap
59. CRT
60. plan position indicator (PPI)
61. phased array
62. 6 MHz
63. amplitude, frequency
64. 4.5 MHz
65. VHF, UHF
66. luminance Y
67. vidicon, CCD
68. scanning
69. red, green, and blue
70. 525
71. 262½
72. 60 (59.94) and 30 (20.97) Hz
73. 15,734 Hz (63.6 μs)
74. sync pulses
75. diplexer
76. red, green, blue
77. I, Q
78. balanced modulator, DSB suppressed in-phase and quadrature carrier AM
79. 3.58 MHz
80. phase
81. Y or luminance
82. frequency synthesizer
83. remote control
84. 589.75 MHz
85. 4 MHz
86. 41.25, 45.75 MHz
87. surface acoustic wave (SAW)
88. vestigial sideband AM
89. 4.5 MHz
90. quadrature detector and differential peak detector
91. balanced
92. color burst
93. sync separator
94. 15,734 Hz
95. 59.94 Hz, integrating
96. linear sawtooth
97. yoke
98. cathode, electron gun
99. magnetic fields, yoke coils
100. three
101. convergence
102. horizontal sweep output
103. flyback
104. headend
105. trunk, feeder
106. drops, 75
107. mixer
108. amplifiers
109. mixer, frequency synthesizer channel selector
110. 3, 4, antenna jack
111. less noise and signal attenuation for a better picture, greater reliability of reception, premium cable programming, and no need for an antenna

Index

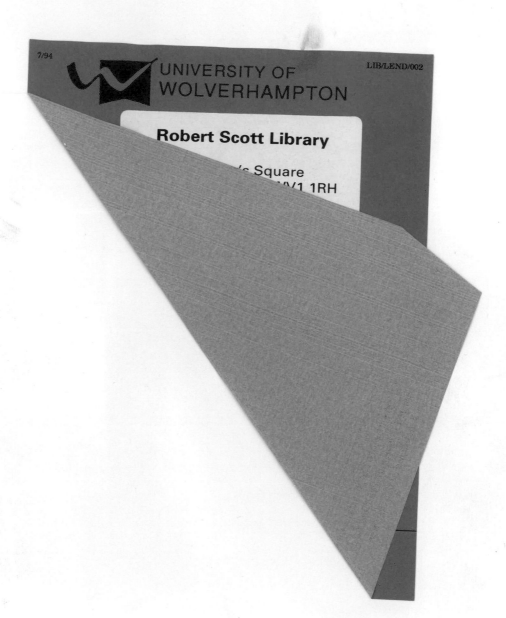